£3.99

11

Flatfishes

Flatfishes

Biology and Exploitation

Edited by

Robin N. Gibson
Scottish Association for Marine Science

Blackwell
Science

© 2005 by Blackwell Science Ltd, a Blackwell Publishing company
Editorial offices:
Blackwell Science Ltd, 9600 Garsington Road, Oxford OX4 2DQ, UK
   Tel: +44 (0) 1865 776868
Blackwell Publishing Professional, 2121 State Avenue, Ames, Iowa 50014-8300, USA
Tel: +1 515 292 0140
Blackwell Science Asia Pty, 550 Swanston Street, Carlton, Victoria 3053, Australia
   Tel: +61 (0)3 8359 1011

The right of the Author to be identified as the Author of this Work has been asserted in accordance with the Copyright, Designs and Patents Act 1988.

All rights reserved. No part of this publication may be reproduced, stored in a retrieval system, or transmitted, in any form or by any means, electronic, mechanical, photocopying, recording or otherwise, except as permitted by the UK Copyright, Designs and Patents Act 1988, without the prior permission of the publisher.

First published 2005

Library of Congress Cataloging-in-Publication Data
Flatfishes : biology and exploitation / edited by Robin N. Gibson.
      p. cm.
   Includes bibliographical references and indexes.
ISBN 0-632-05926-5 (hardback : alk. paper)
1. Flatfishes. 2. Flatfish fisheries. I. Gibson, Robin N.

   QL637.9.P5F58 2004
   597'.69--dc22

2004017649

ISBN 0-632-05926-5

A catalogue record for this title is available from the British Library

Set in 10/13 pt Times New Roman
by Sparks, Oxford, UK – http://www.sparks.co.uk
Printed and bound in India by Gopsons Paper Limited, New Dehli

The publisher's policy is to use permanent paper from mills that operate a sustainable forestry policy, and which has been manufactured from pulp processed using acid-free and elementary chlorine-free practices. Furthermore, the publisher ensures that the text paper and cover board used have met acceptable environmental accreditation standards.

For further information on Blackwell Publishing, visit our website: www.blackwellpublishing.com

Contents

Series foreword xiii
Preface xvii
Acknowledgements xix
List of contributors xxi

1 Introduction 1
Robin N. Gibson
   1.1   The fascination of flatfishes 1
   1.2   A brief history of flatfish research and its contribution to fish biology and fisheries science 2
   1.3   Scope and contents of the book 3
   1.4   Nomenclature 7
   Acknowledgements 8
   References 8

2 Systematic diversity of the Pleuronectiformes 10
Thomas A. Munroe
   2.1   Introduction 10
   2.2   Systematic profile of the Pleuronectiformes 14
   2.3   Intrarelationships of the Pleuronectiformes 15
   2.4   Brief synopses of the suborders and families 17
   2.5   Diversity of the Pleuronectiformes 21
       2.5.1   Overview 21
       2.5.2   Flatfish species diversity 21
       2.5.3   Diversity of species within families 23
       2.5.4   Standing diversity estimate for species of Pleuronectiformes 23
       2.5.5   Relative diversity of the Pleuronectiformes 25
   2.6   Patterns of species discovery among Pleuronectiform families 26
       2.6.1   History 26
       2.6.2   Factors contributing to new species discovery among the Pleuronectiformes 28
           2.6.2.1   Systematic activities 28
           2.6.2.2   Geographic region 30
           2.6.2.3   Depth 31

		2.6.2.4	Size	32
	2.7	Conclusions		34
	Acknowledgements			36
	References			36

3 Distributions and biogeography — 42
Thomas A. Munroe

3.1	Introduction		42
3.2.	Geographic distribution of Pleuronectiform lineages		45
3.3	Global patterns of species richness for the Pleuronectiformes		49
	3.3.1	Latitudinal gradients in species richness	49
	3.3.2	Tropical and subtropical regions	50
	3.3.3	Temperate regions	51
	3.3.4	Species richness on continental shelves	52
	3.3.5	Insular versus continental regions	52
	3.3.6	Continental versus oceanic islands	54
3.4	Species richness in specific environments		54
	3.4.1	Freshwater environments	54
	3.4.2	Antarctic Ocean	55
	3.4.3	Arctic Ocean	57
	3.4.4	Shallow-water versus deep-sea habitats	57
3.5	Historical biogeography		58
	3.5.1	Pleuronectidae	59
	3.5.2	Achiridae	59
	3.5.3	Paralichthyidae	60
	3.5.4	New World tropical flatfishes	60
	3.5.5	Indo-West Pacific region	61
Acknowledgements			63
References			63

4 Ecology of reproduction — 68
Adriaan D. Rijnsdorp and Peter R. Witthames

4.1	Introduction		68
4.2	Spawning		68
	4.2.1	Spawning behaviour	68
	4.2.2	Spawning mode	68
	4.2.3	Egg size	69
	4.2.4	Spawning season	70
	4.2.5	Duration of spawning	72
4.3	Gonad development		73
	4.3.1	Testis	73
	4.3.2	Ovary	73
	4.3.3	Fecundity	75
	4.3.4	Geographical pattern in fecundity	77
	4.3.5	Batch spawning	78

		4.3.6 Egg and sperm quality: maternal and paternal effects	79
	4.4	Age and size at first maturation	79
	4.5	Energetics	81
		4.5.1 Energetics of reproduction and growth	81
		4.5.2 Non-annual spawning	84
		4.5.3 Spawning fast	84
		4.5.4 Sexual dimorphism in reproduction and growth	84
	4.6	Contaminants and reproduction	85
		References	87
5	**The planktonic stages of flatfishes: physical and biological interactions in transport processes**		**94**
	Kevin M. Bailey, Hideaki Nakata and Henk W. Van der Veer		
	5.1	Introduction: the problem	94
	5.2	Flatfish eggs and larvae in the plankton: variations in form and function, time and space	96
		5.2.1 Variations in form and function	96
		5.2.2 Variations in time and space in the plankton	98
	5.3	Physical mechanisms of transport and retention	99
		5.3.1 Dispersal/retention mechanisms	99
		5.3.1.1 Wind-forcing/Ekman transport	99
		5.3.1.2 Tidal currents/selective tidal stream transport	100
		5.3.1.3 Estuarine circulation	100
		5.3.1.4 Fronts and eddies	103
		5.3.2 Models	103
	5.4	Adaptations to transport conditions: geographical and species comparisons	105
		5.4.1 Comparisons among species within a geographic region	106
		5.4.2 Congeneric comparisons in different regions	109
		5.4.3 Conspecific comparisons in different geographic areas	110
		5.4.4 Local adaptations	111
	5.5	Transport and population biology	112
		5.5.1 Population genetics	112
		5.5.2 Recruitment	112
		References	114
6	**Recruitment**		**120**
	Henk W. Van der Veer and William C. Leggett		
	6.1	Introduction	120
	6.2	Range of distribution	121
	6.3	Average recruitment levels	123
	6.4	Recruitment variability	125
		6.4.1 Processes influencing recruitment variability	127
		6.4.2 Recruitment variability in flatfishes relative to other marine fish species	130
		References	132

7	**Age and growth**		**138**
	Richard D.M. Nash and Audrey J. Geffen		
	7.1	Introduction	138
	7.2	Age estimation	139
		7.2.1 Larvae and juveniles	139
		7.2.2 Adults	140
	7.3	Growth of larvae	141
		7.3.1 Variation in growth	141
		7.3.2 Factors affecting larval growth	141
	7.4	Growth during metamorphosis	142
	7.5	Growth on the nursery grounds	146
		7.5.1 Growth models/growth experiments	146
		7.5.2 Maximum achievable growth and evidence for deviations from maximum growth	147
		7.5.3 Growth compensation/depensation	148
	7.6	Growth of adults	149
		7.6.1 Factors affecting adult growth rates	149
		7.6.2 Trade-off between growth and reproduction	150
	7.7	Longevity	150
	References		153
8	**Ecology of juvenile and adult stages of flatfishes: distribution and dynamics of habitat associations**		**164**
	Kenneth W. Able, Melissa J. Neuman and Hakan Wennhage		
	8.1	Introduction	164
	8.2	Definitions	164
	8.3	Distribution and ontogeny	165
		8.3.1 Juvenile habitat associations	168
		8.3.2 Late juvenile and adult habitat associations	170
		8.3.3 Dynamics of habitat associations	171
		8.3.3.1 Tidal, diel and seasonal cycles	171
	8.4	Future emphasis	175
	Acknowledgements		176
	References		176
9	**The trophic ecology of flatfishes**		**185**
	Jason S. Link, Michael J. Fogarty and Richard W. Langton		
	9.1	Introduction	185
	9.2	Major flatfish feeding groups	186
		9.2.1 Polychaete and crustacean eaters	186
		9.2.2 Piscivores	193
		9.2.3 Specialists	195
		9.2.4 Other considerations	196
		9.2.4.1 Anthropogenically produced food	196

		9.2.4.2	Seasonality	196
		9.2.4.3	Ontogeny	196
		9.2.4.4	Spatial factors	197
9.3	Flatfish predators			197
9.4	Flatfish competitors			198
9.5	Flatfish trophic dynamics: a case study of Georges Bank			199
	9.5.1	Shifts in abundance and species composition		199
	9.5.2	Potential competitive interactions		201
	9.5.3	Predation by flatfishes		202
	9.5.4	Have changes in flatfish populations influenced the Georges Bank ecosystem?		203
9.6	Summary and conclusions			203
Acknowledgements				204
References				204

10 The behaviour of flatfishes — 213
Robin N. Gibson

10.1	Introduction			213
10.2	Locomotion and related behaviour			213
	10.2.1	Locomotion		213
	10.2.2	Rheotaxis and station holding		215
	10.2.3	Burying		215
10.3	Colour change			216
10.4	Reproduction			216
	10.4.1	Spawning behaviour		216
10.5	Feeding			219
	10.5.1	Flatfish feeding types		219
	10.5.2	Feeding behaviour		219
		10.5.2.1	Search	220
		10.5.2.2	Encounter/detection	220
		10.5.2.3	Capture	221
		10.5.2.4	Ingestion	221
	10.5.3	Factors modifying feeding behaviour		222
		10.5.3.1	External factors	222
		10.5.3.2	Internal factors	223
10.6	Predation and reactions to predators			223
	10.6.1	Avoidance of predators		223
		10.6.1.1	Reducing encounter and detection probability	223
		10.6.1.2	Burial and substratum selection	224
	10.6.2	Escape from predators following attack		225
	10.6.3	Effect of size on vulnerability and avoidance of ingestion		225
10.7	Movements, migration and rhythms			226
10.8	Behaviour in relation to fishing, aquaculture and stock enhancement			228
	10.8.1	Fishing		228
	10.8.2	Aquaculture		230

	10.8.3	Stock enhancement	230
Acknowledgements			230
References			231

11 Atlantic flatfish fisheries 240
Richard Millner, Stephen J. Walsh and Juan M. Díaz de Astarloa

11.1	Introduction		240
11.2	Main species and nature of the fisheries		240
	11.2.1	Northwest Atlantic	240
		11.2.1.1 Shallow dwellers	243
		11.2.1.2 Mid-shelf dwellers	243
		11.2.1.3 Deep-slope dwellers	243
	11.2.2	Northeast Atlantic	244
	11.2.3	Southern Atlantic	248
		11.2.3.1 Southwest Atlantic	248
		11.2.3.2 Southeast Atlantic	250
11.3	History of exploitation		251
	11.3.1	Northwest Atlantic	251
		11.3.1.1 West Greenland	252
		11.3.1.2 Canada	253
		11.3.1.3 USA	254
	11.3.2	Northeast Atlantic	255
	11.3.3	Southern Atlantic	257
11.4.	Economic importance		258
	11.4.1	Northwest Atlantic	258
		11.4.1.1 West Greenland	259
		11.4.1.2 Canada	259
		11.4.1.3 USA	259
		11.4.1.4 Employment	260
	11.4.2	Northeast Atlantic	261
		11.4.2.1 Employment	262
	11.4.3	Southern Atlantic	262
11.5	Management		263
	11.5.1	Northwest Atlantic	263
		11.5.1.1 International Commission for the Northwest Atlantic Fisheries (ICNAF)	263
		11.5.1.2 Canada	264
		11.5.1.3 Northwest Atlantic Fisheries Organization (NAFO)	264
		11.5.1.4 USA	265
	11.5.2	Northeast Atlantic	265
	11.5.3	Southern Atlantic	267
Acknowledgements			269
References			269

12	**Pacific flatfish fisheries**	**272**
	Thomas Wilderbuer, Bruce Leaman, Chang Ik Zhang, Jeff Fargo and Larry Paul	
	12.1 Introduction	272
	12.2 Main species and nature of fisheries	272
	12.3 History of exploitation	278
	12.3.1 General account	278
	12.3.2 Republic of Korea	279
	12.3.3 Japan	280
	12.3.4 Russia (including the former Soviet Union)	280
	12.3.5 Canada	280
	12.3.6 USA	281
	12.3.7 New Zealand	282
	12.3.8 Australia	283
	12.4 Economic importance	283
	12.5 Management	284
	12.5.1 Western North Pacific	284
	12.5.2 Eastern North Pacific	285
	12.5.3 Australia and New Zealand	286
	12.5.4 Data collection	287
	References	288
13	**Tropical flatfish fisheries**	**292**
	Thomas A. Munroe	
	13.1 Introduction	292
	13.2 Main species and nature of the fisheries	293
	13.2.1 Habitats	293
	13.2.2 Commercially important species and/or taxa	296
	13.2.3 Nature of the fisheries	297
	13.2.4 Types of gear employed	298
	13.2.5 Harvest on spawning concentrations, migrating stocks and impacts on recruitment	299
	13.2.6 Industrial versus artisanal characteristics of the fisheries	300
	13.3 History of exploitation	301
	13.3.1 Commercial landings	301
	13.3.2 Geographic occurrence and historical landings	303
	13.4 Importance	308
	13.4.1 Economic importance	308
	13.4.2 Human importance	309
	13.5 Management and conservation	310
	13.5.1 Fishery conflicts, regulations and management	310
	13.5.2 Conservation	312
	Acknowledgements	313
	References	313

14 Assessment and management of flatfish stocks — 319
Jake Rice, Steven X. Cadrin and William G. Clark

- 14.1 Concepts and terms — 319
- 14.2 Population dynamics, assessment and management — 321
  - 14.2.1 Stock and recruitment — 323
  - 14.2.2 Recruitment, environment, assessment and management — 332
  - 14.2.3 Assessment, management and uncertainty — 334
- 14.3 Assessment and management summary — 335
  - 14.3.1 Northeast Pacific — 335
  - 14.3.2 Northwest Atlantic — 335
  - 14.3.3 Northeast Atlantic — 340
- 14.4 Conclusions — 341
- Acknowledgements — 341
- References — 342

15 Aquaculture and stock enhancement — 347
Bari R. Howell and Yoh Yamashita

- 15.1 Introduction — 347
- 15.2 Hatchery production of larvae and juveniles — 348
  - 15.2.1 Egg production — 348
    - 15.2.1.1 Manipulation of spawning time — 348
    - 15.2.1.2 Egg quality — 350
  - 15.2.2 Food and feeding — 350
    - 15.2.2.1 Live food — 350
    - 15.2.2.2 Formulated feeds — 351
  - 15.2.3 Microbial environment — 352
  - 15.2.4 Juvenile quality — 352
  - 15.2.5 Genetic improvement — 353
- 15.3 Intensive farming — 354
- 15.4 Stock enhancement — 356
  - 15.4.1 Stock enhancement in Europe and the USA — 357
  - 15.4.2 Stock enhancement in Japan — 357
    - 15.4.2.1 Background — 357
    - 15.4.2.2 Objectives and effectiveness of stock enhancement — 358
    - 15.4.2.3 Technology for stock enhancement — 363
  - 15.4.3 Future perspectives — 365
- 15.5 Conclusions — 365
- Acknowledgements — 366
- References — 366

Appendix 1 — 373
Appendix 2 — 377
Index of scientific and common names — 379
Subject index — 387

Series foreword

... flatfish, thirsting, trawled by grief

('Falling on Grass' by Elizabeth Biller Chapman)

Fish researchers (a.k.a. fish freaks) like to explain, to the bemused bystander, how fish have evolved an astonishing array of adaptations; so much so that it can be difficult for them to comprehend why anyone would study anything else. Yet, at the same time, fish are among the last wild creatures on our planet that are hunted by humans for sport or food. As a consequence, today we recognize that the reconciliation of exploitation with the conservation of biodiversity provides a major challenge to our current scientific knowledge and expertise. Even evaluating the trade-offs that are needed is a difficult task. Moreover, solving this pivotal issue calls for a multidisciplinary consilience of fish physiology, biology and ecology with social sciences, such as economics and anthropology, in order to probe the frontiers of applied science. In addition to food, recreation (and inspiration for us fish freaks), it has, moreover, recently been realized that fish are essential components of aquatic ecosystems that provide vital services to human communities. Sadly, virtually all sectors of the stunning biodiversity of fishes are at risk from human activities. In freshwater, for example, the largest mass extinction event since the end of the dinosaurs has occurred as the introduced Nile perch in Lake Victoria eliminated over 100 species of endemic haplochromine fish. But, at the same time, precious food and income from the Nile perch fishery was created in a miserably poor region. In the oceans, we have barely begun to understand the profound changes that have accompanied a vast expansion of human fishing over the past 100 years. The *Blackwell Publishing Fish and Aquatic Resources Series* is an initiative aimed at providing key, peer-reviewed texts in this fast-moving field.

Flatfish, ubiquitous from poles to tropics, are instantly recognizable; yet some biologists regard them as just another advanced teleost, so why, we may ask, do they deserve a monograph and conferences all to themselves? In fact, flatfish, defined as members of the monophyletic order of Pleuronectiformes, are endowed with a number of special and unique features (they underlie that instant recognition), and so your Series Editor is pleased to host this volume, the 9th in the *Blackwell Publishing Fish and Aquatic Resources Series*. The book grew out of a series of six international conferences on flatfish held since 1990. It is the brainchild of, and edited by, Robin Gibson, a leading biologist from the venerable Scottish Marine Biological Association (recently renamed the Scottish Association for Marine Science, located in a stunning landscape just outside Oban in the western highlands of Scotland). The 15 chapters are authored by 27 internationally recognised experts in the field of flatfish biology.

When the oldest fossils appear in the Eocene (53 My BP), flatfish already exhibit a diversity that suggests the order evolved in the Paleocene from perch-like marine ancestors. Today there are about 720 flatfish species in 123 genera, but relations within the group remain obscure. In this book systematics and biogeography are covered in two chapters by Thomas Munroe. Common and Linnaean names from the international fish database FishBase are used, except where recently revised.

During development each flatfish metamorphoses from a rounded, symmetrical fish-like larva to the characteristic flattened adult, with head skewed to remain horizontal as the fish lies on either its left or right side; one eye recapitulates early flatfish evolution and migrates to the upper side. Left- and right-handed eye position appears to be no guide to flatfish taxonomic relationships and the functional advantage of eye handedness remains a mystery.

The fascinating behaviour of flatfish is reviewed by Robin Gibson himself. Flatfish employ a 'swim-and-glide' energy-saving locomotion by undulations of the body. Their 90 degree rotation in body position means that flatfishes control vertical direction by changing the angle of the body, tail and median fins; horizontal direction is altered, rather clumsily, by using the pectoral fin on the eyed side as a rudder. The characteristic flatfish camouflage serves both against detection by their prey and by their predators. Camouflage is achieved partly by very rapid second-by-second colour changes, images falling on the upper part of retina causing three types of chromatophore to expand and contract, and partly by slower changes over a number of weeks to the number and pigmentation of chromatophores, thereby altering the pattern spots and flecks. All flatfish can bury themselves rapidly in the substrate to aid camouflage, and when this is done a muscular sac forces fluid into the orbit, causing the eyes to protrude above ground. Indeed this key feature of flatfish functional anatomy is diagnostic for the order. Flatfish eyes are independently mobile, providing 360 degree vision. Flatfish are ambush predators, which may leave the bottom to capture prey in the water column and then become vulnerable to being eaten themselves. Some species avoid this problem by luring prey with a waggling pectoral fin, while a few species, like angler fish, have evolved specialised food attracting lures. Prey are engulfed and sucked in by the back-pressure from the highly protrusible advanced perch-pattern jaws.

The trophic ecology of flatfishes is covered by Jason Link, Michael Fogarty and Richard Langton; flatfish are largely piscivorous or eat benthic invertebrates like crustaceans and polychaete worms, so they generally have a trophic level of three or more. Some species specialise in eating the tips of bivalve siphons, especially when juvenile, while a few concentrate on echinoderms such as sand dollars and urchins. Flatfish provide a tangible 'ecosystem service' for human benefit because they convert benthic production into a form suitable for consumption by higher predators and humans. Flatfish are sought out as food by specialised predators such as sharks that can detect electrical impulse from their nervous systems.

Flatfish have been eaten by humans for millennia; flatfish bones are found in ancient middens, flatfish are portrayed in rock carvings and paintings from Europe to Australia, and clever re-curved hooks for catching large flatfish are found among aboriginal peoples in the Pacific Northwest and in Northern Australia. Since the 1960s, flatfish comprise almost 25% of all groundfish landings worldwide, and, in this book, the major Pacific, Atlantic and tropical flatfish fisheries each have their own chapters. Overall, the status of most fisheries is very poor and they are generally exploited way beyond levels of maximum production.

Atlantic flatfish fisheries, reviewed by Richard Millner, Stephen J. Walsh and Juan M. Díaz de Astarloa, paint a miserable story of over-exploitation. The huge plaice fisheries of the early 20th Century have been severely depleted, Atlantic halibut is almost extinct, and valuable sole, turbot and brill fisheries are reduced to shadows of their former glory. Species such as the thin and watery-tasting Greenland 'halibut' (also misleading marketed as 'turbot') now dominate catches. The sad litany of stocks experiencing severe declines is accompanied by the realization that, over 80 years, these fisheries have been assessed and managed by some of the most advanced fisheries science in the world. As with other fish in the Atlantic, we see a total failure of fisheries agencies to fulfill their mandate. In the south Atlantic, flatfish stocks have been greatly overexploited, yet almost no good data exits with which to assess and manage the resources. An important exception to the generally miserable state of temperate flatfish resources is the Pacific halibut fishery, which has been managed conservatively by the International Pacific Halibut Commission for almost 100 years (reviewed by Thomas Wilderbuer, Bruce Leaman, Chang Ik Zhang, Jeff Fargo and Larry Pau). Tropical flatfish caught in fisheries tend to be small, taxonomically diverse and poorly known (described by Thomas Munroe); small individuals and species caught by commercial trawlers generate huge and unreported amounts of discards.

In a chapter on reproduction by A. Rijnsdorp and P. Witthames, we learn that, since flatfish live in intimate contact with sediments, the effects of human pollutants on flatfish ecology are large. For example, oil spills reduce ovary development and fecundity, aromatic hydrocarbons and polychlorinated biphenyls in sediments produce smaller eggs, and some pesticides, polychlorinated biphenyls, phthalates and alkyl phenols mimic oestrogen and disrupt female reproductive cycles or feminise males.

Aquaculture and stock enhancement is covered by Bari Howell and Yoh Yamashita. Flatfish aquaculture mainly consists of Japanese flounder (over 40 000 tonnes), turbot (over 5000 t), halibut, sole (both still less than 2000 t). Some newcomers are the Australian greenback flounder, US summer and southern flounders. An insoluble problem that will limit flatfish culture, as with most other fish farming in the developed world, is that most flatfish species that are commercially marketable are generally piscivores high in the trophic web and cannot make many of their own essential amino acids, and so valuable forage fish have to be caught and used to make the food used in farms. Although sea ranching has been attempted for plaice and turbot in Europe, Japanese flounder, marbled sole and six other species in Japan, the success of these ventures in both economic and ecological terms is far from evident. Moreover, sea ranching has recently come into question as genetic consequences, now proving serious in salmonids, have rarely been considered.

This volume presents timely and comprehensive 'state of the art' reviews of flatfish biology, ecology and flatfish fisheries, and readers will find the elements of a fresh synthesis concerning the role of flatfish in today's depleted marine ecosystems. It therefore provides a unique single source of reference on this group for all practitioners and students of marine biology, fish ecology, fishery science and fisheries management.

Professor Tony J. Pitcher
Editor, Blackwell Publishing Fish and Aquatic Resources Series
Fisheries Centre, University of British Columbia, Vancouver, Canada
October 2004

Preface

Instantly recognisable, flatfishes have long fascinated scientist and layman alike. The fascination stems principally from the group's unique asymmetric body form developed as an adaptation to a bottom-living lifestyle following metamorphosis from a bilaterally symmetrical larva. They are a relatively diverse group, over 700 extant species are currently recognised, and they are distributed from the Arctic to Australasia and beyond. This wide distribution, and because they make important contributions to commercial and recreational fisheries in many parts of the world, means that they are familiar to most people and have been the objects of research for more than a century. In recent years there has been a surge of interest in all aspects of the biology of this group. The interest has found expression in, and been greatly stimulated by, the International Symposia on Flatfish Ecology held triennially in the Netherlands (1990, 1993, 1996), the USA (1999) and the UK (2002). A symposium on North Pacific flatfishes was also held in Alaska in 1994. It now seems timely for this rapidly growing body of information to be synthesised into an authoritative account of the biology of this intriguing and economically important group of fishes.

The book brings together accounts written by internationally recognised experts in the field of flatfish biology. The chapters cover systematics and distribution; reproduction and recruitment; ecology and behaviour of the main life history stages; the major fisheries and their management; and the latest developments in flatfish aquaculture and stock enhancement. The volume therefore represents a comprehensive review of the current 'state of the art' in flatfish biology and will be an invaluable source of reference for fish biologists, fisheries scientists and managers, and students of marine biology alike.

Notes

Proceedings of the International Symposia were published as follows. First Symposium: *Netherlands Journal of Sea Research* 1991, **27** (3–4); 1992, **29** (1–3). Second Symposium: *Netherlands Journal of Sea Research* 1994, **32** (2–4); 1995, **34** (1–3). Third Symposium: *Journal of Sea Research* 1997, **37** (3–4); 1998, **40** (1–2). Fourth Symposium: *Journal of Sea Research* 2000, **44** (1–2); 2001, **45** (3–4). Fifth Symposium: *Journal of Sea Research* 2003, **50** (2–3); 2004, **51** (3–4). *Proceedings of the International Symposium on North Pacific Flatfish*: Alaska Sea Grant College Program Report No. 95–04, University of Alaska Fairbanks (1995).

Acknowledgements

This book was conceived well in advance of its final appearance and it is a pleasure to acknowledge the enthusiasm of the authors for the initial idea and for their subsequent encouragement during its preparation. Their willingness to comply with suggestions, requests and questions relating to their contributions made the editor's task an enjoyable one. It is also a pleasure to thank the Director of the Dunstaffnage Marine Laboratory for providing space and facilities during the editor's tenure of a Scottish Association for Marine Science Honorary Research Fellowship. Thanks are also due to Nigel Balmforth of Blackwell Publishing for his support and patience, and for his confidence that the book would one day appear on his list.

List of contributors

Kenneth W. Able
Rutgers University, Institute of Marine and Coastal Sciences, Marine Field Station, 800 c/o 132 Great Bay Blvd, Tuckerton, NJ 08087–2004, USA
able@imcs.rutgers.edu

Kevin M. Bailey
Alaska Fisheries Science Center, 7600 Sand Point Way NE, Seattle, WA 98115, USA
kevin.bailey@noaa.gov

Steven X. Cadrin
Northeast Fisheries Science Center, 166 Water Street, Woods Hole, MA 02543, USA
Steven.Cadrin@noaa.gov

William G. Clark
International Pacific Halibut Commission, P.O. Box 95009, Seattle, WA 98145–2009, USA
Bill@iphc.washington.edu

Juan M. Díaz de Astarloa
Departamento de Ciencias Marinas, Facultad de Ciencias Exactas y Naturales, Universidad Nacional de Mar del Plata, Funes 3350, B7602AYL Mar del Plata, Argentina, CONICET
asterloa@mdp.edu.ar

Jeff Fargo
Pacific Biological Station, Fisheries and Oceans Canada, Nanaimo, BC, Canada V9R 5K6
fargoj@pac.dfo-mpo.gc.ca

Michael J. Fogarty
National Marine Fisheries Service, Northeast Fisheries Science Center, Food Web Dynamics Program, 166 Water Street, Woods Hole, MA 02543, USA
michael.fogarty@noaa.gov

Audrey J. Geffen
Department of Fisheries and Marine Biology, University of Bergen, Postbox 7800, 5020 Bergen, Norway
audrey.geffen@ifm.uib.no

Robin N. Gibson
Scottish Association for Marine Science, Dunstaffnage Marine Laboratory, Oban PA37 1QA, Scotland
Robin.Gibson@sams.ac.uk

Bari R. Howell
CEFAS, Weymouth Laboratory, Barrack Road, The Nothe, Weymouth, Dorset DT4 8UB, UK
Bari.howell@ntlworld.com

Richard W. Langton
Buccoo Reef Trust, TLH Office Building, Unit 29–30, Milford Road, Scarborough, Tobago, West Indies
RLangton@buccooreef.org

Bruce Leaman
International Pacific Halibut Commission, PO Box 95009, Seattle, WA 98145, USA
bruce@iphc.washington.edu

William C. Leggett
Queen's University at Kingston, 74 University Avenue, Kingston, ON, Canada, K7L 3N6
wleggett@post.queensu.ca

Jason S. Link
National Marine Fisheries Service, Northeast Fisheries Science Center, Food Web Dynamics Program, 166 Water Street, Woods Hole, MA 02543, USA
jlink@whsun1.wh.whoi.edu or Jason.Link@noaa.gov

Richard Millner
CEFAS, Lowestoft Laboratory, Pakefield Road, Lowestoft, NR33 OHT, UK
r.s.millner@cefas.co.uk

Thomas A. Munroe
National Marine Fisheries Service National Systematics Laboratory, NMFS/NOAA, National Museum of Natural History, Washington, DC 20560–0153, USA
munroe.thomas@nmnh.si.edu

Hideaki Nakata
Faculty of Fisheries, Nagasaki University, 1–14 Bunkyo-machi, Nagasaki 852–8521, Japan
nakata@net.nagasaki-u.ac.jp

Richard D.M. Nash
Port Erin Marine Laboratory, University of Liverpool, Port Erin, Isle of Man IM9 6JA, UK
rdmnash@liverpool.ac.uk

Melissa J. Neuman
National Oceanic and Atmospheric Administration/National Marine Fisheries Service/SWR Protected Resources Division, 501 W. Ocean Blvd, Suite 4200, Long Beach, CA 90802–4213, USA
Melissa.Neuman@noaa.gov

Larry Paul
National Institute of Water and Atmospheric Research, Greta Point, PO Box 14–901, Kilbirnie, Wellington, New Zealand

Jake Rice
Canadian Science Advisory Secretariat, Department of Fish and Oceans, 200 Kent Street, Ottawa, ON, Canada K1A 0E6
ricej@dfo-mpo.gc.ca

Adriaan D. Rijnsdorp
RIVO, Netherlands Institute for Fisheries Research, PO Box 68, 1970 AB Ijmuiden, The Netherlands
adriaan.rijnsdorp@wur.nl

Henk W. Van der Veer
Netherlands Institute for Sea Research, PO Box 59, 1790 AB Den Burg, The Netherlands
veer@nioz.nl

Stephen J. Walsh
Northwest Atlantic Fisheries Centre, Department of Fisheries and Oceans, 1 White Hills Rd, PO Box 5667, St John's, Newfoundland, Canada A1C 5X1
walshs@dfo-mpo.gc.ca

Hakan Wennhage
Department of Marine Ecology, Kristineberg Marine Research Station, Goteborg University, S-450 34 Fiskebackskil, Sweden
hakan.wennhage@kmf.gu

Thomas Wilderbuer
Alaska Fisheries Science Center, National Marine Fisheries Service, 7600 Sand Point Way NE, Seattle, WA 98115, USA
tom.wilderbuer@noaa.gov

Peter R. Witthames
CEFAS, Centre for Environment Fisheries and Aquaculture Science, Lowestoft Laboratory, Pakefield Road, Lowestoft NR33 0HT, UK
p.r.witthames@cefas.co.uk

Yoh Yamashita
Fisheries Research Station, Kyoto University, Nagahama, Maizuru, Kyoto 625–0086, Japan
yoh@kais.kyoto-u.ac.jp

Chang Ik Zhang
Pukyung National University, Daeyeon-dong, Nam-gu, Pusan 608–737, South Korea
cizhang@dolphin.pknu.ac.kr

Chapter 1
Introduction

Robin N. Gibson

1.1 The fascination of flatfishes

Most people's first encounter with flatfishes is on a fishmonger's slab where their unusual shape makes them instantly recognisable. Flatfishes have certainly featured in the human diet for millennia. They appear in prehistoric rock carvings (Muus & Nielsen 1999), their remains are found in ancient middens (Nicholson 1998; Barrett *et al.* 1999) and they continue to make up a significant proportion of the world groundfish catch today. Gastronomy apart, the interested layman's curiosity is aroused not only by the presence of both eyes on the same side of the head and their flattened shape, but also by the remarkable ability of flatfishes to match the colour and pattern of their background and to bury themselves in the sediment. The last three characters are present in some other bottom-living fishes (e.g. skates and rays, anglerfishes) but together with eye migration in the larva and the less obvious features of protrusible eyes and a dorsal fin that continues onto the head, they make the flatfishes unique.

A question often asked by layman and biologist alike is why do some flatfishes lie on their right side and others on their left? Examination of the occurrence of left and right 'eyedness' within the Order Pleuronectiformes shows that, although some families are predominantly left- or right-eyed (Chapter 2), the trait for a particular direction of asymmetry does not reflect relationships within the order. Eye position is therefore not a useful systematic character. This conclusion holds true whether morphological or molecular evidence is used to deduce inter-relationships (Berendzen & Dimmick 2002). Furthermore, in some species, notably the flounder (*Platichthys flesus*) and starry flounder (*P. stellatus*), 'reversed' individuals are common. Starry flounders off the west coast of the USA, for example, are about half left-eyed and half right-eyed, whereas in Alaska and Japan the percentage of left-eyed fish is 70% and almost 100%, respectively (Policansky 1982a, b). Breeding experiments with this species have demonstrated that the direction of asymmetry is predominantly under genetic control although there may also be some environmental influence (Polikansky 1982a; Boklage 1984). The exact mechanism involved is unclear and remains a subject of debate (McManus 1984; Morgan 1991). To return to the original question, inheritance of eye position suggests that there should be some selective advantage of having eyes on one or other side of the head. It seems intuitively reasonable to assume that it would be advantageous for all members of the population to have the same eye position (Policansky 1982a) and in most species this is indeed the case. During mating, for example, when males and females are in close contact during the release of gametes, pre-mating display and fertilisation are likely to be more effective if both individuals have the same laterality. However, a convincing explanation for eye position has yet to be forthcoming and any explanation will have to account for the existence of left- and right-eyed species in the same habitat. Unless the asymmetry of the eyes is associated with some other as yet unrecognised adaptive character(s), the present conclusion must be that there is no adaptive difference between left- and right-eyedness (Policansky 1982a).

The ability of flatfishes to camouflage themselves against the seabed on which they lie is also a source of fascination for many. Background matching is the result of rapid nervous and slower hormonal responses to visual stimuli received by the eyes and is achieved by differential responses of the chromatophores in the skin. In this way flatfishes can match not only the general colour of their background but also its pattern, even to the extent that the sizes of the spots on a spotted background can be mimicked (Ramachandran *et al.* 1996; Healey 1999).

The variety of flatfishes and their adaptations to a benthic existence also make them intriguing subjects for study by fish biologists. Flatfishes vary in adult size from a few centimetres up to 2 m or more. They are widely distributed in cold, temperate and tropical seas in depths from the intertidal zone to the continental slope. This variation in size and habitat means that they display a considerable range of patterns in ecology and life history and in physiological and behavioural adaptations to life on and in the bottom. Their value as food has also resulted in numerous investigations of these patterns and adaptations in relation to growth, feeding, reproduction and population structure, and the application of the results to management. Yet the role of flatfishes in benthic ecosystems as predators, competitors and prey is still largely unresolved, even though they may account for around a quarter of groundfish species richness and biomass in some areas such as the North Sea (Daan *et al.* 1990). In some coastal nurseries, juvenile flatfishes may numerically dominate the benthic fish fauna (e.g. Gibson *et al.* 1993).

1.2 A brief history of flatfish research and its contribution to fish biology and fisheries science

Although flatfishes feature in many early descriptive zoological treatises and several common species were given their scientific names by Linnaeus in 1758, the first detailed articles describing research on flatfishes appear in the scientific literature at the end of the nineteenth century. Much of this early research was stimulated by the need for information on the biology of the common food fishes and was fuelled by a concern for the state of their fisheries and why catches fluctuated (e.g. Petersen 1894; Holt 1895). At that time, fluctuation in catches was considered to be due principally to changes in migration patterns but also to the possibility that stocks were being over-fished, challenging the earlier assertion by T.H. Huxley that the sea was an inexhaustible resource. It was realised that basic information was lacking and this lack led to the development of numerous research programmes to collect data on age, growth and size at maturity and to examine whether fishing did, in fact, have any effect on populations. It rapidly became evident that fishing could have significant effects and Holt (1895), for example, recommended the imposition of a size limit for plaice (*Pleuronectes platessa*) and common sole (*Solea solea*) in the North Sea. He also considered the possibility of protected areas, close seasons, mesh restrictions and artificial propagation. He dismissed stock enhancement using reared young stages as impractical and uneconomic even though the development of rearing techniques for fishes on a large scale both in North America and Europe had been in progress for some time (Ewart 1885; Dannevig 1897; Blaxter 1975; Smith 1994). Subsequent trials indicated that Holt's opinion was correct and the emphasis in the North Sea moved to the transplantation of wild fish with some success (see Blaxter 2000 for review). Flatfishes played a significant part in the development of these conclusions following experiments in Scotland and the 'Great Fishing Experiments' resulting from cessation of fishing in the North Sea during the two World Wars (summarised by Smith 1994). In these 'Experiments' it was clearly demonstrated that the population structure of North Sea plaice could be greatly altered by fishing but was also capable of recovery when fishing pressure was released.

The early studies in Europe and the USA represented the beginnings of fisheries research and contributed to the formation of bodies such as the International Council for the Exploration of the Sea (ICES) (Rozwadowski 2002) and the International Pacific Halibut Commission (Smith 1994). Much of this work is summarised in subsequent chapters in this book and it has made significant contributions to fish biology and fisheries science. Particular mention can be made of the classic early works on tagging (Petersen 1894) and colour change (e.g. Mast 1914). Beverton & Holt's (1957) seminal treatise on the dynamics of exploited populations incorporated the results of many flatfish studies and intensive investigations of flatfish movements in the North Sea (summarised by Harden Jones 1968) added greatly to our understanding of migration, a topic that continues to produce novel insights into fish behaviour (e.g. Metcalfe & Arnold 1997). The development of ageing techniques for fishes owes much to studies of flatfish species (Chapter 7) and the renewed interest in mass rearing to the juvenile stage pioneered by Shelbourne (Shelbourne 1964) provided material for studies of larval behaviour and physiology that would not have been possible using wild caught individuals (see, for example, Blaxter 1986). These mass rearing techniques also paved the way for further evaluation of the feasibility of flatfish stock enhancement, particularly in Japan, using juveniles rather than eggs and larvae. The Flatfish Symposia (see Preface) have provided a platform for the presentation and discussion of the most recent studies and significant contributions to the study of recruitment processes have been made as a result of discussions at these symposia.

1.3 Scope and contents of the book

The book is an overview of the biology and exploitation of flatfishes. Although necessary constraints on length mean that the coverage of each topic is not fully comprehensive, each chapter does represent a succinct summary of the 'state of the art' in its own field. Furthermore, as Hensley (1997) and several authors in this volume point out, current detailed knowledge is based on only a few, mostly north temperate, species of economic interest.

The contents can be roughly divided into three parts with numerous links between them. The first part deals with systematics and distribution, the second part deals with biology in the widest sense and covers development, ecology, growth and behaviour. The final five chapters describe and discuss aspects of exploitation including the major fisheries and current trends in aquaculture and stock enhancement.

The book starts with chapters on systematics and biogeography that review our current understanding of the evolution and taxonomic diversity of flatfishes. Flatfish fossils are rare but the oldest known date from at least 50 million years ago when many lineages had already diversified. The Order Pleuronectiformes is considered to be monophyletic and over 700 species in 15 family level taxa are presently recognised but the total species diversity for the order is unknown. The flatfish fauna of north temperate regions is generally well known but those of the tropics and deeper water are not. Tropical flatfishes are small and difficult to identify, and many tropical habitats have not been well sampled. These factors, together with the growing realisation that taxa formerly considered to be widespread single species may actually be species complexes, indicate that many flatfish species still await discovery. The companion Chapter 3 provides an overview of flatfish distributions by describing the global occurrence of the flatfish families and their patterns of species richness in terms of geographical region and specific environments. Although flatfishes have a virtually worldwide distribution, this distribution is not uniform; the East China Sea, for example, is particularly speciose but freshwaters, the deeper parts of the sea and high latitudes in the Southern Ocean are comparatively species-poor. Consideration of the historical biogeography of the group provides an explanation of some of these patterns but, here again, incomplete knowledge of systematics and distribution prevents a full picture being obtained as yet. Clearly, much remains to be done in this field of flatfish biology.

Chapter 4 describes the reproduction of flatfishes and focuses on those characteristics that affect the production of offspring and their survival, i.e. egg size, spawning, gonad development and fecundity, onset of sexual maturity, and the energetics of reproduction and growth. The authors explore the adaptive significance of patterns of reproduction from the viewpoint that reproductive characteristics have evolved, and continue to evolve, in response to environmental conditions. They discuss these characteristics in relation to the geographical distribution of species and their implications for population dynamics and the resilience to perturbations caused by exploitation and pollution. Most flatfish eggs and all larvae that hatch from them are planktonic. Consequently their rate and direction of dispersal after spawning are largely dependent on the characteristics of local water movements. Chapter 5 describes the types of water movement to which eggs and larvae are exposed and the physical mechanism by which they are transported to, or conversely retained in, their appropriate nursery grounds. Most transport is assumed to be passive, especially as flatfish larvae are relatively feeble swimmers, but the ability of larvae to migrate vertically enables them to exert some control over their net direction and speed of movement. A comparison of species and genera in the same and different locations reveals remarkable variety in transport patterns, all of which are adapted to local hydrographic conditions and which link spawning grounds to nursery areas. The extent of variation in the success of transport to suitable nursery grounds has considerable implications for population biology and recruitment. The latter topic is examined in more detail in Chapter 6, which reviews the data and hypotheses relating to the generation, regulation and variability of recruitment and analyses three factors relevant to these processes; namely a species' range, and its average level of, and annual variability in, recruitment. Temperature is considered to be the predominant factor determining range but it is also important in determining the duration of the egg and larval stages and hence the critical distance between spawning and nursery grounds. With respect to recruitment level, the authors conclude that level is governed by two distinct processes: the effect of food availability on adult condition at spawning time and density-dependent mortality of juveniles on the nursery grounds. This density-dependent mortality, which results from the concentration of juveniles in two dimensions after settlement, may also be an important contributing factor to the lower recruitment variability of flatfishes compared with other groups.

A knowledge of growth rates and patterns is essential for many areas of fish and fisheries biology and flatfish growth is summarised in Chapter 7. The range of longevity within the group is large (<2 to 60 years) and a complex range of factors affect growth throughout life. Some of the earliest studies seeking methods for ageing fish were carried out on flatfishes and led to the recognition of the value of otoliths as records of past growth. Otoliths are now routinely used for ageing all stages from larva to adult, but our knowledge of flatfish growth patterns has arisen from a combination of both laboratory and field studies. These studies have demonstrated the vital importance of temperature in determining growth rate but numerous other factors come into play with differing importance throughout the life history. Food supply is particularly critical in the larval and juvenile stages and there has been much discussion of the question of density-dependent growth. In the adults, growth patterns and their interpretation are strongly affected by reproduction and the effects of exploitation of populations. The next chapter describes the changes in the distribution of flatfishes as they grow and mature. In many species there are considerable differences between the distribution of the juvenile and adult stages in that the inshore nursery grounds are separated from the offshore spawning grounds. Transfer of eggs and larvae to these nursery grounds, outlined in Chapter 5, is subsequently balanced by a gradual movement back to the adult feeding and spawning grounds. Superimposed on this gradual ontogenetic change in distribution, however, are shorter frequency variations in habitat occupation varying from tidal to annual timescales. The factors controlling these movements are for the most part unknown and although numerous studies have identified abiotic variables such as depth, salinity, temperature and substratum type that correlate with distribution, there may not necessarily be any causal relationship between them. Instead, or in addition, gradients of abiotic factors may be used to locate areas that fulfil the requirements of the particular life history stage. Consideration of scale is also relevant. Whereas large- and small-scale differences in distribution may be defined by depth, salinity and temperature, it is likely that biotic factors such as food availability and predator avoidance are important at smaller scales. Nevertheless, substratum type may be of greater importance in determining habitat occupation for flatfishes than for many other groups. It is certainly true that the substratum is the source of food for the great majority of flatfishes and whereas some species are piscivorous, most prey on invertebrates living in and/or on sediments. Chapter 9 reviews the trophic ecology of flatfishes first by defining the main feeding types and the factors affecting their diet and then by examining the range and effects of flatfish predators and competitors. The evidence for intra- and inter-specific competition is ambiguous but in some cases is strongly suspected. Predation on flatfishes, on the other hand, is intense – particularly in the youngest stages – and predators range from crustaceans through birds and mammals, not the least of which is man. These various aspects of trophic ecology are combined in a case study of Georges Bank in which the authors demonstrate shifts in abundance and species composition, potential competitive interactions and the extent of predation by and on flatfishes, which may be considerable.

Chapter 10 deals with the neglected topic of flatfish behaviour. It includes summaries of locomotion and colour change but concentrates on the key activities of spawning, feeding and reactions to predators that enable fishes to reproduce, grow and survive. These activities take place at a range of spatial scales from localised foraging to long-distance migrations and the chapter also includes a brief account of movement patterns that have evolved to make the most effective use of the environment. Much remains to be learnt about flatfish behaviour and recent studies have demonstrated that they spend much more of their time off the bottom than was previously realised. Finally, the role of behaviour in the capture and culture of flatfishes is discussed and its importance in attempts to augment and conserve wild flatfish populations through stock enhancement.

The last section of the book deals with the exploitation of flatfishes. Three chapters written in a similar format to allow comparison describe flatfish fisheries in the Atlantic and Pacific Oceans and the tropics. Each first describes the main species involved, their relative importance and the nature of the fisheries in the region. Following a short history of flatfish exploitation, an assessment is made of their economic importance and each chapter ends with a description and discussion of management strategies, results and problems for each region. The major fisheries are in the northern hemisphere where the larger species of righteye flounders (Pleuronectidae), soles (Soleidae) and lefteye flounders (Bothidae, Scophthalmidae) are the main targets and include the largest of all flatfishes, the halibuts. In the southern hemisphere catches are smaller and consist mostly of Soleidae, Bothidae, large-tooth flounders (Paralichthyidae) and some tonguefishes (Cynoglossidae). Tropical flatfishes, in contrast to those at higher latitudes, are mainly caught as by-catch rather than in targeted fisheries and the catches are much more diverse. Here, American soles (Achiridae), tonguefishes and psettodids (*Psettodes* spp.) also feature in the catches together with smaller numbers of species from other families. In terms of total fish catches, flatfishes form a small but significant proportion in the North Pacific and Atlantic (~10–30%) but much less in the southern hemisphere (<2% in the South Pacific) and the tropics. Nevertheless, they can command a high price and their economic value is often relatively much higher than landings statistics would suggest.

The increasing exploitation of flatfishes led to the realisation that management strategies would need to be put into place to protect stocks. Such strategies were first mooted in Europe at the beginning of the last century, somewhat later in the North Pacific, and are still largely lacking in the southern fisheries and tropics. For the most part, such strategies are overseen by international bodies but in some areas (e.g. Japan) may be under more local control. Regrettably, and for a variety of reasons discussed in the three chapters, many of these management plans have not prevented stock declines. There are success stories, however, as in yellowtail flounder (*Limanda ferruginea*) and Pacific halibut (*Hippoglossus stenolepis*), where a long history of research and management by the International Pacific Halibut Commission has resulted in one of the most successful fishery management programmes in the world. Successful management requires data and these are often lacking, particularly in the southern hemisphere and the tropics. In tropical regions, landings are gross underestimates of the total catch because by-catch and the products of artisanal fisheries are not included. In addition, landings are mostly not identified, so that catches of individual species are impossible to estimate. Locally, tropical flatfishes are becoming more important as a food source as catches of other groundfishes have declined. In these regions, solutions to over-fishing and habitat destruction will not be prevented by traditional approaches to fisheries management that attempt to regulate only resources. Rather, management will have to focus on people, not fish, to find solutions.

The penultimate chapter takes an overview of assessment and management in general but necessarily concentrates on those areas (the northeastern Pacific and the North Atlantic) where most data are available. The authors first consider aspects of flatfish population dynamics that are important for assessment and management, particularly the relationship between stock size and reproductive success which they consider to be the historic basis for evaluating stock dynamics and conservation targets and limits. They then illustrate how traditional approaches are changing as the exploration of the effects of environmental forcing on growth and recruitment proceeds. Finally, they present a summary of stock assessment results and harvesting policies currently in use in Europe and North America and conclude that although many stocks continue to decline some are stable or increasing. It seems that future developments will require greater use of the precautionary approach and an emphasis on co-operation between scientists, managers and fishermen, all of whom are ultimately seeking the same goal.

Apart from the need to regulate exploitation, the concern over the decline of wild stocks stimulated two other approaches to maintaining the supply of flatfishes for human consumption. First, the enhancement of wild stocks by supplementation with reared individuals and second, intensive farming. Both these approaches required the development of mass culture techniques. The final chapter in the book summarises the history of these two approaches and gives an overview of the current position and likely future developments. Although hatching eggs and rearing early larvae of some species had been successfully achieved by the beginning of the twentieth century, stocking the seas with these early stages failed and it was not until juveniles could be produced in very large numbers that stock enhancement became a more realistic possibility. The key to successful production of juveniles began with advances in larval rearing technology, particularly the use of *Artemia* as food, and the understanding of the importance of the microbial environment on larval survival. In combination with improvement in rearing techniques, the increasing ability to manipulate reproductive cycles of broodstock decreased reliance on the natural spawning cycle and allowed eggs to be produced over a longer period with a consequent increase in the numbers of reared individuals that could be produced. The consistent availability of millions of juveniles led to large stock enhancement programmes in Japan and on a smaller scale elsewhere. Such programmes are currently not economic without subsidy but cost-effectiveness may improve. Much remains to be done to understand the factors affecting the survival of released fishes. In contrast, intensive farming of several species is now well under way and many others are close to commercial exploitation. The exploitation of flatfishes is thus gradually moving into its 'agricultural' phase.

1.4 Nomenclature

Finally, a note on nomenclature. Like most groups the scientific nomenclature of flatfishes is in continual flux and, as with many other commercial species, their common names are confusing. A few examples will make the problem clear. 'Flounder' is widely used as a general term for all flatfishes and is applied with epithets to a great variety of species. 'Sole' is similarly but more restrictively used. Common names in the English language vary notoriously from place to place as in 'American plaice' and 'long rough dab' for *Hippoglossoides platessoides*. Furthermore, species described as a 'flounder' or a 'sole' may not necessarily reflect their systematic position. The 'English sole' (*Parophrys vetula*), for example, is neither English nor a sole. The question then arises of which system of nomenclature should be used in a book on flatfishes that is both unambiguous and reflects current systematic thinking. One approach would be to use the names used by the authors of individual chapters without any attempt to adopt a uniform system. On this basis, reference to the original publications cited would make the identity of species clear. However, this approach could lead to confusion, particularly as there have been major changes in flatfish nomenclature in the past few years, some of which are reflected in recent publications whereas others are not. The approach adopted here will hopefully minimise ambiguity and maximise the readability of the text. In each chapter a species is defined at first mention by its scientific and common name. Thereafter only the common name is used. The scientific names used are those given in FishBase (www.fishbase.org) except for those members of the Pleuronectidae whose names have been changed recently following the extensive taxonomic revision by Cooper & Chapleau (1998). Common names are those given in FishBase although in some cases no common names are available. In such cases the common name used by the chapter author has been used. Two appendices are provided at the end of the book that list the scientific and common names used throughout the text and the pleuronectid scientific names used before and after Cooper & Chapleau's (1998) revision.

Acknowledgements

I am grateful to Kevin Bailey, Andrew Cooper, S.J. de Groot, Tom Munroe and Tom Wilderbuer for their help and advice in writing this chapter.

References

Barrett, J.H., Nicholson, R.A. & Cerón-Carrasco, R. (1999) Archaeo-ichthyological evidence for long-term socioeconomic trends in northern Scotland: 3500BC to AD 1500. *Journal of Archaeological Science,* **26**, 353–388.

Berendzen, P.R. & Dimmick, W.W. (2002) Phylogenetic relationships of Pleuronectiformes based on molecular evidence. *Copeia,* **2002**, 642–652.

Beverton, R.J.H. & Holt, S.J. (1957) On the dynamics of exploited fish populations. *Fishery Investigations, Series II, Marine Fisheries, Great Britain Ministry of Agriculture, Fisheries and Food,* **19**, 1–533.

Blaxter, J.H.S. (1975) The eggs and larvae of fish: a century of experimental research with special reference to Scotland. In: *Development of Fisheries Research in Scotland* (eds R.S. Bailey & B.B. Parrish), pp. 112–122. Fishing News Books Ltd, Farnham, England.

Blaxter, J.H.S. (1986) Development of sense organs and behaviour of teleost larvae with special reference to feeding and predator avoidance. *Transactions of the American Fisheries Society,* **115**, 98–114.

Blaxter, J.H.S. (2000) The enhancement of marine fisheries. *Advances in Marine Biology,* **38**, 1–54.

Boklage, C.E. (1984) On the inheritance of directional asymmetry (sidedness) in the starry flounder, *Platichthys stellatus*: additional analyses of Policansky's data. *Behavioral and Brain Sciences,* **7**, 725–762.

Cooper, J.A. & Chapleau, F. (1998) Monophyly and intrarelationships of the family Pleuronectidae (Pleuronectiformes) with a revised classification. *Fishery Bulletin,* **96**, 686–726.

Daan, N., Bromley, P.J., Hislop, J.R.G. & Nielsen, N.A. (1990) Ecology of North Sea fish. *Netherlands Journal of Sea Research,* **26**, 343–386.

Dannevig, H. (1897) On the rearing of the larval stages of the plaice and other flatfishes. *Annual Report of the Fishery Board for Scotland 1896,* 175–193.

Ewart, J.C. (1885) Report on the progress of fish culture in America. *Annual Report of the Fishery Board for Scotland 1884–1885*, Appendix F, (8), 78–91.

Gibson, R.N., Ansell, A.D. & Robb, L. (1993) Seasonal and annual variations in abundance and species composition of fish and macrocrustacean communities on a Scottish sandy beach. *Marine Ecology Progress Series,* **98**, 89–105.

Harden Jones, F.R. (1968) *Fish Migration*. Edward Arnold Ltd, London.

Healey, E.G. (1999) The skin pattern of young plaice and its modification in response to graded changes in background tint and pattern. *Journal of Fish Biology,* **55**, 937–971.

Hensley, D.A. (1997) An overview of the systematics of flatfishes. *Journal of Sea Research,* **37**, 187–194.

Holt, E.W.L. (1895) An examination of the present state of the Grimsby trawl fishery, with especial reference to the destruction of immature fish. *Journal of the Marine Biological Association of the United Kingdom,* **3**, 339–448.

McManus, I.C. (1984) The inheritance of asymmetries in man and flatfish. *Behavioral and Brain Sciences,* **7**, 731–733.

Mast, S.O. (1914) Changes in shade, color, and pattern in fishes, and their bearing on the problems of adaptation and behavior, with especial reference to the flounders *Paralichthys* and *Ancylopsetta*. *Bulletin of the United States Bureau of Fisheries,* **34**, 177–238.

Metcalfe, J.D. & Arnold, G.P. (1997) Tracking fish with electronic tags. *Nature,* **387**, 665–666.

Morgan, M.J. (1991) The asymmetrical genetic determination of laterality: flatfish, frogs and human handedness. In: *Biological Asymmetry and Handedness. Ciba Foundation Symposium 162*, pp. 234–250. Wiley, Chichester.

Muus, B.J. & Nielsen, J.G. (1999) *Sea Fish*. Scandinavian Fishing Yearbook, Hedehusene, Denmark.

Nicholson, R.A. (1998) Fishing in the Northern Isles: a case study based on fish bone assemblages from two multi-period sites on Sanday, Orkney. *Environmental Archaeology,* **2**, 15–28.

Petersen, C.G.J. (1894) On the biology of our flatfishes and on the decrease of our flatfish fisheries. *Report of the Danish Biological Station 1893*, 1–146.

Policansky, D. (1982a) The asymmetry of flounders. *Scientific American,* **246**, 96–102.

Policansky, D. (1982b) Flatfishes and the inheritance of asymmetries. *Behavioral and Brain Sciences,* **5**, 263–265.

Ramachandran, V.S., Tyler, C.W., Gregory, R.L., *et al.* (1996). Rapid adaptive camouflage in tropical flounders. *Nature,* **379**, 815–818.

Rozwadowski, H.M. (2002) *The Sea Knows No Boundaries: A Century of Marine Science under ICES*. ICES Copenhagen and University of Washington Press, Seattle.

Shelbourne, J.E. (1964) The artificial propagation of marine fish. *Advances in Marine Biology,* **2**, 1–83.

Smith, T.D. (1994) *Scaling Fisheries: The Science of Measuring the Effects of Fishing, 1885–1955*. Cambridge University Press, Cambridge.

Chapter 2
Systematic diversity of the Pleuronectiformes

Thomas A. Munroe

2.1 Introduction

Flatfishes are generally deep-bodied, laterally compressed fishes – easily and immediately recognised anatomically in that juveniles and adults (post-metamorphic individuals) have both eyes present on the same side of the head (Figs 2.1 and 2.2). All flatfishes begin life as pelagic, bilaterally symmetrical fishes. During larval development, however, flatfishes undergo a spectacular ontogenetic metamorphosis where one eye migrates from one side of the head to the other (Brewster 1987). Depending upon the species, either the right or left eye migrates. Some species are indeterminate with respect to which eye migrates; in others eye migration appears to be genetically fixed. The eyes may or may not come to lie in close proximity to each other when eye migration is completed. Further deviations from a bilaterally symmetrical body plan occur in various external and internal structures, including placement of nostrils on the head, differential development of osteological features (especially bones in anterior head skeleton), differences in jaw shape and dentition on either side of the body, degree of development of lateral body musculature, lateral-line development on either side of the body, differential coloration on ocular and blind sides (most species with uniformly whitish or yellowish blind side, others with darkly pigmented blind side), and differences in paired fin development on ocular and blind sides of the body. As a group, flatfishes are the only vertebrates to deviate so radically from a bilaterally symmetrical body plan. Flatfishes share unique morphological specialisations related to this asymmetry, and although earlier hypotheses proposed that flatfishes share a common ancestor with some as yet unidentified perciform group of symmetrical fishes, the origin and sister group of flatfishes remain unknown (Chapleau 1993).

Following eye migration, flatfishes settle out of the water column to assume a benthic lifestyle. Flatfishes generally lie on the bottom on their blind side. They often are found either lying on top of the substratum or partially buried under a fine layer of sand or silt with only their eyes protruding above the substratum. Flatfishes are found on a variety of substrata including silt, mud, sand and sand–shell mixtures, with some species also occurring on rocky or pebbly bottoms. Some species appear to have strong preferences for particular substrata, whereas others are found on a variety of bottom types. In some species, newly settled, juvenile, and adult life stages reside on different substrata (Allen & Baltz 1997; Phelan *et al.* 2001).

Flatfishes represent an interesting and diverse order of marine, estuarine, and to a lesser extent, freshwater euteleostean fishes (Figs 2.1 and 2.2). Flatfishes are well known organisms as they occur in all of the world's oceans, are represented by large numbers of species and genera (Table 2.1), and in some regions their populations are sufficiently large to constitute major fishery resources. They are extremely successful in conducting life on or near the bottom, where they function in pivotal ecological roles as both predator and prey. Within their largely benthic or demersal lifestyles, flatfishes display amazing diversity in size, shape,

Fig. 2.1 Representatives of seven lineages of pleuronectiform fishes. a. Psettodidae; b. Citharidae; c. *Tephrinectes*; d. Scophthalmidae; e. Bothidae; f. Paralichthyidae; g. Pleuronectidae.

12 Flatfishes

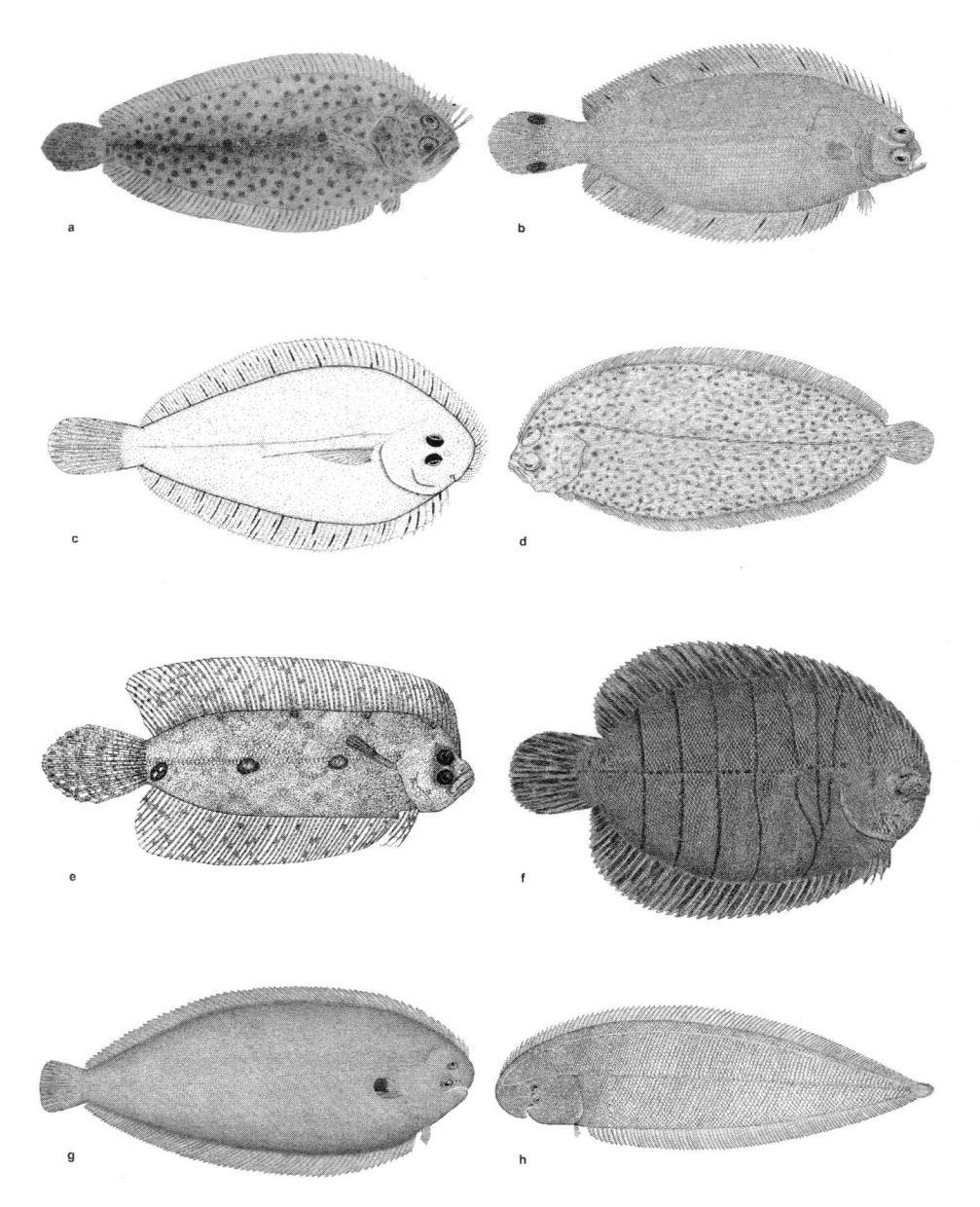

Fig. 2.2 Representatives of eight lineages of pleuronectiform fishes. a. Paralichthodidae; b. Poecilopsettidae; c. Rhombosoleidae; d. Achiropsettidae; e. Samaridae; f. Achiridae; g. Soleidae; h. Cynoglossidae.

habitat and trophic ecologies. How this diversity in morphology, biology and specialised adaptations translates into the evolutionary and ecological success of these fishes is of interest to evolutionary biologists, fisheries ecologists and ichthyologists alike.

Taxonomically, the best known flatfish faunas are those occurring in areas that support large commercial fisheries. These fisheries are primarily located in the northern hemisphere in both Atlantic and Pacific Oceans (Families Pleuronectidae, Scophthalmidae and some

Table 2.1 Summary of taxonomic information, presented as numbers of genera and species by family (except *Tephrinectes*) for pleuronectiform fishes (for comparison purposes, all genera and species of flatfishes listed in earlier studies are arranged according to families recognised in the present study)

Family	Norman (1934) Genera	Norman (1934) Species	Chabanaud (1939) Genera	Chabanaud (1939) Species	Present study[1] Genera	Present study[1] Species
Psettodidae	1	2	1	2	1	2
Citharidae	2	3	2	3	4	7
Scophthalmidae	3	10	4	10	4	9
Paralichthyidae	21	93	21	97	11	95
Bothidae	11	79	11	89	20	145
Pleuronectidae	28	61	27	65	21	60
Tephrinectes	1	1	1	1	1	1
Paralichthodidae	1	1	1	1	1	1
Poecilopsettidae	3	11	3	16	4	30
Rhombosoleidae	8	16	8	16	9	19
Achiropsettidae	2	2	2	2	4	4
Samaridae	4	13	2	15	2	28
Achiridae	–	–	10	31	9	31
Soleidae	–	–	28	86	29	139
Cynoglossidae	–	–	4	117	3	145
Totals	85	292	125	551	121	716

[1] Includes data for nominal species recognised by taxonomists, but not formally described.

representatives of Soleidae and Paralichthyidae). South temperate regions also support commercial fisheries for other groups including the Rhombosoleidae (Australia, New Zealand) and Paralichthyidae (South America; Díaz de Astarloa & Munroe 1998). In tropical areas, flatfishes occur in a variety of habitats including mangrove estuaries and adjacent mud flats, in seagrass beds and on mud bottoms. Within reef-associated habitats, which are broadly distributed across tropical oceans but have a more patchy distribution on a smaller scale, flatfishes are found on reef flats where substrata consist mostly of sand with algae, around coral outcrops or soft corals, and in lagoons associated with reefs. They also occur on back-reef slope areas where sediment deposition occurs and coral development is poorer, as well as on sandy substrata interspersed around reef spurs. The majority of flatfishes inhabiting the Indo-Pacific region, especially species of Bothidae, Samaridae, Poecilopsettidae, Soleidae and Cynoglossidae, are relatively small fishes generally not of commercial importance. Other tropical flatfishes, especially larger species (Psettodidae, and some Paralichthyidae, Cynoglossidae, Soleidae and Bothidae), are captured on a more regular basis in tropical fisheries and, for these, better (although still limited) taxonomic and ecological data are available. The great diversity of difficult-to-identify and usually small species of tropical Indo-Pacific flatfishes has largely gone unstudied by ecologists and ichthyologists, except for the few taxonomists working on these fishes. Consequently, in contrast to the situation for flatfishes occurring in northern areas, the taxonomy and ecology of tropical flatfishes, especially deep-water species, are generally less well known. This disparity in taxonomic and ecological information between tropical flatfishes and those from more northern areas is particularly poignant because it is in tropical waters, especially the Indo-Pacific region, where the greatest diversity of flatfishes is found.

This progress report evaluates our understanding of the evolution and taxonomic diversity of flatfishes. Diversity data are discussed: (1) in the context of evolutionary relationships among flatfishes; (2) in taxonomic diversity as expressed by numbers of families, genera and species recognised by systematists; and (3) with respect to the diversity of habitats (e.g. major trends in geographical and bathymetric distributions) where flatfishes are found. Other aspects of taxonomic diversity, such as intraspecific or population variation, or ecological diversity (feeding morphologies, reproductive strategies, etc.), although interesting and important in their own right, are beyond the scope of this chapter.

2.2 Systematic profile of the Pleuronectiformes

The oldest flatfish fossils, otoliths dating from the Early Eocene some 53–57 million years ago (Mya), indicate the presence of true Pleuronectiformes as far back as the early Tertiary (Schwarzhans 1999). *Eobothus minimus* (Agassiz 1834–1842), a representative of the bothoid lineage with uncertain affinities within the group, is the oldest existing skeleton representative of the Pleuronectiformes, dating at least to the Lutetian (some 45 Mya) in the Eocene (Norman 1934; Chanet 1997, 1999). This species already represents an advanced evolutionary line within the Pleuronectoidei, the main suborder within the Pleuronectiformes (Chanet 1999). The oldest soleids, *Eobuglossus eocenicus* and *Turahbuglossus cuvillieri*, both known from single specimens from the Upper Lutetian of Egypt (Chabanaud 1937; Chanet 1994, 1997), are also among the first known flatfish fossils and they are identical to skeletons of recent soleids. The earliest bothid fossil, a surprisingly 'modern'-looking species of the genus *Arnoglossus,* also dates from the Eocene (Schwarzhans 1999), as does a fossil species of Pleuronectidae (Bannikov & Parin 1997).

The sudden appearance of fossils representing several different lineages of flatfishes, and encompassing nearly all of the structural features and diversity of the Order (Arambourg 1927), indicates that considerable diversification occurred in these fishes earlier than 45 Mya, before the Lutetian (Chanet 1997; Schwarzhans 1999). These early fossils also highlight the fact that in the history of flatfishes, anatomical specialisations including asymmetry of the skull, supracranial extension of the dorsal fin and modifications of the caudal skeleton occurred earlier than these fossil flatfishes.

When flatfishes evolved, and how rapidly they diversified, are still unresolved questions. Fossil flatfishes are relatively rare (only about 250 specimens known worldwide), so knowledge concerning the history of this group is very incomplete (Chanet 1997, 1999). Soleids are the best represented family in the fossil record (Chanet 1999) with a minimum date of origin at about 45 Mya. This time-frame is also the minimum time of divergence between the Soleidae and Cynoglossidae. Fossil pleuronectiform otoliths are unknown from true freshwater sediments (Schwarzhans 1999), which may indicate that the ancestor of this group was a marine fish. And the appearance of several lineages of flatfishes within the same deposits may indicate that diversification was sudden. The gap in fish fossils between the Late Campanian (ca. 75 Mya) and the Late Paleocene (ca. 55 Mya) impedes unravelling the evolutionary history of most acanthomorph fish groups (Patterson 1993), including flatfishes. Pleuronectiform fishes may have existed during this time, but their skeletons may not have been fossilised, or if these fossils exist, they have not been discovered or identified (Chanet 1999). Despite these gaps in information, it is abundantly evident that morphological specialisations characteristic of the Pleuronectiformes evolved long ago and that diversification of major lineages of flatfishes also occurred in the distant past. The pleuronectiform body plan has had a long and successful presence among marine teleost fish assemblages, with a minimum of 50 million years available for diversification of these fishes. For further discussions on fossil flatfishes consult the works of Chanet (1997, 1999), Schwarzhans (1999), and literature cited therein.

2.3 Intrarelationships of the Pleuronectiformes

Intrarelationships among flatfishes are not well resolved. Earlier hypotheses of higher level relationships among the Order Pleuronectiformes were based mainly on the works of Regan (1910, 1929), Norman (1934, 1966) and Hubbs (1945). These hypotheses have undergone considerable changes over time. Amaoka (1969) provided detailed osteological study of sinistral flounders of Japan and sparked renewed interest in re-examining earlier hypotheses regarding flatfish intrarelationships. A phenetic study on right-eyed flounders was conducted by Sakamoto (1984) who identified many useful characters shared by these fishes. Lauder & Liem's study (1983) first applied cladistic methodology to hypothesise relationships of flatfishes as part of a broad-based re-examination of phylogeny of ray-finned fishes. This hypothesis was, however, based on limited data, featuring relatively few characters. Hensley & Ahlstrom (1984) and Ahlstrom et al. (1984) provided a detailed synthesis of knowledge on classification and larval morphology of the Pleuronectiformes. They pointed out weaknesses of earlier classifications, but did not produce a cladogram reflecting their hypotheses of intrarelationships of the flatfishes. First attempts at cladistic hypotheses of relationships were proposed for the Cynoglossidae by Chapleau (1988) and for the Soleidae by Chapleau & Keast (1988).

Chapleau (1993) used available ordered and polarised morphological characters to produce a phylogenetic hypothesis of pleuronectiform intrarelationships. He was able to corroborate that the Order Pleuronectiformes is monophyletic based on three characters: ontogenetic migration of one of the eyes, anterior position of the dorsal fin origin (dorsal fin overlaps the cranium) and the presence of a recessus orbitalis, a muscular sac-like evagination in the membranous wall of the orbit that can be filled with fluid causing protrusion of the eyes to a higher position above the surface of the head (and above substratum when the fish is buried). Chapleau (1993) also showed that many flatfish groups traditionally recognised at the family and subfamily levels could not be interpreted as monophyletic units. His study suggested that convergence, reversal and parallelism are important components of flatfish evolution and that lack of detailed phylogenetic studies for several pleuronectoid taxa hinders the understanding of intrarelationships of flatfishes even at the family level. He identified several areas of investigation necessary for achieving better understanding of evolutionary relationships of flatfishes. Hensley (1997) also discussed recent changes in flatfish classification and further reiterated critical research areas in need of study on systematics and biogeography of pleuronectiform fishes.

Since the publication of Chapleau's (1993) cladistic hypothesis of intrarelationships, other researchers have continued the process of delineating monophyletic taxa within the Order and further developed hypotheses of relationships for various pleuronectiform sub-

```
                    ╱ Psettodidae
                   ╱ Citharidae
                  ╱ Tephrinectes
                 ╱ Scophthalmidae
                ╱ "Paralichthyidae"
               ╱─ Bothidae
                ╲ Pleuronectidae
                ╱ Paralichthodidae
               ╱ Poecilopsettidae
              ╱ Rhombosoleidae
             ╱ Achiropsettidae
            ╱ Samaridae
           ╱ Achiridae
          ╱─ Soleidae
             ╲ Cynoglossidae
```

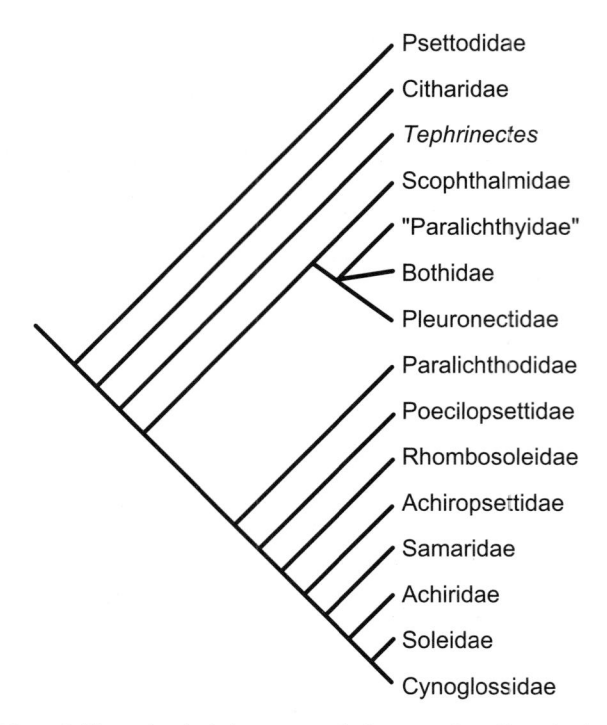

Fig. 2.3 Schematic illustration depicting a composite interpretation of hypothesised relationships of the Pleuronectiformes based on morphological information presented in recently published phylogenetic analyses. Hypothesised relationships generally follow those presented in Chapleau (1993) with additional information incorporated as follows: recognition of Citharidae and placement of *Tephrinectes* following Hoshino (2001), placement of Paralichthodidae following Cooper & Chapleau (1998b), and placement of Achiropsettidae following Evseenko (2000).

groups. Examples of such research are discussed below under the family synopses. Currently, 14 families of flatfishes are recognised, with *Tephrinectes* also representing a distinct lineage within the Order (Fig. 2.3). Because systematic studies on intrarelationships of the Pleuronectiformes are ongoing, knowledge of historical relationships for flatfishes will be a work in progress for some time to come. This active, dynamic area of systematic research will probably necessitate changes in interpretations of higher relationships among flatfishes. In particular, present concepts of some family-level taxa within the Order (e.g. the Paralichthyidae) will need refinement as we learn more about evolutionary relationships of constituent taxa. Molecular evidence incorporated into phylogenetic analyses is just beginning to deliver results that will provide new frameworks for examining intrarelationships of the Pleuronectiformes (Berendzen & Dimmick 2000).

To provide a platform to summarise information on species diversity of flatfishes, some arbitrary groupings of taxa into traditionally recognised categories (usually families) with uncorroborated monophyly were necessary in the present study. A schematic illustration incorporating results from morphological studies presenting recently published hypotheses on intrarelationships of flatfishes (Fig. 2.3) serves as the framework for comparisons among flatfish lineages for the remainder of the chapter. Although the phylogenetic relationships of some families and the monophyly of others (e.g. Paralichthyidae) are in doubt, for the most part, these groups represent well-diagnosed lineages.

2.4 Brief synopses of the suborders and families

Two major lineages of flatfishes are recognised: the Psettoidei, comprising the family Psettodidae, and the Pleuronectoidei, containing all remaining flatfish groups. Detailed discussion of characters used to support hypotheses of relationships proposed in this phylogenetic framework is provided in Chapleau (1993), Cooper & Chapleau (1998a, b), Hoshino & Amaoka (1998), Chanet (1998), Evseenko (2000) and Hoshino (2001).

Psettodidae (Fig. 2.1a). The psettodids, or toothed flounders, are a basal group of flatfishes hypothesised to be the sister group for the Pleuronectoidei. This suborder comprises one family, the Psettodidae, with two species of *Psettodes*. These large flatfishes, with both dextral and sinistral individuals in populations, are characterised by several derived internal features discussed in Chapleau (1993). Externally, these fishes are easily recognised by such plesiomorphic characters as the posterior location of the dorsal fin, which does not advance onto the cranium anterior to the eyes, occurrence of spines in dorsal and anal fins, large mouths with specialised teeth, and nearly rounded bodies without the obvious bilateral asymmetry in lateral musculature development evident in other flatfishes.

The Pleuronectoidei contains all of the more familiar flatfishes. Hensley & Ahlstrom (1984) considered this suborder to comprise all flatfishes except the Psettodidae and soleoid taxa (Cynoglossidae, Achiridae and Soleidae). Chapleau & Keast (1988) determined that the suborder Pleuronectoidei of Hensley & Ahlstrom (1984) was paraphyletic; to become monophyletic it would also have to include all soleoid taxa. Chapleau & Keast (1988) also recommended that the Pleuronectinae, Poecilopsettinae, Rhombosoleinae and Samarinae be raised to family rank. They further suggested that more work be done to define the monophyletic status of these former subfamilies and to determine intrarelationships among the Pleuronectiformes.

Citharidae (Fig. 2.1b). Considerable controversy has surrounded recognition and systematic placement of this family – comprising four genera and seven small to medium-sized species collectively referred to as 'large-scale flounders'. Hubbs (1945) erected this family by regrouping two genera formerly placed in the Bothidae (sinistral taxa) and Pleuronectidae (dextral taxa). Inclusion of genera featuring opposite ocular asymmetries in the same family deviated radically from earlier traditional hypotheses that had grouped flatfish taxa heavily weighted on ocular asymmetry. Hensley & Ahlstrom (1984) doubted the monophyly of the family, and Chapleau's (1993) cladistic analysis of the Order also could not support the monophyly of this family. Recently, Hoshino (2000, 2001) re-examined the status of five genera and six species placed in the Citharidae and concluded that indeed these flatfishes did form a monophyletic group that should be recognised at the family level.

Tephrinectes (Fig. 2.1c). This genus contains only one species, the flower flounder, *Tephrinectes sinensis*, which occurs in coastal seas off China. Populations feature both sinistral and dextral individuals. Hoshino & Amaoka (1998) conducted detailed study of the osteology of this problematic species, formerly placed in a monotypic genus in the Paralichthyinae or Paralichthyidae. These authors concluded that this species was not a member of the Paralichthyidae and recommended removing it from that family. They found characters that warranted recognising *T. sinensis* as a distinct lineage with hypothesised relationships placing it as the sister group of a clade (clade IV of Chapleau 1993) containing the Poecilopsettidae, Rhombosoleidae, Samaridae, Achiridae, Soleidae and Cynoglossidae. Hoshino (2001), based on additional characters, later hypothesised a more basal position for *Tephrinectes*, i.e. following that of the Citharidae, and proposed that this taxon was the sister group to remaining pleuronectoids.

Scophthalmidae (Fig. 2.1d). A small family consisting of four genera with about nine species of small to large-sized sinistral flatfishes characterised by a relatively large mouth and large eyes, two elongated pelvic fin bases (slightly asymmetrical) extending anteriorly to the urohyal, an elongated supra-occipital process forming a bridge with the dorsal margin of the blind-side frontal bone, and caudal vertebrae with asymmetrical transverse apophyses (Hensley & Ahlstrom 1984; Chapleau 1993; Chanet 1998). Historically (Norman 1934), scophthalmid flatfishes were classified as a subfamily within an expanded Bothidae. The larger species have commercial importance and some species are also important in aquaculture. A cladistic appraisal of the Scophthalmidae was recently published (Chanet 2003).

Bothidae (Fig. 2.1e). This large, diverse family of sinistral flatfishes contains about 20 genera and 145 species. Hensley & Ahlstrom (1984) detailed the history of systematic hypotheses regarding classification of the Bothidae. Amaoka (1969) restricted the family by removing the paralichthyines (see above), and elevated the subfamily Bothinae to the rank of family. He also proposed recognising two subfamilies, Taeniopsettinae containing four genera, and the Bothinae containing all other bothids. Hensley & Ahlstrom (1984), however, discussed morphological evidence inconsistent with the monophyly of the Taeniopsettinae. Other genera formerly contained within the Bothidae were removed to the family Achiropsettidae (Evseenko 1984, 2000). Chapleau (1993) concluded that the Bothidae was monophyletic, but that the status of subfamilies needed further substantiation. Many new species of bothid flounders have been discovered recently in Indo-Pacific waters (Amaoka *et al.* 1993, 1997a, b; Amaoka & Mihara 2000) confirming this species-rich family as among the most diverse of the Pleuronectiformes. Larger species enter artisanal and commercial fisheries, especially throughout tropical areas.

Paralichthyidae (Fig. 2.1f). A non-monophyletic family of sinistral (most species; a few species with both sinistral and dextral individuals) flatfishes containing 11 genera and about 95 species. These taxa are regarded herein as a family until further study demonstrates the monophyly and relationships of and within the group, or refines the status of monophyletic subunits presently considered as component taxa of this family. This group of taxa was historically considered to be a subfamily of an expanded Bothidae. Amaoka (1969) proposed elevation of these taxa to family level. Hensley & Ahlstrom (1984) thoroughly discussed changes in composition of this taxon since Norman (1934). Several groups of genera contained in the family were identified as possible monophyletic groups by Hensley and Ahlstrom, but the family itself could not be defined by any synapomorphy. Chapleau (1993) also was unable to establish the monophyly of this family and concluded that further work was needed to clarify relationships of these fishes. Hoshino (1999) provided further evidence that the Paralichthyidae is not monophyletic. Commercial and artisanal fisheries target larger paralichthyid flounders (especially species of *Paralichthys, Pseudorhombus* and *Cyclopsetta*) and some paralichthyids are important aquaculture species.

Pleuronectidae (Fig. 2.1g). Historically (Norman 1934; Sakamoto 1984), the Pleuronectidae included five subfamilies (Pleuronectinae, Paralichthodinae, Rhombosoleinae, Samarinae and Poecilopsettinae). All of these subfamilies have been elevated to family status (Chapleau & Keast 1988; Chapleau 1993). In a phenetic study of 77 species of an expanded

Pleuronectidae, Sakamoto (1984) defined the family based on several plesiomorphic characters (Chapleau 1993). Using information from Sakamoto's study for members of the subfamily Pleuronectinae, as well as new characters, Cooper & Chapleau (1998a) elevated this subfamily to family status and provided evidence corroborating its monophyly. A resolved species-level cladogram for the Pleuronectidae, the first such hypothesis to be derived for a speciose family of flatfishes, was provided by Cooper and Chapleau. This relatively large and diverse family of marine and estuarine, dextral (usually), flatfishes contains some 21 genera and about 60 species including many commercially important species, such as halibuts, the largest species of flatfishes. Several pleuronectid species are considered good candidates for aquaculture.

Paralichthodidae (Fig. 2.2a). This monotypic family contains only the peppered flounder, *Paralichthodus algoensis*, a medium-sized (to about 50 cm), dextral flatfish endemic to the inner continental shelf of South Africa (Heemstra 1986). Classification and relationships of this species have also been problematic. Cooper & Chapleau (1998b) reviewed the history of classification of this species, established its monophyly and determined its phylogenetic position within the Pleuronectiformes as sister group to a clade that unites the Poecilopsettidae, Rhombosoleidae, Samaridae, Achiridae, Soleidae and Cynoglossidae.

Poecilopsettidae (Fig. 2.2b). This relatively small family consists of four genera with approximately 30 species. Most are small-sized, dextral, marine flatfishes of deep-water habitats and rather fragile appearance. Historically (Norman 1934), poecilopsettids were regarded as a subfamily of an expanded Pleuronectidae. Descriptive osteology and hypothesised relationships for members of this family were presented in Sakamoto (1984). A study to demonstrate the monophyly and intrarelationships of this family is in progress (Guibord, unpublished data).

Rhombosoleidae (Fig. 2.2c). This relatively small family of dextral, marine and estuarine flatfishes, some of commercial importance, comprises about nine genera and 19 species that occur primarily in relatively shallow marine waters in the southern hemisphere. Norman (1934) regarded rhombosoleines as a subfamily of an expanded Pleuronectidae. Chapleau & Keast (1988) recommended raising this subfamily to familial rank. Chapleau (1993) briefly discussed the need for more research on this family to define its monophyletic status and to determine its relationships. Such a study examining intrarelationships of these fishes is in progress (Guibord, unpublished data).

Achiropsettidae (Fig. 2.2d). Achiropsettid flounders represent a small family of marine flatfishes that occur primarily in the Southern Ocean. At least nine nominal species of achiropsettid flounders have been described, but a recent review of the family (Evseenko 2000) recognised only four species in four monotypic genera. Many nominal species are known only from a single specimen. Evseenko (2000) noted that several undescribed species were perhaps present in collections, but study material was insufficient to accurately diagnose these nominal species. Three genera now included in the Achiropsettidae (*Achiropsetta, Neoachiropsetta* and *Mancopsetta*) were previously placed in the Bothidae. Evseenko (1984) erected the family Achiropsettidae for these three genera together with *Pseudomancopsetta*. Hensley & Ahlstrom (1984) and Hensley (1986) suggested that these genera (and several others formerly placed in Bothidae) should be removed from the Bothidae and united with the Rhombosoleidae. Evseenko (1984, 1996) suggested closer relationship of these taxa to *Brachypleura* of the Citharidae, or alternatively (Evseenko 1996), that achiropsettids were more closely related to the Poecilopsettidae or Samaridae. Recently, Evseenko (2000) provided data supporting the monophyly of the family and hypothesised the Achiropsettidae as the outgroup to a clade comprising the Samaridae, Achiridae, Soleidae and Cynoglossidae.

Samaridae (Fig. 2.2e). This family of small-sized, dextral flatfishes containing about two genera and some 28+ nominal species occurs in relatively deep tropical and subtropical marine waters of the Indo-Pacific. Samarid flatfishes were formerly regarded as a subfamily of an expanded Pleuronectidae (Norman 1934). Sakamoto (1984) detailed osteological characters of the group that provided a well corroborated hypothesis for its monophyly. Independent studies on species-level systematics of the Samaridae (Alphonso, unpublished data; Amaoka, unpublished data) continue to discover undescribed species in recently collected material, indicating that considerably more work will be needed to assess species diversity accurately within this family.

Achiridae (Fig. 2.2f). A relatively small, but diverse family (about nine genera with some 31 species) of mostly small, dextral flatfishes found in temperate and tropical freshwater, estuarine and coastal marine waters of the Americas. A few achirid species are taken regularly in subsistence or artisanal fisheries of Central and South America, while other species contribute to by-catch in shrimp fisheries. Historically, achirids were regarded as a subfamily of an expanded Soleidae. Chapleau & Keast (1988) established the monophyly of the Achiridae based on six characters. Ramos (1998) also corroborated the monophyly of this family and proposed a phylogenetic hypothesis of intrarelationships of achirid taxa.

Soleidae (Fig. 2.2g). This diverse family of specialised, dextral flatfishes of mostly small to medium size are found worldwide in a variety of marine and estuarine habitats, as well as a few species in freshwater. Chapleau & Keast (1988) established the monophyly of the Soleidae based on five characters. About 29 genera with approximately 139 species are currently recognised within the Soleidae, but the status of many species and genera, and their intrarelationships, remain poorly known. The monophyly and placement of some subgroups within the family have been established (Desoutter 1994; Chapleau & Desoutter 1996; Desoutter-Meninger 1997) and progress has been made in resolving species systematics of these flatfishes (Desoutter 1987, 1994; Desoutter & Chapleau 1997). Much work remains in identifying monophyletic taxa, especially Indo-Pacific genera, within this species-rich family. Those soleid species that reach sufficiently large size and attain large populations are highly desirable food fishes targeted by commercial fisheries. The majority of tropical species, primarily because of small size, low population density and poorly known systematics, remain less well known.

Cynoglossidae (Fig. 2.2h). Tonguefishes are a diverse family of specialised marine, estuarine and freshwater (few species) flatfishes containing about 145 primarily small, sinistral species distributed in about three genera. Chapleau (1988) provided convincing evidence based on 27 derived characters that corroborate the monophyly of this family. Diagnosis of monophyletic genera and their intrarelationships within the family still require further study. Species of *Cynoglossus* (Menon 1977) and *Paraplagusia* (Chapleau & Renaud 1993) have been revised. Various geographic assemblages of species within the species-rich genus *Symphurus* have also been revised (Munroe 1990, 1998) or updated (Munroe 1992; Munroe et al. 1995; Munroe & Marsh 1997; Munroe & Amaoka 1998). Species-level taxonomy of *Cynoglossus* remains problematic, and new species of *Symphurus* continue to be discovered, especially from Indo-Pacific deep-water habitats.

2.5 Diversity of the Pleuronectiformes

2.5.1 Overview

An important step towards understanding the diversity of flatfishes is to evaluate progress being made on the species-level systematics of these fishes. Species numbers were used for these comparisons as they represent the most straightforward measure for comparing diversity between and among families or other taxa. In so doing, a variety of relevant questions can be addressed: How close are we to knowing the world flatfish fauna? How reliable are current estimates for standing diversity (i.e. best current estimate of total diversity) of flatfishes? What progress has been made in discovering the diversity represented among flatfishes? How have recent accumulations in knowledge about flatfish species diversity changed our viewpoints regarding species diversity within families and other taxa? How well known are flatfish assemblages occurring in various geographic regions? What factors contribute to recent changes in estimating the standing diversity of flatfishes? What geographic areas or latitudes are under-sampled? What methodological techniques or approaches, or changes in systematic philosophy, offer the greatest likelihood of providing further information and refinement to these estimates? Some answers to these questions are provided below.

2.5.2 Flatfish species diversity

More than 60 years have elapsed since the publication of Norman's (1934) classic treatise on flatfishes and Chabanaud's (1939) summary checklist of species diversity among flatfishes. Since these works, no ichthyologist has attempted a comprehensive review of systematic and biogeographic information for these fishes. Norman (1934) summarised available systematic information for the flatfishes in a classic work on this group. However, taxonomic information for Soleidae, Achiridae and Cynoglossidae was not included. For families of flatfishes where taxonomic information was considered, Norman recognised 292 species placed in 85 genera (Table 2.1). Later, Chabanaud (1939) summarised taxonomic information for species of Pleuronectiformes he considered valid, including those in families not addressed in Norman's study. Chabanaud's estimate in 1939 for the standing diversity of the Pleuronectiformes totalled 551 species placed in 125 genera. Of families and subfamilies summarised in Norman's study, Chabanaud (1939) recognised 24 additional species and two fewer genera.

More recent estimates of flatfish species diversity include that of Nelson (1994) with approximately 570 species in 123 genera and about 11 families. Hensley (1997) estimated flatfish diversity at 570–620 species, whereas Chapleau & Amaoka (1998) estimated only about 540 species in 117 genera in 7 families. These estimates represent coarse-scale approximations of species diversity for this group because they are unclear as to how the numbers of taxa were estimated. Also unclear is whether these estimates were based on detailed evaluations of the species-description literature. Because Chabanaud's (1939) study represents the only comprehensive estimate of species diversity based on the species-description literature and partitioned by family and genus for all taxa within the Order Pleuronectiformes, it serves as a useful baseline to evaluate progress being made on species discovery within this group of fishes.

To examine current levels of knowledge regarding species-level diversity of the flatfishes, I constructed a database following review and evaluation of a wide variety of primary and secondary literature, including earlier summaries of flatfish diversity (Norman 1934; Chabanaud 1939), a variety of authoritative taxonomic revisions, 100+ regional faunal works, recently published FAO species identification guides (Fischer *et al.* 1995; Carpenter & Niem 2001), and information contained in Eschmeyer (1998) and FishBase (1997). In addition, taxonomists currently conducting systematic investigations on flatfishes were consulted for input regarding published and unpublished information on various groups of flatfishes. These experts provided invaluable assistance that was helpful in estimating the validity and status of nominal species of flatfishes from all major groups of Pleuronectiformes and they also provided estimates of recognised, but undescribed, species among various families under their study.

Of approximately 1339 nominal species of flatfishes described, named or recognised, 716 species are considered valid (e.g. recognised by taxonomic authorities), while another 670 names are regarded as synonyms for pleuronectiform fishes (Table 2.2). Nomenclatural synonyms, i.e. multiple names proposed for the same species and published subsequent to the original description and naming of the species, are not included in data summaries below. Data included in summaries and analyses comprise only valid species and the year in which they were described. Taxa originally described as subspecies or other named subspecific ranks, or earlier treated as synonyms and now recognised as valid species, are listed in summaries for the year in which they were accorded a Linnaean name. Counting only valid taxa in the year they were first recognised as distinct enables us to exclude synonyms that confuse the historical aspect of discovery and also prevents artificially inflating diversity estimates (Patterson 1994, 1996; May & Nee 1995). This approach also clarifies diversity estimates by eliminating problems associated with changing species concepts and modes in systematic biology (Patterson 1994; and discussion below).

Table 2.2 Nominal species of flatfishes described or recognised (e.g. not formally described), with number of species considered valid in the present study and number of synonyms available for nominal species described in each family or genus (names in synonymy based mostly on those in Eschmeyer 1998)

Taxon	Nominal species	Valid species	Names in synonymy
Psettodidae	8	2	6
Citharidae	12	7	5
Scophthalmidae	39	9	30
Paralichthyidae	190	95	95
Bothidae	267	145	124
Pleuronectidae	177	60	117
Tephrinectes	4	1	3
Rhombosoleidae	48	19	29
Achiropsettidae	9	4	5
Poecilopsettidae	34	30	11
Samaridae	30	28	9
Paralichthodidae	2	1	1
Achiridae	56	31	30
Soleidae	227	139	95
Cynoglossidae	236	145	110
Totals	1339	716	670

2.5.3 Diversity of species within families

It is immediately apparent from estimates of the standing species diversity partitioned by flatfish family (Table 2.2) that species are not uniformly distributed among families. Families with low species diversity include the monotypic Paralichthodidae, Psettodidae (2 species), Achiropsettidae (4 species), Citharidae (7 species) and Scophthalmidae (9 species). *Tephrinectes* also is a monotypic lineage within the Order. Families with intermediate levels of species diversity include the Rhombosoleidae (19 species), Samaridae (28), Poecilopsettidae (30) and Achiridae (31). More diverse, species-rich families are the Pleuronectidae (60 species, 8.4% of total flatfish species diversity), Paralichthyidae (95 species, 13.3% of total diversity), Soleidae (139 species, 19.4% of total diversity) and Cynoglossidae and Bothidae, each with 145 species and each representing 20.2% of the total diversity of flatfishes.

2.5.4 Standing diversity estimate for species of Pleuronectiformes

A species accumulation curve reflecting standing diversity of the Pleuronectiformes (Fig. 2.4) plots the cumulative number of valid flatfish species against the year in which they were described or recognised, beginning with the work of Linnaeus (1758) and continuing to 2001. From a historical perspective, descriptions of flatfishes initially began slowly, and remained at a low level from 1758 to about 1832. Beginning in about 1833, however, frequency of discovery of undescribed species of flatfishes increased. And from about 1850 and continuing to the present day, discovery of new flatfish species has increased dramatically, resulting in steady accumulation of recognised diversity within this group of fishes.

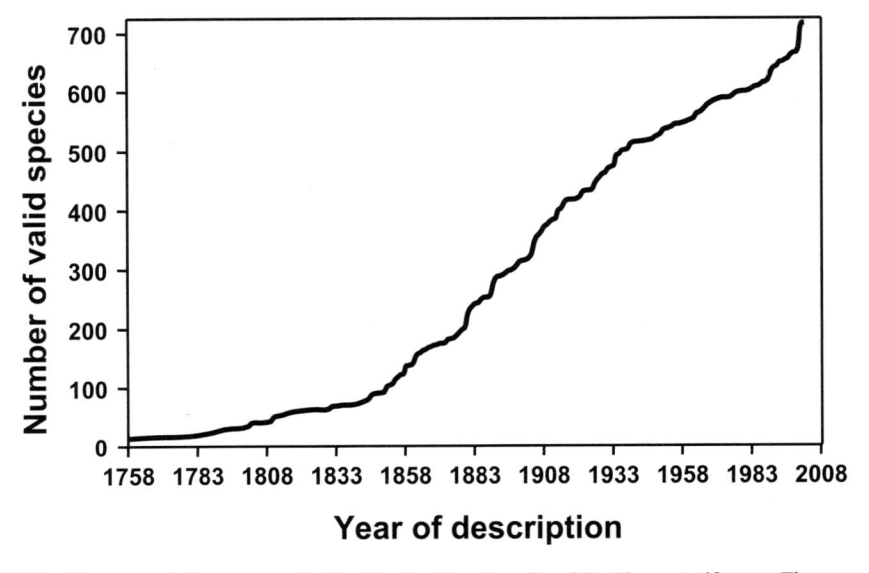

Fig. 2.4 Species accumulation curve reflecting the standing diversity of the Pleuronectiformes. The cumulative number of valid flatfish species is plotted against the year in which the species were first described or recognised (i.e. presently recognised but not yet formally described).

24   Flatfishes

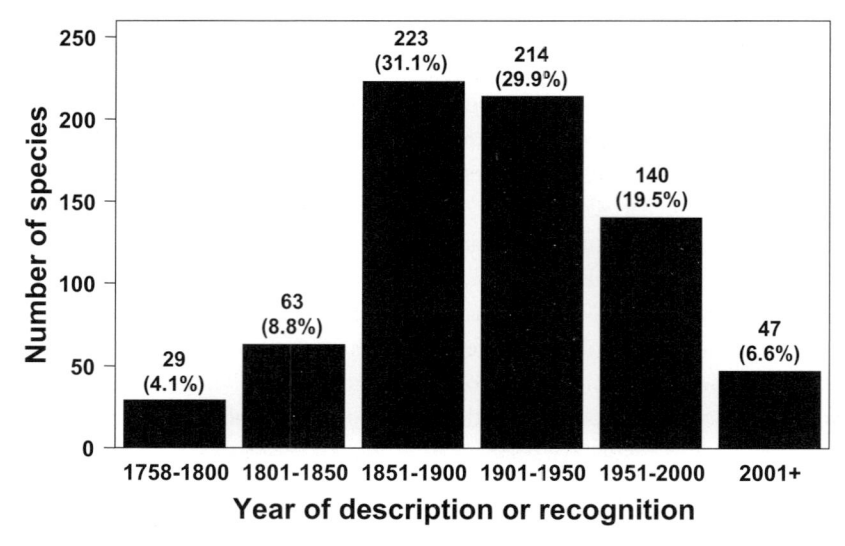

Fig. 2.5 Numbers and relative percent of valid flatfish species described or recognised during approximately 50-year intervals from 1758 to 2000. Data for 2001+ are for species considered valid by taxonomists but not yet formally described.

When viewing numbers of flatfish species described or recognised during approximately 50-year intervals since 1758 (Fig. 2.5), only 29 species were described from 1758 to 1800, while another 63 species (8.8% of the standing diversity) were recognised from 1801 to 1850. The greatest number of newly discovered flatfish species for the 50-year increments (223 species or about 31% of the standing diversity) was described during 1851–1900. Most flatfishes (about 86%) have been described since 1850. Increased levels of taxonomic activity evident during this time period reflect both increases in numbers of practising taxonomists, and the expanding age of scientific exploration and discovery when major collecting trips by researchers of many nations journeyed to various exotic, and previously unexplored, locations. During the twentieth century, discovery of new species also continued at a high rate with 214 additional species (about 30% of standing diversity) discovered from 1901 to 1950, while another 140 species (about 20% of total diversity) were also recognised between 1951 and 2000.

Since Chabanaud's (1939) synopsis, 166 valid species of flatfishes representing 23.2% of the standing diversity of the Order have been described. Another 47 nominal species (= 6.6% of standing diversity), currently recognised by taxonomists and awaiting formal description, were also discovered during this time (Fig. 2.5). Thus, since Chabanaud's (1939) study, approximately 213 nominal species of flatfishes, representing nearly 30% of the standing diversity of the Order Pleuronectiformes, have been described or are currently recognised by taxonomists conducting systematic studies on these fishes.

The number of species described or recognised does not represent the entire taxonomic activity conducted during these time periods. In addition to discovering and describing new species, taxonomists revising various groups of flatfishes have also recognised as valid species other nominal species that were once formerly placed in synonymy. Resurrection of nominal species from synonymy is required as better information becomes available. For flatfishes, nearly as many names reside in synonymy as are currently recognised for valid species (Table 2.2). As Patterson (1996) pointed out, synonyms are not purely errors in applying systematic classifications, but are also by-products of systematic concepts used by investigators. The importance of synonyms for systematic, fisheries and ecological studies of flatfishes is that these names represent a potential source of species currently unrecognised, which, when further evaluation merits it, could represent names for valid species. In the present work, no attempt was made to estimate the number of names resurrected from, or placed into, synonymy during any time interval. However, this type of systematic activity certainly contributed to improved delineation and diagnosis of species, and more importantly, recognition of synonyms as valid species, especially among tropical flatfishes where this likelihood is greatest, could significantly increase future diversity estimates and allow for a more accurate estimation of species diversity of these fishes.

2.5.5 Relative diversity of the Pleuronectiformes

Approximately 716 species of flatfishes (669 named and 47 recognised, but not described) placed in 123 genera, are currently recognised (Table 2.1). The relative diversity of flatfishes can best be appreciated by comparing their diversity with that of other Orders of primarily marine euteleostean fishes (Fig. 2.6). Of marine euteleostean fishes, flatfishes rank as the third most diverse Order both in numbers of species and genera; only the Perciformes and Scorpaeniformes have greater diversity. Given the high degree of morphological specialisation of the pleuronectiforms, and their overall superficial similarity in body plan and benthic lifestyles, such diversity may appear somewhat surprising to those less familiar with this interesting group of fishes.

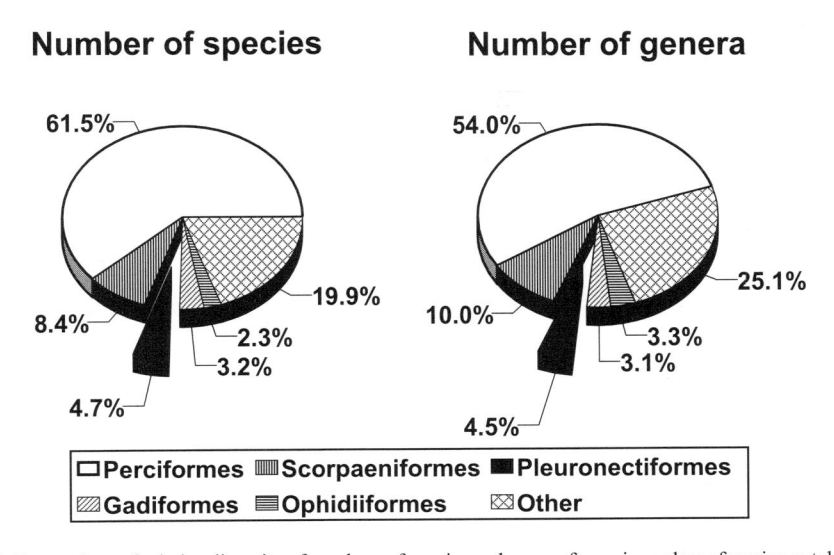

Fig. 2.6 Comparison of relative diversity of numbers of species and genera for major orders of marine euteleostean fishes (non-pleuronectiform data based on Nelson 1999).

2.6 Patterns of species discovery among pleuronectiform families

2.6.1 History

The history of discovery of undescribed species, as reflected in species accumulation curves (Figs 2.7–2.9), varies widely among pleuronectiform families. Generally, flat trajectories of species accumulation curves for groups considered well known are taken to provide robust estimates of total diversity (Soberón & Llorente 1993). For some flatfish families, we appear to have good working knowledge of species-level diversity, as evidenced by the flattened shape of the accumulation curve of species discovery for these families. The period of active species discovery is essentially past for the Pleuronectidae and Scophthalmidae (Fig. 2.7), which occur in the North Atlantic and North Pacific oceans, and in the North Atlantic Ocean, respectively. This is not surprising given that many of these species support commercial fisheries or inhabit areas where other commercially important flatfish species are commonly taken. For some tropical families with low species diversity, such as the Psettodidae and Citharidae (Fig. 2.7), most species were also discovered during the last century. Although new species are still being described in the Paralichthyidae and Rhombosoleidae (Fig. 2.8), relative shapes of species discovery curves for these taxa suggest that we are approaching reasonably accurate assessments of species diversity in these families.

For other families, such as the Achiropsettidae (Fig. 2.7), Samaridae (Fig. 2.8) and Poecilopsettidae (Fig. 2.9), species accumulation curves indicate that an accurate estimate of diversity within these families is not currently available. Samarid and poecilopsettid species

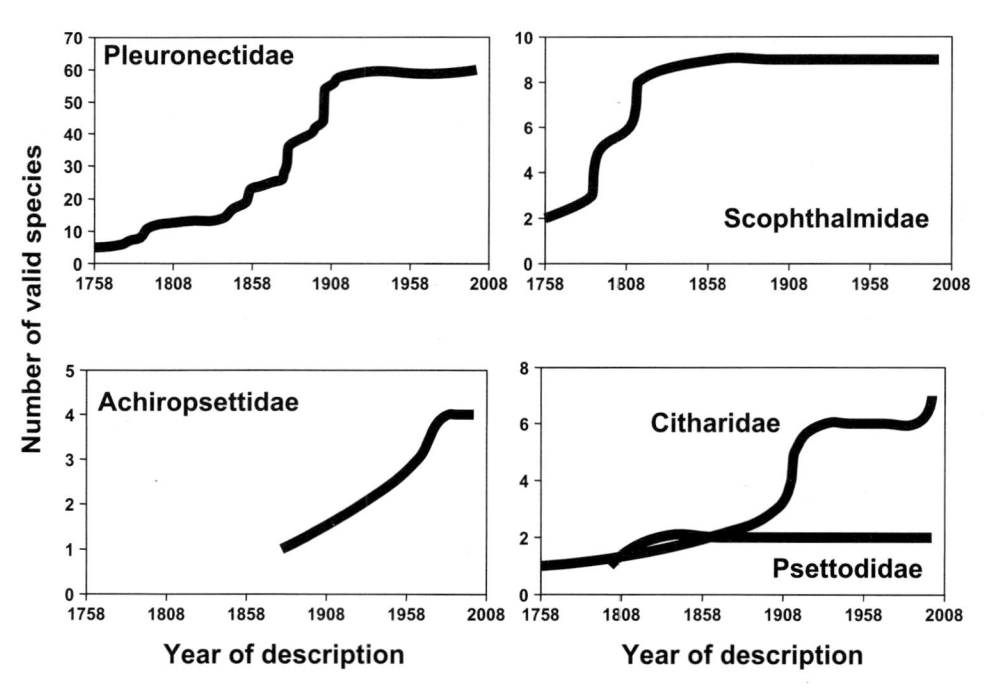

Fig. 2.7 Accumulation curves for valid species of flatfishes described from 1758 to 2000 and for flatfish species currently recognised as valid (2001+) for the pleuronectiform families. Pleuronectidae, Scophthalmidae, Achiropsettidae, Citharidae and Psettodidae.

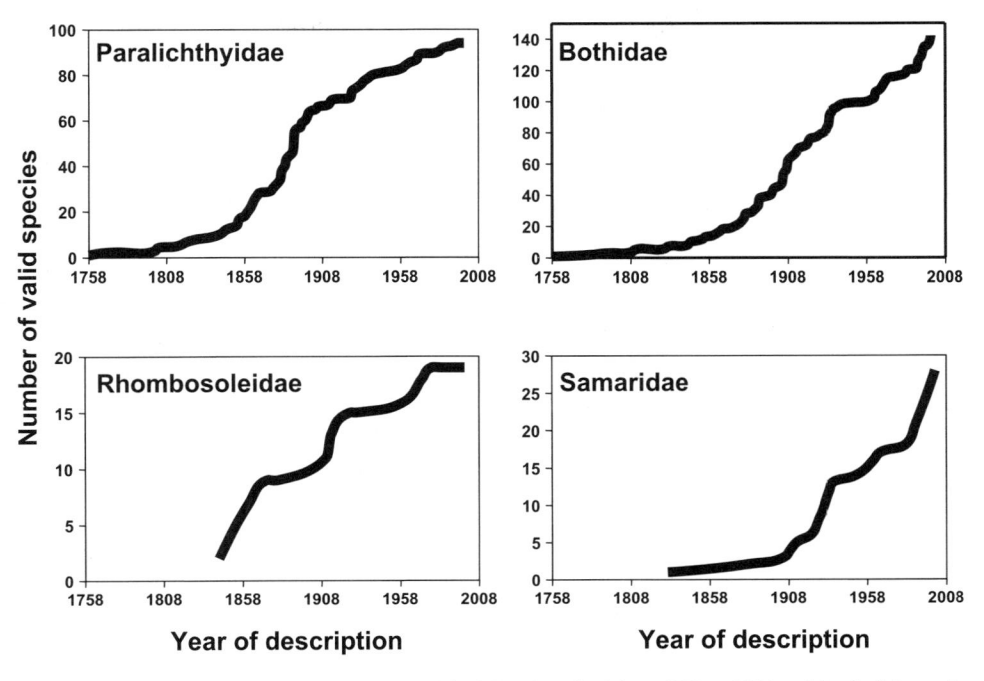

Fig. 2.8 Accumulation curves for valid species of flatfishes described from 1758 to 2000 and for flatfish species currently recognised as valid (2001+) for the pleuronectiform families, Paralichthyidae, Bothidae, Rhombosoleidae and Samaridae.

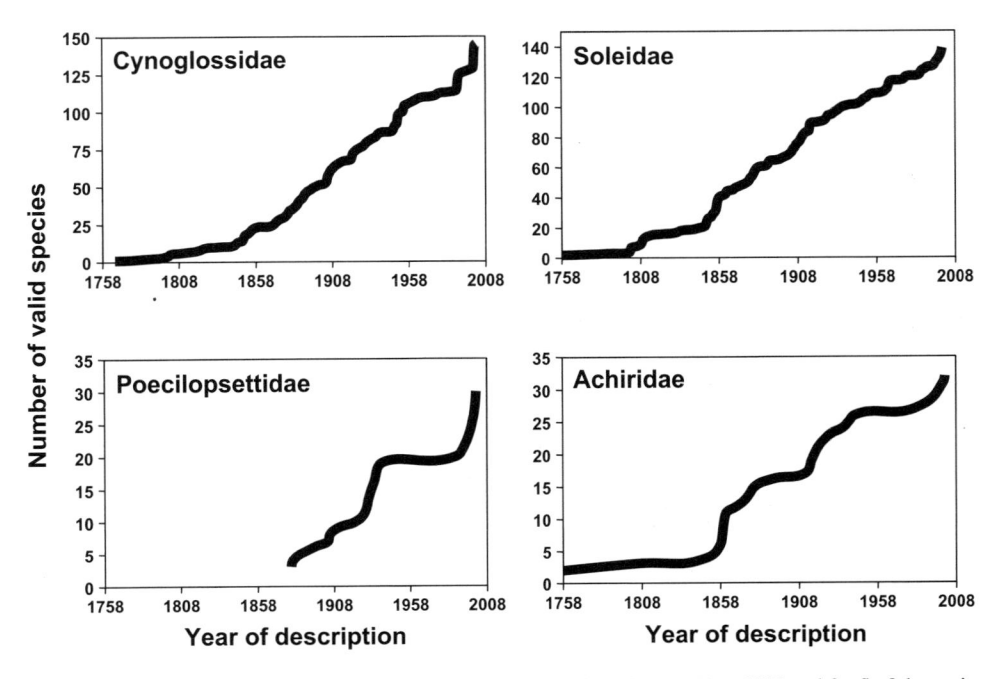

Fig. 2.9 Accumulation curves for valid species of flatfishes described from 1758 to 2000 and for flatfish species currently recognised as valid (2001+) for the pleuronectiform families Cynoglossidae, Soleidae, Poecilopsettidae, and Achiridae.

occur in relatively deep, tropical waters, areas not as rigorously sampled as similar habitats in temperate areas. Achiropsettid flatfishes occur in deep waters of the Southern Ocean, are infrequently collected, and thus remain poorly known taxonomically (Evseenko 2000). For these families, shapes of species accumulation curves may merely reflect sampling heterogeneity and bias in collection intensity, rather than depicting actual levels of species diversity remaining to be discovered in these families. Sampling biases may be inevitable at large taxonomic and spatial scales (Soberón & Llorente 1993), and Patterson (1994) pointed out that even when diversity asymptotes are clearly indicated for taxa, as they are for many vertebrate groups, it does not necessarily follow that existing diversity has been fully uncovered. Recent discoveries of undescribed species in the Samaridae and Poecilopsettidae (Alfonso, unpublished data; Amaoka, unpublished data) from previously poorly sampled areas in the Pacific Ocean would indicate that this is also the case for these flatfishes.

Species discovery is still very active for several species-rich families within the Order. The shape of the species discovery curve for the Bothidae (Fig. 2.8) indicates that we are still actively adding to our knowledge of diversity within this family. Likewise, for the Cynoglossidae, Soleidae and Achiridae (Fig. 2.9), relative shapes of species curves over time suggest that much more work is needed before numbers of species in these families can be reliably estimated. Because species discovery is an ongoing process for these species-rich families, reliable estimates of the total species-level diversity for flatfishes as a whole will not be available for some time. Patterson's (1994) comment that no major group of organisms is known well enough to be simply enumerated also applies to the present level of understanding for species-level diversity of the flatfishes.

2.6.2 Factors contributing to new species discovery among the Pleuronectiformes

Factors influencing the likelihood of discovering a species include such things as proximity to sampling areas frequented by systematists, catchability or morphological uniqueness of the species, species concept employed, and the methodology available to systematists (morphological, histological, genetic and molecular approaches). For flatfishes, factors such as size, shape, activity, abundance, conspicuousness (cryptic vs non-cryptic), habitat, geographic range, size of habitat, availability and recognition of different life stages, and commercial importance all undoubtedly contribute significantly to their probability of discovery. Many factors are correlated, such as size and commercial importance, behaviour and catchability, shape/size and conspicuousness, habitat and sampling frequency. Evaluating the influence of these factors provides insight into the reliability of contemporary estimates of standing diversity for the Pleuronectiformes. Also, better understanding as to why newly discovered species have not been found earlier may also increase our ability to locate additional species yet to be found (Gaston & Blackburn 1994).

2.6.2.1 Systematic activities

Exploration and sampling in new or previously poorly explored geographic regions or in difficult-to-sample habitats has uncovered many undescribed flatfish species. Extension of commercial fishing operations and scientific exploration into previously unexploited deep-water habitats has also resulted in collections of new species. New collecting techniques, including use of submersibles and SCUBA, contribute to discoveries of new flatfishes (Böhlke 1961; Robins & Randall 1965; Munroe et al. 2000; Munroe & McCosker 2001; Munroe & Hashimoto, unpublished observations). New approaches to delineate species, including use of molecular techniques and new morphological characters, or detailed ecological and early life history studies have also resulted in discoveries of new species. The most recently described pleuronectid, northern rock sole, *Lepidopsetta polyxystra*, was first discovered through differences in larvae collected in plankton samples (Orr & Matarese 2000). Developments in molecular studies, in particular, have had substantial impacts in systematics and taxonomy (Avise 1994, 1998; Mayden & Wood 1995; Hillis et al. 1996; Riddle & Hafner 1999), providing important contributions to understanding phylogenetic relationships and species recognition.

Philosophical approaches in conducting taxonomic studies have also changed concepts regarding the number of species that should be recognised. Widespread acceptance of the biological species concept led many taxonomists in the twentieth century to treat earlier names as synonyms of large polytypic species (Patterson 1996). Modern workers, perhaps increasingly motivated by phylogenetic interests, place more emphasis on diagnosis of taxa and less on establishing their reproductive limits (Cracraft 1989). Species concepts play central and under-appreciated roles in all biodiversity matters (Mallet 1995), especially those involving species diversity and synonymy, and within the systematics community, this is an active time for evaluating species concepts (Mallet 1995; Mayden & Wood 1995; Kottelat 1998; Carvalho & Hauser 1999; de Pinna 1999; de Queiroz 1999; Kullander 1999; Nelson 1999; Turner 1999; Pleijel & Rouse 2000).

Potentially influencing diversity estimates for Indo-Pacific (and other tropical) flatfishes is the change in approach regarding 'widespread' species (Gibbs 1986; Gill 1999). New insights into analysing observed patterns of morphological variation have necessitated abandoning conservative approaches regarding the slight morphological variation observed in 'widespread' species. Instead, species with circumglobal or circumtropical distributions, and even those with broad distributions within oceans, are now more often regarded with suspicion that they represent complexes of subtly distinct species. In the Indo-Pacific, approximately 45–85% of the shorefish fauna are considered 'widespread', i.e. with distributions spanning the Indo-West Pacific or greater (Gill 1999). Earlier systematic approaches and more recent studies (Randall 1999) accepted widespread taxa as consisting of one species comprising polytypic populations or subspecies. However, Gill (1999) emphasised that many Indo-Pacific coral reef fishes with apparent widespread distributions were artificial concepts that reflected inconsistent application of criteria for species delineation. He suggested that criteria for recognising such forms at an intraspecific, rather than specific, level were unjustified, and that when sufficient study was conducted on these 'widespread' fish species, they too would be found to actually comprise species complexes. A large number of small flatfishes occur in reef-associated habitats throughout tropical regions and many of these are considered to have widespread distributions. However, most have not been adequately sampled, correctly identified, or thoroughly studied with respect to population variation. Preliminary study of *Cynoglossus* and *Soleichthys*, two genera of relatively small, reef-associated, tropical flatfishes, indicates that several polymorphic 'species' previously considered to be widespread actually represent species complexes (Munroe, unpublished data). Other flatfish taxa where previously 'widespread' species have been shown to consist of species complexes include poecilopsettids (*Poecilopsetta*), bothids (*Arnoglossus* and *Engyprosopon*) and tonguefishes (*Symphurus*). The diversity of species present in these genera is considerably greater than previously recognised, with geographic ranges of individual species much more restricted, and with none of the species being widespread. Roberts & Hawkins (1999) observed that small geographic ranges are common, especially among small-bodied marine organisms. When other small, tropical 'widespread' flatfish species become better studied, how many will be found to comprise species complexes with individual species having narrower endemism than previously suspected?

2.6.2.2 Geographic region

Where flatfishes live has strongly influenced the probability of their discovery. Most flatfishes inhabiting polar and north temperate regions were discovered much earlier than were tropical species (Fig. 2.10). Nearly 80% of species described from polar or north temperate regions were discovered prior to 1900. In recent years, few new species have been discovered in north temperate or polar seas (Fig. 2.11A). As most ichthyologists and fisheries biologists of the past three centuries originated from, and worked mostly in, northern hemisphere areas, the timing of discovery of these northern fishes generally reflects this bias in geographic origin of the investigators. Also relevant to their earlier discovery is that many flatfishes inhabiting northern waters are relatively large, important food fishes that came under intensive study early due to their commercial importance.

The discovery of tropical flatfishes increased dramatically after about 1851 (Fig. 2.10). In fact, more than half of all flatfish species known from tropical regions were discovered only within the past 100 years. In the most recent 30-year period (Fig. 2.11), the majority of newly discovered flatfish species have come from tropical waters, primarily marine waters of the

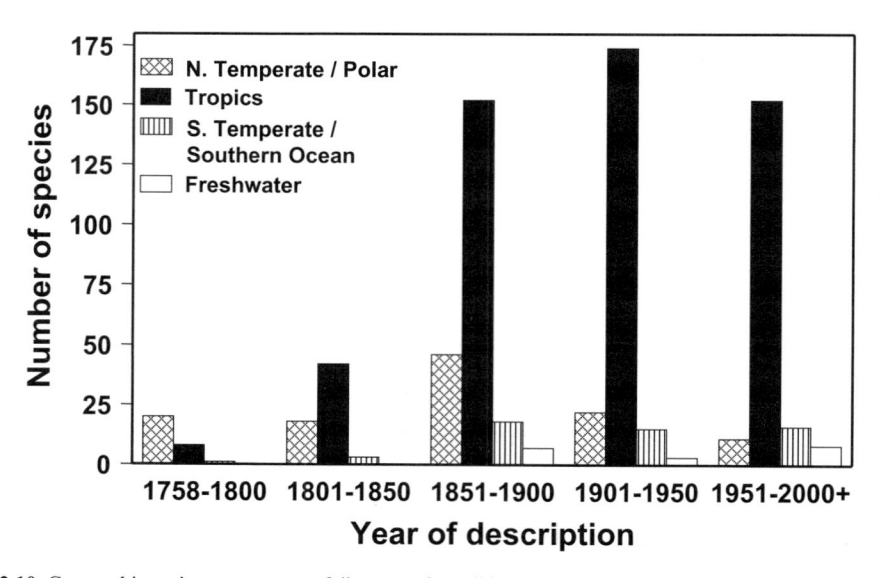

Fig. 2.10 Geographic region versus year of discovery for valid species of flatfishes described or recognised since 1758.

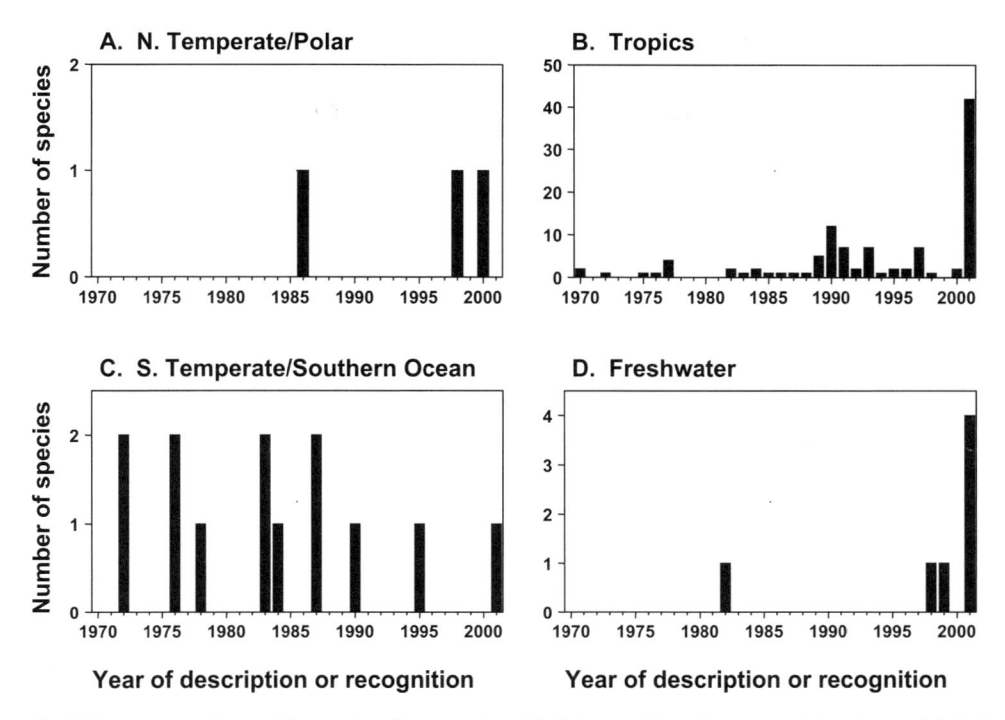

Fig. 2.11 Summary of recent discoveries of new species of flatfishes partitioned by geographic region and plotted against their year of discovery for the most recent 30-year time period. A. North Temperate and Polar Seas. B. Tropical marine waters. C. South Temperate and Southern Ocean regions. D. Freshwater regions.

Indo-West Pacific and tropical South American freshwaters (Fig. 2.11B, D). Marine waters off northern Australian and associated tropical seas, the area with the most diverse marine fish fauna in the world (Briggs 1999), still continue to yield many new species of flatfishes. Geographic differences in discovery rates for tropical flatfishes are generally consistent with patterns of discovery for other tropical marine organisms. Winston (1992), for example, estimated that the marine waters of the Caribbean region were the best studied of tropical areas with an estimated 70% of the total faunal and floral species thought to be known, whereas in contrast, only an estimated 50–60% of the total fauna and flora occurring in non-Caribbean tropical regions are thought to be known.

The overall diversity of flatfishes in south temperate and Southern Ocean waters is relatively low with only 53 species (7.4% of the standing diversity of flatfishes) known from these areas. Most of these flatfishes were described only during the past 100 years. More recently (Fig. 2.11C), a few species have been described, adding to estimates of flatfish diversity from these waters. Several other undescribed species of achiropsettid flounders are also known from the Southern Ocean (Evseenko, personal communication, cited in Heemstra 1990), which will increase diversity estimates for flatfishes from this region when these species are described.

2.6.2.3 Depth

Bathymetric habitats where flatfishes live also influence the probability of their discovery. Species inhabiting more accessible or more easily sampled environments have a greater

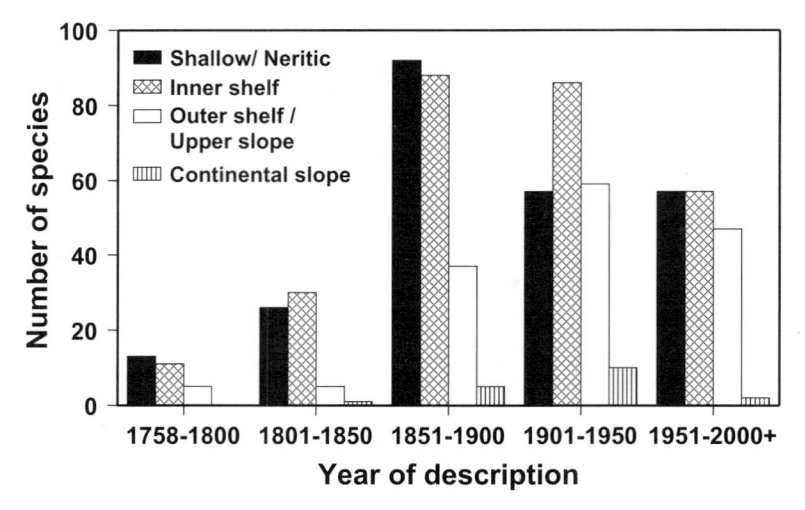

Fig. 2.12 Summarised data on depth of adult habitat versus year of discovery for valid species of flatfishes described or recognised since 1758.

probability of earlier discovery than those found living in inaccessible habitats, or those occurring in habitats that are difficult to sample. Flatfishes occupy diverse depths ranging from shallow marine and freshwaters to deep-water areas on the upper continental slope. Flatfish species living in shallow-water habitats are more likely to be discovered earlier than those living in deep, outer continental shelf or upper continental slope environments. Plots of depth of adult habitat versus year of discovery (Fig. 2.12) clearly indicate that flatfishes from shallow-water environments were indeed discovered frequently during the early years of taxonomic activity. However, lack of significant correlation ($r = 0.14$) between depth of occurrence and year of description indicates that, perhaps somewhat surprisingly, many species of flatfishes continue to be discovered from shallow-water habitats, especially in tropical regions. The majority of flatfishes discovered in the most recent 30-year period were collected in shallow, neritic waters, or were trawled on the inner continental shelf.

Flatfishes inhabiting outer continental shelf and upper continental slope depths still contribute to new discoveries, especially in remote tropical areas of the Indo-Pacific, areas sampled far less frequently and far less thoroughly than shallow-water habitats elsewhere. However, new discoveries of undescribed species of flatfishes deriving from these outer shelf and upper slope habitats are comparatively few relative to those of shallow-water seas. Therefore, if these current trends in species discovery continue, the majority of undiscovered flatfishes can be expected to derive from shallow-water tropical habitats, with species discoveries from deep-water habitats not contributing significantly to future estimates of overall flatfish species diversity.

2.6.2.4 Size

Flatfishes span a size range of about three orders of magnitude ranging from diminutive species, such as shallow-water tonguefishes (*Symphurus*) sexually mature at 2.5–4.0 cm SL (Munroe 1990, 1998) and females in the bothid genus *Tarphops* that are sexually mature

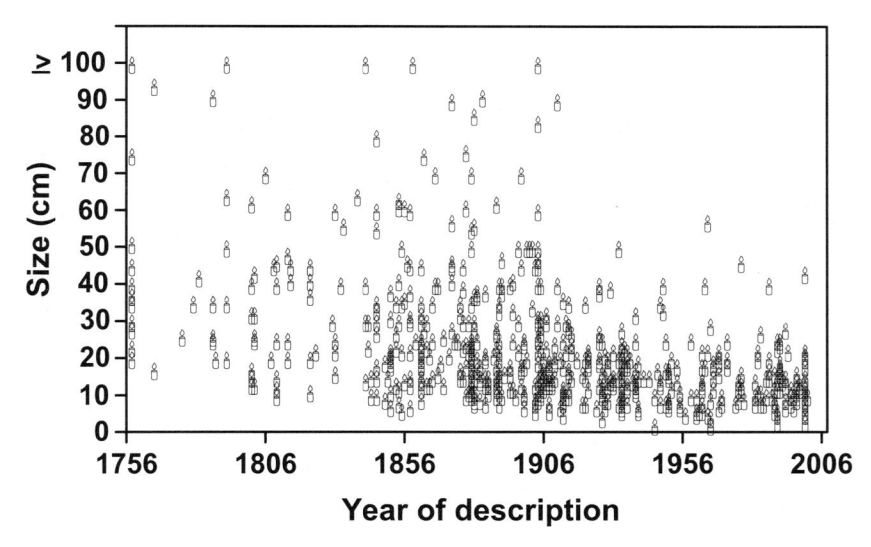

Fig. 2.13 Average total length of adults versus year of discovery for valid species of flatfishes described or recognised since 1758.

at 4.5 cm, to intermediate-sized flatfishes occurring on the continental shelf, such as sand flounders (*Paralichthys* species) that reach to 150 cm, to giant species of halibuts (Pacific halibut, *Hippoglossus stenolepis* reaches nearly 2 m in total length, and may weigh well over 300 kg). The average total length of adults is 30 cm or less for the majority of flatfishes (Fig. 2.13), and few species exceed 80 cm in average adult length.

Intuitively, large-sized species in a community would be expected to be encountered before smaller species, so one might expect that in a given region larger flatfishes would be discovered and described earlier than their smaller-sized counterparts. Among flatfishes, a negative correlation ($r = -0.40$) was found between the average adult size of a species and its year of discovery (Fig. 2.13), indicating that trends between larger body size and increased probability of discovery noted for other groups of organisms (Gaston 1991; Patterson 1994, 2000) are also evident for flatfishes, especially during the early history of species discovery. Most flatfish species larger than 30 cm were discovered before or by the early twentieth century, and for these species it appears that size was an important feature influencing their discovery. For the most recent 50-year interval, the majority (146/173, 84%) of newly discovered flatfish species are 20 cm or less in average adult length (Fig. 2.14). For species discovered during this period, year of discovery is largely independent of species size ($r = -0.04$). Because many small flatfishes (i.e. those <20 cm) are generally poorly known, both taxonomically and ecologically, and are collected irregularly compared with larger-sized species, this finding is not surprising. Probably few, if any, large species (>50 cm) of flatfishes remain to be discovered. Only four of the 190 species discovered in the past 50 years feature adult sizes >30 cm SL. Important in this context is that even the northern rock sole, a pleuronectid species reaching adult sizes to 58 cm SL, and one of the largest flatfish species described in this most recent 50-year time period, was already heavily exploited before its recognition as a distinct species (Orr & Matarese 2000).

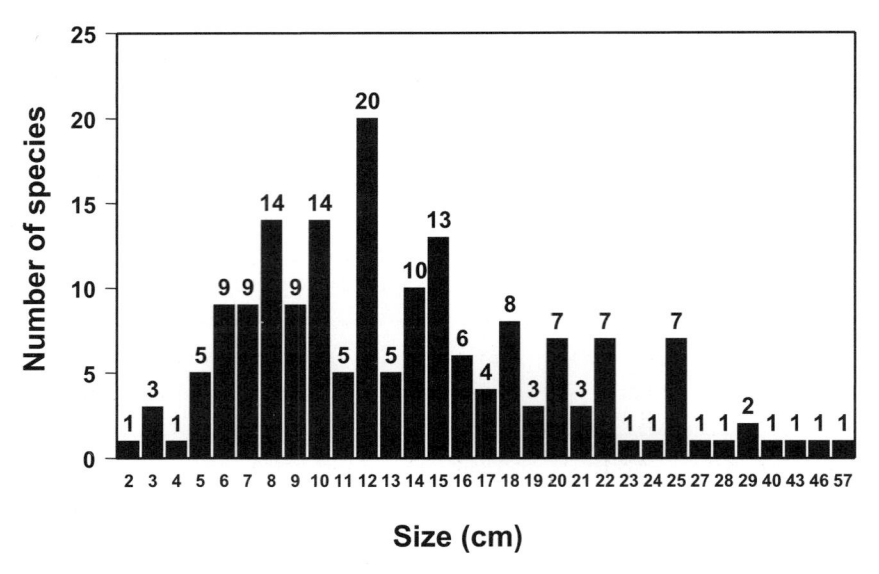

Fig. 2.14 Size frequency distribution of average total length (cm) of adults of 173 valid species of flatfishes described or recognised from 1950 to 2001.

For flatfish species <20 cm, size alone may not necessarily be the primary factor influencing probability of its discovery or recognition. Other species-specific factors such as geographic range size and overall conspicuousness (correlated with population size, geographical range and behaviour) can also influence the probability of their discovery (Diamond 1985; Gaston & Blackburn 1994). Typically, species with more widespread ranges (usually large-bodied forms) (Rapoport 1982) are discovered earlier than those with highly localised distributions. Species with small populations or small geographical ranges, or cryptic or nocturnal species, should have lower probabilities of being discovered, whether large- or small-bodied. At least some of these traits may be correlated with species body size (e.g. species with small geographic ranges tend to be small (Gaston 1991)). However, if body size is of secondary importance to probability of discovery of a species, one obvious prediction is that the relationship between probability of discovery and, for example, geographic range or cryptic behaviour should be stronger than that with body size. Testing whether newly described species are also more cryptic and geographically localised than other species in general will be much harder, although such differences do seem likely (Gaston & Blackburn 1994). As so little is known regarding geographic range sizes, population sizes or behaviours of small-sized flatfishes (and for most flatfishes in general), it is impossible to evaluate important predictors of probability of description for these taxa until better information becomes available.

2.7 Conclusions

Asymmetrical specialisations of the Pleuronectiformes are extremely successful adaptations. Pleuronectiform body plans and lifestyles have been around since at least the late Tertiary, and the contemporary success of these fishes is expressed in the spectacular morphological diversity observed in members of at least 14 different evolutionary lineages. With body sizes spanning three orders of magnitude, flatfishes perform a wide variety of ecological roles and exhibit wide variety in ecological activities. Nocturnal activity is a major adaptation evident in the Soleidae and Cynoglossidae, two of the most species-diverse families in the Order. The ecological success of the Pleuronectiformes can be measured in the diversity of habitats occupied by the 700+ species of flatfishes, with their diverse trophic biologies and their relatively wide bathymetric occurrence. Flatfishes have nearly global occurrence in marine habitats, ranging from Arctic and boreal marine waters to Antarctic and Austral waters. They are broadly distributed throughout the world's temperate marine zones, and are especially speciose in tropical marine regions. Flatfishes are conspicuous elements of most fish assemblages in estuaries, and neritic waters extending from the shoreline to outer margins of the continental shelf and onto the upper continental slope. Only in polar seas, freshwaters, or at bathyal depths (usually absent below about 2000 m) are flatfish diversities substantially reduced compared with that of other marine biotopes.

Considerable advances have been made in our understanding of the systematics of the Pleuronectiformes, including discovering new species, progress in understanding evolutionary relationships within the group, and in uncovering the great biological and ecological diversity represented in these fishes. However, much more work needs to be done. Many new species of flatfishes await discovery and the process of discovering and naming species of flatfishes remains an active, important and viable area of taxonomic research, especially for tropical, subtropical and deep-water species. That 18% of the species diversity of Pleuronectiformes (129 species) was discovered only during the past 30 years highlights the fact that the level of undiscovered diversity among flatfishes is substantial. Sometimes rates of species description in poorly known groups correspond better to the number of specialists studying them rather than to the real magnitude of their undescribed diversity (Gaston & May 1992). For flatfishes, this seems unlikely because many habitats where these fishes are found, especially in remote tropical waters, and deep-water habitats in general, have not been adequately sampled. Difficulties in sampling the majority of marine habitats impede progress on accumulating taxonomic knowledge for many taxa, including flatfishes. These difficulties are especially important for those pleuronectiform groups inhabiting tropical and deep-water habitats where species-level taxonomy still remains poorly known. When these habitats receive concentrated sampling, they continue to yield new species. Therefore, increases in flatfish species diversity recognised over time accurately reflect the magnitude of undescribed diversity present in this group, and not merely any increased activity by systematists studying these fishes. Based on present estimates, we are still not yet close to knowing the total species diversity for many families within the Pleuronectiformes.

Expanded views of flatfish diversity have helped to clarify issues and directions where additional research is needed to better understand the diversity, evolution, biology and biogeography of these fishes. With accumulation of new systematic information – including species discoveries, improved species diagnoses and improved phylogenetic hypotheses – the reliability of information regarding species diversity and geographical distributions will also increase. Improved hypotheses of flatfish relationships using new phylogenetic information will directly impact how species diversity information is interpreted and will also shape the framework for postulating future hypotheses concerning comparative biology of these fishes. Understanding the mode and direction, as well as the history of flatfish evolution, requires robust hypotheses of evolutionary relationships. Evolutionary trends and constraints of evolution can only be uncovered within a phylogenetic context, within which, monophyletic taxa can be identified and non-monophyletic taxa can be re-defined. Discovering patterns of convergent evolution, or unravelling the role of vicariant events in geographic distributions of these fishes, requires knowledge of relationships derived from phylogenetic studies. When conflicting hypotheses exist regarding taxonomy or evolutionary relationships of taxa due to lack of information, or when construction of hypotheses is based on unsuitable characters, then partitioning of knowledge about these poorly known taxa will also be affected (Cotterill & Dangerfield 1997).

Meaningful improvements in flatfish classification and better understanding of the evolutionary history and intrarelationships of these fishes must await well corroborated cladograms for taxa where information is lacking. For flatfish families occurring in north temperate and Arctic regions, species discoveries still occur, but at much slower rates than for those of subtropical and tropical areas. For most flatfish families occurring in northern waters, taxa are sufficiently known so that studies examining phylogenetic relationships are proceeding. For flatfishes inhabiting tropical seas, despite recent progress, considerable diversity is still being discovered and the taxonomy of many tropical flatfishes remains especially problematic. Failure to identify species, and erroneous species identifications, still represent serious impediments to collection of meaningful data for many of these smaller species. Inaccurate identifications and lack of recognition of species diversity, in turn, compromise reliability of information on geographic and ecological distributions, habitat requirements, and trophic and reproductive biology of poorly known flatfishes from tropical regions. Much more systematic work is needed before evolutionary hypotheses can be developed for most tropical flatfishes and their biogeographical history interpreted.

Acknowledgements

Thanks to many colleagues who generously contributed information and specimens that enhanced the scope of this chapter. I am grateful to N. Alfonso, K. Amaoka, B. Chanet, F. Chapleau, A. Cooper, M. Desoutter, A.-C. Guibord, K. Hoshino, M. Nizinski, R. Vari and M. Vecchione for providing information, assistance or critical comment on ideas expressed in this paper. P. Skelton, South African Institute for Aquatic Biodiversity, and the Species Identification and Data Programme (SIDP) of the Marine Resources Service, FAO of the UN, kindly provided permission to use figures of flatfishes. Appreciation is expressed also to the National Marine Fisheries Service for continued support of my systematics research on flatfishes at the National Systematics Laboratory.

References

Agassiz, L. (1834–1842) *Recherches sur les Poissons Fossiles.* Neuchâtel, Switzerland.
Ahlstrom, E.H., Amaoka, K., Hensley, D.A., Moser, H.G. & Sumida, B.Y. (1984) Pleuronectiformes: development. In: *Ontogeny and systematics of fishes. American Society of Ichthyologists & Herpetologists Special Publication Number 1* (eds H.G. Moser, W.J. Richards, D.M. Cohen, M.P. Fahay, A.W. Kendall, Jr. & S.L. Richardson). pp. 640–670. Lawrence, KS.

Allen, R.L. & Baltz, D.M. (1997) Distribution and microhabitat use by flatfishes in a Louisiana estuary. *Environmental Biology of Fishes,* **50**, 85–103.

Amaoka, K. (1969) Studies on the sinistral flounders found in the waters around Japan. Taxonomy, anatomy and phylogeny. *Journal of the Shimonoseki University of Fisheries,* **18**, 65–340.

Amaoka, K. & Mihara, E. (2000) Pisces, Pleuronectiformes: Flatfishes from New Caledonia and adjacent waters – Genus *Arnoglossus.* In: Résultats des Campagnes MUSORSTOM, Vol. 21 (ed. A. Crosnier). *Mémoires du Muséum National d'Histoire Naturelle,* **184**, 783–813.

Amaoka, K., Mihara, E. & Rivaton, J. (1993) Pisces, Pleuronectiformes: Flatfishes from the waters around New Caledonia – A revision of the genus *Engyprosopon.* In: Résultats des Campagnes MUSORSTOM, Vol. 11 (ed. A. Crosnier). *Mémoires du Muséum National d'Histoire Naturelle,* **158**, 377–426.

Amaoka, K., Arai, M. & Gomon, M.F. (1997a) A new species of *Arnoglossus* (Pleuronectiformes: Bothidae) from the southwestern coast of Australia. *Ichthyological Research,* **44**, 131–136.

Amaoka, K., Mihara, E. & Rivaton, J. (1997b) Pisces, Pleuronectiformes: Flatfishes from the waters around New Caledonia. Six species of the bothid genera *Tosarhombus* and *Parabothus.* In: Résultats des Campagnes MUSORSTOM, Vol. 17 (ed. B. Séret). *Mémoires du Muséum National d'Histoire Naturelle,* **174**, 143–172.

Arambourg, C. (1927) Les poissons fossiles d'Oran. *Matériaux pour la carte géologique de l'Algérie, 1ère série, Paléontologie,* **6**, 1–298.

Avise, J.C. (1994) *Molecular Markers, Natural History and Evolution.* Chapman & Hall, London.

Avise, J.C. (1998) The history and purview of phylogeography: a personal reflection. *Molecular Ecology,* **7**, 371–379.

Bannikov, A.F. & Parin, N.N. (1997) The list of marine fishes from Cenozoic (Upper Paleocene-Middle Miocene) localities in southern European Russia and adjacent countries. *Journal of Ichthyology,* **37**, 133–146.

Berendzen, P.B. & Dimmick, W.W. (2002) Phylogenetic relationships of Pleuronectiformes based on molecular evidence. *Copeia,* **2002**, 642–652.

Böhlke, J.E. (1961) Two new Bahaman soles of the genus *Symphurus* (family Cynoglossidae). *Notulae Naturae* (Academy of Natural Sciences, Philadelphia), **344**, 1–4.

Brewster, B. (1987) Eye migration and cranial development during flatfish metamorphosis: a reappraisal (Teleostei: Pleuronectiformes). *Journal of Fish Biology,* **31**, 805–833.

Briggs, J.C. (1999) Modes of speciation: marine Indo-West Pacific. *Bulletin of Marine Science,* **65**, 645–656.

Carpenter, K.E. & Niem, V.H. (2001) *FAO species identification guide for fishery purposes. The living marine resources of the Western Central Pacific,* Vol. 6, *Bony fishes,* Part 4 (*Labridae to Latimeriidae*), *estuarine crocodiles, sea turtles, sea snakes and marine mammals.* FAO, Rome.

Carvalho, G.R. & Hauser, L. (1999) Molecular markers and the species concept: new techniques to resolve old disputes? *Reviews in Fish Biology and Fisheries,* **9**, 379–382.

Chabanaud, P. (1937) Téléostéens dissymétriques du Mokkattam inférieur de Tourah. *Mémoires de l'Institut d'Egypte,* **32**, 1–121.

Chabanaud, P. (1939) Catalogue systématique et chorologique des Téléostéens dyssymétriques du globe. *Bulletin de l'Institut Océanographique Monaco,* **763**, 1–31.

Chanet, B. (1994) *Eobuglossus eocenicus* (Woodward, 1910) from the Upper Lutetian of Egypt, one of the oldest soleids [Teleostei, Pleuronectiformi]. *Neues Jahrbuch für Paläontologie Monatehefte,* **7**, 391–398.

Chanet, B. (1997) A cladistic reappraisal of the fossil flatfishes record consequences on the phylogeny of the Pleuronectiformes (Osteichthyes: Teleostei). *Annales de Sciences Naturelles, Zoologie, Paris* (13éme Série), **18**, 105–117.

Chanet, B. (1998) L'asymétrie des vertèbrae caudales chez les Scophthalmidae [Pleuronectiformes] est-elle une autapomorphie? *Cybium,* **22**, 405–412.

Chanet, B. (1999) Supposed and true flatfishes [Teleostei: Pleuronectiformes] from the Eocene of Monte Bolca, Italy. *Museo Civico di Storia Naturale di Verona, Studi e Ricerche sui Giacimenti Terziari di Bolca,* Vol. VIII, *Miscellanea Paleontologica* (ed. J.C. Tyler), pp. 220–243.

Chanet, B. (2003) Interrelationships of Scophthalmid fishes (Pleuronectiformes: Scophthalmidea). *Cybium,* **27**, 275–286.

Chapleau, F. (1988) Comparative osteology and intergeneric relationships of the tongue soles (Pisces: Pleuronectiformes: Cynoglossidae). *Canadian Journal of Zoology,* **66**, 1214–1232.

Chapleau, F. (1993) Pleuronectiform relationships: a cladistic reassessment. *Bulletin of Marine Science,* **52**, 516–540.

Chapleau, F. & Amaoka, K. (1998) Flatfishes. In: *Encyclopedia of Fishes. A Comprehensive Illustrated Guide by International Experts,* 2nd edn (eds J.R. Paxton & W.N. Eschmeyer). pp. 223–226. Academic Press, San Diego, CA.

Chapleau, F. & Desoutter, M. (1996) Position phylogénétique de *Dagetichthys lakdoensis* (Pleuronectiformes). *Cybium,* **20**, 103–106.

Chapleau, F. & Keast, A. (1988) A phylogenetic reassessment of the monophyletic status of the family Soleidae, with comments on the suborder Soleoidei (Pisces; Pleuronectiformes). *Canadian Journal of Zoology,* **66**, 2797–2810.

Chapleau, F. & Renaud, C.B. (1993) *Paraplagusia sinerama* (Pleuronectiformes: Cynoglossidae), a new Indo-Pacific tongue sole with a revised key to species of the genus. *Copeia,* **1993**, 798–807.

Cooper, J.A. & Chapleau, F. (1998a) Monophyly and intrarelationships of the family Pleuronectidae (Pleuronectiformes), with a revised classification. *Fishery Bulletin,* **96**, 686–726.

Cooper, J.A. & Chapleau, F. (1998b) Phylogenetic status of *Paralichthodus algoensis* (Pleuronectiformes: Paralichthodidae). *Copeia,* **1998**, 477–481.

Cotterill, F.P.D. & Dangerfield, J.M. (1997) The state of biological knowledge. *Trends in Ecology and Evolution,* **12**, 206.

Cracraft, J. (1989) Speciation and its ontology: the empirical consequences of alternative species concepts for understanding patterns and processes of differentiation. In: *Speciation and its Consequences* (eds D. Otte & J.A. Endler). pp. 28–59. Sinauer, Sunderland, MA.

de Pinna, M.C.C. (1999) Species concepts and phylogenetics. *Reviews in Fish Biology and Fisheries,* **9**, 353–373.

de Queiroz, K. (1999) The general lineage concept of species and the defining properties of the species category. In: *Species, New Interdisciplinary Essays* (ed. R.A. Wilson). pp. 49–89. MIT Press, Cambridge, MA.

Desoutter, M. (1987) Statut de *Microchirus boscanion* Chabanaud, 1926 et de *Buglossidium luteum* (Risso, 1810) (Pisces, Pleuronectiformes, Soleidae). *Cybium,* **11**, 427–439.

Desoutter, M. (1994) Révision des genres *Microchirus, Dicologlossa* et *Vanstraelenia* (Pleuronectiformes, Soleidae). *Cybium,* **18**, 215–249.

Desoutter, M. & Chapleau, F. (1997) Taxonomic status of *Bathysolea profundicola* and *B. polli* (Soleidae; Pleuronectiformes) with notes on the genus. *Ichthyological Research,* **44**, 399–412.

Desoutter-Meninger, M. (1997) *Revision systematique des genres de la famille des Soleidae presents sur les cotes de l'Est-atlantique et de la Mediterranee.* DPhil thesis, Museum National d'Histoire Naturelle, Paris.

Diamond, J.M. (1985) How many unknown species are yet to be discovered? *Nature,* **315**, 538–539.

Díaz de Astarloa, J.M. & Munroe, T.A. (1998) Systematics and ecology of commercially important species of paralichthyid flounders (*Paralichthys*: Paralichthyidae: Pleuronectiformes) co-occurring in Uruguayan-Argentine waters: an overview. *Journal of Sea Research,* **39**, 1–9.

Eschmeyer, W.N. (1998) *Catalog of fishes*. Special Publication Number 1 (3 vols). Center for Biodiversity Research and Information, California Academy of Sciences. San Francisco, CA.

Evseenko, S.A. (1984) New genus and species of 'Armless' flounders, *Pseudomancopsetta andriashevi* gen. et sp. nova (Pisces, Pleuronectoidei) and its position in the suborder Pleuronectoidei. *Voprosy Ikhtiologii,* **24**, 709–717 (in Russian). English translation in: *Journal of Ichthyology,* **25** (1985), 1–10.

Evseenko, S.A. (1996) Ontogeny and relationships of flatfishes of the Southern Ocean (Achiropsettidae, Pleuronectoidei). *Voprosy Ikhtiologii,* **36**, 725–752 (in Russian).

Evseenko, S.A. (2000) Family Achiropsettidae and its position in the taxonomic and ecological classifications of Pleuronectiformes. *Journal of Ichthyology,* **40** (Suppl. 1), S110–S138.

Fischer, W., Krupp, F., Schneider, W., Sommer, C., Carpenter, K.E. & Niem, V.H. (1995) *Guía FAO para la identificación de especes para los fines de la pesca. Pacífico centro-oriental*, Vols. 2 & 3. Vertebrados-Parte 1, pp. 647–1200 & Parte 2, pp. 1201–1813. FAO, Rome.

FishBase (1997) *FishBase: a biological database on fish.* International Center for Living Aquatic Resources Management, Makati City, Philippines.

Gaston, K.J. (1991) Body size and probability of description: the beetle fauna of Britain. *Ecological Entomology,* **16**, 505–508.

Gaston, K.J. & Blackburn, T.M. (1994) Are newly described bird species small-bodied? *Biodiversity Letters,* **2**, 16–20.

Gaston, K.J. & May, R.M. (1992) Taxonomy of taxonomists. *Nature,* **356**, 281–282.

Gibbs, R.H., Jr. (1986) The stomioid fish genus *Eustomias* and the oceanic species concept. In: *Pelagic biogeography* (eds A.C. Peirrot-Bults, S. Van der Spoel, B.J. Zahuranec & R.K. Johnson). pp. 98–103. *UNESCO Technical Papers in Marine Science,* **49**.

Gill, A.C. (1999) Subspecies, geographic forms and widespread Indo-Pacific Coral-Reef fish species: a call for change in taxonomic practice. *Proceedings of the 5th Indo-Pacific Fish Conference, Nouméa, New Caledonia, 1997* (eds B. Séret & J.Y. Sire). pp. 79–87. Société Française d'Ichtyologie, Paris.

Heemstra, P.C. (1986) Family No. 260: Pleuronectidae. In: *Smith's Sea Fishes* (eds M.M. Smith & P.C. Heemstra). pp. 863–865. Macmillan South Africa Publishers (Pty), Ltd, Johannesburg.

Heemstra, P.C. (1990) Achiropsettidae, Southern Flounders. In: *Fishes of the Southern Ocean* (eds O. Gon & P.C. Heemstra). pp. 408–413. J.L.B. Smith Institute of Ichthyology, Grahamstown.

Hensley, D.A. (1986) Current research on Indo-Pacific bothids. In: *Indo-Pacific Fish Biology Proceedings of the 2nd International Conference on Indo-Pacific Fishes* (eds T. Uyeno, R. Arai, T. Taniuchi & K. Matsuura). p. 941 (Abstract only). Ichthyological Society, Tokyo.

Hensley, D.A. (1997) An overview of the systematics and biogeography of the flatfishes. *Journal of Sea Research,* **37**, 187–194.

Hensley, D.A. & Ahlstrom, E.H. (1984) Pleuronectiformes: relationships. In: *Ontogeny and systematics of fishes. American Society of Ichthyologists & Herpetologists Special Publication Number 1* (eds H.G. Moser, W.J. Richards, D.M. Cohen, M.P. Fahay, A.W. Kendall, Jr & S.L. Richardson). pp. 670–687. Lawrence, KS.

Hillis, D.M., Moritz, C. & Mable, B.K. (1996) *Molecular Systematics*, 2nd edn. Sinauer Associates, Sunderland, MA.

Hoshino, K. (1999) *Phylogenetic relationships of the paralichthyid flatfishes (Teleostei: Pleuronectiformes) and their positions within the suborder Pleuronectoidei.* DPhil thesis, Hokkaido University, Hakodate, Japan.

Hoshino, K. (2000) *Citharoides orbitalis*, a new sinistral citharid flounder from Western Australia (Pleuronectoidei, Pleuronectiformes). *Ichthyological Research,* **47**, 321–326.

Hoshino, K. (2001) Monophyly of the Citharidae (Pleuronectoidei: Pleuronectiformes: Teleostei) with considerations on pleuronectoid phylogeny. *Ichthyological Research,* **48**, 391–404.

Hoshino, K. & Amaoka, K. (1998) Osteology of the flounder, *Tephrinectes sinensis* (Lacepède) (Teleostei: Pleuronectiformes), with comments on its relationships. *Ichthyological Research*, **45**, 69–77.

Hubbs, C.L. (1945) Phylogenetic position of the Citharidae, a family of flatfishes. *Miscellaneous Publications of the Museum of Zoology, University of Michigan*, **63**, 1–38.

Kottelat, M. (1998) Systematics, species concepts and the conservation of freshwater fish diversity in Europe. *Italian Journal of Zoology*, **65** (Suppl.), 65–72.

Kullander, S.O. (1999) Fish species – how and why. *Reviews in Fish Biology and Fisheries*, **9**, 325–352.

Lauder, G.V. & Liem, K.F. (1983) The evolution and interrelationships of the actinopterygian fishes. *Bulletin of the Museum of Comparative Zoology*, **150**, 95–197.

Linnaeus, C. (1758) *Systema Naturae*, 10th edn, Vol. 1. Laurentii Salviii Holmiae, Stockholm.

Mallet, J. (1995) A species definition for the modern synthesis. *Trends in Ecology and Evolution*, **10**, 294–299.

May, R.M. & Nee, S. (1995) The species alias problem. *Nature*, **378**, 447–448.

Mayden, R.L. & Wood, R.M. (1995) Systematics, species concepts and the evolutionary significant unit in biodiversity and conservation biology. *American Fisheries Society Symposium*, **17**, 58–113.

Menon, A.G.K. (1977) A systematic monograph of the tongue soles of the genus *Cynoglossus* Hamilton-Buchanan (Pisces: Cynoglossidae). *Smithsonian Contributions to Zoology*, **238**, 1–129.

Munroe, T.A. (1990) Eastern Atlantic tonguefishes (*Symphurus*: Cynoglossidae, Pleuronectiformes), with descriptions of two new species. *Bulletin of Marine Science*, **47**, 464–515.

Munroe, T.A. (1992) Interdigitation patterns of dorsal-fin pterygiophores and neural spines, an important diagnostic character for symphurine tonguefishes (*Symphurus*: Cynoglossidae: Pleuronectiformes). *Bulletin of Marine Science*, **50**, 357–403.

Munroe, T.A. (1998) Systematics and ecology of tonguefishes of the genus *Symphurus* (Cynoglossidae: Pleuronectiformes) from the western Atlantic Ocean. *Fishery Bulletin*, **96**, 1–182.

Munroe, T.A. & Amaoka, K. (1998) *Symphurus hondoensis* Hubbs 1915 (Cynoglossidae, Pleuronectiformes), a valid species of western Pacific tonguefish. *Ichthyological Research*, **45**, 385–391.

Munroe, T.A. & McCosker, J.E. (2001) Redescription of *Symphurus diabolicus*, a poorly-known, deep-sea tonguefish (Pleuronectiformes: Cynoglossidae) from the Galápagos Archipelago. *Revista de Biología Tropical*, **49** (Special Suppl. 1), 187–198.

Munroe, T.A. & Marsh, B.S. (1997) Taxonomic status of three nominal species of Indo-Pacific symphurine tonguefishes (*Symphurus*, Cynoglossidae, Pleuronectiformes). *Ichthyological Research*, **44**, 189–200.

Munroe, T.A., Krupp, F. & Schneider, M. (1995) Family Cynoglossidae. In: Vol. 2 *Guía FAO para la identificación de especes para los fines de la pesca. Pacífico centro-oriental* (eds W. Fischer, F. Krupp, W. Schneider, C. Sommer, K.E. Carpenter & V.H. Niem). pp. 1039–1059. FAO, Rome.

Munroe, T.A., Brito, A. & Hernández, C. (2000) *Symphurus insularis*: a new eastern Atlantic dwarf tonguefish (Cynoglossidae: Pleuronectiformes). *Copeia*, **2000**, 491–500.

Nelson, J.S. (1994) *Fishes of the World*, 3rd edn. John Wiley & Sons, New York.

Nelson, J.S. (1999) Editorial and introduction: The species concept in fish biology. *Reviews in Fish Biology and Fisheries*, **9**, 277–280.

Norman, J.R. (1934) *A systematic monograph of the flatfishes (Heterosomata)*, Vol. 1. Psettodidae, Bothidae, Pleuronectidae. British Museum, Natural History, London.

Norman, J.R. (1966) *A draft synopsis of the order, families and genera of recent fishes and fish-like vertebrates.* Unpublished photo offset copies distributed by the British Museum (Natural History), London.

Orr, J.W. & Matarese, A.C. (2000) Revision of the genus *Lepidopsetta* Gill, 1862 (Teleostei: Pleuronectidae) based on larval and adult morphology, with description of a new species from the North Pacific Ocean and Bering Sea. *Fishery Bulletin,* **98**, 539–582.

Patterson, B.D. (1994) Accumulating knowledge on the dimensions of biodiversity: systematic perspectives on Neotropical mammals. *Biodiversity Letters,* **2**, 79–86.

Patterson, B.D. (1996) The 'species alias' problem. *Nature,* **380**, 589.

Patterson, B.D. (2000) Patterns and trends in the discovery of new Neotropical mammals. *Diversity and Distributions,* **6**, 145–151.

Patterson, C. (1993) An overview of the early fossil record of acanthomorphs. *Bulletin of Marine Science,* **52**, 29–59.

Phelan, B.A., Manderson, J.P., Stoner, A.W. & Bejda, A.J. (2001) Size-related shifts in the habitat associations of young-of-the-year winter flounder (*Pseudopleuronectes americanus*): field observations and laboratory experiments with sediments and prey. *Journal of Experimental Marine Biology and Ecology,* **257**, 297–315.

Pleijel, F. & Rouse, G.W. (2000) Least-inclusive taxonomic unit: a new taxonomic concept for biology. *Proceedings of the Royal Society of London,* **B 267**, 627–630.

Ramos, R.T.C. (1998) *Estudo filogenético da família Achiridae (Teleostei: Pleuronectiformes: Pleuronectoidei), com a revisão das formas de água doce da América do Sul cisandina e a reavaliação do monofiletismo de Soleomorpha ('Soleoidei')*. DPhil thesis, Universidade de São Paulo.

Randall, J.E. (1999) Zoogeography of coral reef fishes of the Indo-Pacific region. *Proceedings of the 5th Indo-Pacific Fish Conference, Nouméa, 1997* (eds B. Séret & J.Y. Sire). pp. 23–26. Société Française d'Ichtyologie, Paris.

Rapoport, E.H. (1982) *Areography: geographical strategies of species,* 1st English edn (B. Drausal, translator). Published on behalf of Fundación bariloche by Pergamon Press, New York.

Regan, C.T. (1910) The origin and evolution of the teleostean fishes of the order Heterosomata. *Annals and Magazine of Natural History,* **8**, 484–496.

Regan, C.T. (1929) Fishes. Heterosomata, Vol. IX, pp. 324–325. *Encyclopaedia Britannica,* 14th edn. London.

Riddle, B.R. & Hafner, D.J. (1999) Species as units of analysis in ecology and biogeography: time to take the blinders off. *Global Ecology and Biogeography,* **8**, 433–441.

Roberts, C.M. & Hawkins, J.P. (1999) Extinction risk in the sea. *Trends in Ecology and Evolution,* **14**, 241–245.

Robins, C.R. & Randall, J.E. (1965) *Symphurus arawak*, a new cynoglossid fish from the Caribbean Sea, with notes on *Symphurus rhytisma* and *Symphurus ommaspilus*. *Bulletin of Marine Science,* **15**, 331–337.

Sakamoto, K. (1984) Interrelationships of the family Pleuronectidae (Pisces: Pleuronectiformes). *Memoirs of the Faculty of Fisheries, Hokkaido University,* **31**, 95–215.

Schwarzhans, W. (1999) *Piscium Catalogus: Part Otolithi piscium,* Vol. 2, *A comparative morphological treatise of recent and fossil otoliths of the order Pleuronectiformes*. Verlag Dr Friedrich Pfeil, München.

Soberón, M.J. & Llorente, B.J. (1993) The use of species accumulation functions for the prediction of species richness. *Conservation Biology,* **7**, 480–488.

Turner, G.F. (1999) What is a fish species? *Reviews in Fish Biology and Fisheries,* **9**, 281–297.

Winston, J.E. (1992) Systematics and marine conservation. In: *Systematics, Ecology and the Biodiversity Crisis* (ed. N. Eldredge). pp. 144–168. Columbia University Press, New York.

Chapter 3
Distributions and biogeography

Thomas A. Munroe

3.1 Introduction

Flatfishes have nearly global occurrence in marine habitats (Fig. 3.1), ranging from at least the southern Arctic Ocean to continental seas off Antarctica. They occupy diverse bathymetric environments from shallow marine and freshwater areas to deep-water habitats (to about 2000 m) and are conspicuous elements of most fish assemblages in estuaries and marine waters extending from shorelines to outer margins of the continental shelf. Only in polar seas, most freshwater systems, or in bathyal depths (below about 1500 m) are flatfish diversity values significantly lower compared with those of other aquatic environments (see discussion below).

When examining distributional patterns of taxa, ecologists and systematists often approach these biogeographical data by asking different types of questions. Researchers in these respective disciplines assign different levels of importance to contemporary or historical factors in developing hypotheses that attempt to explain patterns observed in organismal distributions. Consequently, biogeography has come to mean different things to different investigators (Crisci 2001).

Ecological approaches that attempt to explain distributional patterns of taxa, global patterns of community distributions or species richness patterns of particular taxa often focus on contemporary environmental factors in developing their hypotheses. For example, Topp & Hoff (1972) used a modified analysis of faunal coincidence among regions based on relative species abundances to identify different assemblages among Gulf of Mexico flatfishes (arctic-boreal group, warm-temperate to subtropical group, and Caribbean or tropical group). Ecological studies demonstrate that flatfish species distributions within regions are modified by responses of species to various ecological factors including water temperature, salinity, depth, sediment type and its spatial distribution, prey distribution and degree of habitat specialisation of the species (see also Gibson 1994; McConnaughey & Smith 2000). Ecological studies interested in explaining patterns in contemporary distributions of taxa seldom evaluate the contribution of historical events underlying these contemporary distributions. Consequently, they can only provide descriptive interpretations of biogeographic distributions and cannot address questions or test hypotheses related to the origins and evolutionary history of species present within faunal assemblages (Humphries & Parenti 1999).

How and when particular species came to occur in an area requires knowledge not only of evolution of the taxa of interest, but also knowledge concerning the history of the region.

Distributions and biogeography 43

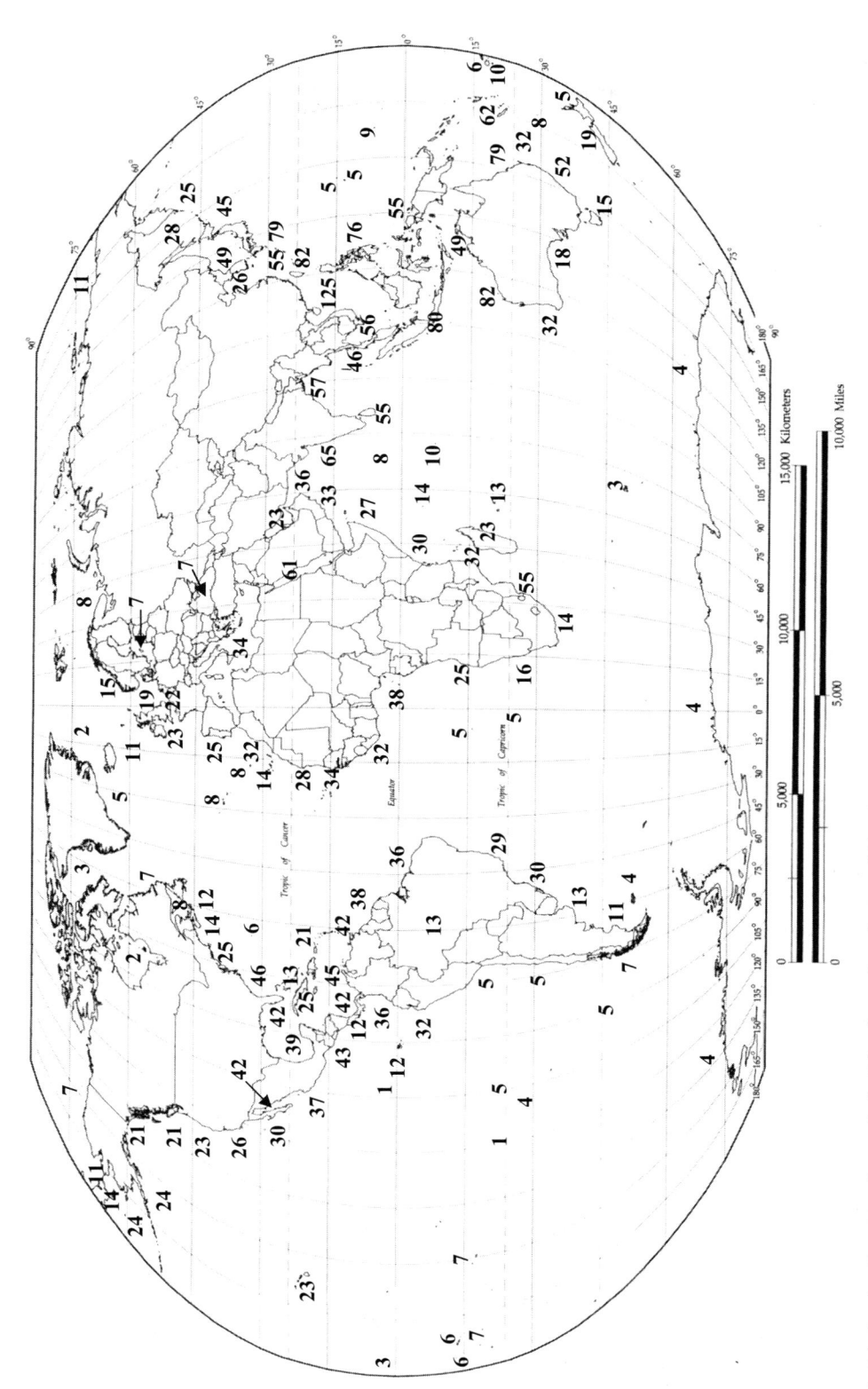

Fig. 3.1 Species diversity estimates for flatfishes from selected geographic regions.

Historical biogeography requires information contained within cladograms, i.e. hypotheses of evolutionary relationships among the organisms based on identification of monophyletic groups, to search for distributional patterns among sister taxa (the most closely related taxa). Systematic information contained within cladograms can then be combined with geological information to develop testable hypotheses structured within a historical framework to explain discontinuities in distributions among related taxa.

Fossils indicate that the origins of flatfishes date back at least to the Early Tertiary some 53–57 million years ago (Mya), and representatives of most flatfish lineages are present in the fossil record at this time (Chanet 1997; Schwarzhans 1999). Contemporary geographic distributions of various flatfish lineages are distinctly different from each other (Norman 1934), indicating that the history of these lineages has differed over time or that members of different lineages have responded differently to commonly experienced historical biogeographic events. Historical geological and oceanographic vicariant events in earth's history, by disrupting gene flow among ancestral populations, are considered to be the primary mechanisms leading to speciation (and extinction) in marine organisms, including flatfishes. These events have shaped and modified the composition of present-day flatfish assemblages occurring in different regions. Only by identifying these historical events and the timing of their occurrences can we begin to understand how, when and why the species and lineages occur where they do. Questions such as these are the substance of historical biogeography (Humphries & Parenti 1999; Crisci 2001).

For most pleuronectiform subgroups (and marine fishes in general), absence of robust hypotheses of intrarelationships precludes reconstruction of their evolutionary history and impedes progress in interpreting the role and timing of historical events influential in their evolution. The monophyly of most flatfish genera inhabiting tropical areas, especially the Indo-West Pacific, has not been demonstrated using phylogenetic analysis. Species discoveries from this area continue to accrue for many genera (see Chapter 2), but for many tropical and deep-sea species, especially, taxonomy remains imprecise (Hensley 1997). Without accurate identifications, distributional information for species is unreliable. Consequently, because data are so limited for many groups of flatfishes, it is premature to propose hypotheses reconstructing their historical biogeography, especially for most tropical flatfishes. It is also premature to describe areas of endemism for most tropical and deep-sea flatfish species, or to discuss area relationships using distributional information based on our imprecise knowledge of species occurring in tropical regions.

To summarise current knowledge regarding the distribution and biogeography of flatfishes, distributional information accumulated in taxonomic revisions and publications describing new flatfish species, as well as that reported for flatfishes in faunal and fishery surveys, was summarised and incorporated into a database. Data for 716 valid species of flatfishes (669 named and 47 recognised, but not described; see Chapter 2) were evaluated for inclusion in the database depending upon reliability of identifications presented in each study. For Indo-West Pacific flatfishes formerly perceived as representing a single 'widespread' species but now recognised as representing a complex of related species, special consideration was taken when evaluating distributional information associated with these specimens from earlier studies. If members of the species complex have allopatric distributions, then previously reported distributional data that could be accurately assigned to individual species were used.

Where members of a species complex have sympatric distributions, distributional information from earlier studies was considered compromised and was not used in data summaries.

In this chapter, distributional information for each major lineage within the Pleuronectiformes (see Chapter 2), as well as for selected genera and species within each lineage, is briefly summarised. Then follow descriptive summaries of global estimates and comparisons of flatfish species richness from different geographic regions and an examination of ecological distributional patterns of flatfishes with respect to three environmental gradients (temperature, depth and salinity). The chapter concludes by discussing information concerning historical biogeography of flatfishes and suggests some interesting directions for future research on this topic.

3.2 Geographic distribution of pleuronectiform lineages

Psettodidae. Two species occur in tropical marine waters, the spottail spiny turbot, *Psettodes belcheri,* found off tropical West Africa and the Indian spiny turbot, *P. erumei*, with widespread distribution throughout the Indo-West Pacific from East Africa to southern China, through Indonesia and northern Australia, and eastward to the Philippines.

Citharidae. Four genera and seven small- to medium-sized species of citharid flatfishes occur in temperate and subtropical seas of Europe and West Africa (*Citharus*); South Africa, throughout the Indian Ocean, the Philippines, Japan and western Australia (*Citharoides*); and widespread in the central and northern Indian Ocean eastward to the Philippines and Australia (*Brachypleura, Lepidoblepharon*) in the West Central Pacific.

Tephrinectes. The single species of this genus, the flower flounder, *T. sinensis*, occurs in coastal seas of China.

Scophthalmidae. Four genera with nine species of scophthalmid flatfishes are known exclusively from the North Atlantic Ocean between the Arctic Circle and Tropic of Cancer. Eight species occur in the eastern North Atlantic Ocean from the southwestern Barents Sea (Andriashev 1954), Scandinavia to Iceland, the Mediterranean, to as far south as Morocco. The windowpane, *Scophthalmus aquosus*, is the only western Atlantic member of this family (Gulf of Maine to northern Florida).

Bothidae. Twenty genera and 145 species of bothid flatfishes occur worldwide primarily in tropical and subtropical waters, with the majority of species occurring in relatively shallow marine waters. A few species in a smaller number of genera (e.g. *Parabothus, Chascanopsetta*) occur on the outer continental shelf and upper continental slope. Bothid flatfishes are most diverse in the tropical Indo-West Pacific, where species occur from the east coast of Africa and Red Sea throughout the Indian Ocean and Indo-Australian Archipelago, Japan, Australia and New Zealand, and across the Central Pacific (Norman 1934). In the western Atlantic, bothids are found from about as far north as the seas off Long Island, NY, to about Rio de Janeiro, Brazil. In the eastern Atlantic, bothids range as far north as southern Scotland, the Kattegat and Christiania fjord (Norman 1934), extend into the Mediterranean and Black Seas, and range along the West African coast to South Africa.

Many bothid genera have widespread geographic distributions with representative species widely distributed throughout temperate and tropical seas. *Arnoglossus* is a speciose genus with members distributed from off the Atlantic coast of Europe and Africa, in the

Mediterranean and Black Seas (scaldback, *Arnoglossus kessleri*), throughout the Indo-West and South Central Pacific to the Nazca Submarine Ridge (*A. multirastris*) in the southeastern Pacific (Parin 1991). *Engyprosopon*, a genus of about 30 nominal species of small-sized flatfishes, has members distributed throughout the Indo-Pacific from South Africa, northern Indian Ocean and Red Sea to the Indo-Australian Archipelago and Japan, the Hawaiian Islands, and with a larva also found at the Sala-y-Gomez Submarine Ridge (Belyanina 1990). Species of *Bothus* occur circumglobally in tropical and warm temperate waters. Several, including the flowery flounder, *B. mancus*, Indo-Pacific oval flounder, *B. myriaster*, and leopard flounder, *B. pantherinus*, have nearly circumglobal distributions throughout tropical waters (not occurring in eastern Atlantic). This species-rich family is among the most diverse of the Pleuronectiformes and many new bothids continue to be discovered in Indo-Pacific waters (Amaoka *et al.* 1993, 1997a; Amaoka & Mihara 2000).

Paralichthyidae. Eleven genera and 94 species of paralichthyid flounders are distributed worldwide in tropical, subtropical and temperate seas. In the Pacific, family members extend from about 45°N to about 35°S (Norman 1934), while in the western Atlantic, nine genera occur from the Gulf of Maine (Bigelow & Schroeder 1953) to southern Argentina (Díaz de Astarloa & Munroe 1998). Paralichthyid flounders do not occur off northwestern Europe (Norman 1934), and only a single species each of *Syacium* and *Citharichthys* represent this family in the eastern Atlantic off West Africa.

Only 2 of 11 paralichthyid genera, *Pseudorhombus* (23 species) and *Tarphops* (2 species), are found entirely in the Indo-West Pacific, with species ranging from East Africa and the Red Sea throughout the Indian Ocean and Indo-Australian Archipelago to the western Pacific, including Korea and Japan (Amaoka 1969). A third genus, *Paralichthys*, is represented in the western Pacific by a single species (Japanese flounder, *P. olivaceus*).

The greatest diversity of genera and species of paralichthyids occurs in seas of the New World, especially the Caribbean Sea and tropical eastern Pacific. Of 15 species of *Paralichthys*, 14 have entirely New World distributions (8 western Atlantic, 6 eastern Pacific). The eight species of *Hippoglossina* also are restricted to the western Atlantic and eastern Pacific oceans.

Pleuronectidae. Twenty-one genera and 60 species, including many commercially important species, such as halibuts, plaice (*Pleuronectes platessa*) and winter flounder (*Pseudopleuronectes americanus*), constitute this relatively large and diverse family. Pleuronectids have an amphi-boreal distribution in North Atlantic and North Pacific marine and estuarine waters and this is the predominant flatfish family found in Arctic-boreal and cold temperate seas of the northern hemisphere (Norman 1934; Cooper & Chapleau 1998). There is an asymmetric distribution of pleuronectid species in the Pacific versus that of the Atlantic Ocean. Taxonomic endemism is greater among Pacific pleuronectids (21 genera and 5 subfamilies present, with 12 genera and 2 subfamilies endemic to this region), but relatively high taxonomic diversity is present among Atlantic pleuronectids (9/21 genera and 3/5 subfamilies, and 12 species that constitute one-fifth of total species diversity for the family).

Most (48/60) pleuronectid species are endemic in the Pacific Ocean off North America and Asia in the region extending from the Bering Strait southward to the Gulf of California in the east (Norman 1934) and to the North China Sea in the west (Shmidt 1950). Eleven Pacific species range into the Arctic Ocean, but none is restricted to these waters. Two pleuronectids, the Greenland halibut (*Reinhardtius hippoglossoides*) and Arctic flounder (*Pleuronectes glacialis*) are wide-ranging with Arctic-boreal distributions in Atlantic and Pacific waters. Ten pleuronectids are endemic to Atlantic waters with geographic ranges extending from the Barents Sea, Iceland and Greenland Strait (Wheeler 1969) southward along the European shelf to the Mediterranean, and from off western Greenland southward along the North American shelf to off North Carolina (Cargenelli *et al.* 1999; Periera *et al.* 1999).

Paralichthodidae. This monotypic family contains only the peppered flounder, *Paralichthodus algoensis*, a medium-sized (to about 50 cm) flatfish known only from the inner continental shelf off South Africa from Mossel Bay to Delagoa Bay in 1–100 m (Heemstra 1986).

Samaridae. These small-sized flatfishes (3 genera and 28+ species) occur in relatively deep waters of the tropical and subtropical Indo-Pacific. New species of samarids continue to be discovered (Alfonso 2001; K. Amaoka, unpublished data) in recently collected material, indicating that considerable work is needed before precise assessments of species diversity and geographic distributions can be obtained for this family. Samarid flatfishes are found from East Africa and the Red Sea to northeastern and northwestern Australia, China, southeastern Japan, the Philippines, Hawaiian Islands, and eastward to the Sala-y-Gomez Submarine Ridge in the eastern Pacific (Parin 1991; Amaoka *et al.* 1997b; Alfonso 2001).

Members of *Samariscus* are the most broadly distributed, with representative species spanning over approximately 60° of latitude (Alfonso 2001) from southeastern Africa and Madagascar to the Hawaiian Islands (Quéro *et al.* 1989), with a juvenile specimen of an unidentified species of *Samariscus* also known from the Sala-y-Gomez Submarine Ridge (Amaoka *et al.* 1997b).

Plagiopsetta glossa, the tongue flatfish, is found from southern Japan to Australia and New Caledonia, whereas species of *Samaris* occur in continental seas off South Africa and along eastern Africa, off Madagascar, Reunion Island, in the Red Sea, and from the northern Indian Ocean, China, the Philippines, Indo-Malay Archipelago, and across northern Australia to New Caledonia (Alfonso 2001).

Poecilopsettidae. Representatives of this family (4 genera with 30+ species) are mostly small-sized, deep-water species. Poecilopsettids occur primarily in tropical and subtropical Indo-Pacific seas, with some species having widespread distributions (A.-C. Guibord, unpublished data), and with some species occurring in the tropical Atlantic. This family is notably absent from the tropical eastern Pacific. Members of *Marleyella* are found in the Indian Ocean from Natal to the Maldives and India. Species of *Nematops* are found from off Luzon Island, Philippines, to coastal waters off Sydney, Australia, and also on the western Australian continental shelf. A single specimen of the long-fin righteye flounder, *Nematops macrochirus*, representing the only extra-limital occurrence of this genus beyond the Indo-Pacific, purportedly was collected at St Helena Island, Central Atlantic.

Species of *Poecilopsetta* are found in deep waters of the Indo-Pacific and Atlantic Oceans. Most occur in the Indo-Pacific from Natal to India and also off western Australia, Honshu Island, Japan, New Caledonia, continental shelf off southeastern Australia to about off Sydney, and in the Hawaiian Archipelago. Two species inhabit the Atlantic, from off New England to Brazil in the western Atlantic, with a single capture of this genus in the eastern Atlantic off the southwestern coast of England.

Rhombosoleidae. This relatively small family of flatfishes, some of commercial importance, comprises nine genera and 19 species that occur primarily in relatively shallow marine and estuarine waters exclusively in the southern hemisphere (Norman 1934; A.-C. Guibord, unpublished data). Most genera and species are found in temperate waters off southern New Zealand and along Australia's southern coast. The Remo flounder, *Oncopterus darwini*, occurs off southeastern South America. Most species inhabit relatively shallow water, except the banded-fin flounders, *Azygopus pinnifasciatus* and *A. flemingi*, which occur in deep waters off southern Australia and the South Island of New Zealand in the Tasman Sea, respectively. The Indonesian ocellated flounder, *Psammodiscus ocellatus*, is found in tropical waters of the Indo-Australian Archipelago.

Achiropsettidae. Four species of achiropsettid flatfishes are endemic to deep continental shelf and continental slope waters in the Southern Ocean usually south of 50°S (except off southern South Africa and off southern New Zealand). All four species have circumpolar and peri-Antarctic distributions (Evseenko 2000), but these distributions are discontinuous with captures separated by enormous distances. Achiropsettid flatfishes have been collected off southern South America and the Falkland Islands, South Africa, New Zealand, and subantarctic and Antarctic shelves and underwater rises in the Southern Ocean. These are the southernmost occurring flatfishes, with the Antarctic armless flounder, *Mancopsetta maculata*, ranging into continental seas south of 65°S off Antarctica (Heemstra 1990; Evseenko 2000). More specimens are needed to better understand the biogeography of these fishes.

Soleidae. About 29 genera with 139+ species are currently recognised within the Soleidae, but the status of many species and genera remain poorly known. This diverse family of mostly small- to medium-sized fishes is found nearly worldwide in temperate and tropical waters in a variety of marine and freshwater habitats. Soles are conspicuously absent in the western Atlantic, and are rare in the eastern Pacific. The majority of soles inhabit shallow-water marine and estuarine environments, but a few species are found deeper (to about 1300 m).

In the eastern Atlantic, soles occur from off southern Iceland (the solenette, *Buglossidium luteum*) and in continental waters off Europe from southern Norway (common sole, *Solea solea*) to Spain and Portugal, throughout the Mediterranean Sea with a few species occurring in the Black Sea, and from North to South Africa. Soles are widespread throughout the Indo-Pacific, with maximum diversity occurring in the Indo-Malayan Archipelago and off northern Australia. Soleids also occur at oceanic islands throughout the Central Pacific extending eastward as far as Hawaii, Easter Island and the Galápagos Archipelago, where this family is represented only by Herre's sole, *Aseraggodes herrei* (Grove & Lavenberg 1997).

Among soleid genera, some have widespread distribution and others only fairly restricted occurrence in particular regions. Members of *Solea* occur from northern Europe and Iceland, throughout the eastern Atlantic and Indian Ocean to the West Central Pacific region. *Aseraggodes* (ca. 19 species) has widespread distribution, with species occurring from East Africa throughout the Indo-West Pacific, to Hawaii, Easter Island and the Galápagos Archipelago. Species of *Brachirus* (11 species), *Soleichthys* (14 species), *Pardachirus* (6 species) and *Zebrias* (19 species) also occur throughout the Indo-West Pacific region, with some species of *Soleichthys* occurring also at oceanic sites in the West Central Pacific. Several soleid genera are endemic in the eastern Atlantic including *Buglossidium, Dicologlossa, Monochirus, Pegusa* and *Vanstralenia*. Three of four species of *Bathysolea* are found in the eastern Atlantic, and six or seven species of *Microchirus* also inhabit this region. Two species of *Austroglossus* occur off South Africa. *Phyllichthys, Paradicula* and *Rendahlia,* each with small numbers of species, are endemic to waters of Australia and New Guinea. Soleid species are found in freshwaters in southern Africa (*Dagetichthys*) and southeast Asia, New Guinea, Borneo and northern Australia (e.g. *Brachirus, Achiroides*). Geographic distributions for the majority of tropical soleid genera and species, primarily because of their small size and unresolved systematics, are poorly known.

Achiridae. A relatively small, but diverse family (9 genera with 31+ species) of mostly small-sized flatfishes found exclusively in temperate and tropical freshwater, estuarine and coastal marine habitats of the Americas. The greatest diversity of achirids is found in tropical and subtropical waters within this region. Species distributions within the family extend from the Gulf of Maine (Bigelow & Schroeder 1953) to northern Argentina in the western Atlantic (Figueiredo & Menezes 2000), and from southern California (Barnhart 1936) to northern Peru (Chirichigno 1974) in the eastern Pacific, including a single species at the Galápagos Islands (Grove & Lavenberg 1997). Approximately 13 species, placed among 3–5 genera, occur in freshwaters of South America (Ramos 1998).

Cynoglossidae. Three genera and 145+ species of tonguefishes are found circumglobally primarily in shallow, warm temperate, subtropical and tropical marine and estuarine waters. Several species of *Cynoglossus* inhabit freshwater rivers of southeast Asia and northern Australia, and some rivers located on larger western Pacific islands, such as Borneo and New Guinea. Species of *Paraplagusia* are restricted to shallow, marine waters of the Indo-West Pacific from South Africa to the Indo-Malay Archipelago, to southern Japan, southwards to northeastern and northwestern Australia, and New Caledonia (Chapleau & Renaud 1993). *Symphurus* species occur throughout the world's temperate, subtropical and tropical oceans (Munroe 1992). Indo-West Pacific symphurine species occur in tidepools and other shallow-water habitats, but the majority of species inhabit relatively deep waters on the continental shelf and upper slope, with some inhabiting depths to about 1500 m (Munroe 1992; Munroe & Amaoka 1998). Atlantic (Munroe 1990, 1998) and eastern Pacific species (Munroe *et al.* 1995; Munroe & McCosker 2001) occur in both shallow and deep-water habitats to about 1000 m. *Cynoglossus* species are found primarily in estuarine and relatively shallow, continental marine waters of the tropical eastern Atlantic and Indo-West Pacific, with some also occurring in coastal habitats of Indo-West Pacific islands featuring extensive areas of soft-bottom habitats and/or estuaries (Menon 1977). In the Western Pacific, species of *Cynoglossus* are found from the Sea of Japan to southern Australia, including the Philippines and eastward to New Caledonia. Species of *Cynoglossus* and *Paraplagusia* are curiously absent in New World habitats.

3.3 Global patterns of species richness for the Pleuronectiformes

3.3.1 Latitudinal gradients in species richness

Although having nearly global occurrence in marine habitats, flatfish species diversity is distinctly asymmetric with respect to geographic location (Fig. 3.1). Among marine regions, some of the lowest species diversity estimates for flatfishes are recorded from habitats located in northern and southern polar waters, the extremes of flatfish distribution. Generally, diversity estimates for marine flatfishes increase as one proceeds along continental shelf regions from polar towards equatorial waters, with maximum diversity of flatfish assemblages usually occurring on tropical and subtropical continental shelves within each ocean.

Although evident for flatfishes in general, this pattern does not hold for individual families, some of which have distinctly different distributional patterns compared with that observed for flatfishes in general.

3.3.2 Tropical and subtropical regions

The greatest diversity of flatfishes occurs in tropical and subtropical marine waters where approximately 528 species representing nearly 74% of the total diversity of the Order are found. Many species continue to be discovered from tropical Indo-West Pacific waters; therefore species richness values for this region are conservative estimates. Species richness estimates are highest for flatfish assemblages occurring in marine waters in the area bordered by northern Australia and New Caledonia to the south and east, Indonesia, Malaysia and the Gulf of Thailand in the west, the Philippines and southern Japan in the northeast, and the South China Sea to the north. This region is also where the greatest diversity of marine fishes is found (Briggs 1974, 1999; Planes 1998).

The South China Sea supports the greatest diversity of flatfish species (125). Other Indo-West Pacific localities with diverse flatfish assemblages include Taiwan (82 species), the Indo-Malay Archipelago (80 species), Philippines (76 species), northwestern (82 species) and northeastern Australia (79 species), southern Japan (79 species) and Gulf of Thailand (56 species). Several Indian Ocean localities also recording high levels of flatfish diversity include the Arabian Sea off southern India (about 65 species), Red Sea (61 species), Bay of Bengal (57 species), Sri Lanka (55 species), Andaman Sea (46 species) and seas off southeastern South Africa (55 species).

Some tropical regions, e.g. the Persian Gulf where only 23 species have been documented, the northern Arabian Sea (36 recorded species), waters around New Guinea (ca. 55 species) and off northern Australia (49 species), have noticeably lower diversity estimates for flatfishes than do adjacent areas. Possibly, flatfish diversity is actually lower in these areas, but such low values may also indicate lack of adequate documentation for many species occurring in these areas.

Among the four major tropical marine regions (Briggs 1974), diversity of flatfishes is highest for the western Pacific (125 species). Flatfish assemblages are considerably less diverse in the tropical western Atlantic off Colombia and Venezuela (45 species), the tropical eastern Pacific off Central America (43 species), and in the eastern Atlantic in the Gulf of Guinea (38 species). Diversity trends observed for flatfishes among these tropical marine regions largely follow those observed for marine fishes in general (Briggs 1974).

Most tropical and subtropical flatfishes have rather restricted geographic distributions, but a few species have more widespread occurrence. Several bothids, the flowery flounder, Indo-Pacific oval flounder and leopard flounder, for example, have nearly circumglobal distributions in tropical waters, while the Indian spiny turbot occurs throughout the Indian Ocean from East Africa to Indonesia, and in the western Pacific from the Indo-Malay Archipelago to Taiwan and southern China, southward to northern Australia and eastward to the Philippines.

3.3.3 Temperate regions

On a global basis, approximately 171 flatfish species are found in temperate waters. Flatfish diversity in northern temperate areas (117 species, 16% of total diversity of the Order) is slightly more than twice that of southern temperate areas (54 species, 7.5% of total flatfish diversity). Among temperate areas, higher diversities of flatfishes occur in the West Pacific including the East China Sea (55 species), off southeastern Australia (52 species), Sea of Japan (49 species) and off northern Honshu Island (45 species). Temperate areas in the Atlantic and eastern Pacific Oceans record far fewer species than their counterparts in the western Pacific Ocean. Only 25 species occur off the eastern United States south of Cape Cod to Cape Hatteras, North Carolina; 23 species are reported in waters south of the British Isles; 34 in the Mediterranean Sea; and 26 species occur off northern California. Comparatively low diversity values are noted for flatfishes off northern Argentina (13 species), Namibia (16 species), southern Australia (18 species), Tasmania (15 species), New Zealand (19 species), the North Sea (19 species) and Chile (5–7 species). Flatfish assemblages in different temperate regions comprise distinctly different species. Most species occurring in temperate regions have relatively restricted distributions within that region or, at most, occur within temperate regions only within the same ocean. Relatively few species exhibit anti-equatorial distributions.

Among cold temperate regions, the highest flatfish diversity occurs in the western North Pacific, specifically the Okhotsk Sea where about 28 species occur. Twenty-five species are also known from around the Kuril Islands. About 24 flatfish species are recorded from the southern Bering Sea compared with only 14 species from the northern Bering Sea. Other North Pacific locations with relatively high species diversity include waters off the Aleutian Islands where 24 species have been reported, and in the Gulf of Alaska and along southern Canada, with 21 species known from each location. By comparison, North Atlantic cold temperate regions record fewer species than North Pacific areas at comparable latitudes. The maximum diversity reported from these areas is the 19 species recorded from the North Sea.

Among North Atlantic cold temperate regions, flatfish diversity is higher in eastern compared with western localities at the same latitude. The highest diversity in the western North Atlantic is the 14 species reported from the Gulf of Maine, located considerably further south than the North Sea (19 species). North of the Gulf of Maine, 12 species have been recorded from the Scotian Shelf, while only 7 are known from the colder waters off Labrador. About 15 species are known to occur off Norway, whereas off southern Iceland only 11 flatfish species have been reported. North Atlantic regions recording distinctly depauperate flatfish assemblages compared with other regions at comparable latitudes are generally those with reduced salinities and/or restricted connections to other bodies of water. Such areas include the Baltic Sea where only 7 species occur (compared with 19 in North Sea), the Black Sea (7 species compared with 34 in the Mediterranean) and Hudson Bay where only 2 species are reported (versus 7 off Labrador).

Flatfish diversities in southern hemisphere temperate provinces are much lower than those recorded from comparable latitudes in the northern hemisphere. For example, 11 species are known from off southern Argentina, whereas flatfish diversity diminishes to only 4 species at the Falkland Islands and southernmost Argentina, about 14 species occur off southernmost South Africa, and in the eastern Pacific, between 5 and 7 species occur in temperate continental shelf waters off northern and southern Chile.

3.3.4 Species richness on continental shelves

Within all geographic regions, flatfish diversity is generally greater in areas featuring extensive continental shelves with complex habitats located in shallow water. Comparable regions at similar latitudes with narrower shelves and/or lower habitat diversity generally record fewer species. Continental shelves in the West Central Pacific, the Indo-Malay Archipelago and the Caribbean Sea are extensive and offer a large variety of complex soft-sediment habitats. Correspondingly, these locations record the highest diversities of flatfishes. In the tropical western Pacific, flatfish diversity is markedly greater (125 species in the South China Sea) than corresponding areas in the Arabian Sea off southern India (about 65 species), the Red Sea (61 species), the Bay of Bengal (57 species), Andaman Sea (46 species), off southeastern South Africa (55 species) and in the tropical eastern Pacific (about 43 species). These other areas have narrower shelves and/or shelf regions with less habitat variability and complexity than those in the western Pacific. Likewise, in the tropical western Atlantic, flatfish diversity (about 45 species) is also higher than that (38 species) recorded at comparable latitudes in the eastern Atlantic where the shelf is narrower and sediment environments less complex (coral reef habitats are generally absent in eastern Atlantic).

Flatfish diversity at continental areas is also higher in geographic regions where components from different faunal provinces intermix (Fig. 3.1). In the western North Atlantic, the 46 species recorded from the region between Cape Hatteras, North Carolina, and West Palm Beach, Florida, is among the highest diversity estimates observed for flatfishes in the entire Atlantic, and is slightly higher than diversity values reported for the Gulf of Mexico (42 species) and tropical West Central Atlantic (45 species). The assemblage occurring off the southeastern United States comprises a mixture of species from the subtropical province to the south and that of a temperate province to the north. Other regions recording higher than expected diversity values for flatfish assemblages resulting from mixing of species of different biogeographic provinces include the Gulf of California, off northwest Africa, off southeast Africa, Taiwan, southeastern and southwestern Australia, and the North Sea (Fig. 3.1).

Lower than expected species richness values for flatfishes are noted on some continental shelf regions where cold-water currents and upwelling occur. Such lower diversity values are evident along the southwestern coast of Africa (25 species in Angolan waters diminishing to 16 off Namibia and 14 off South Africa); along southeastern South America (30 species off Uruguay, while only 13 and 11 species occur off northern and southern Argentina, respectively); and off western South America where 32 species are known from waters off northern Peru, but only 5–7 occur in Chilean waters just to the south. Possibly, lower species diversity (36) reported from the northern Arabian Sea compared with other areas in this region (Red Sea, 61 species) and latitude (57 species, Bay of Bengal) is also influenced by upwelling in this area. However, better documentation of species occurrences in this region is needed before any definite relationships between flatfish species diversity and upwelling can be demonstrated.

3.3.5 Insular versus continental regions

In all oceans, notable differences in flatfish diversity occur between continental and insular locations (Fig. 3.1; Table 3.1). The most dramatic example of this trend is evident in the West

Table 3.1 Flatfish species diversity at selected islands or island groups located in Atlantic, Pacific or Indian Oceans

Atlantic Ocean		West and Central Pacific Ocean		Indian Ocean		Eastern Pacific Ocean	
Location	No. of species	Location	No. of species	Location	No. of species	Location	No. of species
Azores	8	New Caledonia	62	Madagascar	23	Galápagos	12
Madeira	8	Chesterfield Islands	32	Sri Lanka	55	Cocos Islands	12
Canary Islands	14	Marshall	9	Mascarene Islands	13	Clipperton	1
Ascension	5	Gilbert	3	Maldives	8	Juan Fernandez	5
Bermuda	6	Vanuatu	10	Say de Malha Bank	10		
Bahamas	13	Fiji	6	Seychelles	14		
Cuba	25	Tonga	7				
Trinidad & Tobago	42	American Samoa	6				
Falkland Islands	4	French Polynesia	7				
St Helena	5	Pitcairn	1				
		Easter Island	4				
		Hawaiian Islands	23				
		North Marianas	5				
		Caroline Islands	5				
		Kermadec	5				
		Lord Howe	8				
		New Zealand	19				

Pacific. From the South China Sea, where 125 species of flatfishes have been recorded, diversity estimates decline appreciably with increasing distance from this region, especially for areas on the Pacific Plate (see Springer 1982). For example, 76 species are recorded from the Philippines, but only 5 from the northern Mariana Islands, 9 at the Marshall Islands, and 23 species from the Hawaiian Archipelago. Flatfish diversity values are also markedly reduced at progressively more remote islands located along a southeast tangent starting from the Philippines. Along this transect, five species are recorded in the Caroline Islands and six at American Samoa, seven in French Polynesia, one at Pitcairn Island, five from the region around the Sala-y-Gomez Ridge, and only four species at Easter Island. Similarly, islands located progressively eastward off northern Australia (where 79 species occur) generally record fewer flatfishes with increasing distance from continental areas: Chesterfield Islands and adjoining area (32 species), New Caledonia (62 species), only 10 at Vanuatu, 6 at Fiji, 7 at Tonga, 8 at Lord Howe Island, and just 5 species from the Kermadec Islands. Even at New Zealand and surrounding waters, where more extensive soft-sediment habitat occurs and more intensive sampling has been conducted, relatively few species of flatfishes (only 19) have been taken.

Within geographic regions, continental islands (usually larger with heterogeneous soft-sediment habitats) generally support greater diversity of flatfishes than do oceanic islands (usually smaller with less habitat complexity) located in the same region. In the Indian Ocean, flatfish diversity at Sri Lanka (55) is approximately four times as great as that at the Mascarene (13 species) or Maldive (14 species) Islands. In the tropical West Central Atlantic, the greatest diversity of flatfishes found at any island region (42 species; J.D. Hardy, personal communication) is recorded from Trinidad and Tobago, while 21 and 25 species are noted from Puerto Rico and Cuba, respectively. Oceanic islands in this region, in contrast, record fewer flatfishes: 13 from the Bahamas, 14 at the Canary Islands, 8 each at the Azores and Madeira,

6 at Bermuda (Smith-Vaniz *et al.* 1999), and Ascension and St Helena Islands record only 5 species each. Similar trends are observed for Pacific Ocean islands where 62 species are known from New Caledonia compared with 32 from the Chesterfield Islands region, and only 23 species in the Hawaiian Archipelago, 4–10 at Vanuatu, Fiji, Tonga, the Caroline Islands, American Samoa and French Polynesia. In the southern Pacific, similar trends between continental islands and oceanic islands are also apparent (19 flatfish species from New Zealand waters versus 8 at Lord Howe Island, 5 at Kermadec Islands, 4 at Easter Island and 1 species at Pitcairn Island).

3.3.6 Continental versus oceanic islands

More diverse flatfish assemblages at continental compared with oceanic islands in the same geographic region or at similar latitudes result from a variety of factors, some of which are inter-related, including the island's history (age) and history of its fauna, island size, location, remoteness, and its habitat diversity. Islands of continental origin have had a different geological origin and history compared with oceanic islands, and continental islands often possess (or retain) some members of the fauna typical of adjoining continental areas. Continental islands, often because of their larger size and geology, also provide greater amounts and greater diversity of habitats for flatfishes, especially soft-sediment environments typically found in estuaries, coastal lagoons and nearshore coastal regions resulting from deposition of sediments from fluvial discharge.

Sampling intensity and sampling bias are also important factors affecting diversity estimates. The marine fauna of some oceanic islands (Hawaii, New Caledonia, Easter Island, Bermuda) are much better sampled and their flatfish assemblages better known than are most others. Benthic fishes, such as flatfishes, are particularly vulnerable to trawling, and are generally better known from areas with extensive soft-sediment environments where commercial trawling takes place than they are from insular areas with extensive coral or rocky reefs where little or no trawling occurs.

An island's location (relative to source pool of colonising fishes) and degree of remoteness or isolation (likelihood of colonisation) also influence its faunal diversity. Remote oceanic islands in general have relatively few species of flatfishes. Atlantic islands, such as Bermuda (6 species), the Azores (8), Madeira (8), Ascension (5) and St Helena (5) record very few species of flatfishes, as do eastern Pacific islands including the Galápagos (12), Cocos Island (12), Easter Island (4 species), and Pitcairn Island and Clipperton Island (Robertson & Allen 1996) each with only 1 species. In the Indian Ocean, relatively few flatfishes occur at the more remote locations including the Mascarene (13) and Maldive (8) Islands, and this is also true for similar Central West Pacific islands (Fig. 3.1; Table 3.1).

3.4 Species richness in specific environments

3.4.1 Freshwater environments

Overall, diversity of freshwater flatfishes, i.e. those completing their entire life cycle within freshwater, is low compared with that occurring in other biotopes. Nearly all freshwater flat-

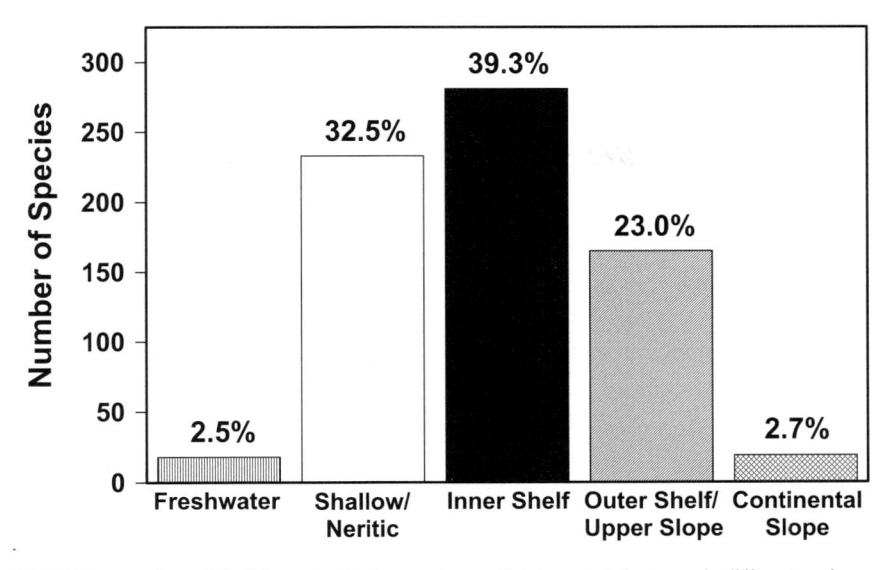

Fig. 3.2 Relative numbers of flatfish species (and percentages of total species) that occur in different environments.

fishes (about 18 species, 2.5% of total diversity of the Order, see Fig. 3.2) occur in the tropics. Adaptation to freshwater evolved independently several times among the more specialised lineages of flatfishes including the Achiridae (several genera), Soleidae (several genera) and Cynoglossidae (*Cynoglossus* spp.). Absence of flatfish fossils from freshwater deposits (Schwarzhans 1999) may indicate that invasion of freshwater by flatfishes occurred later in their evolution. Today, freshwater flatfishes are found in river systems of Central and South America (Achiridae), southern Africa (Soleidae) and in many larger rivers located throughout southeast Asia, Borneo, New Guinea and northern Australia (Soleidae, Cynoglossidae). The greatest diversity of freshwater flatfishes (ca. 13 species) occurs among the Achiridae inhabiting rivers of Central and South America (Ramos 1998). Somewhat surprising is the relatively low number of species recorded from freshwaters of Africa and southeast Asia.

3.4.2 Antarctic Ocean

Only four species of achiropsettid flatfishes occur in the extreme Southern Ocean and the diversity of flatfishes within this region is the lowest recorded for any continental shelf habitat, including those in Arctic seas. This low diversity does not appear to be an artefact of poor sampling (Eastman 2000) and the fish fauna is well characterised taxonomically (Gon & Heemstra 1990). Since the Southern Ocean encompasses about one-tenth of the world ocean and the Antarctic continental shelf constitutes about 7–8% of the total continental shelf area in the world (Kennett 1982), why are there so few lineages and species of flatfishes (and other fishes) in Antarctic waters compared with other geographic regions?

Eastman (2000) proposed that the discrepancy between the large area of the Antarctic shelf and the low species richness of fishes in general is in part attributable to a combination of regional historical and contemporary factors, especially tectonic, oceanographic and climatic events. During the early Miocene (25–22 Mya), seafloor spreading and opening of the Drake Passage to deep water began to separate Antarctica from other Gondwanan landmasses (Kennett 1982). Development of the unrestricted Antarctic Circumpolar Current and initial formation of the Antarctic Polar Front, with somewhat lower temperatures to the south, began to define the Southern Ocean and the Antarctic Zoogeographic Region. These events isolated Antarctica and initiated climatic changes that eventually eliminated some inshore habitats. After the middle Miocene (ca. 16 Ma), expansion of the ice sheet led to destruction of inshore habitats by ice, with repeated grounding of parts of the sheet as far as the shelf break (Anderson 1999). Invertebrate faunas and trophic conditions on the shelf changed. Emigration, regional extirpation and/or extinction reduced the diversity of the fish fauna, and its taxonomic composition was altered through geologic time. A post-Eocene faunal replacement took place with little carryover of families into the modern fauna. By the late Miocene-late Pliocene (6.5–2.0 Mya), low water temperatures, extreme geographic isolation and lack of south flowing surface currents limited immigration of many southern hemisphere epipelagic and mesopelagic groups into waters south of the Antarctic Polar Front. These conditions would also have prevented dispersal of planktonic life stages of benthic fishes from these areas as well. The shelf fauna of Antarctica became increasingly endemic and distinctive, but it is not known when it became modern in taxonomic composition (Eastman 2000). Today, the high Antarctic fauna remains isolated from those of other regions. Isolating factors include thermal isolation by subzero temperatures, geographic isolation by distance and oceanographic isolation by currents (Eastman 2000).

Fossil evidence will be needed to determine when and what flatfishes occurred historically in Antarctic waters. Possibly, the only flatfishes surviving through historical changes occurring in Antarctic waters were the ancestors of modern achiropsettids. Contemporary achiropsettids are deeper-water species compared with other southern hemisphere flatfishes. If ancestral achiropsettids were also deep-water inhabitants, perhaps they were less affected by changes in climatic conditions in Antarctic waters than were other flatfishes dwelling in shallower waters in this region. Alternatively, if arrival of flatfishes in Antarctic waters occurred more recently through dispersal from other regions rather than by differential survival of an earlier fauna, then it would appear that physical factors isolating Antarctic waters from other regions have been sufficiently strong enough to prevent all but a few flatfishes (the achiropsettids) from successfully colonising this region. Contemporary ecological conditions in the high Antarctic eliminate certain habitats such as estuaries, intertidal zones and reefs that are prime sites for fish diversity elsewhere (Eastman 2000). In other regions, flatfishes are very diverse in shallow-water habitats, but the Antarctic continental shelf averages 500 m in depth (vs 130 m elsewhere), thus many inner continental shelf habitats commonly occupied by flatfishes elsewhere are non-existent or not abundant. Inner shelf areas in the Antarctic are subjected to ice scouring and periodic disruption and degradation of inshore habitats and, because of extreme temperatures, there is no river run-off with accompanying erosion and deposition of soft-sediments into nearshore marine habitats in Antarctica. Absence of these major habitats may contribute to the rarity of flatfishes in this region. So even if other flatfish taxa are able to disperse to Antarctic waters, conditions on the continental shelf there may prohibit survival and establishment of viable populations.

3.4.3 Arctic Ocean

Flatfish diversity in Arctic seas, although greater than that observed in Antarctic waters, is also depauperate compared with other shelf habitats. Flatfishes resident in Arctic waters include several pleuronectids and at least two species of Scophthalmidae. Of Arctic species, the Greenland halibut has the northernmost distribution, with captures occurring north of 80°N off the northern end of Ellesmere Island (N. Alfonso, personal communication) and in Smith Sound at 78°N (Crawford 1992). Among Arctic-occurring flatfishes, only the Greenland halibut and Arctic flounder are widely dispersed, with both having circum-Arctic distributions between 70 and 80°N. These species have been taken in Arctic seas off Asia (Andriashev 1954), Europe, Greenland and North America. The most diverse Arctic flatfish assemblages (north of Arctic Circle to about 75°N) are the 11 species found in the Chukchi Sea off northeastern Russia, 9 species occurring along the northwestern Alaskan coast and into the Beaufort Sea (Mecklenburg et al. 2002), and the 8 species reported from the southern Barents Sea. Only Greenland halibut, Arctic flounder and American plaice (*Hippoglossoides platessoides*) have been reported in waters north of 70°N off West Greenland (off Baffin Island, ca. 71°N; N. Alfonso, personal communication).

Some geophysical conditions in the Arctic Ocean are different from those encountered in the Antarctic region. The Arctic Ocean has connections with both North Atlantic and North Pacific Oceans. Although the continental shelf is relatively narrow and restricted to the southern perimeter of the Arctic Ocean, it does not undergo the ice scouring at depth that occurs on the Antarctic shelf. Proceeding north into the Arctic Ocean means leaving the continental shelf and moving to open ocean where water depth increases dramatically. As flatfishes are generally not diverse in either pelagic or deep-water habitats anywhere (see below), this may account in part for why so few species are found in relatively deep waters of the open Arctic Ocean. Extreme low temperatures in the Arctic Ocean may also limit distributions for many flatfishes, as most species reported from Arctic Ocean sites generally occur only along its southern perimeter. However, as sampling under the permanent ice pack is difficult, possibly current information regarding Arctic Ocean flatfishes may merely reflect limitations in sampling access and not actual northern limits in distributions.

Greenland halibut, among boreal flatfishes, is only one of two flatfishes with widespread and common occurrence at high latitudes. This is also one of the most pelagically occurring flatfishes, spending much time in the water column. Some achiropsettids with widespread distribution in the Southern Ocean (Evseenko 2000) also have prolonged pelagic life stages (up to 2–3 years). Does a protracted pelagic stage in their life history convey some advantage to these flatfishes for successful habitation of polar oceans?

3.4.4 Shallow-water versus deep-sea habitats

On a global basis, species richness of flatfishes differs markedly with respect to depth of habitat (Fig. 3.2). Depth of occurrence data (depths where major portion of populations are found) clearly indicate that the greatest diversity of flatfishes is located in habitats from nearshore to depths of about 100 m on the continental shelf. About 251 species worldwide (35% of total diversity of Pleuronectiformes) inhabit shallow depths in freshwater, estuarine or neritic areas, and another 281 species (39% of total diversity) occur on the inner continental

Table 3.2 Some examples of deep-water representatives of pleuronectiform fishes

Family	Species	Depth (m)	Literature
Scophthalmidae	Fourspotted megrim, *Lepidorhombus boscii*	About 800	Nielsen (1986)
Pleuronectidae	Greenland halibut, *Reinhardtius hippoglossoides*	Reaching 2000[1]	Nielsen (1986)
	Atlantic halibut, *Hippoglossus hippoglossus*	Reaching 2000[1]	Nielsen (1986)
	Deepsea sole, *Embassichthys bathybius*	To 1370	Hart (1973)
	Pacific halibut, *Hippoglossus stenolepis*	At least 1100	Hart (1973)
	Witch, *Glyptocephalus cynoglossus*	About 1570	Scott & Scott (1988)
	Dover sole, *Microstomus pacificus*	About 1100	Hart (1973)
Bothidae	*Chascanopsetta* species	About 650	Amaoka & Parin (1990)
	Monolene species	At least 650	Nielsen (1961)
Poecilopsettidae	Several species	At least 1600	Potts & Ramsey (1987)
Achiropsettidae	Finless flounder, *Achiropsetta tricholepis*	To 1020	Heemstra (1990)
	Antarctic armless flounder, *Mancopsetta maculata*	To 1115	Heemstra (1990)
Soleidae	Deepwater sole, *Bathysolea profundicola*	About 1300	Quéro et al. (1986)
Cynoglossidae	*Symphurus* species (few)	About 1500	Munroe (1992)

[1]Usually found much shallower.

shelf. From these two depth regions combined, about 74% of the known flatfish species are found. In contrast, deep-ocean habitats throughout the world do not support very diverse flatfish assemblages. About 184 species (nearly 26% of total diversity of flatfishes) live on the outer continental shelf and upper continental slope. Of these, only about 19 species (3% of all flatfishes) are permanent residents on the continental slope. Although comparatively few flatfishes live on the outer continental shelf and upper continental slope, deep-water representatives are found across a broad spectrum of flatfish lineages, including some basal as well as more specialised lineages, with no readily apparent pattern in taxonomic distribution of deep-water taxa (Table 3.2).

3.5 Historical biogeography

The best available information on historical biogeography among the flatfishes is that for the Pleuronectidae (discussed below). However, other flatfish taxa with restricted geographic distributions, such as the Achiridae in the New World, Scophthalmidae in the North Atlantic, Soleidae primarily in the Old World (only one species of *Aseraggodes* in tropical eastern Pacific), and restricted southern hemisphere distribution of the Achiropsettidae and Rhombosoleidae, undoubtedly also have their explanations in global tectonic history. The flatfish fauna of New Zealand and south temperate Australia is exceptional in the number of species endemic to these areas (Norman 1934; Schwarzhans 1999). Rhombosoleid flatfishes, except for one species, occur in New Zealand or Australian waters. The large number of taxa uniquely occurring in either New Zealand or southern Australian waters may reflect the long-term isolation of these two regions (Schwarzhans 1999), and may provide some insight into the history of these regions. Other flatfish taxa where resolution of relationships should provide interesting data towards unravelling their historical biogeography, and geographical areas where resolution of flatfish intrarelationships could help identify important features in the history of these areas, are discussed below.

3.5.1 Pleuronectidae

The Pleuronectidae are the predominant flatfish family found in Arctic-boreal seas of the northern hemisphere. Hypothesised intrarelationships of taxa within this family were presented in Cooper & Chapleau (1998), and a discussion concerning the biogeographical history of this group is proposed in Cooper (unpublished manuscript). Cooper recognised seven regions of endemism based on the present distribution of the Pleuronectidae. Fossil and phylogenetic evidence both suggest the Pacific Ocean as the centre of diversification for the Pleuronectidae. These data also support the hypothesis that certain elements of several independently derived lineages of the Pacific pleuronectid fauna dispersed into the Atlantic via a trans-Arctic dispersal corridor some time following opening of the Bering Land Strait some 7.0 to 3.5 Mya (Marincovich & Gladenkov 1999). Considerably fewer pleuronectid species occur in the Atlantic (10 species) compared with the Pacific (48 species), suggesting that a dispersal corridor was not necessarily available to all Pacific pleuronectid taxa, or that not all Pacific taxa could successfully disperse to the Atlantic. At higher taxonomic ranks, no differentiation of Atlantic lineages from Pacific relatives is apparent; however, the 10 Atlantic endemic pleuronectid species have sister taxa found in the western North Pacific Ocean. Some time after arriving in the Atlantic, subsequent speciation occurred in all but one of the pleuronectid taxa that dispersed into the Atlantic. An estimated time line of 2.5 Mya has been suggested for dating speciation events between Atlantic and Pacific sister taxa.

3.5.2 Achiridae

The historical biogeography and origin(s) of freshwater achirids in tropical South America are also interesting subjects for further study. The diversity of the Achiridae is centred in tropical South America, where six endemic genera are found, with several restricted entirely to freshwater. Intrarelationships are unresolved for the Achiridae and presently no testable hypotheses regarding origins and historical biogeography of these fishes are available. Of interest though is that a diverse cross-section of the South American river fauna, including dolphins, fishes, crabs and snails, appear to be derived from marine ancestors (Lovejoy *et al.* 1998). Several hypotheses have been advanced to explain the occurrence of marine-derived taxa in South American river faunas. Based on molecular phylogenetic analysis of South American freshwater stingrays (Potamotrygonidae), coupled with reconstructions of Amazonian palaeogeography, Lovejoy *et al.* (1998) hypothesised that marine-derived freshwater stingrays originated as a by-product of massive movements of marine waters from the Caribbean into the upper Amazon region during the Early Miocene epoch, 15–23 Mya. If other apparently marine-derived taxa originated simultaneously with freshwater stingrays, then Miocene marine transgressions of South America will have had a profound effect on the diversification and structuring of neotropical communities. Origin(s) of freshwater achirid taxa are unknown and in need of study. However, the hypothesis advanced by Lovejoy *et al.* (1998) certainly provides an interesting initial comparative framework for hypothesis development and testing regarding origins of South American freshwater achirid flatfishes.

3.5.3 *Paralichthyidae*

Another distributional pattern observed among Atlantic flatfishes of interest and meriting further study are the eastern Atlantic occurrences of a single species in each of the paralichthyid genera, *Citharichthys* and *Syacium,* genera more speciose and widespread in the New World. Assuming that these are monophyletic genera, does the presence of these species in West African waters date back to an earlier time pre-dating opening of the early Atlantic when a common fauna may have occupied shallow seas of the proto-Atlantic, or do they derive from more recent events? As Europe and North American continents drifted apart, did they share a common flatfish fauna? If so, has most of this common fauna become extinct in West African waters, leaving only the species of *Syacium* and *Citharichthys* as surviving descendants? Or does the presence of these species in West African waters result from dispersal of larvae originating from the western Atlantic and establishment of population(s) in the eastern Atlantic, which with isolation over time have undergone subsequent speciation?

Also of interest are questions concerning geographical distributions of members of the paralichthyid genus, *Paralichthys*. All members of this genus, except the Japanese flounder, which occurs off Japan, Korea and China, are found in coastal seas of the New World. Assuming that this genus is monophyletic, did Japanese flounder originate as a result of a North Pacific vicariant event that divided a former widespread northern Pacific species into separate populations on either side of the Pacific which have undergone subsequent speciation? Or did this western Pacific species originate via westward dispersal, establishment and subsequent speciation of a population deriving from a New World species?

Only when sister-group relationships within these taxa are resolved can questions regarding their origins and historical biogeography be adequately addressed.

3.5.4 *New World tropical flatfishes*

One of the most important historical biogeographical events in the New World tropics was splitting of an American tropical marine fauna by formation and emergence of the Isthmus of Panama. Occurrences of sister species of many marine taxa on either side of the isthmus, with one member in the Caribbean Sea and the other occurring in the tropical eastern Pacific Ocean, indicate that this region once shared a common fauna. With emergence of the Isthmus of Panama, communication between these marine environments and gene flow between populations on either side of the isthmus were interrupted and populations subsequently diverged. Final closure of the Isthmian Sea is estimated to have occurred some 3–5 Mya when North and South America were joined together, and this time-frame serves as a minimum time estimate for stoppage of gene flow for marine taxa with sister species occurring on either side of the isthmus. Molecular studies (Knowlton *et al.* 1993) indicate that some speciation events for tropical western Atlantic-eastern Pacific sister taxa actually pre-dated full closure of the Panamanian Isthmus, suggesting that some populations had diverged well before final emergence of the peninsula. Phylogenetic studies of flatfishes in this region have been little explored, but unravelling relationships among these taxa should prove interesting in the development of biogeographical models for the New World tropics, as well as for supplying much needed information regarding the origins of related taxa present in other regions of the Atlantic and Pacific Oceans.

3.5.5 Indo-West Pacific region

The vast tropical and subtropical Indo-Pacific region extends from the Red Sea and east coast of Africa to the Hawaiian Islands, Easter Island (Randall 1999; Briggs 2000) and to the Nazca and Sala-y-Gomez Submarine ridges (Parin 1991). On the east it is bounded by the deep-water eastern Pacific Barrier and on the west by the Old World Land Barrier. Smaller biogeographic subunits within this region are recognised (Springer 1982; Randall 1999). Within the Indo-West Pacific region, some of the highest levels of species diversity for flatfishes, especially for members of the Bothidae, Samaridae, Poecilopsettidae, Soleidae and Cynoglossidae, are also noted. For some of these groups, high diversities are reported from within a comparatively small triangular region (East Indies Triangle) formed by the Philippines, Malay Peninsula and New Guinea. This region is also where the majority of tropical marine families have their greatest concentration of species (Briggs 1999, 2000).

Numerous hypotheses attempting to explain the high levels of species diversity observed in the East Indies Triangle have been proposed (Janzen 1967; Woodland 1983; Jokiel & Martinelli 1992; Randall 1999; Briggs 1999, 2000; Turner *et al.* 2001; Santini & Winterbottom 2002). Much of the debate centres around modes of speciation (allopatric vs sympatric) and timing of speciation events (more recent vs historical). Several hypotheses have invoked a dispersal-colonisation model as the primary mechanism to explain distributional patterns observed for taxa. Other hypotheses (Springer 1982, 1988; Springer & Williams 1990; Randall 1999; Santini & Winterbottom 2002) have proposed that historical vicariant events played major roles in the biogeography of taxa in this region.

How important dispersal has been in the historical distribution and evolution of flatfishes is largely unknown. Schwarzhans (1999) pointed out that based on the fossil record, dispersal of flatfishes across major deep-ocean barriers appears to remain the exception rather than the rule, as many flatfish lineages appear endemic to certain regions and likely have been so during their evolution. Features of oceanographic barriers that would be critical to flatfish larval survival or that are important to dispersal processes are unknown because information concerning environmental tolerances and dispersal capabilities of flatfish larvae are largely unstudied, especially for Indo-West Pacific species. Many tropical species, including fishes with planktonic life history stages, have very restricted distributions within large areas (Springer 1982, 1988) that appear to have relatively uniform physical conditions (compared with temperate regions). Restricted distributions of these taxa, in part at least, may be attributable to their inability to cross barriers of even apparently minor differences in environmental conditions (Janzen 1967). For organisms living in predictable environments, such as tropical seas, perhaps even small changes in the environmental parameters may constitute significant barriers to dispersal. Dispersal capabilities are also dependent upon a variety of factors, many of which are intrinsic to the species, including presence and duration of a pelagic larval stage, larval behaviour, degree of habitat preference of settling stages and adult vagility. Most flatfishes have planktonic eggs and larvae that potentially endow these species with significant dispersal capabilities. Some flatfishes, including bothids, achiropsettids, cynoglossids and poecilopsettids, have large, and presumably long-lived, larvae that potentially could be capable of dispersing over large areas (Evseenko 2000). The fate of dispersing larvae, their success in finding adequate habitat upon settling, or their requirements for particular settlement habitat, remain unknown for the vast majority of flatfish species. Equally unknown and largely uninvestigated is the role that larval swimming competency and other larval behaviours have relative to dispersal (Stobutzki & Bellwood 1994; Leis *et al.* 1996). Larvae, especially large and competent late pelagic stage larvae, may not be passive dispersants, but instead, they may control their position in the water column, direct their swimming, and ultimately to varying extent, control their fate regarding dispersal beyond certain regions (see Chapter 5).

Some distributional patterns observed for flatfishes (and other fishes: Randall 1999) in tropical oceans and elsewhere suggest that various taxa may have been successful in dispersing across deep-ocean barriers to colonise isolated locations. Examples among flatfishes that suggest dispersal may have contributed to distributions include the widespread occurrences throughout broad expanses of the world's oceans of species within genera from several families (Bothidae: *Bothus, Monolene* and *Chascanopsetta*; Poecilopsettidae: *Poecilopsetta*; Soleidae: *Aseraggodes* and *Solea*; Cynoglossidae: *Symphurus*), or the widespread Southern Ocean distribution of the finless flounder, and armless flounder, *Mancopsetta milfordi* (Achiropsettidae), and occurrences of congeneric members of various families (Soleidae: *Aseraggodes*; Bothidae: *Bothus*) at remote oceanic islands such as the Hawaiian Archipelago, Easter Island and the Galápagos Archipelago in the Pacific, and *Symphurus* (Cynoglossidae) at Ascension and St Helena islands in the Atlantic and French Polynesia and Hawaii in the Pacific. Many of these taxa have not been studied using phylogenetic techniques, so sister-group relationships among genera and species are unknown. Phylogeographic and population genetic approaches employing molecular techniques hold much promise in providing new insights into identifying sister-group relationships among these taxa that in turn will assist in revealing the role of dispersal and colonisation in their geographic distributions (Shulman & Bermingham 1995; Hilbish 1996; Planes & Galzin 1997; Shulman 1998; Bonhomme & Planes 2000; Muss *et al.* 2001).

For the majority of marine fishes occurring in the Indo-West Pacific, including most genera of flatfishes, cladistic hypotheses of intrarelationships are wanting. Therefore, biogeographical hypothesis including information from historical vicariant events is still at an early stage for most taxa from this region. Springer (1982) discussed the biogeography of Pacific marine fishes and proposed some interesting hypotheses based on plate tectonics and vicariant events in this region to explain the general patterns of distribution exhibited by many of the fish families occurring on the Pacific Plate. Other researchers have proposed that Pleistocene sea-level changes may have resulted in the proliferation of species in the Indo-West Pacific region via isolation and subsequent divergence of populations (Myers 1989; Springer & Williams 1990; Randall 1999). A testable framework for interpreting historical events in the East Indies region was provided by Santini & Winterbottom (2002), who proposed a hypothesis of vicariant biogeography to the study of coral reef fauna in the Indo-western Pacific region, a region of complex tectonic history. They concluded that, contrary to previous claims, most lineages of coral reef fauna inhabiting the Indo-western Pacific originated through vicariant events associated with, and following, the break-up of Gondwana. They hypothesise that a series of events fragmented an earlier widespread fauna, isolating different components of these earlier lineages. Based on their analysis, the high diversity of taxa in the East Indies region has resulted from the amalgamation of faunas from many continental and volcanic fragments and islands converging onto the region due to tectonic events associated with and following the break-up of Gondwana. Further conclusions arrived at in their study were that taxa now occurring in this region have had different origins and biogeographic histories. And if long-range dispersal and sympatric speciation occurred in taxa there, these were probably not the predominant mechanisms of speciation, at least in clades they examined.

As Indo-West Pacific flatfish species (and other fish species; see Gill 1999) become better recognised, as the monophyly of flatfishes and other taxa occurring in this region are defined and their phylogenetic relationships resolved, several vicariance models proposed to explain the history of faunal distributions in this region will be available as testable frameworks for interpretations of their historical biogeography.

Acknowledgements

Many people contributed information for this chapter, for which I am grateful. Any mistakes in interpretation of this information lie solely with the author. I thank N. Alfonso, K. Amaoka, B. Chanet, A.-C. Guibord, F. Chapleau, A. Cooper, M. Desoutter, J.M. Díaz de Astarloa, K. Hoshino, M. Nizinski, L. Parenti, R. Vari and J. Williams for information, assistance, literature and critical comment on ideas expressed in this paper. I thank the National Marine Fisheries Service for providing me with opportunities to study the evolution and ecology of these interesting fishes.

References

Alfonso, N. (2001) *Revision of the family Samaridae (Pleuronectiformes: Actinopterygii)*. MS thesis, University of Ottawa.

Amaoka, K. (1969) Studies on the sinistral flounders found in the waters around Japan. Taxonomy, anatomy and phylogeny. *Journal of the Shimonoseki University of Fisheries*, **18**, 65–340.

Amaoka, K. & Mihara, E. (2000) Pisces, Pleuronectiformes: Flatfishes from New Caledonia and adjacent waters – Genus *Arnoglossus*. In: Résultats des Campagnes MUSORSTOM, Vol. 21 (ed. A. Crosnier). *Mémoires du Muséum National d'Histoire Naturelle*, **184**, 783–813.

Amaoka, K. & Parin, N.V. (1990) A new flounder, *Chascanopsetta megagnatha*, from the Sala-y-Gomez Submarine Ridge, Eastern Pacific Ocean (Teleostei: Pleuronectiformes: Bothidae). *Copeia*, **1990**, 717–722.

Amaoka, K., Mihara, E. & Rivaton, J. (1993) Pisces, Pleuronectiformes: Flatfishes from the waters around New Caledonia – A revision of the genus *Engyprosopon*. In: Résultats des Campagnes MUSORSTOM, Vol. 11 (ed. A. Crosnier). *Mémoires du Muséum National d'Histoire Naturelle*, **158**, 377–426.

Amaoka, K., Arai, M. & Gomon, M.F. (1997a) A new species of *Arnoglossus* (Pleuronectiformes: Bothidae) from the southwestern coast of Australia. *Ichthyological Research*, **44**, 131–136.

Amaoka, K., Hoshino, K. & Parin, N.V. (1997b) Description of a juvenile specimen of a rarely-caught, deep-sea species of *Samariscus* (Pleuronectiformes, Samaridae) from Sala y Gomez Submarine Ridge, eastern Pacific Ocean. *Ichthyological Research*, **44**, 92–96.

Anderson, J.B. (1999) *Antarctic Marine Geology*. Cambridge University Press, Cambridge.

Andriashev, A.P. (1954) *Fishes of the northern seas of the U.S.S.R. Keys to the fauna of the U.S.S.R.* Zoological Institute. USSR Academy of Sciences, Moscow (English translation published for Smithsonian Institution and National Science Foundation, Washington, DC, by Israel Program for Scientific Translations, Jerusalem, 1974).

Barnhart, P.S. (1936) *Marine Fishes of Southern California*. University of California Press, Berkeley, CA.

Belyanina, T.P. (1990) Larvae and fingerlings of little-known benthic and benthopelagic fishes from the Nazca and Sala y Gomez Ridges. *Journal of Ichthyology*, **30**, 1–11.

Bigelow, H.B. & Schroeder, W.C. (1953) Fishes of the Gulf of Maine. *United States Fish & Wildlife Service, Fishery Bulletin*, **53**, 1–577.

Bonhomme, F. & Planes, S. (2000) Some evolutionary arguments about what maintains the pelagic interval in reef fishes. *Environmental Biology of Fishes*, **59**, 365–383.

Briggs, J.C. (1974) *Marine Zoogeography*. McGraw-Hill, New York.

Briggs, J.C. (1999) Modes of speciation: marine Indo-West Pacific. *Bulletin of Marine Science*, **65**, 645–656.

Briggs, J.C. (2000) Centrifugal speciation and the centres of origin. *Journal of Biogeography*, **27**, 1183–1188.

Cargenelli, L.M., Greisbach, S.J., Packer, D.B., Berrien, P.L., Morse, W.W. & Johnson, D.L. (1999) Essential fish habitat source document: Witch flounder, *Glyptocephalus cynoglossus*, life history and habitat characteristics. *United States Department of Commerce, NOAA Technical Memorandum*, NMFS-NE-139.

Chanet, B. (1997) A cladistic reappraisal of the fossil flatfishes record consequences on the phylogeny of the Pleuronectiformes (Osteichthyes: Teleostei). *Annales des Sciences Naturelles, Zoologie, Paris* (13th Series), **18**, 105–117.

Chapleau, F. & Renaud, C.B. (1993) *Paraplagusia sinerama* (Pleuronectiformes: Cynoglossidae), a new Indo-Pacific tongue sole with a revised key to species of the genus. *Copeia*, **1993**, 798–807.

Chirichigno, N.F. (1974) Lista sistemática de los peces marinos comunes para Ecuador-Perú-Chile. *Conferencia sobre explotación y conservación de las riquezas maritimas del Pacífico Sur*. Chile-Ecuador-Perú. Secretaria General, Lima.

Cooper, J.A. & Chapleau, F. (1998) Monophyly and intrarelationships of the family Pleuronectidae (Pleuronectiformes), with a revised classification. *Fishery Bulletin*, **96**, 686–726.

Crawford, R.E. (1992) Life history of the Davis Strait Greenland Halibut, with reference to the Cumberland Sound fishery. Manuscript Report, *Canadian Journal of Fisheries and Aquatic Sciences*, **230**, 1–19.

Crisci, J.V. (2001) The voice of historical biogeography. *Journal of Biogeography*, **28**, 157–168.

Díaz de Astarloa, J.M. & Munroe, T.A. (1998) Systematics, distribution and ecology of commercially important species of paralichthyid flounders occurring in Argentinean-Uruguayan waters (*Paralichthys*: Paralichthyidae): an overview. *Journal of Sea Research*, **39**, 1–9.

Eastman, J.T. (2000) Antarctic notothenoid fishes as subjects for research in evolutionary biology. *Antarctic Science*, **12**, 276–287.

Evseenko, S.A. (2000) Family Achiropsettidae and its position in the taxonomic and ecological classifications of Pleuronectiformes. *Journal of Ichthyology*, **40** (Suppl. 1), S110–S138.

Figueiredo, J.L. & Menezes, N.A. (2000) *Manual de Peixes Marinhos do Sudeste do Brasil. VI. Teleostei* (5). Museu de Zoologia, Universidade de São Paulo, Brazil.

Gibson, R.N. (1994) Impact of habitat quality and quantity on the recruitment of juvenile flatfishes. *Netherlands Journal of Sea Research*, **32**, 191–206.

Gill, A.C. (1999) Subspecies, geographic forms and widespread Indo-Pacific Coral-Reef fish species: a call for change in taxonomic practice. *Proceedings of the 5th Indo-Pacific Fish Conference, Nouméa, 1997* (eds B. Séret & J.Y. Sire). Paris: Société Française d'Ichtyologie, 1999, pp. 79–87.

Gon, O. & Heemstra, P.C. (1990) *Fishes of the Southern Ocean*. J.L.B. Smith Institute of Ichthyology, Grahamstown, South Africa.

Grove, J.S. & Lavenberg, R.J. (1997) *The Fishes of the Galápagos Islands*. Stanford University Press, Stanford, CA.

Hart, J.L. (1973) Pacific fishes of Canada. *Bulletin of the Fisheries Research Board of Canada*, **180**, 1–740.

Heemstra, P.C. (1986) Family No. 260: Pleuronectidae. In: *Smith's Sea Fishes* (eds M.M. Smith & P.C. Heemstra). pp. 863–865. Macmillan South Africa Publishers (Pty) Ltd, Johannesburg.

Heemstra, P.C. (1990) Achiropsettidae, Southern Flounders. In: *Fishes of the Southern Ocean* (eds O. Gon & P.C. Heemstra). pp. 408–413. J.L.B. Smith Institute of Ichthyology, Grahamstown.

Hensley, D.A. (1997) An overview of the systematics and biogeography of the flatfishes. *Journal of Sea Research*, **37**, 187–194.

Hilbish, T.J. (1996) Population genetics of marine species: the interaction of natural selection and historically differentiated populations. *Journal of Experimental Marine Biology and Ecology*, **200**, 67–83.

Humphries, C.J. & Parenti, L.R. (1999) *Cladistic Biogeography: Interpreting Patterns of Plant and Animal Distributions*, 2nd edn. Oxford University Press, Oxford.

Janzen, D.H. (1967) Why mountain passes are higher in the tropics. *American Naturalist*, **101**, 233–249.

Jokiel, D.L. & Martinelli, F.J. (1992) The vortex model of coral reef biogeography. *Journal of Biogeography*, **19**, 449–458.

Kennett, J.P. (1982) *Marine Geology*. Prentice Hall, Englewood Cliffs, NJ.

Knowlton, N., Weigt, L.A., Solórzano, L.A., Mills, D.K. & Bermingham, E. (1993) Divergence in proteins, mitochondrial DNA, and reproductive compatibility across the Isthmus of Panama. *Science*, **260**, 1629–1632.

Leis, J.M., Sweatman, H.P.A. & Reader, S.E. (1996) What the pelagic stages of coral reef fishes are doing out in blue water: daytime field observations of larval behavioural capabilities. *Marine and Freshwater Research*, **47**, 401–411.

Lovejoy, N.R., Bermingham, E. & Martin, A.P. (1998) Marine incursion into South America. *Nature*, **396**, 421–422.

McConnaughey, R.A. & Smith, K.R. (2000) Associations between flatfish abundance and surficial sediments in the eastern Bering Sea. *Canadian Journal of Fisheries and Aquatic Sciences*, **57**, 2410–2419.

Marincovich, L., Jr. & Gladenkov, A.Y. (1999) Evidence for an early opening of the Bering Strait. *Nature*, **397**, 149–151.

Mecklenburg, C.W., Mecklenburg, T.A. & Thorsteinson, L.K. (2002) *Fishes of Alaska*. American Fisheries Society, Bethesda, MD.

Menon, A.G.K. (1977) A systematic monograph of the tongue soles of the genus *Cynoglossus* Hamilton-Buchanan (Pisces: Cynoglossidae). *Smithsonian Contributions to Zoology*, **238**, 1–129.

Munroe, T.A. (1990) Eastern Atlantic tonguefishes (*Symphurus*: Cynoglossidae, Pleuronectiformes), with descriptions of two new species. *Bulletin of Marine Science*, **47**, 464–515.

Munroe, T.A. (1992) Interdigitation patterns of dorsal-fin pterygiophores and neural spines, an important diagnostic character for symphurine tonguefishes (*Symphurus*: Cynoglossidae: Pleuronectiformes). *Bulletin of Marine Science*, **50**, 357–403.

Munroe, T.A. (1998) Systematics and ecology of western Atlantic tonguefishes (*Symphurus*: Cynoglossidae: Pleuronectiformes). *Fishery Bulletin*, **96**, 1–182.

Munroe, T.A. & Amaoka, K. (1998) *Symphurus hondoensis* Hubbs 1915 (Cynoglossidae, Pleuronectiformes), a valid species of western Pacific tonguefish. *Ichthyological Research*, **45**, 385–391.

Munroe, T.A. & McCosker, J.E. (2001) Redescription of *Symphurus diabolicus*, a poorly-known, deep-sea tonguefish (Pleuronectiformes: Cynoglossidae) from the Galápagos Archipelago. *Revista de Biología Tropical*, **49** (Special Suppl. 1), 187–198.

Munroe, T.A., Krupp, F. & Schneider, M. (1995) Family Cynoglossidae. In: *Guía FAO para la Identificación de Especes para los Fines de la Pesca. Pacífico Centro-Oriental* (eds W. Fischer et al.). pp. 1039–1059. FAO, Rome.

Muss, A., Robertson, D.R., Stepien, C.A., Wirtz, P. & Bowen, B.W. (2001) Phylogeography of *Ophioblennius*: the role of ocean currents and geography in reef fish evolution. *Evolution*, **55**, 561–572.

Myers, R.S. (1989) *Micronesian Reef Fishes*. Coral Graphics, Guam.

Nielsen, J.G. (1961) Psettoidea and Pleuronectoidea (Pisces, Heterosomata). *Atlantide Report*, **6**, 101–127.

Nielsen, J.G. (1986) Pleuronectidae. In: *Fishes of the North-Eastern Atlantic and Mediterranean/ Poissons de l'Atlantique du Nort-Est et de la Méditerranée* (eds P.J.P. Whitehead et al.). pp. 1299–1307. UNESCO, Paris.

Norman, J.R. (1934) *A Systematic Monograph of the Flatfishes (Heterosomata)*, Vol. 1, *Psettodidae, Bothidae, Pleuronectidae*. British Museum of Natural History, London.

Parin, N.V. (1991) Fish fauna of the Nazca and Sala y Gomez submarine ridges, the easternmost outpost of the Indo-West Pacific zoogeographic region. *Bulletin of Marine Science*, **49**, 671–683.

Periera, J.J., Goldberg, R., Zikowski, J.J., Berrien, P.L., Morse, W.W. & Johnson, D.L. (1999) Essential fish habitat source document: Winter flounder, *Pseudopleuronectes americanus*, life history and habitat characteristics. *United States Department of Commerce NOAA Technical Memorandum*, NMFS-NE-138.

Planes, S. (1998) Biogeography of coral reef fish in the Pacific. *Océanus*, **24**, 197–206.

Planes, S. & Galzin, R. (1997) New perspectives in biogeography of coral reef fish in the Pacific using phylogeography and population genetic approaches. *Vie et Milieu*, **47**, 375–380.

Potts, D.T. & Ramsey, J.S. (1987) *A Preliminary Guide to Demersal Fishes of the Gulf of Mexico Continental Slope (100 to 600 Fathoms)*. Alabama Sea Grant Extension, Auburn University, Mobile, AL.

Quéro, J.C., Desoutter, M. & Lagardère, F. (1986) Soleidae. In: *Fishes of the North-Eastern Atlantic and Mediterranean/Poissons de l'Atlantique du Nort-Est et de la Méditerranée* (eds P.J.P. Whitehead et al.). pp. 1308–1324. UNESCO, Paris.

Quéro, J.C., Hensley, D.A. & Maugé, A.L. (1989) Pleuronectidae de l'ile de la Réunion et de Madagascar. II. Genres *Samaris* et *Samariscus*. *Cybium*, **13**, 105–114.

Ramos, R.T.C. (1998) *Estudo filogenético da família Achiridae (Teleostei: Pleuronectiformes: Pleuronectoidei), com a revisão das formas de água doce da América do Sul cisandina e a reavaliação do monofiletismo de Soleomorpha ('Soleoidei')*. DPhil thesis, Universidade de São Paulo.

Randall, J.E. (1999) Zoogeography of coral reef fishes of the Indo-Pacific region. *Proceedings of the 5th Indo-Pacific Fish Conference, Nouméa, 1997* (eds B. Séret & J.Y. Sire). Paris, Société Française d'Ichtyologie, 1999, pp. 23–26.

Robertson, D.R. & Allen, G.R. (1996) Zoogeography of the shorefish fauna of Clipperton Atoll. *Coral Reefs*, **15**, 121–131.

Santini, F. & Winterbottom, R. (2002) Historical biogeography of Indo-western Pacific coral reef biota: is the Indonesian region a centre of origin? *Journal of Biogeography*, **29**, 189–205.

Schwarzhans, W. (1999) *Piscium Catalogus: Part Otolithi piscium*, Vol. 2. *A Comparative Morphological Treatise of Recent and Fossil Otoliths of the Order Pleuronectiformes*. Verlag Dr Friedrich Pfeil, München.

Scott, W.B. & Scott, M.G. (1988) Atlantic fishes of Canada. *Canadian Bulletin of Fisheries and Aquatic Sciences*, **219**, 1–731.

Shmidt, P. Yu. (1950) *Fishes of the Sea of Okhotsk* (ed. P.V. Ushakov). Academy of Sciences of the USSR, *Transactions of the Pacific Committee*, Volume VI. (Translated from Russian in 1965 by O. Ronen, Israel Program for Scientific Translations, Jerusalem). (Available from United States Department of Commerce Clearinghouse for Federal Scientific and Technical Information, Springfield, VA.)

Shulman, M.J. (1998) What can population genetics tell us about dispersal and biogeographic history of coral-reef fishes? *Australian Journal of Ecology*, **23**, 216–225.

Shulman, M.J. & Bermingham, E. (1995) Early life histories, ocean currents, and the population genetics of Caribbean reef fishes. *Evolution*, **49**, 897–910.

Smith-Vaniz, W.F., Collette, B.B. & Luckhurst, B.E. (1999) Fishes of Bermuda: History, biogeography, annotated checklist, and identification keys. *American Society of Ichthyologists and Herpetologists Special Publication*, **4**, 1–424. Allen Press Inc., Lawrence, KS.

Springer, V.G. (1982) Pacific Plate biogeography, with special reference to shorefishes. *Smithsonian Contributions to Zoology*, **367**, 1–182.

Springer, V.G. (1988) The Indo-Pacific blenniid fish genus *Ecsenius*. *Smithsonian Contributions to Zoology*, **465**, 1–134.

Springer, V.G. & Williams, J.T. (1990) Widely distributed Pacific Plate endemics and lowered sea-level. *Bulletin of Marine Science*, **47**, 631–640.

Stobutzki, I. & Bellwood, D.K. (1994) An analysis of the sustained swimming abilities of pre- and post-settlement coral reef fishes. *Journal of Experimental Marine Biology and Ecology*, **175**, 275–296.

Topp, R.W. & Hoff, F.H., Jr (1972) Flatfishes (Pleuronectiformes). *Memoir of the Hourglass Cruises*, **4**, 1–135. (Florida Department of Natural Resources, St Petersburg, FL).

Turner, H., Hovenkamp, P. & Van Welzen, P.C. (2001) Biogeography of Southeast Asia and the West Pacific. *Journal of Biogeography*, **28**, 217–230.

Wheeler, A.W. (1969) *The Fishes of the British Isles and North West Europe*. Macmillan, London.

Woodland, D.J. (1983) Zoogeography of the Siganidae (Pisces): an interpretation of distribution and richness patterns. *Bulletin of Marine Science*, **33**, 713–717.

Chapter 4
Ecology of reproduction

Adriaan D. Rijnsdorp and Peter R. Witthames

4.1 Introduction

Reproduction represents the phase in the life cycle that links the new generation of offspring to the adult population. As such, successful reproduction is vital for the continued existence of populations in their natural environment. This chapter focuses on those reproductive characteristics that affect the production of offspring and their survival: spawning, gonad development and fecundity, onset of sexual maturity, and the energetics of reproduction and growth. Because reproductive characteristics have evolved, and continue to evolve, in response to environmental conditions, the adaptive significance of emergent patterns will be explored in relation to the geographical distribution of the species and their implications for the population dynamics of the species, and the resilience to perturbations caused by exploitation and pollution will be discussed.

4.2 Spawning

4.2.1 Spawning behaviour

The spawning behaviour of flatfishes is reviewed in Chapter 10. Although field observations are scarce, the evidence indicates that flatfishes may show complex spawning behaviour, especially in relation to the assortive mating in relation to size, which may have important consequences for the reproductive success of a commercially exploited population where the abundance of larger fish in the population as well as the sex ratio of these larger fish is reduced substantially (Rijnsdorp 1994).

4.2.2 Spawning mode

Most flatfish species produce pelagic eggs. In the northeast Atlantic all of the 21 species are known to spawn pelagic eggs (Russell 1976). In the northwest Atlantic area, 3 of 18 species (winter flounder *Pseudopleuronectes americanus*, American smooth flounder *Pleuronectes putnami*, hogchoker *Trinectes maculatus*; Miller *et al.* 1991), and in the north Pacific, 5 of 120 species (marbled flounder *Pseudopleuronectes yokohamae*, cresthead flounder *P. schrenki*, dusky sole *Lepidopsetta mochigarei*, rock sole *L. bilineata* and kurogarei *Pseudopleuronectes*

Table 4.1 Egg size (mm) in flatfishes compared with egg size in other teleost species in the northeast Atlantic

Group	Egg type	Mean	SD	Minimum	Maximum	n
Pacific flatfishes	Pelagic	1.37	0.64	0.83	3.06	16
	Demersal	0.84	–	–	–	1
Atlantic flatfishes	Pelagic	1.44	0.79	0.50	4.25	38
	Demersal	1.00	0.22	0.85	1.25	3
Other teleosts	Pelagic	1.21	0.39	0.73	2.70	39
	Demersal	1.49	0.63	0.75	3.25	30

obscurus; Minami & Tanaka 1992; Grigorev 1995; Shuozeng 1995) are known to spawn demersal eggs. In flatfishes, demersal eggs appear to be restricted to estuarine or shallow coastal species where they may reduce the risk of advective losses of offspring.

4.2.3 Egg size

Egg sizes in flatfishes range between 0.50 and 4.25 mm and do not differ from those in other teleosts (Table 4.1). No differences in egg size are apparent across ocean basins. Comparison of egg size across flatfish families showed that the mean and range of egg sizes in Pleuronectidae was larger than in other families. There is a tendency among teleosts for demersal eggs to be larger than pelagic eggs (Chambers 1997) but there is no indication for such a pattern in flatfishes. Minami & Tanaka (1992) showed that egg size in northwest Pacific Pleuronectidae was higher in offshore spawning species compared with inshore spawning species. Such a relationship also exists across flatfish families over a wide range of species according to an analysis of the relationship between egg size and maximum depth of the distribution (Fig. 4.1b). Egg size also increases with latitude within the family of the Pleuronectidae and Soleidae (Fig. 4.1a). However, no consistent latitudinal pattern in egg size emerges within species. Egg size in common sole, *Solea solea*, decreased with latitude (Rijnsdorp & Vingerhoed 1994), whereas in American plaice, *Hippoglossoides platessoides*, no pattern has been found (Walsh 1994). The interpretation of the latitudinal patterns is complicated by the general decline in egg size with the time of spawning. This pattern exists within species

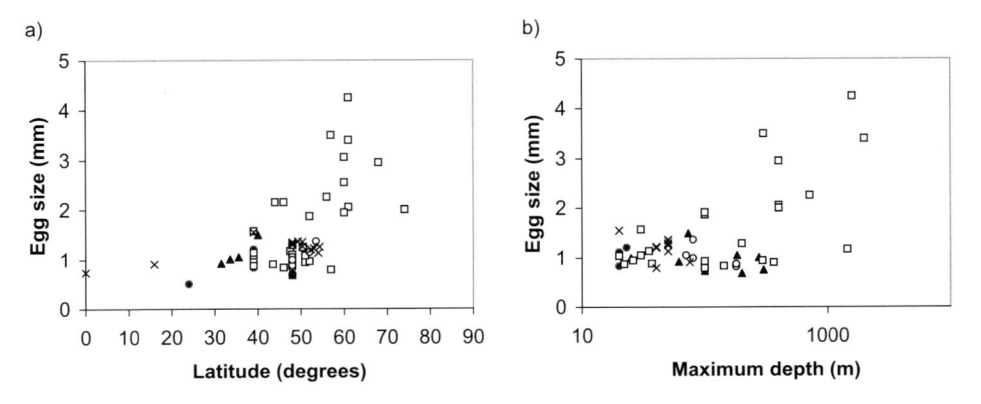

Fig. 4.1 Egg size in flatfish in relation to (a) latitude and (b) maximum depth: ▲ Bothidae, ● Cynoglossidae, □ Pleuronectidae, ○ Scophthalmidae, × Soleidae.

as well as among species (Russell 1976; Rijnsdorp & Vingerhoed 1994) and is consistent with the general pattern in teleosts (Chambers 1997). Chambers (1997) showed that egg size also co-varies with temperature and salinity. Low salinity is positively related to egg size in flounder, *Platichthys flesus*, although this explained only the difference between egg size off Norway and in the Baltic (Solemdal 1967, 1973).

Egg size is correlated with several ecologically relevant traits such as size at hatching, size of mouth parts and burst swimming speed that may affect feeding success. As such, there may be a positive feedback from a large initial size to the survival opportunities later in life. Egg size is viewed as adaptive to the feeding conditions of the larvae (Bagenal 1971). Also, it has been shown that larger eggs experience a lower mortality rate, both among and within species (Rijnsdorp & Jaworski 1990; Pepin 1991).

4.2.4 Spawning season

In a review of life histories of northwest Pacific flatfishes, Minami & Tanaka (1992) showed that spawning season varies among flatfish families in relation to latitude. In Pleuronectidae and Paralichthyidae, reproduction is restricted to summer in high latitude waters, while it shifts to spring and winter in temperate waters. Soleidae and Cynoglossidae are distributed at lower latitudes and spawn mainly during summer.

In most species spawning at a time when day length increases, there is a tendency to spawn later in the year at higher latitudes (Fig. 4.2). The regression of the time of peak spawning against latitude indicates a delay in spawning of 2.7 days (SE = 0.77) per degree latitude, although species vary considerably in their response (Table 4.2). An opposite trend is indicated for the northwest Atlantic summer flounder, *Paralichthys dentatus*, which spawns at decreasing day length: spawning starts in September in the northern end of its range and progresses southward during autumn and winter (Smith 1973).

The timing of spawning in temperate fishes is generally stable (Cushing 1969). Peak spawning in plaice, *Pleuronectes platessa*, as observed from plankton surveys, occurred on average on 19 January with a standard deviation of 7 days (Simpson 1959). In the spring

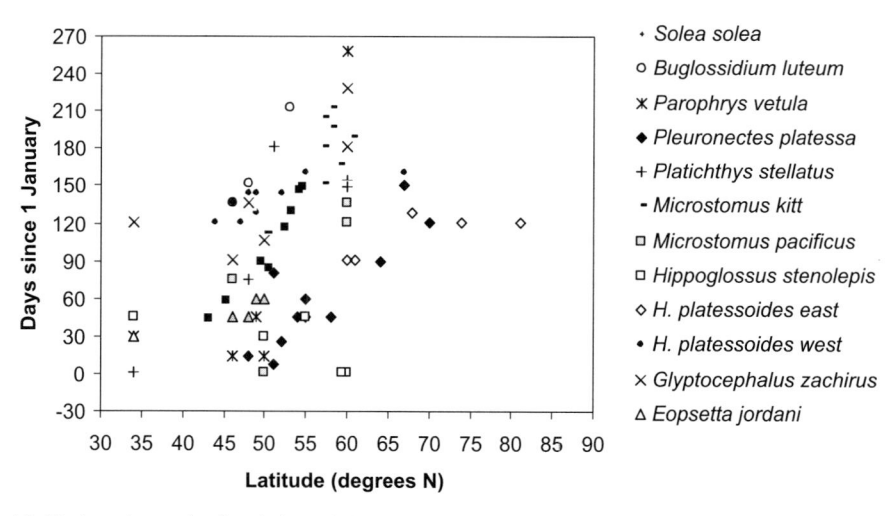

Fig. 4.2 Timing of spawning in relation to latitude in different populations of flatfish.

Table 4.2 Regression parameters of the linear regression of the time of peak spawning against latitude

Species	Slope	Correlation	n	P	Reference
Solenette	11.1	1.00	3	*	Whitehead et al. 1986
Buglossidium luteum					
Petrale sole	1.7	0.90	5	*	Castillo 1995
Eopsetta jordani					
Rex sole	3.8	0.72	6	*	Castillo 1995
Glyptocephalus zachirus					
American plaice	1.3	0.65	4	ns	Walsh 1994
Hippoglossoides platessoides (1)					
Hippoglossoides platessoides (2)	1.7	0.79	10	**	Walsh 1994
Pacific halibut	−3.5	−0.74	4	ns	Castillo 1995
Hippoglossus stenolepis					
Lemon sole	0.4	0.02	6	ns	Rae 1965
Microstomus kitt					
Dover sole	3.3	0.65	5	ns	Castillo 1995
Microstomus pacificus					
Starry Flounder	5.9	0.86	5	*	Castillo 1995
Platichthys stellatus					
Plaice	5.3	0.86	11	**	Deniel 1981; Simpson 1959; Wimpenny 1953
Pleuronectes platessa					
English sole	18.7	0.96	4	*	Castillo 1995
Parophrys vetula					
Common sole	9.2	0.97	9	**	Rijnsdorp & Vingerhoed 1994; Whitehead et al. 1986
Solea solea					

* $P<0.05$; ** $P<0.01$; ns, not significant. For *Hippoglossoides platessoides* the western (1) and eastern (2) Atlantic populations are analysed separately.

spawning common sole, however, peak spawning was more variable and appeared to be related to the winter and spring temperature (Van der Land 1991). An analysis of the proportion of spawning stages in market samples taken throughout the spawning period for North Sea plaice and common sole (Rijnsdorp 1989) revealed that the onset and end of spawning were negatively correlated with the water temperature at the start of the spawning period (plaice: December–January; common sole: February–March) and during the spawning period (plaice: 1st quarter; common sole: 2nd quarter) (Fig. 4.3). Common sole showed a

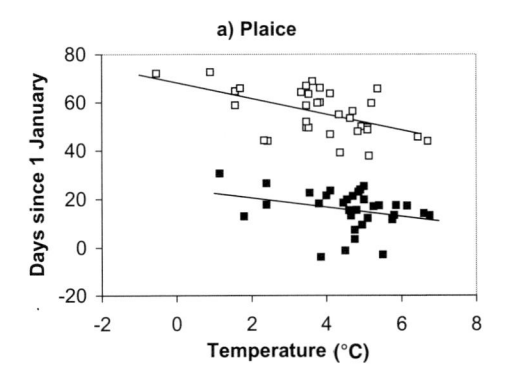

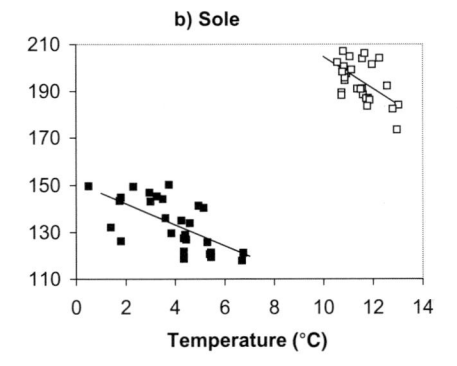

Fig. 4.3 The relationship between the onset (■) and end (□) of the spawning period (days since 1 January) in relation to the mean water temperature at the start of the spawning period and during the spawning period in (a) North Sea plaice and (b) common sole, as determined from Dutch market samples taken between 1957 and 1996.

stronger response to temperature than plaice. The end of the spawning season appeared to be more strongly influenced by temperature than the onset of spawning.

Although the majority of species are characterised by having one single spawning period during the year, a bimodal periodicity has been reported for some flatfish species. In Greenland halibut (*Reinhardtius hippoglossoides*), spawning was observed during winter as well as during summer (Junquera & Zamarro 1994; Albert *et al.* 2001). In Wakasa Bay (Japan), two spawning periods were reported for *Pseudorhombus oligolepis* (April–June and October–December (Kamisaka *et al.* 1999) and largescale flounder *Engyprosopon grandisquama* (May and November; Minami and Tanaka 1992). The bimodal spawning in *Pseudorhombus oligolepis* is related to a small size and the 6 months required for maturation (Kamisaka *et al.* 1999).

The patterns observed in flatfishes are consistent with the view that the timing of spawning is synchronised with the productivity of the pelagic system (Cushing 1969; Bagenal 1971). In high latitude areas, pelagic productivity is high during summer and flatfishes spawn well in advance so that their eggs hatch in the most productive period. In temperate waters, productivity shows a distinct peak in spring and a smaller secondary peak in autumn. Flatfishes mainly utilise the spring peak and only few species spawn in autumn. The autumn spawning of summer flounder coincides with the breakdown of the thermal stratification on the continental shelf and the timing of the autumn plankton bloom (Smith 1973). In tropical waters, productivity is low throughout the year. Spawning may occur at any time (Garcia Abad *et al.* 1992; Rajaguru 1992), but may be locally linked to the patterns of rainfall or wind (Ramanathan *et al.* 1977).

4.2.5 Duration of spawning

The spawning season is commonly described in terms of the month at which spawning starts and ends (Russell 1976; Whitehead *et al.* 1986; Minami & Tanaka 1992). This allows a crude estimate of the duration of spawning. At high latitudes, spawning generally lasts between 2 and 4 months (Miller *et al.* 1991; Minami & Tanaka 1992). At low latitudes spawning periods may extend over up to 10 months (Rajaguru 1992), although shorter spawning periods are reported as well (Ramanathan *et al.* 1990; Garcia Abad *et al.* 1992).

Spawning of individual fish takes less time. In North Sea plaice, females were in spawning condition for 3–6 weeks and the length of this stage increased with age. In recruit spawners, spawning starts later, while the end of the spawning season appeared to be similar for all females. Males are in spawning condition for at least 11 weeks and arrive at the spawning grounds well before the spawning season starts. Young males tended to finish spawning earlier than older males (Rijnsdorp 1989).

In the sand sole *Solea lascaris* and Adriatic sole *S. impar*, females spawn several batches during each of two periods within a prolonged spawning period of 4.5 and 3.5 months, respectively. The eggs spawned during the second period had recruited from the stock of previtellogenic oocytes at the beginning of the spawning period (Deniel *et al.* 1989).

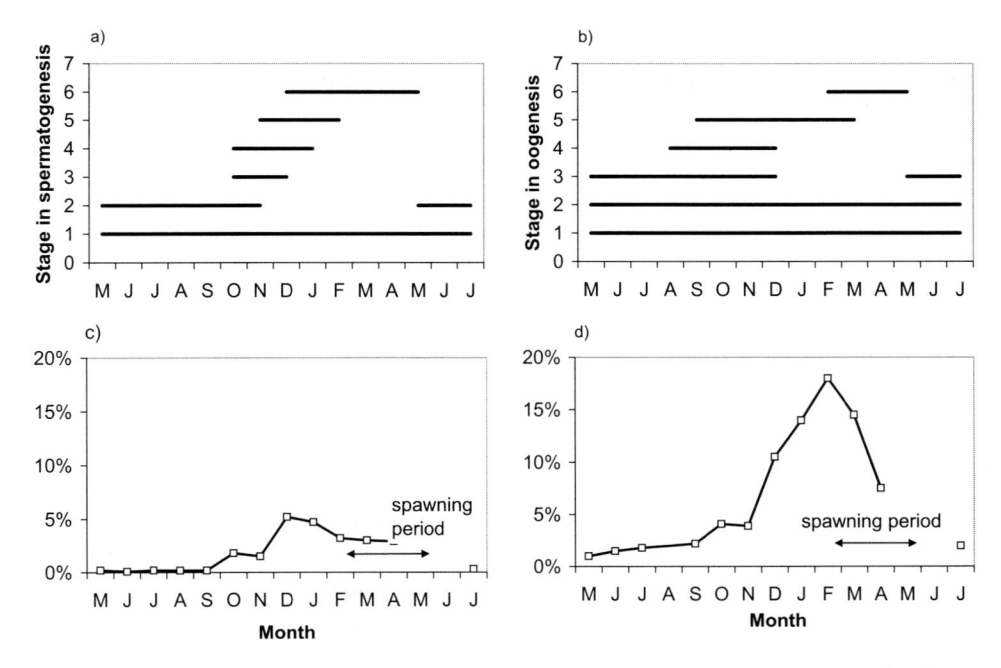

Fig. 4.4 Schematic representation of the gonad development in plaice: (a) spermiogenesis; (b) oogenesis; (c) %GSI testis; (d) %GSI ovary. Spermatogenesis codes denote: 1, spermatogonia; 2, mitotic increase in spermatogonia; 3, primary spermatocytes; 4, secondary spermatocytes; 5, spermatids; 6, spermatozoa. Oogenesis codes denote: 1, oogonia; 2, primary oocytes; 3, yolk precursors; 4, early vitellogenesis; 5, advanced vitellogenesis; 6, ripe ova. Modified from Barr (1963a, b).

4.3 Gonad development

4.3.1 Testis

There are few publications on testis structure in flatfishes (plaice, Barr 1963a; greenback flounder *Rhombosolea tapirina,* Barnett & Pankhurst 1999). The seasonal development in plaice is illustrated in Fig. 4.4. The terminology follows the comprehensive review of testis in teleosts (Grier 1981). Spermatogenesis starts with mitotic cell proliferation of spermatogonia (5–7 μm in size), which become aggregated in cysts. A meiosis phase follows as the spermatogonia turn into primary and then secondary spermatocytes that undergo spermiogenesis to form spermatids. Spermatids are transformed into sperm during spermiation. Unlike the large female gamete (0.7–5 mm depending on the species), sperm are very small and less than 1 μm in size. In contrast to plaice, most greenback flounder males collected were only partially spermiated and milt was not freely running. In this respect they closely resemble common sole, which is rarely found to be running copious quantities of milt. These differences are reflected in the gonadosomatic indices (common sole <1%, Deniel 1981; plaice 5%, Barr 1963a) and suggest a different spawning strategy.

4.3.2 Ovary

As in other telosts (Tyler & Sumpter 1996) the production of gametes (oogenesis) in flatfishes starts when the ovary develops during sexual differentiation, probably during the larval phase. A few months after metamorphosis, a small pair of ovaries have formed projecting back from the abdominal cavity towards the tail and oogenesis to produce previtellogenic oocytes from stem cells (oogonia) surrounded by follicle cells is well under way (Barr 1963b; Deniel 1981; Morse 1981; Hunter *et al.* 1992; Witthames & Greer Walker 1995; Rideout *et al.* 1999). This juvenile stage lasts until puberty when the fish switches to an annual cycle culminating in the uptake of large amounts of yolk (vitellogenin) into developing oocytes. The ovarian development receives cues from the environment via the hypothalamus pituitary axes mediated by the secretion of neural and peptide hormones (gonadotropins), regulating the production of sex steroids (Tyler *et al.* 1999). Maturation is controlled by a combination of photoperiod and temperature signals as in most teleosts (Norberg *et al.* 1999). In winter and spring spawning species like plaice and common sole, vitellogenesis starts in August when the largest previtellogenic oocytes enter the cortical alveoli stage (Fig. 4.4). After an initial burst in the recruitment of oocytes into vitellogenesis the number of developing oocytes is fairly stable. Final maturation and the first meiotic division occur in the leading oocyte cohort following stimulation by gonadotropin hormone 2 produced by the pituitary. These oocytes swell in size due to an influx of water following breakdown of yolk reserves and then ovulate by bursting out of the follicle ready for spawning (Barr 1963b; Witthames & Greer Walker 1995).

A proportion of the developing oocytes abort their development towards final maturation and become atretic. This atresia can be identified by hypertrophy in the follicular layer, which is accompanied by the disintegration of the zona pellucida (outer boundary of the oocyte) and leakage of cytoplasm and yolk granules into the ovary lumen (Wood & Van Der Kraak 1999). Both previtellogenic and vitellogenic oocytes may become atretic but in the latter case this involves large metabolic costs associated with the synthesis and breakdown of complex macromolecules. The dynamics of the process in relation to the annual reproductive cycle of a wild population has only been published for common sole (Witthames *et al.* 1999), where it was most prevalent just prior to and during the first part of spawning. This species has an asynchronous pattern of oocyte development even just before spawning and atresia is more prevalent in the smallest vitellogenic size classes (200–450 µm; Witthames & Greer Walker 1995). The total number of oocytes lost from the ovary during spawning (60 days for an individual female) was estimated to be 28% of the annual fecundity (Armstrong *et al.* 2001). In plaice, characterised by a more synchronous population of large 1000-µm vitellogenic oocytes, atresia in wild spawning fish was almost negligible (Nash *et al.* 2000) although advanced stage atretic oocytes were found sporadically in spent ovaries. Plaice may well be similar to herring (a species with the same capital spawning strategy, see section 4.5, and large synchronous population of vitellogenic oocytes) in that atresia is most prevalent well before (October) the start of spawning in February (Kurita *et al.* 2003). No data are available for flatfishes with an indeterminate spawning strategy but, by analogy with other teleosts having such a strategy, atresia probably occurs at the end of spawning or when the fish is exposed to unfavourable environmental conditions (Hunter & Macewicz 1985). Atresia may be particularly pronounced during the first year of maturation. This may be a general case but firm evidence has only been reported for common sole (Ramsay & Witthames 1996) and turbot *Scophthalmus maximus* (Bromley *et al.* 2000).

Different patterns of developing oocytes in the ovary arise from the asynchrony of vitellogenic oocyte recruitment from the previtellogenic population. One group of species shows only one mode of vitellogenic oocytes (unimodal type) that is carried through vitellogenesis and increases in size throughout the development period. In the final stage vitellogenic oocytes may hydrate synchronously and be released in a single batch, or hydrate asynchronously and be released in several successive batches (e.g. plaice). The second type is the multi-modal type, where several development stages of vitellogenic oocytes can be recognised. During development there may be a continuous recruitment of previtellogenic oocytes into vitellogenesis. The eggs in each mode will be spawned as separate batches of eggs. Within the multi-modal type fish, a hiatus in the size distribution of vitellogenic and previtellogenic oocytes may or may not develop. Hiatus formation prior to spawning may indicate that the number of oocytes to be released that season is determined (e.g. common sole).

Within the stock of previtellogenic oocytes several cohorts may be distinguished. These are believed to represent the cohorts that will develop in consecutive years (Dunn & Tyler 1969). Burton & Idler (1984) inferred that oocyte recruitment occurred close to spawning and that the pattern of oocyte development was compatible with a 3-year cycle in winter flounder.

Gametogenesis in Greenland halibut from the northwest Atlantic is not synchronous between individuals within a population, suggesting that the spawning season is not well defined. Differences in oocyte size-frequency distributions in prespawning, spawning and spent conditions suggest that Greenland halibut are capable of de novo vitellogenesis prior to and during spawning, which would mean that the spawning pattern is not determinate. This species may be capable of 'fast-tracking' oocytes to maturity, whereby oocyte batches may be brought quickly through vitellogenesis during the spawning season so as to increase the fish's yearly reproductive output (Rideout et al. 1999). A rapid maturation of early stage oocytes has also been suggested for plaice and American plaice to increase reproductive output (Horwood 1990; Maddock & Burton 1998). Reproductive output, however, can only be enhanced if the developmental time of oocytes is much shorter than the duration of the spawning season. In common sole from the eastern English Channel, Witthames & Greer Walker (1995) showed that the maximum developmental rate allowed little scope for previtellogenic oocytes to complete maturation during the spawning period of about 60 days.

During development the weight of the ovary, expressed as a percentage of the total body weight (gonadosomatic index, GSI), increases from a low of about 1% after the spawning season to a peak value just prior to spawning. The timing of ovary growth may differ between age groups. In plaice, the ovarian development of recruit spawners lagged behind that of older females by about 1 month (Rijnsdorp 1989).

Peak GSI shows large inter-specific differences between about 5% in species such as sand sole to a maximum of about 35% in the flounder. In individual flounder, maximum GSI values as high as 54% have been observed (Deniel 1981). Maximum GSI values for species that spawn in winter are higher than those for species spawning later in the year (Fig. 4.5).

4.3.3 Fecundity

Fecundity is defined as the standing stock of advanced yolked eggs. In determinate spawners, this number equals the annual fecundity of a female. In indeterminate spawners, the annual fecundity will be higher because primary oocytes recruit to the developing vitellogenic oocytes during the spawning season.

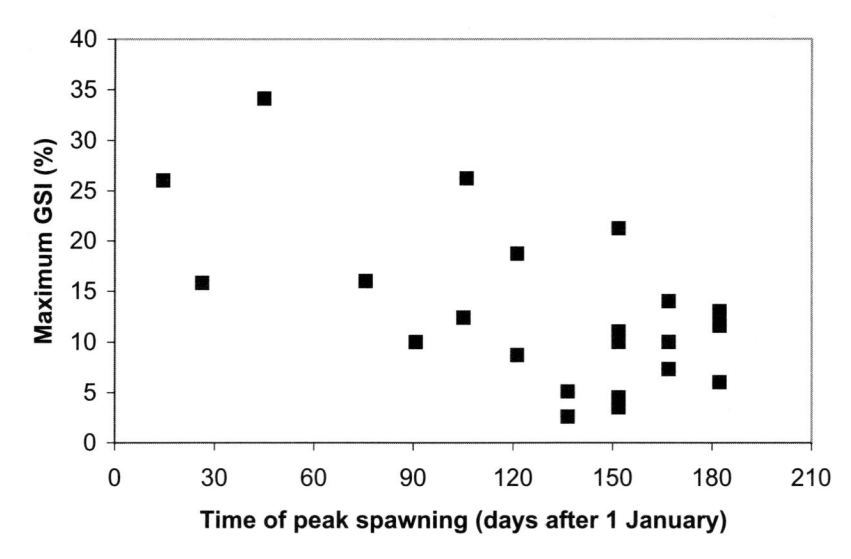

Fig. 4.5 The relationship between the mean maximum gonadosomatic index (%GSI) and the date of peak spawning (days since 1 January).

Fecundity is positively related to body size but is highly variable among individuals and may differ by a factor of two in similar-sized individuals. A trade-off between egg size and egg number is suggested by the inter-specific comparison of fecundity and egg size revealing that the variation in egg number among species is strongly reduced if egg size is taken into account (Roff 1982, 1983). Within plaice, a trade-off is suggested by the decrease in size-specific fecundity with age, and the lack of such an effect on the size-specific ovary weight (Rijnsdorp 1991). Howell & Scott (1989) showed that individual turbot produced consistently relatively large or small eggs in successive seasons.

The total number of eggs produced annually is regulated by body condition and nutritional status by affecting both the recruitment of previtellogenic oocytes into vitellogenesis and the loss of vitellogenic oocytes through follicular atresia (Witthames & Greer Walker 1995; Bromley *et al.* 2000). More direct evidence that vitellogenic oocyte recruitment is limited to early maturation stages in determinate spawners has been provided by unilateral ovariectomy in rainbow trout *Oncorhynchus mykiss*. If the operation is carried out early during the annual cycle, fecundity in the remaining ovary will be enhanced. Later in the season there was no effect on recruitment although the operation increased the investment in the standing stock of oocytes by increasing their mean size (Tyler *et al.* 1994). Oocyte recruitment during the early maturation phase is strongly influenced by the extent to which each female has recovered body condition following the previous spawning season. Large females that have largely switched energy allocation from growth to egg production are particularly prone to regulation at this stage compared with smaller females that are still prioritising growth, because the annual variation in somatic condition is larger in the former. This explains why studies of individual fecundity in relation to somatic growth, back-calculated from annual growth patterns of otoliths, failed to find a positive correlation between size-specific fecundity and somatic growth in plaice and common sole (Rijnsdorp 1990; Millner *et al.* 1991). However, body condition just prior to spawning was positively correlated with variation in fecundity and this may be caused by atresia regulation of fecundity after oocyte recruitment has been completed. More generally in flatfish populations, fecundity-size relationships are rather stable under highly different growth regimes. Inter-annual variation in size-specific fecundity varies among species between 6% and 67% (mode = 12%) (Rijnsdorp 1994). Within North Sea plaice populations, the inter-annual variation in fecundity of various size classes ranged between 7% and 15% over a 7-year study period and correlated with the variation in the condition factor, but not with variations in somatic growth (Rijnsdorp 1990, 1991). These observations may be interpreted as the result of a mechanism whereby fecundity and somatic growth are fine-tuned to the rate of energy acquisition during the feeding season (Rijnsdorp 1990). Under favourable feeding conditions fish grow fast, have a high oocyte recruitment and will build up energy reserves in proportion to their body size. Under poor feeding conditions, fish may still build up the energy stores necessary for reproduction at the expense of a reduced growth and oocyte recruitment. Only in very poor feeding conditions does the model predict a reduction in size-specific fecundity due to increased atresia. This model applies to species that cover their reproductive expenses from the energy reserves laid down in the body during the previous feeding season (capital spawners such as plaice, section 4.5).

In species that cover their reproductive expenses from energy acquired during the spawning season (income spawners), fecundity may be more loosely related to the size of the female and mainly determined by the recruitment and growth rate of vitellogenic oocytes (Deniel 1981). In intermediate spawners, where reproduction is partly subsidised by energy acquired during the spawning season, such as in some common sole populations, fecundity may depend on the energy reserves, but fine-tuned to the rate of energy acquisition through variations in atresia just prior to and during the spawning season.

4.3.4 Geographical pattern in fecundity

Fecundity may differ among populations of the same species. In a well-known study, Bagenal (1966) showed that size-specific annual fecundity of plaice showed a geographical pattern with values increasing from a low value in the centre (North Sea) to the edges of the distribution. This pattern was strongly dependent on the low annual fecundity in the southern North Sea estimated by Simpson (1951) in 1948 and 1949. Recent studies have indicated that these fecundity estimates may be atypical for the population. Fecundity around 1980 was much higher than observed by Simpson (Horwood *et al.* 1986), and close to the fecundity estimates for the years around 1900 that were re-analysed by Rijnsdorp (1991). In addition, the low fecundity estimates of Simpson (1951) are not supported by the ovary weights that were only 10% lower than in the 1980s (Rijnsdorp 1991). We therefore conclude that fecundity is fairly constant over most of the distribution range with the exception of a much higher value in the Baltic and a slightly lower value in the Barents Sea.

In common sole, fecundity of a 35-cm female increases with latitude from 200 000 eggs off Portugal to 450 000 eggs in the southeastern North Sea (Table 4.3). Superimposed on the latitudinal trend, there is a clear difference in fecundity between the western part of the North Sea (Flamborough) and the eastern part (German Bight) situated at the same latitude. Mean egg size shows a reversed trend but to a lesser degree. High fecundity populations produced smaller eggs than low fecundity populations. These data do not support the suggestion that the reproductive investment, expressed as the dry weight of eggs per gram body weight, is

Table 4.3 Geographic pattern in reproductive parameters of the common sole

Area	Latitude (°N)	Longitude (°E)	Fecundity at 35 cm (1000 eggs)	Relative fecundity (eggs. body weight^{-1})	Peak spawning date (days since 1 January)	Mean egg size (mm)	Reproductive investment (g dry weight eggs. g wet body weight^{-1})
German Bight	54.0	5	440[4]	1038[4]	148[5]	1.132[5]	0.064
Flamborough	54.3	0	258[4]	701[4]	151[5]	1.257[5]	0.059
Irish Sea	53.3	–4	342[4]	888[4]	–	–	–
Texel	53.0	3	432[4]	1040[4]	131[5]	1.226[5]	0.081
Southern Bight	52.3	3	–	–	118[5]	1.207[5]	–
English Channel east	50.3	0.3	325[4]	825[4]	111[5]	1.264[5]	0.071
Bristol Channel	50.3	–5	221[1]	416[2]	86[5]	1.360[5]	0.044
English Channel west	49.3	–4.3	240[4]	597[4]	91[5]	1.369[5]	0.065
Bay de Douarnenez	48.1	–4	275[3]	–	~45[3]	–	–
Bay of Biscay	47	–3	221[4]	641[4]	60[5]	–	–
Portugal	38	–9	205[4]	579[4]	–	–	–

Data sources: [1]Horwood & Greer Walker 1990; [2]Horwood 1993; [3]Deniel 1981; [4]Witthames *et al.* 1995; [5]Rijnsdorp & Vingerhoed 1994.

higher in the northeastern populations characterised by a higher variability in reproductive success (Rijnsdorp & Vingerhoed 1994).

The geographic pattern in fecundity and egg size is also related to spawning time. Spawning starts in the south where populations show a low fecundity and a large egg size. At higher latitudes, spawning starts later, fecundity is higher and egg size is lower. The current high fecundity in the eastern part of the North Sea contrasts with the lower fecundity reported in the early 1960s (quoted in Witthames *et al.* 1995). In the 1960s and 1970s, growth rate, gonadosomatic index and condition factor of North Sea common sole increased (de Veen 1978; Millner & Whiting 1996), suggesting that the low fecundity could be related to poor feeding conditions in the 1960s.

4.3.5 Batch spawning

Most flatfish species are batch spawners, which means that they release several successive batches of eggs (McEvoy 1984; Houghton *et al.* 1985; Zamarro 1991; Manning & Crim 1998; Nagler *et al.* 1999; Smith *et al.* 1999). For few species such as winter flounder (Burton & Idler 1984), stone flounder *Platichthys bicoloratus* (Park 1995) and three spotted flounder *Pseudorhombus triocellatus* (Ramanathan *et al.* 1990), a single spawning is reported.

Batch spawning is related to an ovulatory periodicity of 1–4 days as reported in Atlantic halibut (*Hippoglossus hippoglossus*), plaice, yellowtail flounder (*Limanda ferruginea*) and turbot (McEvoy 1984; Howell & Scott 1989; Norberg *et al.* 1991; Koya *et al.* 1994; Manning & Crim 1998). Under constant temperature and light conditions, individual turbot had predictable rhythms which differed between individuals (McEvoy 1984).

4.3.6 Egg and sperm quality: maternal and paternal effects

There is little published information on these topics for flatfishes and frequently the observations have been in the context of aquaculture where captivity is associated with poor quality eggs and sperm from broodstock. Captive turbot showed peaks in fertilisation success that coincided with peaks in egg production (Howell & Scott 1989) and milt from captive plaice has been shown to be much more viscous (hermatocrit >85%) compared with wild plaice (hermatocrit <60%; Vemeirssen *et al.* 1998). In both these cases the low quality gametes are thought to arise from depressed steroid levels induced by stress in captivity. Whether these observations have any relevance to wild populations is unknown but similar situations might arise from stress induced by either disease or pollution.

Egg weight and egg size have been shown to be proportional to female size or female age in several flatfish species and this variation could have a direct consequence for larval viability (Chambers *et al.* 1988; Buckley *et al.* 1991; Witthames, unpublished data).

4.4 Age and size at first maturation

Maturation is a developmental process that is coupled to the physiology of energy acquisition and hormone kinetics. According to the conventional view, a fish becomes mature after it has passed some fixed size or age threshold (Roff 1982). However, in a more general representation, fish mature according to a trajectory in the length-age space (Stearns & Koella 1986; Rijnsdorp 1993a). In this model, which includes the fixed size or age threshold as special cases, both length and age at first maturity are functions of growth rate and environmental conditions during the juvenile phase. Length and age at first maturation estimated at the time of spawning may be used as proxies of the average juvenile-adult transition points (Fig. 4.6).

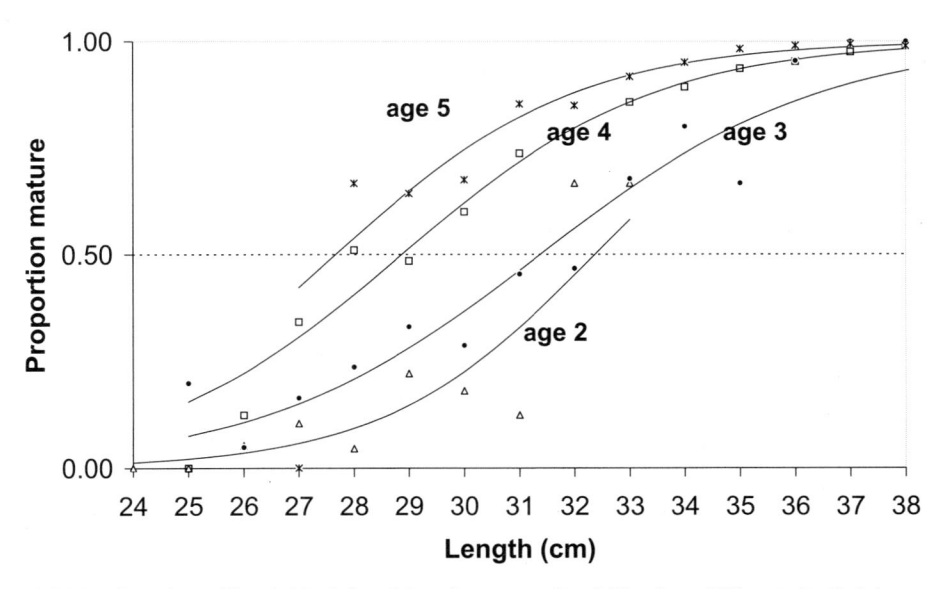

Fig. 4.6 Maturity ogives of female North Sea plaice of age groups 2 to 5. The size at 50% maturity (L_{50}) decreases with age from about 32 cm at age 2 to 27.6 cm at age 5 (data from Dutch market samples 1990–1999).

Because immature and mature fish may differ in their spatial distribution and/or behaviour, it may be difficult to obtain a representative sample of the population to estimate maturation characteristics (Rijnsdorp 1989; Horwood 1993; Sampson & Al-Jufaily 1999). Nevertheless, samples from commercial fisheries are often assumed to give a reasonably accurate estimate of the maturation characteristics of the population (de Veen 1978; Beacham 1983; Rijnsdorp 1989; Junquera & Zamarro 1994; Rickey 1995).

Because the distinction between mature and immature males is often not easy (see section 4.3), data on the onset of first maturation in male flatfishes are less well documented than in females. However, for species for which data are available, males become mature at a smaller size and younger age than females, although the difference may only be minor for species maturing at a young age and small size (Shuozeng 1995). There are some reports of very small males caught in spawning condition. Berestovskij (1992) reported two 3-year-old mature male American plaice of 9 cm in the Barents Sea, compared with the normal age of 6–7 years and size of 17–19 cm.

Maturation is generally expressed as the length (L_{50}) or age (A_{50}) at which 50% of the population has become mature. In an across-species comparison, female L_{50} appears to increase with latitude ($r = 0.40$, $P<0.05$, $n = 68$), but this pattern is related to the pattern of overall body size in flatfish species being larger at high latitudes (Pauly 1994). A comparison of the L_{50} among populations within a species does not reveal a consistent relationship with latitude. In plaice and English sole, L_{50} increases with latitude (Rijnsdorp 1989; Sampson & Al-Jufaily 1999), whereas the L_{50} of common sole in the Irish Sea was lower than in the more southern Bristol Channel population (Horwood 1993). In American plaice, the size at maturation increased with latitude in the eastern Atlantic, whereas size at maturity increased in the western Atlantic at the mid-latitudes, but decreased further north (Walsh 1994). Also in witch, *Glyptocephalus cynoglossus*, no latitudinal trend in L_{50} and A_{50} was apparent (Beacham 1983; Bowering 1987, 1989).

The latitudinal pattern in the onset of maturation may be (partly) related to the temperature conditions during the juvenile phase. In American plaice, the L_{50} and A_{50} declined with increasing temperature (Walsh 1994). A study in the Irish Sea has shown large differences in plaice from the western Irish Sea and Liverpool Bay with A_{50} of 4.4 and 2.7 years, respectively (Nash *et al.* 2000). These differences are larger than those observed within the North Sea (Rijnsdorp 1989).

Size and/or age at maturation may change over time (Beacham 1983; Bowering 1987, 1989; Rijnsdorp 1989; Bowering & Brodie 1991). In North Sea plaice populations, the percentage of mature females has changed periodically superimposed on an increasing trend, whereas the L_{50} in the population decreased by 0.7 cm per decade (Fig. 4.7). A similar change has been observed in other exploited flatfish species. Changes are likely to be related to the observed increase in juvenile growth (reviewed in Van der Veer *et al.* 1994), although in English sole L_{50} decreased and A_{50} increased while juvenile growth rate decreased (Sampson & Al-Jufaily 1999). Rijnsdorp (1993a) showed that growth rate affected the size at maturation 3 years before the onset of sexual maturity, consistent with a maturation period of more than 1 year (Dunn & Tyler 1969; Burton & Idler 1984). The decrease in the L_{50} and A_{50} as observed in several exploited flatfish stocks may also be the result of a recent evolutionary change in response to the increase in fishing mortality. As the current fishing mortality rates are well above the natural mortality rates, reproductive lifespan has decreased considerably. Under

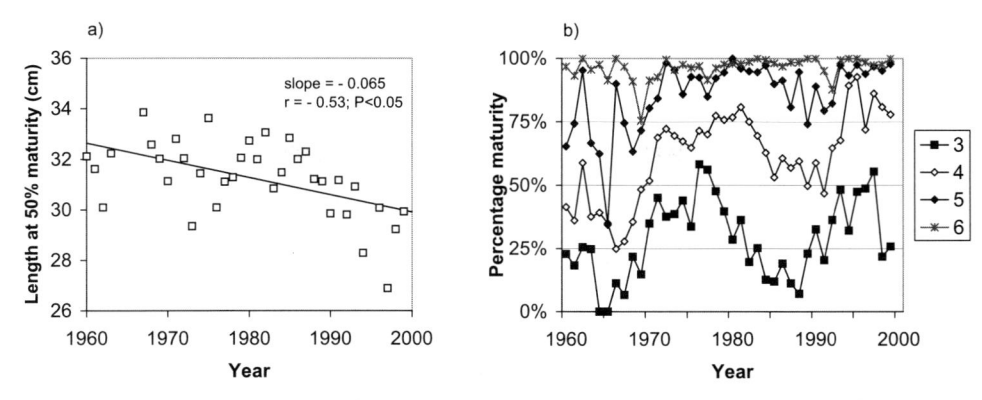

Fig. 4.7 Changes in the L_{50} of 4-year-old female plaice (a) and changes in the percentage mature females of age groups 3 to 6 between 1960 and 1999 (b), as observed in Dutch market samples.

these conditions, fish that start to reproduce at a younger age or smaller size will increase their lifetime reproductive output as compared with fish that mature at a higher age or larger size. It has been inferred that the decrease in L_{50} and A_{50} observed in plaice in the North Sea was partly due to a genetic change due to fisheries selection for early maturing fish, because only part of the observed decrease in L_{50} could be ascribed to the increase in juvenile growth and the slight increase in temperature (Rijnsdorp 1993b; Grift et al. 2003).

4.5 Energetics

4.5.1 Energetics of reproduction and growth

The timing of reproduction and the timing of energy acquisition are decoupled. Spawning is confined to a relatively short period during the year, whereas the energy acquisition follows a seasonal cycle with a peak in summer and a low during winter (MacKinnon 1972; Dygert 1990; Rijnsdorp 1990; Hanson & Courtenay 1997). The extent to which energy acquisition and reproduction are decoupled varies. In capital spawners, reproduction is fully subsidised from the energy reserves. On the other side of the spectrum, income spawners meet the cost of reproduction from acquired resources directly.

Capital spawners mainly spawn in winter or early spring when the rate of energy uptake is low. Figure 4.8 shows the seasonal pattern in the energetics of growth and reproduction for a 6-year-old female plaice in the Irish Sea (Dawson & Grimm 1980). In the feeding period the energy content increases gradually until reaching a peak in November. During the winter fast between December and February, the overall energy content decreases because of metabolism and the transposition of energy reserves from body into ovarian tissue. Between February and March, energy content drops as a consequence of extruding the eggs. Energy loss during spawning was estimated at 38%, about half of which was used for egg production. In spring and summer the energy is mainly stored in the carcass, but from September onwards energy is stored in both the carcass and the ovary. Ovary growth continues after feeding has ceased in December and about 50% of the ovary growth is re-allocated from the body energy reserves (Dawson & Grimm 1980; Rijnsdorp 1990). Although males exhibit a similar seasonal cycle,

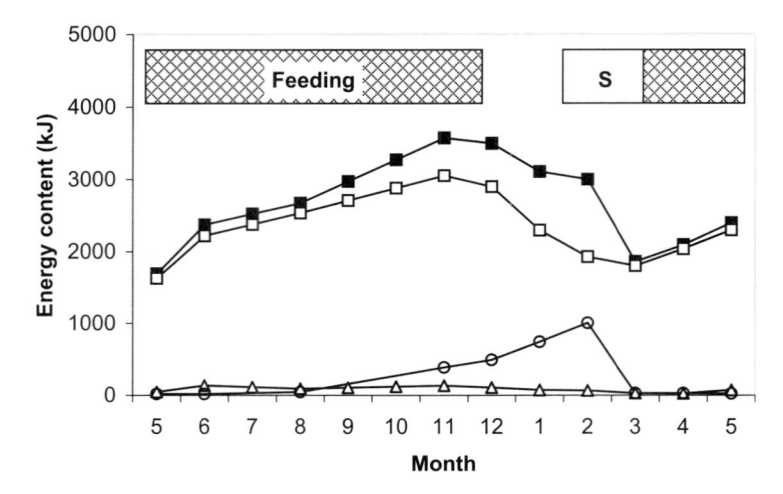

Fig. 4.8 Energy content by month of ovaries (○), liver (Δ), carcass (□) and total body (■) of a female temperate flatfish species plaice. Spawning (S) occurs in February and March, feeding between April and November. Body energy content increases during the feeding season due to an increase in the carcass, ovaries and liver. Ovary growth between December and February is realised through re-allocation of energy from the carcass to the ovary. The decrease in energy content between December and February is due to winter metabolism, and between February and March due to the shedding of eggs and spawning metabolism (based on Dawson & Grimm 1980).

the energy stored is less than in females, because spawning appears to be less demanding. In North Sea plaice, the energy loss in males was 27% compared with 44% in females (Rijnsdorp & Ibelings 1989). Comparable energy losses over winter and spawning were observed in females of English sole (45%; Dygert 1990) and American plaice (39%; MacKinnon 1972). In plaice, energy for reproduction and fasting metabolism is delivered by lipids (50–75%) and proteins (25–50%) (Dawson & Grimm 1980; Rijnsdorp & Ibelings 1989).

For a summer spawning flatfish, no detailed studies on the energetics of reproduction and growth are available, but the study of the sand sole by Deniel (1981) and Deniel *et al.* (1989) illustrates the main features. Vitellogenesis starts in February, whereas the first hyaline oocytes appear in April when a second batch of small oocytes enter primary vitellogenesis. GSI increases during this period from a low of 2% in February to a peak of 7–9% between May and July. Although there is continuous spawning activity from late April till August, individual females produce their batches mainly during two periods in May and July, respectively. There are thus two breeding cycles, which are related to the recruitment of new batches of previtellogenic oocytes in February and April, respectively. The condition factor showed only small seasonal variation with a low of 0.58–0.62 between May and September, and a high of 0.68–0.74 between October and April (Deniel 1981).

The different level of decoupling of energy acquisition and energy expenditure in reproduction results in differences in reproductive characteristics (Fig. 4.9). Winter spawning species tend to be determinate spawners such as plaice. Summer spawning species such as sand sole are indeterminate spawners. Species spawning in spring, such as common sole, may be intermediate with early spawning, determinate spawning populations and late spawning indeterminate populations. These differences correlate with the GSI (Fig. 4.5) and with the amplitude of the seasonal cycle in body condition. Because small bodies have a smaller capacity to build

Ecology of reproduction 83

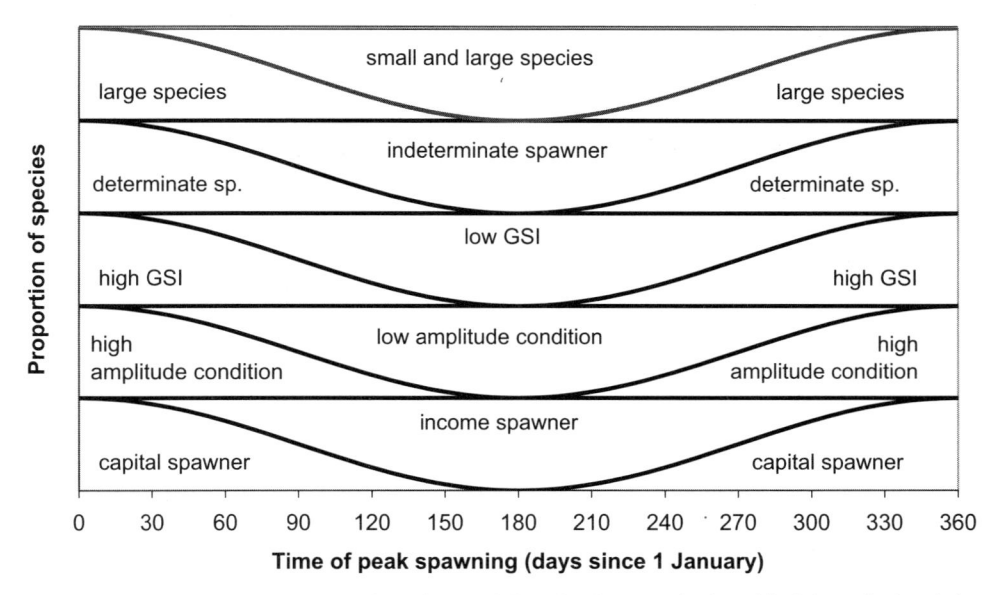

Fig. 4.9 Conceptual relation between various characteristics related to reproduction of flatfish species in relation to the timing of spawning. The curves show the gradual change in characteristics between the two ends of the continuum. Flatfishes that spawn in winter reproduce at the expense of the energy reserves laid down during the previous feeding season (capital spawners), show a large seasonal variation in body condition, show a high GSI at spawning, are determinate spawners and belong to species with large body sizes. In contrast, summer spawning flatfishes reproduce from energy assimilated during the spawning season (income spawners), show a low seasonal variation in body condition, show a low GSI at spawning, are indeterminate spawners and belong to species with either small or large body size.

up energy reserves for reproduction relative to their metabolic demands compared with larger individuals, small-sized species are not able to spawn in winter or in early spring (Fig. 4.10).

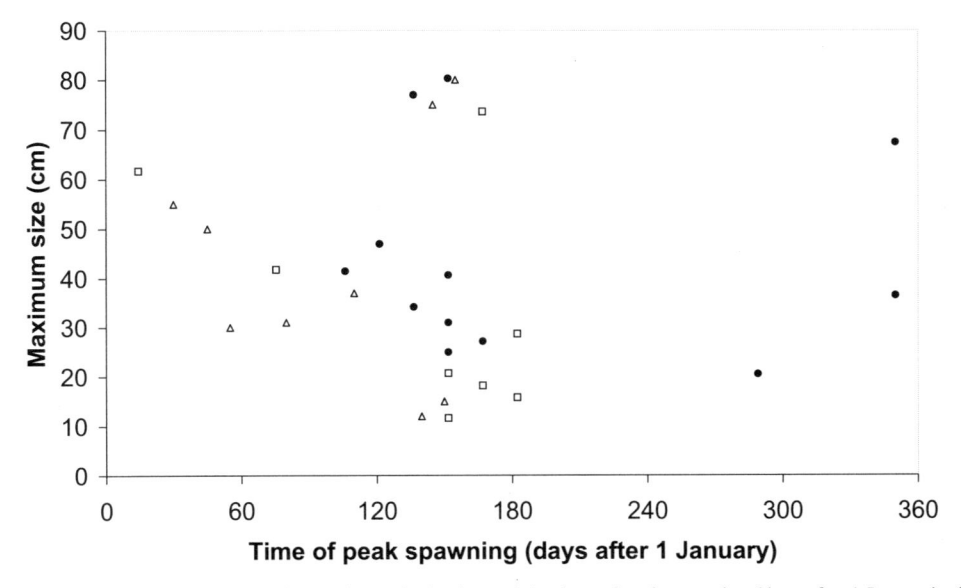

Fig. 4.10 Relationship between the maximum body size and the time of peak spawning (days after 1 January) of flatfish species from the North Sea (△), Bay of Biscay (□) and Bohai Sea (●).

4.5.2 Non-annual spawning

Non-annual spawning is reported for some flatfish species living in cold waters: American plaice, winter flounder and Arctic flounder *Pleuronectes glacialis* (Templeman & Andrews 1956; Burton & Idler 1984; Burton & Maddock 1995). The non-spawning state coincides with a poor body condition (jellied muscle) and seems to be restricted to larger adults. An extreme example was reported by Templeman & Andrews (1956) who recorded a fillet sample of American plaice with 96.2% water, 0.06% fat and 2.8% protein. This jellied condition, which is also reported for American smooth flounder and Dover sole (Burton & Maddock 1995), is probably due to the inability to restore its energy reserves after spawning losses. The risk of not being able to restore the energy reserves may be particularly large in areas where surplus production is highly variable or where surplus production is relatively low.

4.5.3 Spawning fast

In some flatfish species, feeding ceases during the spawning period (MacKinnon 1972; McEvoy 1984; Rijnsdorp 1989; Ntiba & Harding 1993). Although cessation of feeding may be partly related to a low availability of benthic food, there is clear evidence that feeding incidence of flatfishes is related to the maturity stage: flatfishes in spawning condition do not feed whereas co-occurring juveniles or spent fishes do feed (Roff 1982; Rijnsdorp 1989; Stoner *et al.* 1999). Cessation of feeding of spawning fishes may be caused by competition for the limited amount of oxygen among spawning activity and other processes such as growth, feeding and predator avoidance (Rijnsdorp & Ibelings 1989).

4.5.4 Sexual dimorphism in reproduction and growth

The proportion of the energy available for growth and reproduction (surplus energy) increases with body size. After maturation, the proportion of the available energy used for reproduction increases at the expense of the energy available for somatic growth. Close to the maximum size, a fish will use almost the total amount of energy laid down in the body during the feeding period for reproduction and fasting metabolism.

Roff (1983) proposed that the energy allocation schedule of flatfishes was a trade-off between the delay in reproduction to increase future reproductive resources and the increase of a mortality risk related to feeding activities. While the reproductive output of females (fecundity) increases with body size, reproductive output of males does not increase beyond a certain size because the energy needed for production of semen and behavioural cost will be less than the female investment. Comparison of the surplus energy–body size relationship between male and female plaice in Fig. 4.11 indeed shows that surplus energy in males fall behind the surplus energy of females beyond a size that is higher than the size at first maturation (Rijnsdorp & Ibelings 1989). The lower surplus energy acquisition in males may be related to the reduced feeding as observed in dab *Limanda limanda* (Lozan 1992). It is hypothesised, therefore, that the sexual dimorphism in growth of flatfishes is the combined result of the earlier start of reproduction in males and the reduction in surplus energy acquisition above a certain size. As natural mortality in the ocean is strongly related to size, this

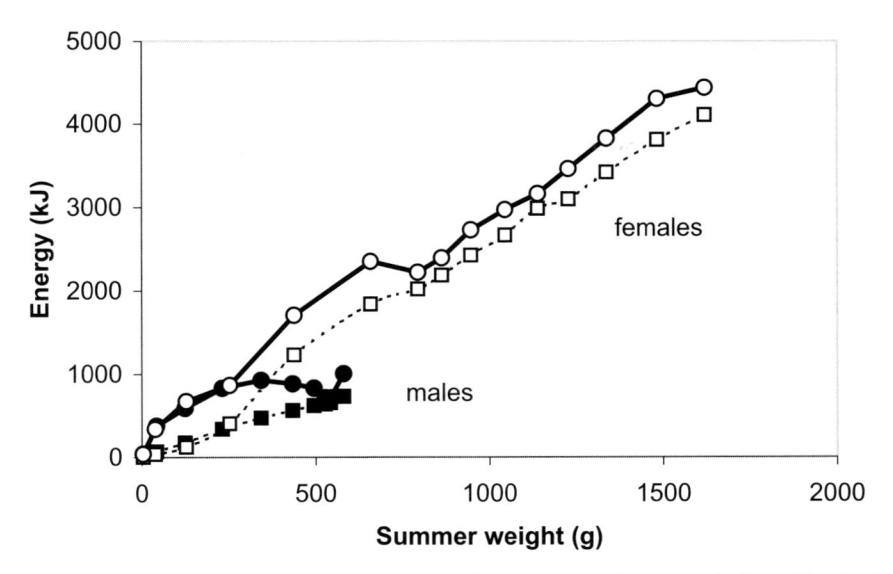

Fig. 4.11 Relation of surplus energy (circles) and reproduction (squares) to the summer body weight of male and female plaice. Modified from Rijnsdorp & Ibelings (1989).

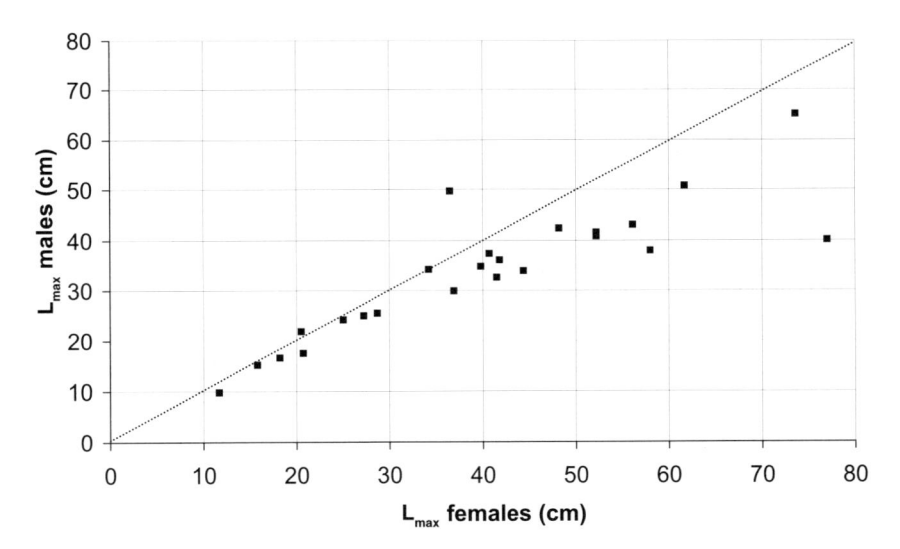

Fig 4.12 Relationship between the maximum size (L_{max}) of male and female flatfish. The dashed line represents the line of equal sizes.

hypothesis predicts that the degree of sexual dimorphism in flatfish will be largest in large-bodied flatfish as indicated in Fig. 4.12.

4.6 Contaminants and reproduction

The effects of contaminants on flatfish reproduction have been studied almost exclusively in coastal and estuarine populations of five species, with the primary aim of assessing the environmental impact of organic chemicals. As in all fishes, exposure to these chemicals in the water may lead to direct uptake through the skin or gills, but flatfishes also come into contact with contaminated sediments because of their behaviour and habitat preference. Organic contaminants contain a chain of carbon molecules that readily become adsorbed to sediments and thus increase the local concentration experienced by bottom-dwelling flatfishes that bury in the sediment or rest on the surface layers. The other major route of uptake is by ingestion of food following a process of bio-accumulation (Moles *et al.* 1994).

Reproduction could be affected by exposure to contaminants in many ways during the production of offspring including sex determination, stem cell proliferation, gamete development and, finally, spawning or subsequent embryo viability. However, most published work to date has focused on effects on development of female gametes in the ovary and feminisation of males. Laboratory and field surveys have been instigated following major environmental disasters or in response to restoration of habitats damaged previously by unregulated industrial activity.

Several reports (cited in Brule 1987) have described the effects of the oil spill from the *Amoco Cadiz* on the reproductive potential of plaice around the Normandy coast. Studies on ovaries reported a decrease in maturity at age and length and in the gonadosomatic index. A high incidence of large empty or atretic follicles and the appearance of connective tissue masses were also seen in the ovarian structure. After the Prudoe Bay oil spill, effects on the reproduction of several flatfish species were examined in controlled laboratory studies (reviewed in Johnson *et al.* 1993). These included intramuscular injections of the crude oil into gravid female rock sole and flathead sole (*Hippoglossoides elassodon*) and in vitro incubations of ovary fragments with the contaminant. In both cases, the concentration of oestradiol declined both in the plasma and in the incubation medium following a dose-dependent negative response to the contaminant. This hormone would affect the GSI and maturity because of its central role in inducing the liver to secrete vitellogenin. However, not all species responded in the same way. Populations of English sole were affected in respect of their ovary development, plasma oestradiol, egg weight/size, spawning and fertilisation success, larval size and viability, but fecundity remained the same. In contrast, winter flounder was only affected in respect of egg weight/size and larval size response. The authors were unable to identify the causes of the differences but suggested that whole animal factors such as diet, migration and behaviour could affect exposure. An alternative, but less convincing explanation, may lie in the species having different metabolic responses, for example, through the ability to induce and create more active detoxification enzyme systems. In an environmental study, fecundity and egg weight in English sole from four localities around Puget Sound were compared in relation to the levels of aromatic hydrocarbons and polychlorinated biphenyls found in the sediments (Johnson *et al.* 1997). Contaminant concentrations were significantly higher in both the liver and the ovary and followed the same rank order observed in local sediments. Fish from the most polluted locality produced notably more and smaller eggs after accounting for differences in size. These differences also co-varied with circulating levels of vitellogenin, oestradiol and follicular atresia, suggesting abnormally high recruitment of oocytes and control of elevated fecundity by atresia. Because the fish originate from larvae advected from the same offshore spawning area at all four localities, a genetic cause may be discounted and a phenotypic response of the adult to the environment seems likely.

Temperature regimes of the different localities were very similar and therefore contaminant levels, possibly in combination with nutrition, offer the only explanation.

Some contaminants have been shown to affect flatfishes through endocrine disruption and this has been a major research topic on the molecular mechanism causing population-level effects (Allen *et al.* 1999a, b). The range of organic compounds and their oestrogenic potency are very diverse (some pesticides, polychlorinated biphenyls, phthalates, alkylphenols and synthetic oestrogens) and their effect in oestrogenic activity may be positive or negative. At the molecular level, these contaminants mimic the structure, at least in some key part, of a steroid hormone (oestrogens) that controls some aspect of the female reproductive cycle. Consequently it binds to the steroid receptors and triggers the same response as the steroid hormone on the cellular chemistry, causing feminisation of males. In the laboratory studies flounder were taken from a clean control site and exposed for 3 weeks to a synthetic oestrogen (ethynyloestradiol) and oestrogen mimic *p*-nonylphenol (Allen *et al.* 1999a). The synthetic oestrogen required a minimum threshold dose (10 ng l^{-1}) to induce vitellogenin production in the plasma of males and to raise vitellogenin levels in females significantly. The oestrogen mimic did have these effects but also significantly affected liver weight. In an earlier mesocosm experiment, no feminisation of males was observed but ovaries showed less synchronous oocyte development and plasma levels of oestradiol and testosterone (a male hormone) were twice the control levels (Lambert & Jannsen 1995).

In the extensive population survey of Allen *et al.* (1999b) elevation of plasma vitellogenin, indicating oestrogenic stimulation of the liver, was found to different degrees in the ten estuaries studied. Concentrations were all higher than at the clean control sites. Some individuals in two estuaries showed intersex characteristics with a mixed ovotestis. In extreme cases, the testis was completely filled with advanced stage vitellogenic oocytes. No observations were made on the vas deferens to discover if the feminisation blocked sperm release and rendered the male completely impotent. The sex ratio was not affected in the same way as the vitellogenin induction but other factors outside the estuary may be involved. Sexual differentiation during ontogeny is more likely controlled during the egg and larval phase during onshore drift to nursery beaches from the offshore spawning grounds. Some flounders were also collected in offshore areas of the North Sea and Irish Sea, also showing some degree of feminisation (higher in the Irish Sea than in the North Sea). All these fish also had higher plasma levels of vitellogenin than the fish from the control site. Contaminants may also be deposited in the lipid/protein reserves of the egg and thus affect development during ontogeny when these reserves are mobilised. However, no major problem in recruitment seems to exist because flounder were abundant in all estuaries.

References

Albert, O.T., Nilssen, E.M., Stene, A., Gundersen, A.C. & Nedreaas, K.H. (2001) Maturity classes and spawning behaviour of Greenland halibut (*Reinhardtius hippoglossoides*). *Fisheries Research*, **51**, 217–228.

Allen, Y., Scott, A.P., Matthiessen, P., Haworth, S., Thain, J.E. & Feist, S. (1999a) Survey of estrogenic activity in United Kingdom estuarine and coastal waters and its effects on gonadal development of the flounder *Platichthys flesus*. *Environmental Toxicology and Chemistry*, **18**, 1791–1800.

Allen, Y., Matthiessen, P., Scott, A.P., *et al.* (1999b) The extent of oestrogenic contamination in the UK estuarine and marine environments – further surveys of flounder. *Science of the Total Environment*, **233**, 5–20.

Armstrong, M.J., Connolly, P., Nash, R.D.M., *et al.* (2001) An application of the annual egg production method to estimate the spawning biomass of cod (*Gadus morhua* L.), plaice (*Pleuronectes platessa* L.) and sole (*Solea solea* L.) in the Irish Sea. *ICES Journal of Marine Science*, **58**, 183–203.

Bagenal, T.B. (1966) The ecological and geographical aspects of the fecundity of the plaice. *Journal of the Marine Biological Association of the United Kingdom*, **46**, 161–186.

Bagenal, T.B. (1971) The interrelation of the size of fish eggs, the date of spawning and the production cycle. *Journal of Fish Biology*, **3**, 207–219.

Barnett, C.W. & Pankhurst, N.W. (1999) Reproductive biology and endocrinology of greenback flounder *Rhombosolea tapirina* (Gunther 1862). *Marine and Freshwater Research*, **50**, 35–42.

Barr, W.A. (1963a) The endocrine control of the sexual cycle in the plaice, *Pleuronectes platessa* (L). III. The endocrine control of spermatogenesis. *General and Comparative Endocrinology*, **3**, 216–225.

Barr, W.A. (1963b) The endocrine control of the sexual cycle in the plaice (*Pleuronectes platessa* L.). I. Cyclical changes in the normal ovary. *General and Comparative Endocrinology*, **3**, 197–204.

Beacham, T.D. (1983) Variability in size and age at sexual maturity of witch flounder *Glyptocephalus cynoglossus*, in the Canadian maritime region of the northwest Atlantic Ocean. *Canadian Field Naturalist*, **97**, 409–422.

Berestovskij, E.G. (1992) The size at sexual maturation and age of males of the Barents Sea long rough dab, *Hippoglossoides platessoides limandoides*. *Journal of Ichthyology*, **32**, 160–162.

Bowering, W.R. (1987) Distribution of witch flounder in the southern Labrador and eastern Newfoundland area and changes in certain biological parameters after 20 years of exploitation. *Fishery Bulletin*, **85**, 611–629.

Bowering, W.R. (1989) Witch flounder distribution off southern Newfoundland, and changes in age, growth, and sexual maturity patterns with commercial exploitation. *Transactions of the American Fisheries Society*, **118**, 659–669.

Bowering, W.R. & Brodie, W.B. (1991) Distribution of commercial flatfishes in the Newfoundland-Labrador region of the Canadian Northwest Atlantic and changes in certain biological parameters since exploitation. *Netherlands Journal of Sea Research*, **27**, 407–422.

Bromley, P.J., Ravier, C. & Witthames, P.R. (2000) The influence of feeding regime on sexual maturation, fecundity and atresia in first-time spawning turbot. *Journal of Fish Biology*, **56**, 264–278.

Brule, T. (1987) The reproductive biology and the pathological changes of the plaice *Pleuronectes platessa* (L.) after the 'Amoco Cadiz' oil spill along the North-West coast of Brittany. *Journal of the Marine Biological Association of the United Kingdom*, **67**, 237–247.

Buckley, L.J., Smigielski, A.S., Halavik, T.A., Caldarone, E.M., Burns, B.R. & Laurence, G.C. (1991) Winter flounder *Pseudopleuronectes americanus* reproductive success. II. Effects of spawning time and female size on size, composition and viability of eggs and larvae. *Marine Ecology Progress Series*, **74**, 125–135.

Burton, M.P. & Idler, D.R. (1984) The reproductive cycle in winter flounder, *Pseudopleuronectes americanus* (Walbaum). *Canadian Journal of Zoology*, **62**, 2563–2567.

Burton, M.P. & Maddock, D. (1995) Reproduction, muscle hydration, and condition cycle variation in northern pleuronectids. In: Proceedings of the International Symposium on North Pacific Flatfish. pp. 73–87. *Alaska Sea Grant College Program Report No. 95–04*. University of Alaska, Fairbanks, AK.

Castillo, G.C. (1995) Latitudinal patterns in reproductive life history traits of Northeast Pacific flatfish. In: Proceedings of the International Symposium on North Pacific Flatfish. pp. 51–72. *Alaska Sea Grant College Program Report No. 95–04*. University of Alaska, Fairbanks, AK.

Chambers, R.C. (1997) Environmental influences on egg and propagule sizes in marine fishes. In: *Early Life History and Recruitment in Fish Populations* (eds R.C. Chambers & E.A. Trippel). pp. 63–102. Chapman and Hall, London.

Chambers, R.C., Leggett, W.C. & Brown, J.A. (1988) Variations in and among early life history traits of laboratory-reared winter flounder *Pseudopleuronectes americanus*. *Marine Ecology Progress Series*, **47**, 1–15.

Cushing, D.H. (1969) The regularity of spawning seasons of some fish. *Journal du Conseil International pour l'Exploration de la Mer*, **33**, 81–92.

Dawson, A.S. & Grimm, A.S. (1980) Quantitative seasonal changes in the protein, lipid and energy content of the carcass, ovaries and liver of adult female plaice, *Pleuronectes platessa* L. *Journal of Fish Biology*, **16**, 493–504.

Deniel, C. (1981) *Les poissons plats [Teleostei, Pleuronectiformes] en Baie de Douarnenez. Reproduction, croissance en migration des Bothidae, Scophthalmidae, Pleuronectidae and Soleidae.* PhD thesis, Universite de Bretagne Occidentale.

Deniel, C., Le Blanc, C. & Rodriguez, A. (1989) Comparative study of sexual cycles, oogenesis and spawning of two Soleidae, *Solea lascaris* (Risso, 1810) and *Solea impar* (Bennet, 1831), on the western coast of Brittany. *Journal of Fish Biology*, **35**, 49–58.

de Veen, J.F. (1978) Changes in North Sea sole stocks (*Solea solea* (L.)). *Rapports et Proces-Verbaux des Reunions Conseil International pour l'Exploration de la Mer*, **172**, 124–136.

Dunn, R.S. & Tyler, A.V. (1969) Aspects of the anatomy of the winter flounder ovary with hypotheses on oocyte maturation time. *Journal of the Fisheries Research Board of Canada*, **26**, 1943–1947.

Dygert, P.H. (1990) Seasonal changes in energy content and proximate composition associated with somatic growth and reproduction in a representative age-class of female English sole. *Transactions of the American Fisheries Society*, **119**, 791–801.

Garcia Abad, M.C., Yanez Arancibia, A., Sanchez Gil, P. & Tapia Garcia, M. (1992) Distribution, reproduction and feeding of *Syacium gunteri* Ginsburg (Pisces: Bothidae) in the Gulf of Mexico. *Revista de Biologia Tropical*, **40**, 27–34.

Grier, H.J. (1981) Cellular organisation of the testis and spermatogenesis in fishes. *American Zoologist*, **21**, 345–357.

Grift, R.E., Rijnsdorp, A.D., Barot, S., Heino, M. & Dieckmann, U. (2003) Fisheries-induced trends in reaction norms for maturation in North Sea plaice. *Marine Ecology Progress Series,* **257**, 247–257.

Grigorev, S.S. (1995) Characteristics of flatfish eggs and larva samples near eastern Kamchatka and in the western Bering Sea during June–July 1991. In: Proceedings of the International Symposium on North Pacific Flatfish. pp. 101–116. *Alaska Sea Grant College Program Report No. 95–04.* University of Alaska, Fairbanks, AK.

Hanson, J.M. & Courtenay, S.C. (1997) Seasonal distribution, maturity, condition, and feeding of smooth flounder (*Pleuronectes putnami*) in the Miramichi Estuary, southern Gulf of St Lawrence. *Canadian Journal of Zoology*, **75**, 1226–1240.

Horwood, J.W. (1990) Fecundity and maturity of plaice (*Pleuronectes platessa*) from Cardigan Bay. *Journal of the Marine Biological Association of the United Kingdom*, **70**, 515–529.

Horwood, J.W. (1993) The Bristol Channel sole (*Solea solea* (L.)): a fisheries case study. *Advances in Marine Biology*, **29**, 215–367.

Horwood, J.W. & Greer Walker, M. (1990) Determinacy of fecundity in sole (*Solea solea*) from the Bristol Channel. *Journal of the Marine Biological Association of the United Kingdom*, **70**, 803–813.

Horwood, J.W., Bannister, R.C.A. & Howlett, G.J. (1986) Comparative fecundity of North Sea plaice (*Pleuronectes platessa* L.). *Proceedings of the Royal Society, London Series B,* **228**, 401–431.

Houghton, R.G., Last, J.M. & Bromley, P.J. (1985) Fecundity and egg size of sole (*Solea solea* (L.)) spawning in captivity. *Journal du Conseil International pour l'Exploration de la Mer*, **42**, 162–165.

Howell, B.R. & Scott, A.P. (1989) Ovulation cycles and post-ovulatory deterioration of eggs of the turbot (*Scophthalmus maximus* L.). *Rapports et Proces-Verbaux des Reunions Conseil International pour l'Exploration de la Mer*, **191**, 21–26.

Hunter, J.R. & Macewicz, B.J (1985) Rates of atresia in the ovary of captive and wild northern anchovy, *Engraulis mordax*. *Fishery Bulletin*, **83**, 119–136.

Hunter, J.R., Macewicz, B.J., Lo, N.-H. & Kimbrell, C.A. (1992) Fecundity, spawning, and maturity of female Dover sole *Microstomus pacificus*, with an evaluation of assumptions and precision. *Fishery Bulletin*, **90**, 101–128.

Johnson, L., Casillas, E., Sol, S., Collier, T., Stein, J. & Varanasi, U. (1993) Contaminant effects on reproductive success in selected benthic fish. *Marine Environmental Research*, **35**, 165–170.

Johnson, L.L., Sol, S.Y., Lomax, D.P., Nelson, G.M., Sloan, C.A. & Casillas, E. (1997) Fecundity and egg weight in English sole, *Pleuronectes vetulus*, from Puget Sound, Washington – influence of nutritional status and chemical contaminants. *Fishery Bulletin*, **95**, 231–249.

Junquera, S. & Zamarro, J. (1994) Sexual maturity and spawning of Greenland halibut (*Reinhardtius hippoglossoides*) from Flemish Pass area. *NAFO Scientific Council Studies*, **20**, 47–52.

Kamisaka, Y., Tagawa, M. & Tanaka, M. (1999) Semi-annual reproductive cycle of a small flounder *Tarphops oligolepis* in Wakasa Bay. *Fisheries Science*, **65**, 98–103.

Koya, Y., Matsubara, T. & Nakagawa, T. (1994) Efficient artificial fertilisation method based on ovulation cycle in barfin flounder *Verasper moseri*. *Fisheries Science*, **60**, 537–540.

Kurita, Y., Meier, S. & Kjesbu, O.S. (2003) Oocyte growth and fecundity regulation by atresia of Atlantic herring (*Clupea harengus*) in relation to body condition throughout the maturation cycle. *Journal of Sea Research*, **49**, 203–219.

Lambert, J.G.D. & Janssen, P.A.H. (1995) A long term study of the effects of polluted sediment on the annual reproductive cycle of the female flounder, *Platichthys flesus*. In: *Proceedings of the Fifth International Symposium on the Reproductive Physiology of Fish, University of Texas at Austin, 2–8 July 1995* (eds F.W. Goetz & P. Thomas). pp. 176–178. Austin, TX.

Lozan, J.L. (1992) Sexual differences in food intake, digestive tract size, and growth performance of the dab, *Limanda limanda* L. *Netherlands Journal of Sea Research*, **29**, 223–227.

McEvoy, L.A. (1984) Ovulatory rhythms and over-ripening of eggs in cultivated turbot, *Scophthalmus maximus* L. *Journal of Fish Biology*, **24**, 437–448.

MacKinnon, J.C. (1972) Summer storage of energy and its use for winter metabolism and gonad maturation in American plaice (*Hippoglossoides platessoides*). *Journal of the Fisheries Research Board of Canada*, **29**, 1749–1759.

Maddock, D.M. & Burton, M.P.M. (1998) Gross and histological observation of ovarian development and related condition changes in American plaice (*Hippoglossoides platessoides*). *Journal of Fish Biology*, **53**, 928–944.

Manning, A.J. & Crim, L.W. (1998) Maternal and interannual comparison of the ovulatory periodicity, egg production and egg quality of the batch-spawning yellowtail flounder. *Journal of Fish Biology*, **53**, 954–972.

Miller, J.M., Burke, J.S. & Fitzhugh, G.R. (1991) Early life history patterns of Atlantic North American flatfish: likely (and unlikely) factors controlling recruitment. *Netherlands Journal of Sea Research*, **27**, 261–275.

Millner, R.S. & Whiting, C.L. (1996) Effects of fishing and environmental variation on long-term changes in growth and population abundance of sole in the North Sea from 1940 to the present. *ICES Journal of Marine Science*, **53**, 1185–1195.

Millner, R.S., Whiting, C.L., Greer Walker, M. & Witthames, P.R. (1991) Growth increment, condition and fecundity in sole (*Solea solea* (L.)) from the North Sea and eastern Channel. *Netherlands Journal of Sea Research*, **27**, 433–439.

Minami, T. & Tanaka, M. (1992) Life history cycles in flatfish from the northwestern Pacific, with particular reference to their early life histories. *Netherlands Journal of Sea Research*, **29**, 35–48.

Moles, A., Rice, S. & Norcross, B.L. (1994) Non avoidance of hydrocarbon laden sediments by juvenile flatfishes. *Netherlands Journal of Sea Research*, **32**, 361–367.

Morse, W.W. (1981) Reproduction of the summer flounder, *Paralichthys dentatus* (L.). *Journal of Fish Biology*, **19**, 189–203.

Nagler, J.J., Adams, B.A. & Cyr, D.G. (1999) Egg production, fertility, and hatch success of American plaice held in captivity. *Transactions of the American Fisheries Society*, **128**, 727–736.

Nash, R.D.M, Witthames, P.R., Pawson, M. & Alesworth, E. (2000) Regional variation in the dynamics of reproduction and growth of Irish Sea plaice, *Pleuronectes platessa* L. *Journal of Sea Research*, **44**, 55–64.

Norberg, B., Valkner, V., Huse, J., Karlsen, I. & Leroy-Grung, G. (1991) Ovulatory rhythms and egg viability in the Atlantic halibut (*Hippoglossus hippoglossus*). *Aquaculture*, **97**, 365–371.

Norberg, B., Kjesbu O.S., Tarangere G.L., Andersson, E. & Stefansson, S.O. (1999) Natural environmental influences on reproduction. *Proceedings of the 6th International Symposium on the Reproductive Physiology of Fish.* Jon Greg AS, Bergen.

Ntiba, M.J. & Harding, D. (1993) The food and feeding habits of long rough dab, *Hippoglossoides platessoides* (Fabricius 1780) in the North Sea. *Netherlands Journal of Sea Research*, **31**, 189–199.

Park, J.S. (1995) Maturity and spawning season of the stone flounder *Kareius bicoloratus* in approaches to Kyongyolbiyolto of Yellow Sea Korea. *Fisheries Science Research Kunsan*, **11**, 51–57.

Pauly, D. (1994) A framework for latitudinal comparisons of flatfish recruitment. *Netherlands Journal of Sea Research*, **32**, 107–118.

Pepin, P. (1991) Effect of temperature and size on development, mortality and survival rates of the pelagic early life history stages of marine fishes. *Canadian Journal of Fisheries and Aquatic Sciences*, **48**, 503–518.

Rae, B.B. (1965) *The Lemon Sole.* Fisheries News (Books) Ltd, London.

Rajaguru, A. (1992) Biology of two co-occurring tonguefishes, *Cynoglossus arel* and *C. lida* (Pleuronectiformes: Cynoglossidae), from Indian waters. *Fishery Bulletin*, **90**, 328–367.

Ramanathan, N., Vijaya, P., Ramaiyan, V. & Natarajan, R. (1977) On the biology of the large-scaled tongue sole *Cynoglossus macrolepidotus* (Bleeker). *Indian Journal of Fisheries Ernakulam*, **24**, 1–2.

Ramanathan, N., Venkataramani, V.K. & Venkataramanujam, K. (1990) Breeding biology of flatfish *Pseudorhombus triocellatus* (Bloch and Schn.) from Tuticorin waters, east coast of India. *Indian Journal of Marine Sciences*, **19**, 151–152.

Ramsay, K. & Witthames, P.R. (1996) Using oocyte size to assess seasonal ovarian development in *Solea solea* (L.). *Journal of Sea Research*, **36**, 275–283.

Rickey, M.H. (1995) Maturity, spawning, and seasonal movement of arrowtooth flounder, *Atheresthes stomias*, off Washington. *Fishery Bulletin*, **93**, 127–138.

Rideout, R.M., Maddock, D.M. & Burton, M.P. (1999) Oogenesis and the spawning pattern in Greenland halibut from the North-west Atlantic. *Journal of Fish Biology*, **54**, 196–207.

Rijnsdorp, A.D. (1989) Maturation of male and female North Sea plaice (*Pleuronectes platessa* L.). *Journal du Conseil International pour l'Exploration de la Mer*, **46**, 35–51.

Rijnsdorp, A.D. (1990) The mechanism of energy allocation over reproduction and somatic growth in female North Sea plaice, *Pleuronectes platessa* L. *Netherlands Journal of Sea Research*, **25**, 279–290.

Rijnsdorp, A.D. (1991) Changes in fecundity of female North Sea plaice (*Pleuronectes platessa* L.) between three periods since 1900. *ICES Journal of Marine Science*, **48**, 253–280.

Rijnsdorp, A.D. (1993a) Relationship between juvenile growth and the onset of sexual maturity of female North Sea plaice, *Pleuronectes platessa*. *Canadian Journal of Fisheries and Aquatic Sciences*, **50**, 1617–1631.

Rijnsdorp, A.D. (1993b) Fisheries as a large-scale experiment on life-history evolution: disentangling phenotypic and genetic effects in changes in maturation and reproduction of North Sea plaice, *Pleuronectes platessa* L. *Oecologia*, **96**, 391–401.

Rijnsdorp, A.D. (1994) Population regulating processes during the adult phase in flatfish. *Netherlands Journal of Sea Research*, **32**, 207–223.

Rijnsdorp, A.D. & Ibelings, B. (1989) Sexual dimorphism in the energetics of reproduction and growth of North Sea plaice, *Pleuronectes platessa* L. *Journal of Fish Biology*, **35**, 401–415.

Rijnsdorp, A.D. & Jaworski, A. (1990) Size-selective mortality in plaice and cod eggs: a new method in the study of egg mortality. *Journal du Conseil International pour l'Exploration de la Mer*, **47**, 256–263.

Rijnsdorp, A.D. & Vingerhoed, B. (1994) The ecological significance of geographical and seasonal differences in egg size in sole *Solea solea* (L.). *Netherlands Journal of Sea Research*, **32**, 3–4.

Roff, D.A. (1982) Reproductive strategies in flatfish: a first synthesis. *Canadian Journal of Fisheries and Aquatic Sciences*, **39**, 1686–1698.

Roff, D.A. (1983) An allocation model of growth and reproduction in fish. *Canadian Journal of Fisheries and Aquatic Sciences*, **40**, 1395–1404.

Russell, F.S. (1976) *The Eggs and Planktonic Stages of British Marine Fishes*. Academic Press, London.

Sampson, D.B. & Al-Jufaily, S.M. (1999) Geographic variation in the maturity and growth schedules of English sole along the U.S. west coast. *Journal of Fish Biology*, **54**, 1–17.

Shuozeng, D. (1995) Life history cycles of flatfish species in the Bohai Sea, China. *Netherlands Journal of Sea Research*, **34**, 195–210.

Simpson, A.C. (1951) The fecundity of plaice. *Fishery Investigations London Series II*, **17**, 1–27.

Simpson, A.C. (1959) The spawning of plaice *Pleuronectes platessa* in the North Sea. *Fishery Investigations, London Series II*, **22**, 1–111.

Smith, T.I.J., McVey, D.C., Jenkins, W.E. *et al.* (1999) Broodstock management and spawning of southern flounder, *Paralichthys lethostigma*. *Aquaculture*, **176**, 1–2.

Smith, W.G. (1973) The distribution of summer flounder, *Paralichthys dentatus*, eggs and larvae on the continental shelf between Cape Cod and Cape lookout, 1965–1966. *Fishery Bulletin*, **71**, 527–548.

Solemdal, P. (1967) The effect of salinity on buoyancy, size and development of flounder eggs. *Sarsia*, **29**, 431–442.

Solemdal, P. (1973) Transfer of Baltic flatfish to a marine environment and long term effects on reproduction. *Oikos* (*Supplement*), **15**, 268–276.

Stearns, S.C. & Koella, J.E. (1986) The evolution of phenotypic plasticity in life-history traits: predictions of reaction norms for age and size at maturity. *Evolution*, **40**, 893–913.

Stoner, A.W., Bejda, A.J., Manderson, J.P., Phelan, B., Stehlik, L.L. & Pessutti, J.P. (1999) Behavior of winter flounder, *Pseudopleuronectes americanus*, during the reproductive season: laboratory and field observations on spawning, feeding, and locomotion. *Fishery Bulletin*, **97**, 999–1016.

Templeman, W. & Andrews, G.L. (1956) Jellied condition in the American plaice *Hippoglossoides platessoides* (Fabricius). *Journal of the Fisheries Research Board of Canada*, **13**, 147–182.

Tyler, C.R. & Sumpter, J.P. (1996) Oocyte growth and development in teleosts. *Reviews in Fish Biology and Fisheries*, **6**, 287–318.

Tyler, C.R., Nagler, J.J. & Pottinger, T.G. (1994) Effects of unilateral ovariectomy on recruitment and growth of follicles in the rainbow trout, *Oncorynchus mykiss*. *Fish Physiology and Biochemistry*, **13**, 309–316.

Tyler, C.R., Santos, E.M. & Prat, F. (1999) Unscrambling the egg – cellular, biochemical molecular and endocrine advances in oogenesis. In: *Proceedings of the 6th International Symposium on the Reproductive Physiology of Fish, Bergen 1999* (eds B. Norberg, O.S. Kjesbu, G.L. Tarangere, E. Andersson & S.O. Stefansson). pp. 273–280. Jon Greg AS, Bergen.

Van der Land, M.A. (1991) Distribution of flatfish eggs in the 1989 egg surveys in the southeastern North Sea, and the mortality of plaice and sole eggs. *Netherlands Journal of Sea Research*, **27**, 277–286.

Van der Veer, H.W., Berghahn, R. & Rijnsdorp, A.D. (1994) Impact of juvenile growth on recruitment in flatfish. *Netherlands Journal of Sea Research*, **32**, 153–152.

Vemeirssen, E.L.M., Scott, A.P., Costadinos, C.M. & Zohar, Y. (1998) Gonadotrophin-releasing hormone agonist stimulates milt fluidity and plasma concentrations of 17,20 β-dihydroxylated and 5β reduced, 3α-hydroxylated C21 steroids in male plaice (*Pleuronectes platessa*). *General and Comparative Endocrinology*, **112**, 163–177.

Walsh, S.J. (1994) Life history traits and spawning characteristics in populations of long rough dab (American plaice) *Hippoglossoides platessoides* (Fabricius) in the North Atlantic. *Netherlands Journal of Sea Research*, **32**, 241–254.

Whitehead, P.J.P., Bauchot, M.L., Hureau, J.C., Nielsen, J. & Tortonese, E. (1986) *Fishes of the North-eastern Atlantic and the Mediterranean.* United Nations Educational, Scientific and Cultural Organisation, Paris.

Wimpenny, R.S. (1953) *The Plaice.* Arnold, London.

Witthames, P.R. & Greer Walker, M. (1995) Determination of fecundity and oocyte atresia in sole (*Solea solea*) (Pisces) from the Channel, the North Sea and the Irish Sea. *Aquatic Living Resources*, **8**, 91–109.

Witthames, P.R., Greer Walker, M., Dinis, M.T. & Whiting, C.L. (1995) The geographic variation in the potential annual fecundity of Dover sole *Solea solea* (L.) from European shelf waters during 1991. *Netherlands Journal of Sea Research*, **34**, 45–58.

Witthames, P.R., Greenwood, L. & Lyons, B. (1999) Ovarian atresia in *Solea solea* (L.). In: *Proceedings of the 6th International Symposium on the Reproductive Physiology of Fish, Bergen 1999* (eds B. Norberg, O.S. Kjesbu, G.L. Tarangere, E. Andersson & S.O. Stefansson). p. 106. Jon Greg AS, Bergen.

Wood, A.W. & Van Der Kraak, J. (1999) Apoptotic cell death in Rainbow trout (*Oncorhynchus mykiss*) and goldfish (*Carassius auratus*) ovarian follicles in *vivo* and *vitro*. *Proceedings of the 6th International Symposium on the Reproductive Physiology of Fish, Bergen 1999* (eds B. Norberg, O.S. Kjesbu, G.L. Tarangere, E. Andersson & S.O. Stefansson). pp. 295–297. Jon Greg AS, Bergen.

Zamarro, J. (1991) Batch fecundity and spawning frequency of yellowtail flounder (*Limanda ferruginea*) on the Grand Bank. *NAFO Scientific Council Studies*, **15**, 43–51.

Chapter 5
The planktonic stages of flatfishes: physical and biological interactions in transport processes

Kevin M. Bailey, Hideaki Nakata and Henk W. Van der Veer

5.1 Introduction: the problem

Marine fishes spawn from tens of thousands to millions of eggs per female every year in order to leave sufficient offspring to replace themselves. Temperate species like plaice (*Pleuronectes platessa*) and flounder (*Platichthys flesus*) put up to 44% of their body's energy capital into the spawning process (Rijnsdorp 1994). A strategy of broadcasting many small planktonic eggs to drift in the ocean is a high risk gamble with a low probability of return given that the average survival rate to the age of first reproduction is 0.00001–0.000001%. The obvious risks of this strategy are arriving at an unsuitable habitat, or arriving at the right habitat at the wrong time. Given the natural variability in current speeds and direction, the potential for drift into nursery environments of varying quality might be expected to result in high recruitment variability. There are also some more subtle effects on population genetics caused by the potential mixing of different spawning groups, including outbreeding depression or loss of local adaptive advantage by interbreeding. Surely, however, there must be some overwhelming benefits to a planktonic life stage? One benefit would be the potential for dispersal and colonisation of new habitats (Strathmann 1974). In addition, gene flow between breeding groups might be enhanced, giving rise to the salutary effect of hybrid fitness.

Another advantage of a planktonic stage is to minimise inter-specific competition. Young feeding larvae harvest the numerous prey at the small end of the plankton size spectrum in the sea. Small fish larvae have slow swimming speeds and therefore prey abundance is a critical factor in determining encounter rates and feeding success. As larvae grow and their swimming speeds increase, a shift to larger but less numerous prey becomes possible because speed and improved vision now dominate encounter rates. Prey shifts allowed by ontogenetic development of prowess and by life history changes alleviate competition with other size groups of fish of the same species.

The relative importance of the many causes of mortality of marine fish eggs and larvae varies over a wide range of space and time scales. During planktonic stages there is a remarkable culling of individuals (Fig. 5.1) to leave the most fit for the given set of environmental conditions, or alternatively, the most lucky to have been spawned in the right place at the right time. However, surviving the larval stage is no guarantee of living long enough to reproduce. Thus, drift in the plankton has another critical consequence; larvae have to end their drift period in an area that is suitable as a nursery for the juvenile stage. The question is whether flatfishes do anything to maximise their probability of arriving there, or whether it is a random process.

Mean cohort declines

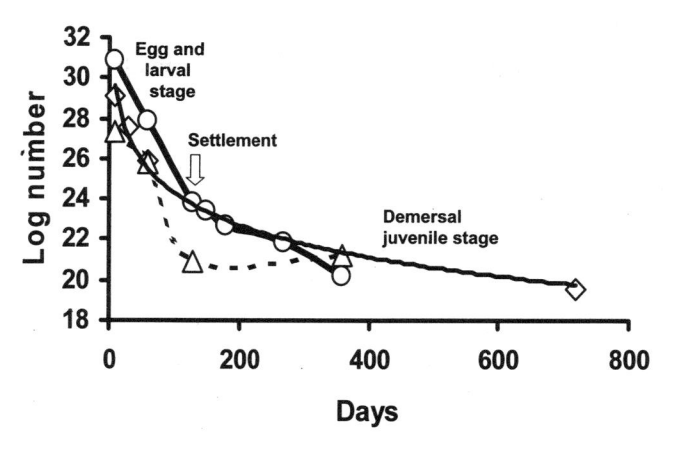

Fig. 5.1 Cohort declines for plaice over the first 2 years of life. Data plotted from Van der Veer (1986) (diamonds), Beverton & Iles (1992) (circles) and Nash (1998) (triangles). The strength of the year-class may be regulated during the early pelagic and post-larval demersal stage and mortality rates may vary on a geographic and inter-annual basis.

Year-class strength may be largely determined by the end of the larval stage (Fig. 5.1; Pihl 1990; Horwood *et al.* 2000) but can be modified (Van der Veer *et al.* 1990) or even controlled in the juvenile stages (Bailey 2000). According to competing concepts about what controls the recruitment of marine fishes, several involve larval drift. According to the 'supply-side hypothesis' (e.g. Connell 1985; Milicich *et al.* 1992), it is the abundance of larvae that arrive in their nursery areas that controls the abundance of fishes at a later stage. Likewise the 'nursery size hypothesis' (e.g. Rijnsdorp *et al.* 1992) states that each nursery sustains a population in proportion to the size of the nursery area, but that supply to the nurseries or sometimes retention (Iles & Sinclair 1982; Sinclair 1988) is often a limiting factor. In the 'match-mismatch' hypothesis (Cushing 1972), fish spawning times and survival may be related to the zooplankton production cycle, but spatial aspects are also important, as Heath (1996) states: 'interactions between dispersal patterns and the spatial and temporal dynamics of plankton biomass are of critical importance for survival, and that the timing of larval production is closely coupled to these factors, implying a degree of adaptation in fish spawning strategies'.

The problem of larval drift and finding a nurturing environment for later stages is particularly acute for flatfishes. Flatfishes generally spawn in water deeper than their juvenile nurseries, and most species have a juvenile stage with fairly specific habitat (such as sediment grain size and temperature preferences) and prey requirements. Different flatfish populations may have different strategies to arrive at, and/or maintain themselves in suitable nurseries. Thus nearshore spawners may take advantage of retention features to maintain themselves in an appropriate habitat, whereas offshore spawners need to get inshore, or even into specific estuaries. Depending on the relative positions of spawning and nursery habitats, they may have local adaptations to spawn in areas where there are retention features to maintain larvae in favourable habitats; or alternatively, they may spawn in areas where there are transport steering features such as local topography or prevalent winds that accomplish a goal of targeted movement.

The goal in this chapter is to outline how different species, or even subpopulations within species of flatfishes, have adapted to transport conditions, to discuss the physical mechanisms and habitat-specific variability in transport, and to outline the consequences to population biology of flatfishes. Other organisms with planktonic life stages, including invertebrates and coral reef fishes, share many of these same problems with flatfishes: eggs and larvae drift in the plankton but juveniles need to recruit to specific habitats. Therefore, comparisons of how quite different organisms have solved similar problems would be illuminating as to how flatfishes have adapted. The cues that may elicit specific behaviours are not discussed in depth here, as there are other excellent reviews of these topics (Boehlert & Mundy 1988; Gibson 1997). This synthesis is largely restricted to a few commercially exploited species from the northern hemisphere, representing a small percentage of the roughly 700 known species of flatfishes, three-quarters of which are tropical (see Chapter 2); unfortunately data on the majority of these other species are sparse.

5.2 Flatfish eggs and larvae in the plankton: variations in form and function, time and space

Various aspects of egg and larval morphology and ontogenetic development affect the process of transport; some of these include size, shape, density and swimming ability. Where and when eggs and larvae are distributed in the plankton will also play a crucial role in the transport process.

5.2.1 Variations in form and function

As a group, flatfish produce a wide variety of egg and larval forms in the plankton. Eggs vary in diametre from 0.5–0.8 mm for the Pacific sand dab (*Citharichthys sordidus*) to 4.0–4.5 mm for the Greenland halibut (*Reinhardtius hippoglossoides*). Variations in egg size may involve adaptations to minimise predation, provide greater maternal investment in individuals, or properties to confer some other benefit to offspring, such as longer egg stage duration whereby hatching larvae are more developmentally advanced. Egg size may vary considerably among congeneric species. For example, for *Pleuronichthys* spp. inhabiting waters off the coast of California, egg size may vary from 1.84–2.08 mm for the curlfin sole (*P. decurrens*) to 0.94–1.08 mm for the spotted turbot (*P. ritteri*) (Sumida *et al.* 1979). Even within a species like plaice, egg size may vary considerably with season (see references in Gibson 1999).

Shell architectures and egg densities also vary considerably. Some flatfishes have intricately sculpted eggs, especially from the genus *Pleuronichthys*. The chorions of different flatfish species have been described as striated, reticulated, rugose, adhesive or covered with a polygonal network (Moser 1996), whereas others are remarkably smooth. Robertson (1981) suggested that egg size of marine fishes and sculpturing of the chorion may influence the ascent rates from depths where eggs are spawned. Based on measuring sinking rates of preserved eggs, he inferred that small eggs and sculpted eggs would have a slower rate of ascent than larger or smooth eggs. Ascent rates will influence the relative duration that eggs spend in bottom and surface currents and affect their ultimate fate.

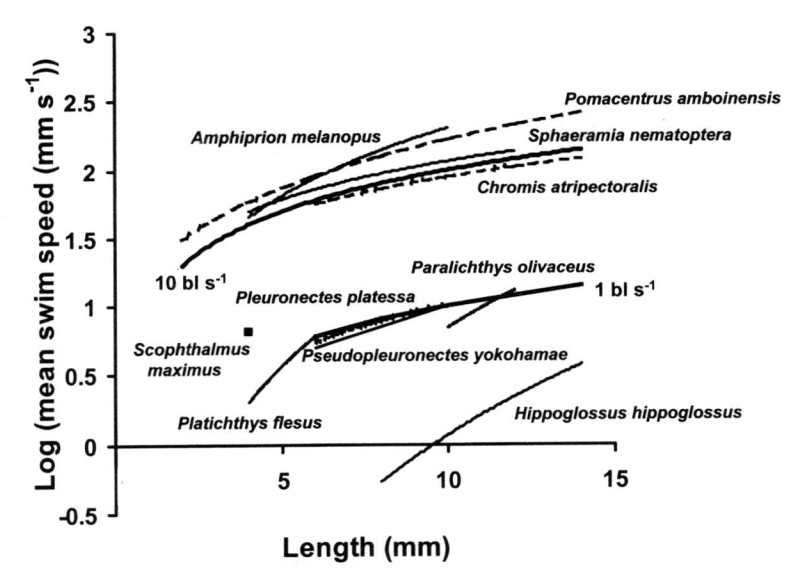

Fig. 5.2 The log of mean swimming speed (mm s^{-1}) plotted against larval length for tropical reef fishes (the species clustering at the top of the graph around 10 BL s^{-1}) and for six temperate flatfish species (clustering around 1 BL s^{-1}). Data are from numerous published sources and use a variety of methods to measure speed over different durations. Differences may be partly due to temperature.

The size of larval flatfishes is also quite diverse. Length-at-hatching can vary from about 2 mm for Pacific sand dab to 10–16 mm for the Greenland halibut. Size-at-transformation varies from about 8 mm for starry flounder (*Platichthys stellatus*) to 49–72 mm for the rex sole (*Glyptocephalus zachirus*). Moser (1996) cites extremes of <5 mm for archirids to >100 mm for *Chascanopsetta*. In general, deeper dwelling species spawn larger eggs and have a longer pelagic life and larger size at metamorphosis than shallower dwelling species (Minami & Tanaka 1992). Some flatfish species are fairly flexible in the size at transformation, whereas others are relatively fixed.

Larval size is closely coupled to swimming speed (Fig. 5.2). Generally, flatfish larvae are weak swimmers in the plankton. Fukuhara (1988) noted that the swimming speeds in body lengths per second (BL s^{-1}) of flatfish larvae are low compared with larvae of other temperate and subarctic pelagic fishes. He suggested that flatfishes have some structural deficiencies of the fin rays that cause them to be feeble. He also noted that swimming speeds of flatfishes decrease at metamorphosis in contrast with speeds for pelagic fishes which tend to increase at metamorphosis.

The decrease in swimming speed at metamorphosis for flatfishes is similar to observations of coral reef fishes, that show decreased swimming speeds after settling onto their reef habitat, but reef fishes also show flexible behaviour (Leis & Carson-Ewart 1997, 1999). In spite of some common traits between flatfishes and coral reef fishes (for example, the necessity of transforming larvae to recruit to specific target habitats), there are some remarkable differences that include swimming abilities of pelagic larvae. Compared with flatfish larvae, many reef fishes have larvae that are notably strong swimmers (Leis & Carson-Ewart 1997; Fisher *et al.* 2000) and thus larvae may play an active role in searching for habitat. Most reef fishes are perciforms, which have fully developed fins at a small size. Leis & Carson-Ewart found that larvae of the pomacentrid *Chromis atripectoralis* only 7 mm in length averaged a sustained swimming speed of 34 bl s^{-1}, and most taxa tested swam at about 10–15 BL s^{-1} (Fig. 5.2). By contrast, flatfish larvae swim at a relatively lethargic 1 BL s^{-1} (Blaxter 1986). Leis & Carson-Ewart suggested that swimming performance may be favoured in a coral reef system, as there would be selective pressure on larvae to find a reef 'target'. Current speeds in the area where they made their measurements were 10–15 cm s^{-1}, so virtually all of the late stage larvae were able to swim faster than currents for sustained periods, and therefore the inference is that they can control their position and distribution. They further suggested that flatfishes may have a larger 'target' for their destination. Alternatively, flatfishes may have adapted to the transport problem in their environment in other ways. For example, by vertically migrating only a short distance, they may preferentially select their residence in currents moving in different directions, thus influencing the direction of transport in currents whose speeds far exceed their own swimming speeds. Where the currents are tidal, such behaviour is called selective tidal stream transport (STST) (Forward & Tankersley 2001).

5.2.2 Variations in time and space in the plankton

Are spawning times and locations of flatfish species distributed such that the arrival of offspring in their juvenile nurseries is maximised? In the case of plaice, one of the most studied species, the spawning grounds are close to the nursery in areas where currents are weak or nondirectional. In areas where currents are strong, eggs appear to be released in an optimum direction and distance from the settlement areas to guarantee successful colonisation (Gibson 1999). This is an essential criterion for most flatfishes, because juveniles have specific nursery and resource requirements. Using plaice as an example, the larvae of this species cannot delay metamorphosis, so enforced settlement can occur in unsuitable habitats. However, late-stage plaice larvae may demonstrate considerable behavioural flexibility in the settlement process before metamorphosis, such as 'pseudo-settlement' (Tanaka *et al*. 1989) when larvae are transported into shallow areas. Juveniles have specific habitat requirements for optimal growth and survival, including food, temperature and avoidance of predation, as those factors are influenced by depth, sediment type, exposure and predator abundance. While it is not believed that settled juveniles starve, predation after settlement is believed to be critical and is strongly influenced by the growth-dependent predator-prey size ratio (Bailey 1994).

Since juveniles have fairly specific habitat and prey requirements, the target for larval transport is not necessarily geographically narrow (as compared with a coral reef), but is restricted in terms of habitat-defining parameters. Juvenile habitat is discussed more thoroughly in Chapter 8, but specifically, different species groups tend to have specific feeding patterns as they settle (e.g. Minami & Tanaka 1992). Other fishes with transport 'targets', for example reef fishes, also tend to have specific feeding requirements at settling (Harmelin-Vivien 1989).

The timing of spawning and rate of egg development in the plankton is likely to be an adaptation in part to the timing of prey production, seasonal and geographic abundance of predators, and seasonal changes in transport conditions. Spawning time may be variable for some species (e.g. common sole *Solea solea* in the North Sea, Bolle *et al*. 1999; English sole *Parophrys vetula* off the Oregon coast, Boehlert & Mundy 1987), and could signal a change in currents, or be related to plankton production cycle, or even be a purely physiological response to temperature, i.e. through gonadal development. Other species may have consistent spawning times; for example, plaice in the North Sea have a mean spawn date of 19 January with a standard error of ± 2 days (Cushing 1990).

Deep-water species spawn earlier in the year, generally in winter, and tend to have larger larvae (Minami & Tanaka 1992). The offshore spawning species tend to spend a longer time in the plankton, and metamorphose at a larger size (Minami & Tanaka 1992). Larvae with a metamorphic size >25 mm may spend more than 3 months in the plankton. In fact, Dover sole (*Microstomus pacificus*) and rex sole larvae may spend a year or longer in the plankton (Pearcy *et al.* 1977). Larvae with a long planktonic life have a more variable size-at-settling (Minami & Tanaka 1992). This variation in size-at-settling reflects the variability in transport and time needed to find a successful clue for settlement to a suitable habitat. Larvae originating from warm water and nearshore spawners are smaller and undergo metamorphosis at a smaller size as well. Species with a metamorphic length <10 mm have a pelagic duration <1 month.

In general, flatfishes tend to move to shallow water to spawn, and nursery grounds for juveniles are shallower than the spawning grounds (Minami & Tanaka 1992). Many species show remarkable consistency from year to year in the location of spawning (Nash & Geffen 1999; Fox *et al.* 2000). Modelling studies have shown that the specific location of spawning, as well as depth and directed swimming, play an important role in retention of fish larvae over offshore banks (Werner *et al.* 1993).

Almost all flatfishes have pelagic eggs or eggs that are deep in the water column offshore. However, five species in the Pacific Ocean (marbled flounder *Pseudopleuronectes yokohamae*, rock sole *Lepidopsetta bilineata*, the newly discovered northern rock sole *L. polyxystra* (Orr & Matarese 2000), dusky sole *Lepidopsetta mochigarei* and kurogarei *Pseudopleuronectes obscurus*) and one in the Atlantic (winter flounder *Pseudopleuronectes americanus*) have demersal eggs. These species spawn in late winter and spring in shallow coastal waters <20 m depth. Demersal eggs in these waters may be an adaptation to prevent their offshore dispersal by surface currents (Pearcy 1962).

5.3 Physical mechanisms of transport and retention

5.3.1 Dispersal/retention mechanisms

Transport and retention are physical processes responsible for moving early pelagic life stages toward an appropriate habitat, or for keeping them within an appropriate habitat (see Norcross & Shaw 1984, for a review). Because Werner *et al.* (1997) have reviewed the physical processes by region on the continental shelf and discussed their effect on pelagic stages of marine species, here the physical mechanisms of larval dispersal/retention are briefly summarised, mainly focusing on flatfish examples.

5.3.1.1 Wind-forcing/Ekman transport

Wind conditions during larval development for plaice showed significant correlations with the 0-group abundance along the Danish coast (Nielsen *et al.* 1998) and with year-class strength on the Swedish west coast (Pihl 1990), implicating the role of wind-induced transport for recruitment. Van der Veer *et al.* (1998) also showed that wind-induced variability in circulation and larval dispersal patterns might be a key factor in determining subsequent year-class strength. In addition to the winter temperature at the spawning grounds, which has an inverse relationship with year-class strength of the plaice (Van der Veer 1986), residual currents induced by persistent westward winds in cold winters could also have an effect on recruitment in a specific year (Van der Veer & Witte 1999).

Many flatfish species have their major spawning period in winter-early spring, when strong storm-related winds predominate, so there could be adaptations to wind-induced circulation. In this regard, it is noteworthy that on/offshore wind-induced currents often produce vertical shear structure in water circulation of shallow coastal seas. In Sendai Bay, Japan, wind-induced circulation could be most responsible for transport of stone flounder (*Platichthys bicoloratus*) eggs and larvae from the spawning site to the vicinity of estuarine nurseries. The spawning site located at the northernmost part of this bay is apparently adapted to southward Ekman transport induced by westerly winds (Nakata *et al.* 1999a). In a shelf region of the Sea of Japan, on the other hand, strong westerly winds predominate during the main spawning period of littlemouth flounder (*Pseudopleuronectes herzensteini*), and larval retention and settlement could be enhanced by onshore drift at the surface. In fact, a significant negative correlation was found between the frequency of strong westerly blows and fishery catch 2 years after the westerly blows (Nakata *et al.* 2000).

English sole, an estuarine-dependent species with eggs and larvae spawned offshore on the US Pacific coast, may have a dual strategy for immigrating to estuarine nurseries: onshore Ekman transport for newly transforming larvae and selective tidal stream transport for older larvae once they have settled on nearshore nurseries outside the estuary (Boehlert & Mundy 1987). This strategy may enhance the ability of this species to use the relatively small, isolated Pacific coast estuaries as juvenile nursery areas.

5.3.1.2 *Tidal currents/selective tidal stream transport*

Tidal currents are important for larval transport in tidal inlets and estuaries if associated with vertical migration behaviour (active STST). This behaviour usually develops in later-stage larvae as they approach the settlement zone. STST has been shown (whether active or passive) in several species (see Table 5.1). Bergman *et al.* (1989) hypothesised that plaice larvae accumulate in the inshore nursery area of the Dutch Wadden Sea from the coastal zone due to a passive process through tidal water exchange and settlement occurs in suitable nurseries during flood tide (passive STST).

5.3.1.3 *Estuarine circulation*

Another physical process establishing a two-layered circulation system is freshwater discharge into estuaries. In this case again, complex interactions between vertical migration/movement and horizontal advection often greatly contribute to the transport/retention of pelagic eggs and larvae. Vertical migration is used by larvae to select either the seaward outflow at the surface or the inflow near the bottom for the purposes of entering, leaving or remaining in estuaries (e.g. Epifanio 1988). This may be a behavioural adaptation to avoid

Table 5.1 Differences in spawning and transport-related characteristics of flatfishes

Species	Area	Spawning	Pelagic duration (days)	Nursery	Transport characteristics	Authors
Family Pleuronectidae						
Pleuronectes platessa	North Sea	30–60 km offshore	30–90	Nearshore, inland seas	Transport with bottom currents, then STST	Van der Veer et al. 1998; Cushing 1990; Rjinsdorp et al. 1985
	Irish Sea	Nearshore		Nearshore	Retention	Nash & Geffen 1999
	Kattegat/Belt Sea	Offshore		Shallow water ~5 m	Wind-driven currents, selective transport	Neilsen et al. 1998
P. quadrituberculatus	Gulf of Alaska	Nearshore		Nearshore	Retention and selective transport	Bailey, unpubl. data; Norcross, pers. comm.
Pseudopleuronectes herzensteini	Coastal Japan	Coastal waters ~50m		Nearshore, coastal	Drift north and inshore retention	Nakata 1996
P. americanus	NW Atlantic	Estuaries	28–42	Estuaries, coves	Tidal accumulation, retention	Chant et al. 2000
P. yokohamae	Japan			STST		Takahashi et al. 1986, cited in Tanaka et al. 1989
Limanda aspera	Gulf of Alaska	Inshore	30–60	Inshore, inner bays	Apparent retention	Bailey, unpubl. data
L. ferruginea	Grand Banks, Newfoundland	Offshore banks	90–120	Offshore banks	Retention	Neilson et al. 1988; Walsh 1992
L. limanda	North Sea	Offshore		Inshore	Offshore settlement, inshore migration	Bolle et al. 1994
Glyptocephalus cynoglossus	Newfoundland	Offshore banks	120–360	Offshore banks	Retention	Neilson et al. 1998;
G. zachirus	Oregon	Offshore 100–300 m	365	Offshore outer shelf	Long larval life, Ekman onshore transport	Pearcy et al. 1977
Platichthys flesus	East English Channel		30–60	Coastal estuaries	Drift north, then after flexion STST towards coast	Grioche et al. 2000
	SE North Sea			Coastal estuaries	STST	Campos et al. 1994
	North Sea	Offshore		Coastal estuaries, tidal flats	STST	Jager 1998, 1999
Parophrys vetula	Oregon	Offshore		Estuaries	Onshore Ekman	Boehlert and Mundy 1987

Table 5.1 (Continued.)

Species	Area	Spawning	Pelagic duration (days)	Nursery	Transport characteristics	Authors
Hippoglossoides platessoides	North Atlantic	Offshore banks	30–120	Offshore banks	Retention	Walsh 1994; Neilson et al. 1988
Microstomus pacificus	Oregon	Offshore >400 m	365–550	Offshore, outer shelf	Ekman onshore and post settlement migration	Pearcy et al. 1977; Toole et al.1997
Eopsetta jordani	Oregon	Offshore, deep water	180	Inner shelf	Ekman onshore	Pearcy et al. 1977
Hippoglossus stenolepis	Gulf of Alaska	Continental slope >200 m	90–180	Inshore bays	Directed transport	Bailey & Picquelle 2002
Rheinhardtius stomias	Gulf of Alaska	Continental slope	90–180	Inshore bays	Directed transport	Bailey & Picquelle 2002
R. hippoglossoides	Norway	Continental slope 600–900 m	150	Slope, banks, coastal waters	Transport northwards onto banks and coastal waters	Haug et al. 1986
Family Achiridae						
Trinectes maculatus	NW Atlantic coast	Estuaries		Estuaries	Presumed retention, movement up estuaries	Miller et al. 1991
Family Paralichthyidae						
Paralichthys dentatus	Onslow Bay, North Carolina	Shelf, offshore	30–70	Nearshore, estuaries	Cross shelf, STST	Burke et al. 1998
P. olivaceus	Wakasa Bay, Japan	Nearshore, near nursery	30–60	Nearshore	Near bottom retention	Burke et al. 1998
P. olivaceus	Shijiki Bay, Japan	Offshore		Nearshore	STST	Tanaka et al. 1989
Family Soleidae						
Solea solea	Bay of Biscay, France	40–80 km offshore	30–60	Inshore bays, estuaries	STST?	Amara et al. 1998
					Retention offshore, then Ekman transport, diffusion inshore	Koutsikopoulos et al. 1991
	Eastern English Channel	Nearshore		Inshore coastal	Retention, tidal migrations	Grioche et al. 2000

Authors for most information are cited, some information was also obtained from Miller et al. 1991.

the effect of high flushing rates of the estuaries and to maintain horizontal position (Heath 1992). Ontogenic changes in vertical migration behaviour possibly lead to longitudinal gradients in larval size, age and species composition within the estuary. In the two-layered Cape Fare River estuary, it was demonstrated that behavioural responses of the larvae of flounder (*Paralichthys* sp.), primarily to tide and photoperiod, facilitates their transport to appropriate habitats (Weinstein *et al.* 1980).

5.3.1.4 Fronts and eddies

Reproduction of migratory species is often concentrated in geographic areas with relatively stable long-term hydrographic characteristics, such as fronts and eddies (Norcross & Shaw 1984). Recent examples include: Atlantic herring *Clupea harengus* (Iles & Sinclair 1982), Atlantic cod *Gadus morhua* (Ellertsen *et al.* 1990; Taggart *et al.* 1996), Alaska pollock *Theragra chalcogramma* (Bailey *et al.* 1997), Japanese anchovy *Engraulis japonicus* (Nakata 1996) and others (see Heath 1992).

In the case of flatfishes, the main spawning grounds of the littlemouth flounder are consistently found in a shelf region near the Sado Strait in the Sea of Japan. This location is characterised as the depth range from 50 m to 100 m at the most upstream region of relatively wide shelf, where current speeds are appreciably low and coastal eddies are often observed (Nakata *et al.* 2000), suggesting that littlemouth flounder may be adapted to the current system of this region. Dover sole larvae may be prevented from inshore transport at time of settling and kept in suitable outer continental shelf water by oceanic fronts associated with winter convergence (Hayman & Tyler 1980). Chant *et al.* (2000) indicated that a small-scale eddy generated during flood tides appeared to contribute to advection of winter flounder larvae to a cove in a southern New Jersey estuary, resulting in high numbers of the settled juveniles in the cove.

In addition to providing a possible mechanism of egg and larval retention within the coastal nurseries, fronts and eddies potentially play significant roles in accumulation and production of prey organisms, thus contributing to survival and subsequent recruitment of the larvae retained in their vicinity (Nakata 1996). Munk *et al.* (1999) recently demonstrated the variability in frontal zone formation at the shelf break in relation to the distribution and abundance of five species of gadoid larvae, and pointed out that frontal zone variability had a diverse influence on the larval populations.

5.3.2 Models

Numerical models of water circulation are useful to explore egg and larval transport mechanisms because of the complicated dynamic interactions between physical and biological processes, such as coupling of water circulation and larval vertical migration which are inherent in transport mechanisms (Bartsch *et al.* 1989; Werner *et al.* 1993; Lough *et al.* 1994). Van der Veer *et al.* (1998) applied a 2D circulation model of the southern North Sea to simulate the inter-annual variability in dispersal of plaice eggs and larvae from the spawning area in the Southern Bight towards the Dutch coastal nursery areas. An interesting finding from the model simulations was that inter-annual variability in transport is quite large and of the same order of magnitude as that in larval abundance observed near the nursery areas, suggesting that the variability in circulation patterns during the early pelagic stages in the open sea might

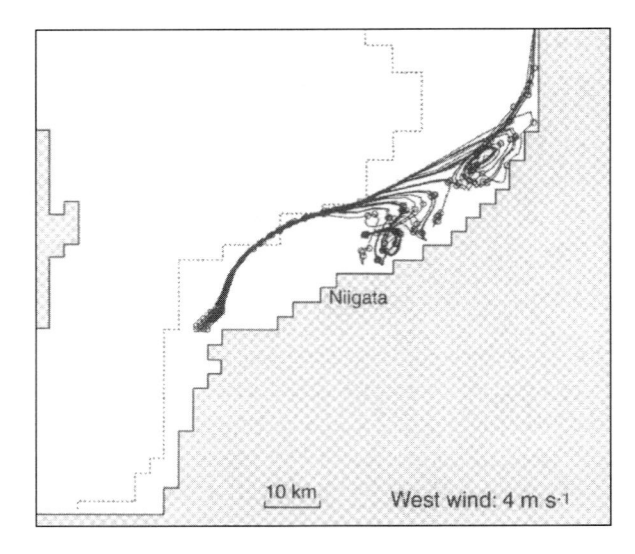

Fig. 5.3 Computed trajectories of particles, simulating littlemouth flounder eggs and larvae, released from the spawning location for 30 days under the condition of constant westerly winds. The broken line is the 100-m contour (from Nakata *et al.* 2000).

be a key factor in determining year-class strength of plaice. However, plaice larvae could exhibit some active behaviour such as settling on the bottom by late larvae, increasing the role of currents near the bottom in their transport.

Transport of littlemouth flounder eggs and larvae has been modelled in a shelf region of the Sea of Japan using a 3D Euler-Lagrangian model, with special interest in the effect of wind on inshore retention (Nakata *et al.* 1999b, 2000). The general pattern of egg and larval transport from the spawning habitat on the shelf to adjacent coastal nurseries was reproduced under a constant westerly wind, assuming vertical movement of the eggs and larvae (upward in the earlier phase and downward in the later phase of the drift period), and the effect of the westerly wind speed on the egg and larval retention in the inner shelf was evaluated (Fig. 5.3). It was shown that the retention rate could be rapidly reduced when the wind speed exceeded a critical value (Fig. 5.4). Supporting the model results, the number of juveniles collected in the nursery area (1991–1998) showed a significant negative correlation with the frequency of strong westerlies in April. Although field data on the vertical distribution and movement of littlemouth flounder eggs and larvae are limited, the most realistic pattern of transport was obtained only in the case where sequential upward and downward movement was assumed (Nakata *et al.* 1999b). This may indicate that upward movement in the earlier phase could be an adaptation to avoid wind-induced offshore drift at the spawning depth (below 30 m), and to facilitate entrainment into coastal eddies observed in the shelf region.

Considerable advances in the ability to explore the couplings of physics and biology in the models are very encouraging. The examples of flatfish species are still limited, but more research should be encouraged in combination with well-designed field surveys and laboratory experiments. Comparison between species under the same physical setting, or a comparison between different physical settings for the same species using models, would also be useful. Some of these types of comparisons are presented below, but it should be noted that usually

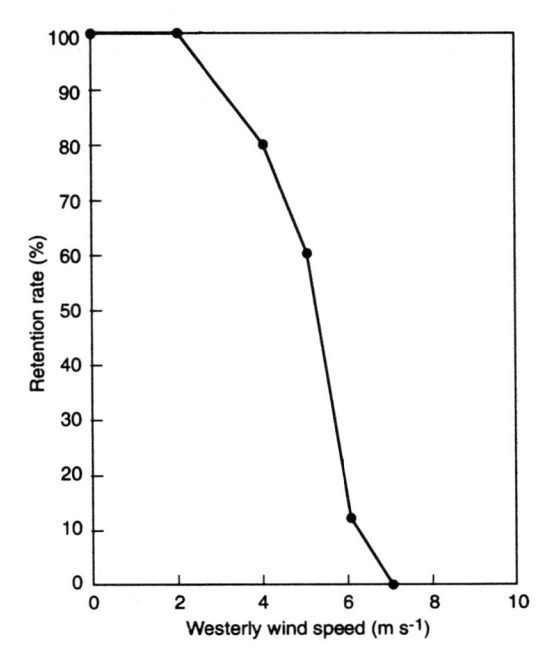

Fig. 5.4 From a 3D model simulating transport of littlemouth flounder eggs, the estimated retention rates (%) of particles released from the spawning grid after a 30-day drift under various westerly wind speeds (from Nakata *et al.* 2000).

the knowledge of the physical or biological parameters is incomplete. For this purpose, precise field data on the behaviour of target species – especially in the vertical water column – are indispensable, as well as data on circulation in the spawning and nursery habitats.

5.4 Adaptations to transport conditions: geographical and species comparisons

Across the taxonomic range, flatfishes have a variety of transport requirements during the larval to juvenile stage (Miller *et al.* 1991; Fig. 5.5). As noted earlier, they are relatively feeble swimmers, so they must have other adaptations to reach nearshore targets. Spawners in estuaries generally have their nurseries in estuaries, so retention is necessary. Species that spawn on the coastal shelf/slope with coastal shelf/slope nurseries also require retention. These species may spawn in areas where fronts, eddies or other retention features may prevail. On the other hand, many species require cross-shelf migration into shallow water, into estuaries or even transport up estuaries into freshwater. In this case, spawning in areas that favour directed transport, or behaviours that take advantage of currents, such as STST, are common. This section presents more detailed examples of potential adaptations that flatfishes have developed in order to get to their nurseries.

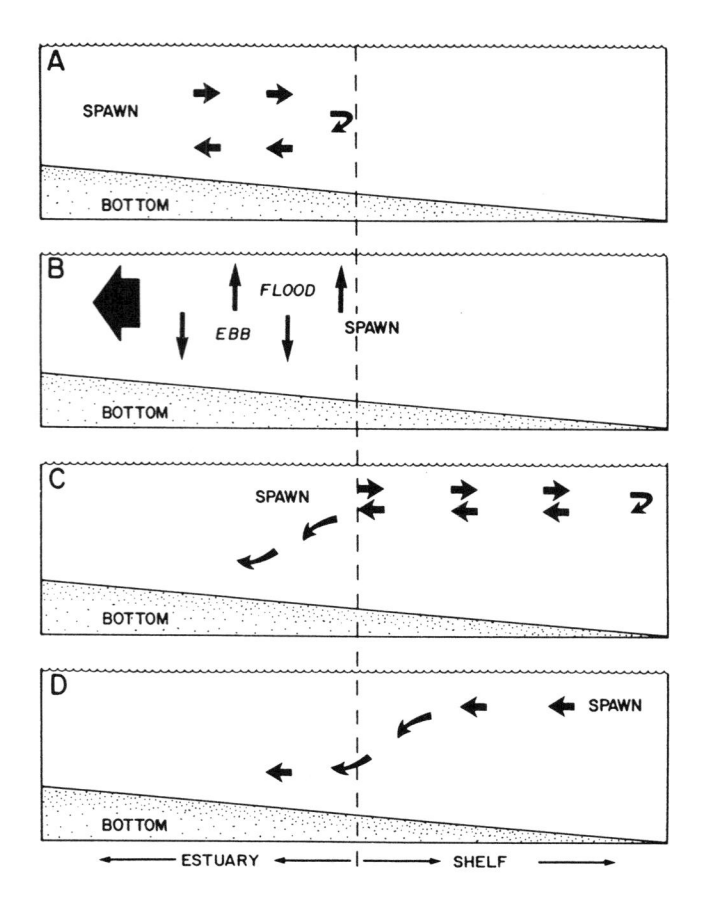

Fig. 5.5 Comparison of four mechanisms for dispersal and recruitment of estuarine larvae. A. Spawning occurs in the estuary and young larvae are transported downstream by surface flows; older larvae sink and are transported upstream in the residual bottom flow. B. Spawning occurs near the estuary mouth; larvae migrate up in the water column on flood tides and sink on ebb tides, resulting in net upstream transport. C. Larvae are spawned near the estuary mouth and are flushed onto the shelf; larvae are retained near the estuary mouth by current patterns. D. Larvae are spawned on the shelf and are transported onshore by current; post-larvae sink and are transported into the estuary by residual bottom flow (from Epifanio 1988).

5.4.1 Comparisons among species within a geographic region

Within a relatively small geographic area, species may be transported in different directions based on their distribution and behaviour in the local current system. In Shelikof Strait, located in the Gulf of Alaska between Kodiak Island and the Alaska Peninsula (Fig. 5.6), there is an estuarine type circulation with the surface Alaska Coastal Current (ACC) flowing towards the southwest, countered by a deep current flowing up the strait. A weak nearshore current flows down the strait hugging the coastline. At about the same time and within the same region, different species of larvae are moving in different directions and at different rates. For example, Pacific halibut (*Hippoglossus stenolepsis*) larvae (eggs 300–400 m deep, larvae in the upper 50–100 m) are transported from the slope region at the exit of the strait in an up-strait direction northward to their nursery grounds in shallow water. Flathead sole (*Hippoglossoides elassodon*) (eggs spawned at 150–250 m, later-stage eggs and larvae in the

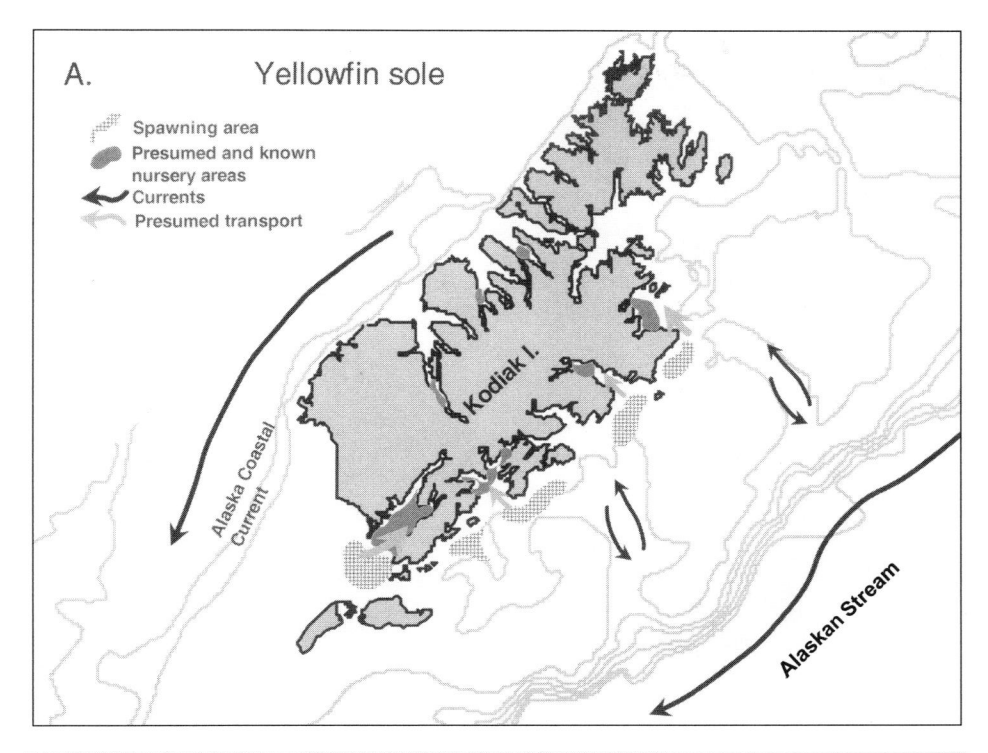

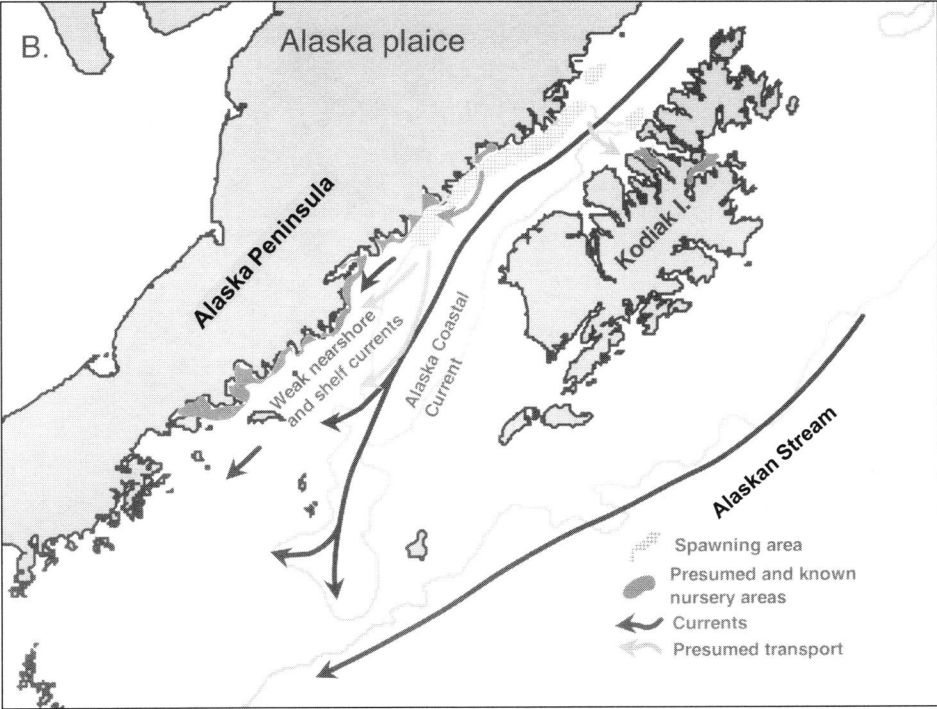

Fig. 5.6 Potential schematic for different transport patterns of four species of flatfish larvae in the Shelikof Strait, Alaska. A. Yellowfin sole are spawned in shallow water over banks on the outer side of Kodiak Island. Their eggs and larvae are retained in the current system and juveniles recruit inshore to local bays. B. Alaska plaice are spawned in shallow water on either side of the sea valley, are transported downstream and are retained near nursery areas in bays.

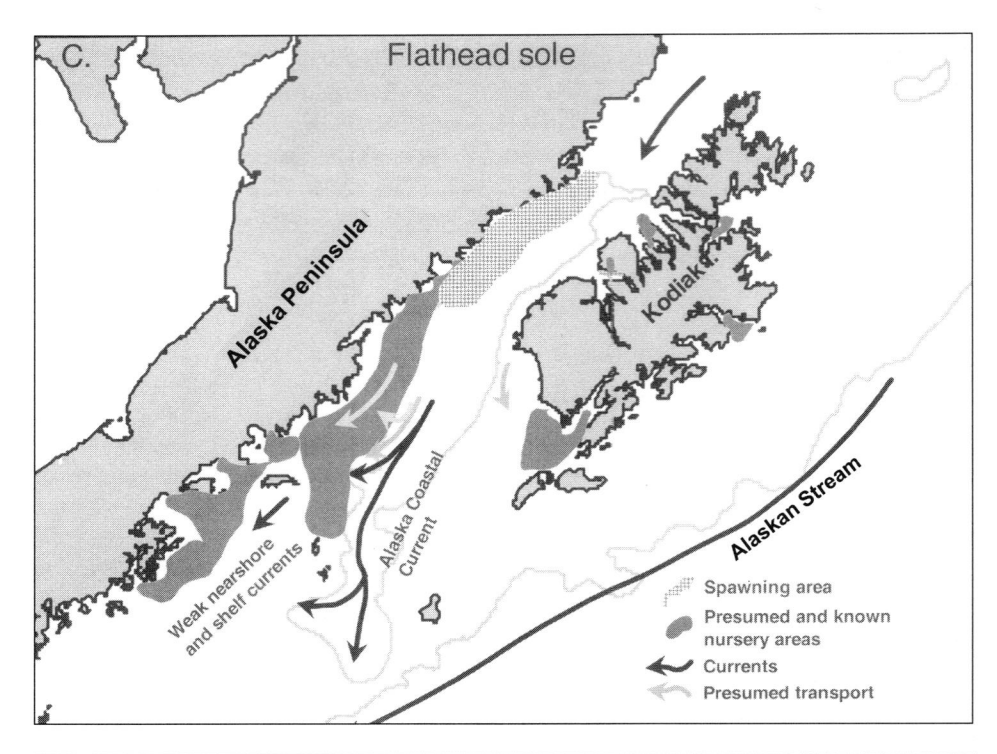

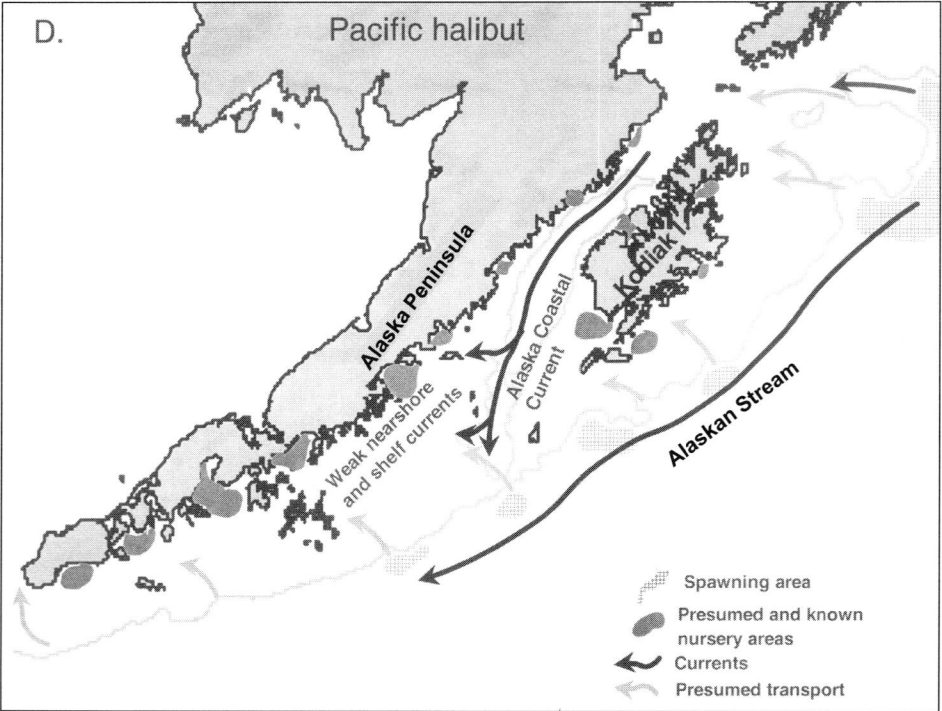

Fig. 5.6 (*Continued.*) C. Flathead sole are spawned over deep water in the sea valley and are transported downstream in the Alaska Coastal Current towards nursery areas on the continental shelf. D. Pacific halibut are spawned offshore over deep water and are carried inshore by topographically steered currents up the sea valley; halibut nurseries are in coastal bays. Information on larval flathead sole was provided by S. Porter (Alaska Fisheries Science Center, Seattle, WA, USA) and information on juvenile flatfishes by B. Norcross (University of Alaska, Fairbanks, AK, USA) and in Norcross *et al.* (1999).

upper 40 m) are spawned near the centre of the strait and are transported southwestward in the ACC, while Alaska plaice (*Pleuronectes quadrituberculatus*) (eggs spawned in shallow water 50–100 m and larvae shallow) are transported downstream and are retained in the shallow coastal waters in the vicinity of their nurseries in coastal bays. Yellowfin sole (*Limanda aspera*) are spawned on offshore banks, are retained nearby and recruit inshore to bays.

Species in the same vicinity may demonstrate different transport characteristics through variations in vertical migration behaviour within a system of currents. In the eastern English Channel, Grioche *et al.* (2000) showed that flounder larvae drifted northward until they neared transformation. During their drift, flounder larvae did not migrate vertically and were not retained until flexion when they migrated to near bottom on ebb flow, which advected them towards the coast. On the other hand, common sole remained in the nearshore coastal waters and its retention was attributed to tidal and diurnal vertical migrations undertaken by the larvae.

Amara *et al.* (1998) compared the transport characteristics of common sole with the thickback sole (*Microchirus variegatus*) in the Bay of Biscay. In this case, both species spawn 40–80 km offshore. Late-stage common sole larvae and early-stage juveniles are transported to their nurseries in bays and estuaries after about 60 days, possibly using STST. On the other hand, thickback sole larvae have no swimbladder, show limited vertical migration and settle in the offshore habitat. Their larvae are not caught above 40 m depth in the water column and may make an early shift to near bottom, where currents are weaker, effectively retaining them in the offshore habitat. Using the development of accessory primordia in otoliths, the authors inferred that thickback sole shift to lateral swimming orientation associated with settling before transformation, whereas common sole show orientation to the bottom in late metamorphosis.

5.4.2 Congeneric comparisons in different regions

There are differences in the spawning location and transport characteristics of closely related species in different areas. For example, Burke *et al.* (1998) compared transport characteristics of two species of *Paralichthys* in different oceans. In Wakasa Bay, Japan, the Japanese flounder (*P. olivaceus*) stays near bottom in spite of the tidal cycle. The tidal currents in this region of coastal Japan are weak and estuarine habitat is limited by a narrow and steep continental shelf. In contrast, the Atlantic coast of the USA has a strong tide, a wide shallow continental shelf and extensive estuarine habitat. In Beaufort Inlet, the summer flounder (*P. dentatus*) abundance varied with the tidal stage; high densities of flounder were sampled at the bottom during ebb tide and in the water column during the flood. Late-stage larvae appear to use vertical migration and tidal streams to migrate from offshore into the estuarine nursery, and settlement occurs in the intertidal. By comparison, in Wakasa Bay, Japanese flounder larvae may be passively transported into the bay. However, in another area of Japan where the tides are stronger, Japanese flounder larvae seem to be using STST to migrate inshore (Tanaka *et al.* 1989). Burke *et al.* (1998) showed from laboratory experiments with wild-caught and laboratory-reared larvae that the degree of development of an endogenous tidal rhythm in the larvae appears to depend on the regional coherence and strength of tidal variation, suggesting some behavioural flexibility in paralichthyids in different areas.

Other congeneric species in different oceanic systems show remarkable differences in transport characteristics (Table 5.1). Species in the genus *Limanda* show a great variety of mechanisms for reaching their shallow nurseries. Yellowfin sole (*L. aspera*) in the Gulf of

Alaska spawn in shallow water where currents are weak and larvae are retained. Yellowfin flounder (*L. ferruginea*) that spawn on offshore banks in the western North Atlantic are apparently retained there. Dab (*L. limanda*) in the North Sea spawn offshore, and settlement occurs offshore followed by an inshore migration of juveniles. Finally, marbled flounder (*Pseudopleuronectes Limanda yokohamae*) are believed to use STST to reach their nearshore nurseries.

Two species in the genus *Glytocephalus* also demonstrate remarkable differences in transport strategies. Rex sole (*G. zachirus*) off Oregon has an extremely long planktonic life, believed to be about a year. Spawning is in late winter. Presumably larvae take advantage of the high production associated with prevalent spring and summer upwelling, and utilise wintertime onshore Ekman currents nearly a year later to reach nurseries over the shelf. Witch (*G. cynoglosus*) also has a long planktonic life, but is spawned on offshore banks and retained there.

5.4.3 Conspecific comparisons in different geographic areas

Within the family Pleuronectidae there are also some remarkable differences in transport characteristics varying from retention, to selective transport using Ekman currents, to STST (Table 5.1). However, different subpopulations of plaice also have different mechanisms to arrive in nearshore nurseries (Fig. 5.7). In the Irish Sea, plaice spawn nearshore in close proximity to nursery areas in areas of reduced tidal flow and larvae are apparently retained there. In the Kattegat/Belt Sea area of Scandinavia where tidal currents are weak, spawning occurs

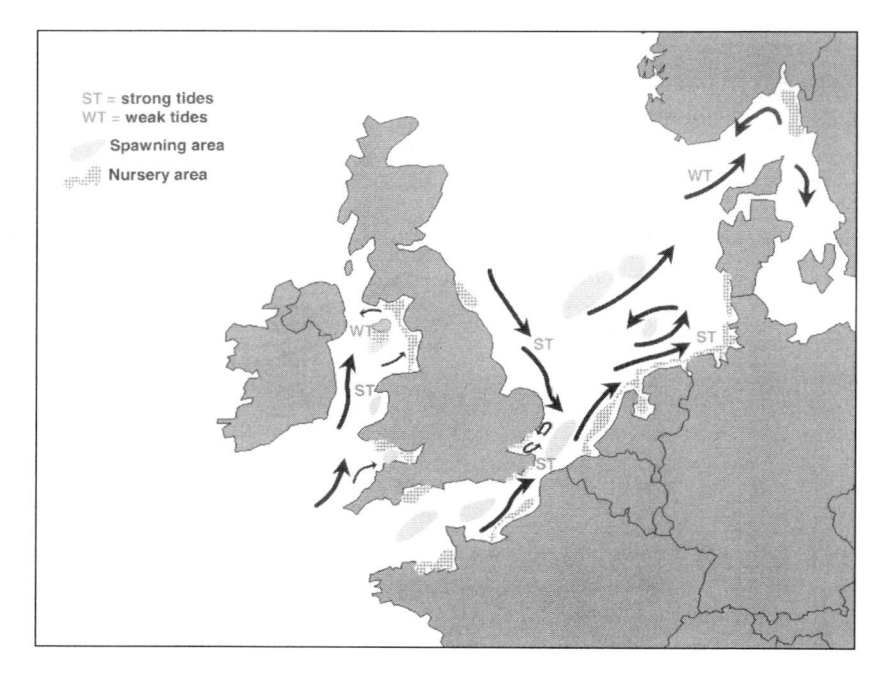

Fig. 5.7 Transport conditions for the plaice in three major spawning areas. In the Irish Sea near the Isle of Man, tidal and residual currents are weak and plaice spawn near their nursery area. In the English Channel and North Sea, residual currents are strong towards the northeast and tidal currents are strongly shoreward. Larvae drift with residual currents and may use tidal currents to arrive at coastal nurseries. In the Skagerrak-Kattegat area, tidal currents are weak and larvae are transported to the Swedish coastal region by wind-driven currents.

offshore and larvae ride wind-driven Ekman currents to reach shallow nurseries. Finally in the North Sea, larvae ride bottom currents into near coastal waters and perhaps use STST to reach nearshore and inland sea nurseries. Further research is needed to determine whether these adaptations are inherent characteristics of local populations or arise from behavioural flexibility and responses to different cues in the different areas.

5.4.4 Local adaptations

Although the above inter-species comparisons show differences in transport mechanisms, species within a common geographic region dominated by different types of currents may have prevailing transport characteristics. For example, in the North Sea where tidal currents are strong, numerous flatfish species use STST to reach inshore nurseries including plaice, flounder and common sole. It is recognised that generalisations about tidal currents over broad regions of the shelf are dangerous with regard to life history adaptations of local fish populations, as tidal currents can be strongly affected by local features, such as topography. In the California Current system, an upwelling system dominated by Ekman currents, many species either have adaptations to minimise transport offshore, such as spawning demersal eggs or viviparity (Parrish *et al*. 1981). They may spawn during periods of downwelling, or in areas characterised by onshore transport. In addition, some species have an unusually long planktonic period with a flexible age-at-settlement, favouring the eventual finding of suitable nursery habitat. In other areas dominated by specific and predictable current patterns, there may be a similar predilection for specific local adaptations. Given the future availability of data on many more species, it will be of interest to discover whether life history patterns are dominated by taxonomic similarities or whether different species in different areas have converged on similar patterns of local adaptation, a trend that the limited amount of available data tend to support.

According to Gibson (1999), with reference to plaice, 'the selection of settlement area is determined by the hydrographic relationship between the spawning ground and the nursery area. In areas where currents are only weakly directional, spawning grounds are situated close to the nursery ground. Where currents are strong, the choice of location of the spawning ground has presumably evolved so that eggs are released at an optimum direction and distance from the settlement areas. Stocks are thus "hydrographically contained" (Cushing 1990) within a limited area'. Likewise, according to Boehlert & Mundy (1988) 'behaviours associated with shoreward movement are likely related to distribution in the water column and have evolved to take advantage of mean current conditions in the species habitat'. The above studies, which range from comparisons across the flatfish order, to studies of congeneric species, to studies within a species, support the concept that there is local adaptation of transport characteristics in different oceanic systems to favour transport towards or retention in nearshore nurseries. Similar observations have been made for invertebrate species (e.g. LeFevre & Bourget 1992). As those authors summarised, different species have distinct vertical distributions in a given environment and a single species may have different distributions in different environments, affecting how they may be transported or retained.

5.5 Transport and population biology

Variations in transport of flatfish eggs and larvae have important consequences for the dynamics of local populations, the genetic structure of populations, metapopulation dynamics and the recovery of local populations after depletions.

5.5.1 Population genetics

In theory, species with short larval duration should show more genetic heterogeneity among subpopulations because there would be less gene flow through planktonic dispersal (Doherty *et al.* 1995). However, gene flow can be limited even for species with long pelagic stages as a consequence of larval retention features (Palumbi 1995). For example, geographical differentiation of populations of Dover sole along the continental slope of the northeastern Pacific Ocean is consistent with retention of larvae, despite extended pelagic periods, and is inconsistent with long-distance dispersal of adults (Stepien 1999). Witch also has a long larval stage, but shows significant stock structuring that may be related to larval retention (Fairbairn 1981; Table 5.1). Other species, such as Pacific halibut, have a long planktonic life, and little apparent evidence for subpopulation structure (Grant *et al.* 1984).

Gene flow should vary inversely with dispersal rates, but even species with potentially high dispersal can show genetic population structure that reflects both the historical demography and present dispersal patterns (Rocha-Olivares & Vetter 1999). Furthermore, present patterns of genetic structure may reflect highly pulsed dispersal patterns that may have occurred in past global climate changes or due to shifts in currents over the last 1–3 million years. Even though present currents may 'connect' them, the persistence of genetic differences indicates a lack of effective contemporary gene exchange (Benzie 1999).

The metapopulation concept may be useful for assessing gene flow, risk of extinction, and potential for recolonisation of depleted populations. Given that the major potential for population dispersal is in the planktonic stage, metapopulations for many flatfishes may be organised around larval drift patterns (Bailey 1997). In support of this concept is the apparent small degree of population structure of several flatfish species in the northeast Atlantic except among major basins like the Mediterranean and North Sea (Exadactylos *et al.* 1998; Hoarau *et al.* 2004). More structure may be found where retention is common (northwest Atlantic: e.g. witch, Atlantic cod). Cushing (1990) observed that the spawning and feeding grounds of different plaice stocks are located near different tidal streamlines, and that these pathways may prevent extensive mixing of populations. Exceptions to metapopulations forming around drift patterns may be where long-distance migrations occur, for example Pacific halibut.

5.5.2 Recruitment

Many factors have an influence on recruitment (see Chapter 6). However, certain controls may be more important at different latitudes, or among groups of species with similar life history traits (Miller *et al.* 1991). Because of their dependence on transport to nearshore nurseries, variability in this process may be of particular importance to flatfishes.

Numerous studies point out the importance of transport of eggs and larvae to the recruitment process of flatfishes (e.g. Boehlert & Mundy 1987; Nakata *et al.* 2000). Modelling results show that variations in circulation patterns in the North Sea might be a key factor in determining year-class strength of plaice (Van der Veer *et al.* 1998). This is further supported by empirical field studies of plaice recruitment (Van der Veer 1986; Van der Veer *et al.* 1990). Year-class strength of plaice depends on successful larval delivery, but variability can be dampened on the nursery grounds. Larval transport and sea temperature may both be impacted similarly by wind conditions, and may also interact by the effect of temperature on development rate and therefore the duration of the drift period (Van der Veer & Witte 1999).

A comparison of plaice recruitment among stocks has some intriguing implications. In the Kattegat/Skaggerak where tidal currents are weak, wind-driven currents dominate larval transport (Neilson *et al.* 1998). Pihl (1990) found that variations in onshore winds are related to the abundance of plaice in their nearshore nursery and Pihl *et al.* (2000) considered that concentration of larvae in the water column and exchange of water in the nursery ground determine the rate of larval delivery, and may explain some patterns of newly settled plaice larvae along the coastline of the Swedish Skagerrak archipelago.

In the North Sea, where the larval transport is long distance, larval density near the nursery is correlated with the number of settling age-0s, suggesting that transport is a key factor in the recruitment process. But in the Irish Sea, where the nursery is in close proximity to spawning, the settled numbers are correlated with egg abundance, suggesting that variability in larval drift is not a critical factor. In fact, year-class strength may be determined in the juvenile nursery (Nash & Geffen 2000).

Rijnsdorp *et al.* (1992) suggested that for common sole, the period in which recruitment level is determined is the pelagic or early juvenile stage based on correlations between age-0 abundance and recruitment levels. The factors determining recruitment vary over a scale of 100–200 km, and similarities in recruitment patterns of 0-group common sole are restricted to nursery areas which have a similar direction of coastline. The authors suggested that hydrographic conditions involved in transport of larvae to coastal nurseries could be important. The coefficient of variation in recruitment was highest in the North Sea (127%), lower in the Irish Sea (97%), and significantly less in the eastern and western English Channel (34% and 55%, respectively). Recruitment variability is interesting to compare with transport characteristics; in the North Sea, post-larval common sole are carried to nursery ground by tides (Berghahn 1984, cited in Rijnsdorp 1992), whereas tidal currents in the Irish Sea are relatively weak and common sole may spawn near nurseries to minimise transport distance.

Recruitment of Pacific halibut has been correlated with transport and cross-shelf transport during its 6-month pelagic larval phase. For example, Parker (1989) found that recruitment is density-dependent but is also influenced by strong winter winds, which favour production of strong year-classes. It was suggested that alongshore and cross-shelf winds generated transport conditions favourable for survival. Parker (1989) and Bailey & Picquelle (2002) also suggested bathymetric steering of currents carrying offshore larvae into coastal nurseries. Accelerated coastal currents could entrain offshore waters up troughs which could be a key avenue for directing larvae in the Gulf of Alaska towards inshore nurseries.

In summary, losses may be substantial when eggs are spawned or larvae are transported beyond typical habitat boundaries. The geographical locations and the times of fish spawning represent evolutionary adaptations to the climatological mean water circulation pattern (Bakun 1985). The deviations from the transport pathway under the mean water circulation may be a cause of recruitment variations. Various physical mechanisms have been proposed to explain the variations in transport pathways as a causal factor for recruitment success/failure. It should also be recognised that the relative importance of individual factors may change from year to year and may also vary over the range of a species because species life history traits vary over their ranges (Miller et al. 1991). In the case of many flatfishes and estuarine-dependent species whose juvenile nursery habitats are spatially distinct from spawning locations, physical processes affecting the transport in the pelagic stages are of great importance to the formation of year-class strength.

References

Amara, R., Poulard, J., Lagardere, F. & Desaunay, Y. (1998) Comparison between the life cycles of two Soleidae, the common sole, *Solea solea*, and the thickback sole, *Microchirus variegatus*, in the Bay of Biscay (France). *Environmental Biology of Fishes*, **53**, 193–209.

Bailey, K.M. (1994) Predation on juvenile flatfish and recruitment variability. *Netherlands Journal of Sea Research*, **32**, 175–189.

Bailey, K.M. (1997) Structural dynamics and ecology of flatfish populations. *Journal of Sea Research*, **37**, 269–280.

Bailey, K.M. (2000) Shifting control of recruitment of walleye pollock *Theragra chalcogramma* after a major climatic and ecosystem change. *Marine Ecology Progress Series*, **198**, 215–224.

Bailey, K.M. & Picquelle, S. (2002) Larval distribution of offshore spawning flatfish in the Gulf of Alaska: potential transport pathways and enhanced onshore transport during ENSO events. *Marine Ecology Progress Series*, **236**, 205–217.

Bailey, K.M., Stabeno, P.J. & Powers, D.A. (1997) The role of larval retention and transport features in mortality and potential gene flow of walleye pollock. *Journal of Fish Biology*, **51**, 135–154.

Bakun, A. (1985) Comparative studies and the recruitment problems: searching for generalizations. *California Cooperative Oceanic Fisheries Investigations Reports*, **26**, 30–40.

Bartsch, J., Brander, K., Heath, M., Munk, P., Richardson, K. & Svendsen, E. (1989) Modelling the advection of herring larvae in the North Sea. *Nature*, **340**, 632–636.

Benzie, J.A. (1999) Genetic structure of coral reef organisms: ghosts of dispersal past. *American Zoologist*, **39**, 131–145.

Berghahn, R. (1984) Untersuchungen an Plattfischen und Nordseegarnelen (*Crangon crangon*) im Eulitoral des Wattenmeeres nach dem Ubergang zum Bodenleben. *Helgoländer Meeresuntersuchungen*, **43**, 461–477.

Bergman, M.J.N., Van der Veer, H.W., Van der Stam, A. & Zuidema, D. (1989) Transport mechanisms of larval plaice (*Pleuronectes platessa* L.) from the coastal zone into the Wadden Sea nursery area. *Rapports et Procès-Verbaux des Réunions Conseil International pour l'Exploration de la Mer*, **191**, 43–49.

Beverton, R.J.H. & Iles, T.C. (1992) Mortality rates of 0-group plaice (*Pleuronectes platessa* L.), dab (*Limanda limanda* L) and turbot (*Scophthalmus maximus* L.) in European waters. II. Comparison of mortality rates and construction of life table for 0-group plaice. *Netherlands Journal of Sea Research*, **29**, 49–59.

Blaxter, J.H.S. (1986) Development of sense organs and behaviour of teleost larvae with special reference to feeding and predator avoidance. *Transactions of the American Fisheries Society*, **115**, 98–114.

Boehlert, G.W. & Mundy, B.C. (1987) Recruitment dynamics of metamorphosing English sole, *Parophrys vetulus*, to Yaquina Bay, Oregon. *Estuarine, Coastal and Shelf Science*, **25**, 261–281.

Boehlert, G.W. & Mundy, B.C. (1988) Roles of behavioral and physical factors in larval and juvenile fish recruitment to estuarine nursery areas. *American Fisheries Society Symposium,* **3**, 51–67.

Bolle, L.J., Dapper, R., Witte, J.IJ. & Van der Veer, H.W. (1994) Nursery grounds of dab (*Limanda limanda* L.) in the southern North Sea. *Netherlands Journal of Sea Research,* **32**, 299–307.

Bolle, L.J., Rijnsdorp, A.D., Clerck, R.D., Damm, U., Millner, R.S. & Van Beek, F.A. (1999) The spawning dynamics of sole *Solea solea* (L.) in the North Sea. In: *Book of Abstracts, Fourth International Symposium on Flatfish Ecology, Atlantic Beach, NC,* p. 6.

Burke, J.S., Ueno, M., Tanaka, Y., *et al.* (1998) The influence of environmental factors on early life history patterns of flounders. *Journal of Sea Research,* **4**, 19–32.

Campos, W.L., Kloppmann, M. & von Westernhagen, H. (1994) Inferences from the horizontal distribution of dab *Limanda limanda* (L.) and flounder *Platichthys flesus* (L.) larvae in the southeastern North Sea. *Journal of Sea Research,* **32**, 277–286.

Chant, R.J., Curran, M.C., Able, K.W. & Glenn, S.M. (2000) Delivery of winter flounder (*Pseudopleuronectes americanus*) larvae to settlement habitats in coves near tidal inlets. *Estuarine, Coastal and Shelf Science,* **51**, 529–541.

Connell, J.H. (1985) The consequences of variation in initial settlement vs. post-settlement mortality in rocky intertidal communities. *Journal of Experimental Marine Biology and Ecology,* **93**, 11–45.

Cushing, D.H. (1972) The production cycle and the numbers of marine fish. *Symposia of the Zoological Society of London,* **29**, 213–232.

Cushing, D.H. (1990) Hydrographic containment of a spawning group of plaice in the Southern Bight of the North Sea. *Marine Ecology Progress Series,* **58**, 287–297.

Doherty, P.J., Planes, S. & Mather, P. (1995) Gene flow and larval duration in seven species of fish from the Great Barrier Reef. *Ecology,* **76**, 2373–2391.

Ellertsen, B., Fossum, P., Solemdal, P., Sundby, S. & Tilseth, S. (1990) Environmental influence on recruitment and biomass yields in the Norwegian Sea ecosystem. In: *Large Marine Ecosystems: Patterns, Processes and Yields* (eds K. Sherman, L.M. Alexander & B.D. Gold). pp. 19–35. American Association for the Advancement of Science.

Epifanio, C.E. (1988) Transport of invertebrate larvae between estuaries and the continental shelf. *American Fisheries Society Symposium,* **3**, 104–114.

Exadactylos, A., Geffen, A.J. & Thorpe, J.P. (1998) Population structure of the Dover sole, *Solea solea* L., in a background of high gene flow. *Journal of Sea Research,* **40**, 117–130.

Fairbairn, D.J. (1981) Which witch is which? A study of the stock structure of witch flounder (*Glyptocephalus cynoglossus*) in the Newfoundland region. *Canadian Journal of Fisheries and Aquatic Sciences,* **38**, 782–794.

Fisher, R., Bellwood, D.R. & Suresh, D.J. (2000) Development of swimming abilities in reef fish larvae. *Marine Ecology Progress Series,* **202**, 163–173.

Forward, R.B. & Tankersley, R.A. (2001) Selective tidal-stream transport of marine animals. *Oceanography and Marine Biology: An Annual Review,* **39**, 305–353.

Fox, C.J., O'Brien, C.M., Dickey-Collas, M. & Nash, R.D.M. (2000) Patterns in the spawning of cod (*Gadus morhua* L.), sole (*Solea solea* L.) and plaice (*Pleuronectes platessa* L.) in the Irish Sea as determined by generalized additive modelling. *Fisheries Oceanography,* **9**, 33–49.

Fukuhara, O. (1988) Morphological and functional development of larval and juvenile *Limanda yokohamae* (Pisces: Pleuronectidae) reared in the laboratory. *Marine Biology,* **99**, 271–281.

Gibson, R.N. (1997) Behaviour and the distribution of flatfishes. *Journal of Sea Research,* **37**, 241–256.

Gibson, R.N. (1999) The ecology of the early life stages of the plaice, *Pleuronectes platessa* L.: a review. *Bulletin of the Tohoku National Fisheries Research Institute,* **62**, 17–50.

Grant, W.S., Teel, D.J. & Kobayashi, T. (1984) Biochemical population genetics of Pacific halibut (*Hippoglossus stenolepsis*) and comparison with Atlantic halibut (*H. hippoglossus*). *Canadian Journal of Fisheries and Aquatic Sciences,* **41**, 1083–1088.

Grioche, A., Harlay, X., Koubbi, P. & Fraga Lago, L. (2000) Vertical migrations of fish larvae: Eulerian and Lagrangian observations in the Eastern English Channel. *Journal of Plankton Research,* **22**, 1813–1828.

Harmelin-Vivien, M.L. (1989) Implications of feeding specialization on the recruitment processes and community structure of butterflyfishes. *Environmental Biology of Fishes,* **25**, 101–110.

Haug, T., Kjorsvik, E. & Solemdal, P. (1986) Influence of some physical and biological factors on the density and vertical distribution of Atlantic halibut *Hippoglossus hippoglossus* eggs. *Marine Ecology Progress Series,* **33**, 207–216.

Hayman, R.A. & Tyler, A.V. (1980) Environment and cohort strength of Dover sole and English sole. *Transactions of the American Fisheries Society,* **109**, 54–70.

Heath, M.R. (1992) Field investigations of the early life stages of marine fish. *Advances in Marine Biology,* **28**, 1–174.

Heath, M.R. (1996) The consequences of spawning time and dispersal patterns of larvae for spatial and temporal variability in survival to recruitment. In: *Survival Strategies in Early Life Stages of Marine Resources* (eds Y. Watanabe, Y. Yamashita & Y. Oozeki). pp. 175–207. A.A.Balkema, Rotterdam.

Hoarau, G., Piquet, M.T., Van der Veer, H.W., Rijnsdorp, A.D., Stam, W.T. & Olsen, J.L. (2004) Population structure of plaice (*Pleuronectes platessa* L.) in northern Europe: a comparison of resolving power between microsatellites and mitochondrial DNA data. *Journal of Sea Research,* **51**, 183–190.

Horwood, J., Cushing, D. & Wyatt, T. (2000) Planktonic determination of variability and sustainability of fisheries. *Journal of Plankton Research,* **22**, 1419–1422.

Iles, T.D. & Sinclair, M. (1982) Atlantic herring: stock discreteness and abundance. *Science,* **215**, 627–633.

Jager, Z. (1998) Accumulation of flounder larvae (*Platichthys flesus* L.) in the Dollard (Ems estuary, Wadden Sea). *Journal of Sea Research,* **40**, 43–58.

Jager, Z. (1999) Selective tidal stream transport of flounder larvae (*Platichthys flesus* L.) in the Dollard (Ems estuary). *Estuarine, Coastal and Shelf Science,* **49**, 347–362.

Koutsikopoulos, C., Fortier, L. & Gagne, J.A. (1991) Cross-shelf dispersion of Dover sole (*Solea solea*) eggs and larvae in Biscay Bay and recruitment to inshore nurseries. *Journal of Plankton Research,* **13**, 923–945.

Le Fevre, J. & Bourget, E. (1992) Hydrodynamics and behaviour: transport processes in marine invertebrate larvae. *Trends in Ecology and Evolution,* **7**, 288–289.

Leis, J.M. & Carson-Ewart, B.M. (1997) In situ swimming speeds of the late pelagic larvae of some Indo-Pacific coral-reef fishes. *Marine Ecology Progress Series,* **159**, 165–174.

Leis, J.M. & Carson-Ewart, B.M. (1999) In situ swimming and settlement behaviour of larvae of an Indo-Pacific coral-reef fish, the coral trout *Plectropomus leopardus* (Pisces: Serranidae). *Marine Biology,* **134**, 51–64.

Lough, R.G., Smith, W.G., Werner, F.E., *et al.* (1994) Influence of wind-driven advection on interannual variability in cod eggs and larval distributions on Georges Bank: 1982 vs 1985. *ICES Marine Science Symposium,* **198**, 356–378.

Milicich, M.J., Meekan, M.G. & Doherty, P.J. (1992) Larval supply: a good predictor of recruitment of three species of reef fish (Pomacentridae). *Marine Ecology Progress Series,* **86**, 153–166.

Miller, J.M., Burke, J.S. & Fitzhugh, G.R. (1991) Early life history patterns of Atlantic North American flatfish: likely (and unlikely) factors controlling recruitment. *Netherlands Journal of Sea Research,* **27**, 261–275.

Minami, T. & Tanaka, M. (1992) Life history cycles in flatfish from the northwestern Pacific, with particular reference to their early life histories. *Netherlands Journal of Sea Research,* **29**, 35–48.

Moser, H.G. (1996) Pleuronectiformes. *California Cooperative Oceanic Fisheries Investigations, Atlas,* **33**, 1323–1324.

Munk, P., Larsson, P.O., Danielssen, D.S. & Moksness, E. (1999) Variability in frontal zone formation and distribution of gadoid fish larvae at the shelf break in the northeastern North Sea. *Marine Ecology Progress Series,* **177**, 221–233.

Nakata, H. (1996) Coastal fronts and eddies: their implications for egg and larval transport and survival processes. In: *Survival Strategies in Early Life Stages of Marine Resources* (eds Y. Watanabe, Y. Yamashita & Y. Oozeki). pp. 227–241. A.A.Balkema, Rotterdam.

Nakata, H., Yamashita, Y. & Yamada, H. (1999a) Wind-induced advection of stone flounder (*Kareius bicoloratus*) eggs and larvae in Sendai Bay, northern Japan, and its effect on the larval settlement in coastal nurseries. In: *Book of Abstracts, Fourth International Symposium on Flatfish Ecology 1999, Atlantic Beach, NC*, p. 49.

Nakata, H., Suenaga, Y. & Fujihara, M. (1999b) Wind-induced drift of brown sole eggs and larvae in the shelf region near Sado Strait in relation to the recruitment mechanism. *Bulletin of Tohoku National Fisheries Research Institute,* **62**, 51–60.

Nakata, H., Fujihara, M., Suenaga, Y., Nagasawa, T. & Fujii, T. (2000) Effect of wind blows on the transport and settlement of brown sole (*Pleuronectes herzensteini*) larvae in a shelf region of the Sea of Japan: numerical experiment with an Euler-Lagrangian model. *Journal of Sea Research,* **44**, 91–100.

Nash, R.D.M. (1998) Exploring the population dynamics of Irish Sea plaice (*Pleuronectes platessa* L.) through the use of Paulik diagrams. *Journal of Sea Research,* **40**, 1–18.

Nash, R.D.M. & Geffen, A.J. (1999) Variability in stage I egg production and settlement of plaice *Pleuronectes platessa* on the west side of the Isle of Man, Irish Sea. *Marine Ecology Progress Series,* **189**, 241–250.

Nash, R.D.M. & Geffen, A.J. (2000) The influence of nursery ground processes in the determination of year-class strength in juvenile plaice *Pleuronectes platessa* L. in Port Erin Bay, Irish Sea. *Journal of Sea Research,* **44**, 101–110.

Nielsen, E., Bagge, O. & MacKenzie, B.R. (1998) Wind-induced transport of plaice (*Pleuronectes platessa*) early life-history stages in the Skagerrak-Kattegat. *Journal of Sea Research,* **39**, 11–28.

Norcross, B.L. & Shaw, R.F. (1984) Oceanic and estuarine transport of fish eggs and larvae: a review. *Transactions of the American Fisheries Society,* **113**, 153–165.

Norcross, B.L., Blanchard, A. & Holladay, B.A. (1999) Comparison of models for defining nearshore flatfish nursery areas in Alaskan waters. *Fisheries Oceanography,* **8**, 50–67.

Orr, J.W. & Matarese, A.C. (2000) Revision of the genus *Lepidopsetta* Gill, 1862 (Teleostei: Pleuronectidae) based on larval and adult morphology, with a description of a new species from the North Pacific Ocean and Bering Sea. *Fishery Bulletin,* **98**, 539–582.

Palumbi, S.R. (1995) Using genetics as an indirect estimator of larval dispersal. In: *Marine Invertebrate Larvae* (ed. L. McEdward). pp. 369–387. CRC Press, New York.

Parker, K.S. (1989) Influence of oceanographic and meteorological processes on the recruitment of Pacific halibut, *Hippoglossus stenolepis*, in the Gulf of Alaska. In: Effects of ocean variability on recruitment and an evaluation of parameters used in stock assessment models (eds R.J. Beamish & G.A. McFarlane). *Canadian Special Publication in Fisheries and Aquatic Science* No. 108, 221–237.

Parrish, R.H., Nelson, C.S. & Bakun, A. (1981) Transport mechanisms and reproductive success of fishes in the California Current. *Biological Oceanography,* **1**, 175–203.

Pearcy, W.G. (1962) Distribution and origin of demersal eggs within the order Pleuronectiformes. *Journal du Conseil International Exploration de la Mer,* **27**, 233–235.

Pearcy, W.G., Hosie, M.J. & Richardson, S.L. (1977) Distribution and duration of pelagic life of larvae of Dover sole, *Microstomus pacificus*; rex sole, *Glyptocephalus zachirus*; and petrale sole, *Eopsetta jordani*, in waters off Oregon. *Fishery Bulletin,* **75**, 173–183.

Pihl, L. (1990) Year-class strength regulation in plaice (*Pleuronectes platessa* L.) on the Swedish west coast. *Hydrobiologia,* **195**, 79–88.

Pihl, L., Modin, J. & Wennhage, H. (2000) Spatial distribution patterns of newly settled plaice (*Pleuronectes platessa* L.) along the Swedish Skagerrak archipelago. *Journal of Sea Research,* **44**, 65–80.

Rijnsdorp, A.D. (1994) Population-regulating processes during the adult phase in flatfish. *Netherlands Journal of Sea Research,* **32**, 207–223.

Rijnsdorp, A.D., Van Stralen, M. & Van der Veer, H.W. (1985) Selective tidal transport of North Sea plaice larvae *Pleuronectes platessa* in coastal nursery areas. *Transactions of the American Fisheries Society,* **114**, 461–470.

Rijnsdorp, A.D, Van Beek, F.A., Flatman, S., *et al.* (1992) Recruitment of sole stocks, *Solea solea* (L.), in the Northeast Atlantic. *Netherlands Journal of Sea Research,* **29**, 173–192.

Robertson, D.A. (1981) Possible functions of surface structure and size in some planktonic eggs of marine fishes. *New Zealand Journal of Marine and Freshwater Research,* **5**, 147–153.

Rocha-Olivares, A. & Vetter, R.D. (1999) Effects of oceanographic circulation on the gene flow, genetic structure, and phylogeography of the rosethorn rockfish (*Sebastes helvomaculatus*). *Canadian Journal of Fisheries and Aquatic Sciences,* **56**, 803–813.

Sinclair, M. (1988) *Marine Populations: An Essay on Population Regulation and Speciation.* Washington Sea Grant Press, Seattle, WA.

Stepien, C.A. (1999) Phylogeographical structure of the Dover sole *Microstomus pacificus*: the larval retention hypothesis and genetic divergence along the deep continental slope of the northeastern Pacific Ocean. *Molecular Ecology,* **8**, 923–939.

Strathmann, R. (1974) The spread of sibling larvae of sedentary marine invertebrates. *American Naturalist,* **108**, 29–44.

Sumida, B.Y., Ahlstrom, E.H. & Moser, H.G. (1979) Early development of seven flatfishes of the eastern north Pacific with heavily pigmented larvae (Pisces, Pleuronectiformes). *Fishery Bulletin,* **77**, 105–145.

Taggart, C.T., Thompson, K.R., Maillet, G.L., Lochmann, S.E. & Griffin, D.A. (1996) Abundance distribution of larval cod (*Gadus morhua*) and zooplankton in a gyre-like water mass on the Scotian Shelf. In: *Survival Strategies in Early Life Stages of Marine Resources* (eds Y. Watanabe, Y. Yamashita & Y. Oozeki). pp. 155–173. A.A. Balkema, Rotterdam.

Tanaka, M., Goto, T., Tomiyama, M., Sudo, H. & Azuma, M. (1989) Lunar-phased immigration and settlement of metamorphosing Japanese flounder larvae into the nearshore nursery ground. *Rapports et Procès-Verbaux des Réunions Conseil International pour l'Exploration de la Mer,* **191**, 303–310.

Toole, C.L., Markle, D.F. & Dorion, D. (1997) Settlement timing, distribution, and abundance of Dover sole (*Microstomus pacificus*) on an outer continental shelf nursery area. *Canadian Journal of Fisheries and Aquatic Sciences,* **54**, 531–542.

Tsuruta, Y. (1978) Field observations on the immigration of larval stone flounder into the nursery ground. *Tohoku Journal of Agricultural Research,* **29**, 136–145.

Van der Veer, H.W. (1986) Immigration, settlement and density-dependent mortality of a larval and early post-larval 0-group plaice (*Pleuronectes platessa*) population in the western Wadden Sea. *Marine Ecological Progress Series,* **29**, 223–236.

Van der Veer, H.W. & Witte, J.IJ. (1999) Year-class strength of plaice *Pleuronectes platessa* in the Southern Bight of the North Sea: a validation and analysis of the inverse relationship with winter seawater temperature. *Marine Ecology Progress Series,* **184**, 245–257.

Van der Veer, H.W., Pihl, L. & Bergman, M.J.N. (1990) Recruitment mechanisms in North Sea plaice *Pleuronectes platessa*. *Marine Ecology Progress Series,* **64**, 1–12.

Van der Veer, H.W., Ruardij, P., Van den Berg, A.J. & Ridderinkhof, H. (1998) Impact of interannual variability in hydrodynamic circulation on egg and larval transport of plaice *Pleuronectes platessa* L. in the southern North Sea. *Journal of Sea Research,* **39**, 29–40.

Walsh, S.J. (1992) Factors influencing distribution of juvenile yellowtail flounder (*Limanda ferruginea*) on the Grand Bank of Newfoundland. *Netherlands Journal of Sea Research,* **29**, 193–203.

Walsh, S.J. (1994) Life history traits and spawning characteristics in populations of long rough dab (American plaice) *Hippoglossoides platessoides* (Fabricius) in the North Atlantic. *Netherlands Journal of Sea Research,* **32**, 241–254.

Weinstein, M.P., Weiss, S.L., Hodson, R.G. & Gerry, L.R. (1980) Retention of three taxa of post larval fishes in an intensively flushed tidal estuary, Cape Fear River, North Carolina. *Fishery Bulletin,* **78**, 419–436.

Werner, F.E., Page, F.H., Lynch, D.R., *et al.* (1993) Influences of mean advection and simple behavior on the distribution of cod and haddock early life stages on Georges Bank. *Fisheries Oceanography,* **2**, 43–64.

Werner, F.E., Quinlan, J.A., Blanton, B.O. & Luettich, R.A., Jr. (1997) The role of hydrodynamics in explaining variability in fish populations. *Journal of Sea Research,* **37**, 195–212.

Yamashita, Y., Tsuruta, Y. & Yamada, H. (1996) Transport and settlement mechanisms of larval stone flounder, *Kareius bicoloratus*, into nursery grounds. *Fisheries Oceanography,* **5**, 194–204.

Chapter 6
Recruitment

Henk W. Van der Veer and William C. Leggett

6.1 Introduction

At the beginning of the last century Hjort (1914, 1926) advanced the hypothesis that year-class strength in marine fishes is controlled during a 'critical phase' in the early life history. The mechanisms involved were thought to be a combination of density-independent processes related to fluctuations in the physical environment and density-dependent processes caused by either predation or food competition. After almost a century, this picture has altered very little: year-class strength in marine fishes appears to be primarily determined by mortality processes operating during the pre-juvenile stage of the life history. This process appears to result from a combination of coarse control during the period of egg and/or larval drift, followed by a second interval of finer-scale regulation later in the early life history (for review see Leggett & DeBlois 1994).

While this general pattern of year-class regulation remains largely unaltered by a century of research, the intervening period has produced a large volume of excellent studies, primarily on recruitment processes in single species. A continuing debate reflected in all these studies is the extent to which the processes determining recruitment are species- and/or area-specific or are part of a more general pattern affecting more than one species or species group. Arguments in support of the presence of general patterns are founded on evidence of a link between the factors controlling recruitment and species- or group-specific early life history patterns in fishes (Roff 1982; Rothschild & DiNardo 1987). Further support for this view is provided by the fact that adjacent populations often show synchrony in year-class strength over spatial scales of hundreds of kilometres (Walsh 1994b; Myers *et al*. 1997; Fox *et al*. 2000). Additional evidence is provided by the observation that flatfishes, as a group, are characterised by a relatively low recruitment variability (Beverton 1995). This implies the existence of flatfish-specific life history characteristics that moderate recruitment variability.

Miller *et al*. (1991) were among the first to examine the early life history patterns specific to flatfishes in the context of factors controlling recruitment. Their study, which was restricted to a qualitative analysis of the juvenile (demersal) phase of a subset of North American species, led to a series of explicit predictions which have subsequently been shown to be inconsistent with current knowledge about latitudinal variability in recruitment in these species (Leggett & Frank 1997; Philippart *et al*. 1998). To date, a quantitative framework for analysing and interpreting the relationship between flatfishes' early life histories and factors regulating recruitment remains elusive. In this chapter the data and hypotheses relating to the generation and regulation of recruitment and recruitment variability in flatfishes are reviewed, and the ideas of Miller *et al.* (1991) extended to the pelagic life stage of flatfishes during which the coarse-grained determination of recruitment success appears to occur.

The starting point for the analysis is the assumption that recruitment variability will be caused either by inter-annual variability in population egg production, by inter-annual variability in survival of these eggs and the resulting larvae or by a combination of both (Cushing 1995; Rickman *et al.* 2000). The approach taken is to identify relationships between flatfish life history traits and recruitment. More traditional overviews of recruitment variation in flatfishes can be found in Iles & Beverton (2000), Van der Veer *et al.* (2000), in the proceedings of five Flatfish Symposia, published by the (*Netherlands*) *Journal of Sea Research* (see Preface for details) and in Chapter 5 of this volume.

From this perspective, three factors require analysis in the context of an understanding of recruitment processes in flatfishes: the distributional range of the species; the average level of recruitment achieved over time; and the annual variability in recruitment.

6.2 Range of distribution

A general requirement for the development of a stable population by any species is the ability to close the life cycle. In flatfishes, the pelagic dispersal phase appears to be the most critical period, because settling occurs in specified nursery areas and therefore the duration of the egg and larval stage must conform to the length of the period of transport required to ensure metamorphosis. Otherwise, larvae will metamorphose and settle in suboptimal habitats and, as a consequence, experience reduced survival. Because development rate is strongly influenced by temperature conditions during drift (see for example, plaice (*Pleuronectes platessa*): Harding *et al.* 1978; Van der Veer & Witte 1999), this delicate balance between development time and settlement location is subject to temporal and spatial variation. Consequently, the time window for settlement, determined by the size and location of the nursery zone relative to the spawning site and the environmental conditions experienced en route, becomes critical. It is also probable that this temporal 'window of opportunity' will become progressively constrained toward the poles, a product of the protraction of spawning and the temperature-induced reduction in development rate that occurs with latitude (Minami & Tanaka 1992). Should this reduction occur, the 'window of opportunity' could limit the species' ability to close the life cycle, and hence its distribution.

One solution to this latitudinal trend would be to increase the duration of drift (i.e. increase the distance between spawning and nursery sites) at higher latitudes in response to the declining average temperature during drift. However, the available evidence (Walsh 1994a) suggests that the period of larval drift does not increase meaningfully with latitude. A second solution, involving physiological adaptation to the decreased average temperatures experienced at higher latitudes, might involve counter-gradient growth compensation during the larval stage (Conover 1992). To date, there has been no systematic investigation of this possibility. The prolonged development times of eggs and larvae spawned at higher latitudes should also increase the potential for mixing between populations, perhaps destroying their integrity or negating population-specific adaptations. The increase in egg and larval development time with latitude could also serve to limit the geographic distribution of the species and to induce variability in recruitment. The data currently available are insufficient to assess this hypothesis.

Overall, there is little doubt that the most important factor governing the distributional range of the flatfishes is temperature. Dominating this regulating influence is the absolute temperature tolerance limit of each species, and the effect of temperature on the abundance of the key prey of one or more life stages. This influence may vary between species depending on which of the many life stages is the most sensitive to these direct or indirect temperature effects. The distribution of many flatfish species (e.g. Atlantic halibut (*Hippoglossus hippoglossus*), plaice), extends to, or close to, the poles (Wheeler 1978; see also Chapter 3). In these species, therefore, latitudinal distribution appears to be limited only by temperatures at the warm-water limits of their ranges. Adult body size in flatfishes declines toward the equator (Pauly 1994; Van der Veer *et al.* 2003), a response generally attributed to the decline in ocean productivity and the resulting limitation of food supply at lower latitudes (Gross *et al.* 1988). The warm-water limit of distribution of flatfishes may, therefore, be defined by the point at which energy uptake can no longer compensate for metabolic costs.

In species where this energetic constraint operates at the adult level, compensation may be achieved through migration which takes the adult to more favourable feeding environments during some portion of the year, a behaviour exhibited by plaice (Harden Jones 1968; Greer Walker *et al.* 1978). An alternative, and perhaps more likely, limitation on distribution may be imposed by the effects of energy limitations on egg size. Egg size, like adult body size, declines with decreasing latitude in cold water-adapted flatfishes (see Miller *et al.* 1991; and for a general overview see Chambers 1997). When energy available for egg production, or the effects of limitations in the energy available to embryos and early stage larvae become limiting, the life cycle can no longer be closed. This limitation could be imposed by a paucity of energy for egg production, or the interacting effects of low egg energy and higher metabolic rates during development imposed by the higher ambient temperatures experienced at lower latitudes. In these situations, too, behavioural adaptations such as a shift in the timing of spawning to earlier in the seasonal cycle of temperature could compensate, in part, but ultimately behavioural adaptations will also prove ineffective. The data currently available on patterns in the timing of spawning at lower latitudes are inadequate to assess the possibility of this trend.

Temperate zone flatfish species appear to experience both warm and cold water limits to their distribution. The limiting processes likely to be operating at low latitudes are described above. Temperate species living near the northern limit of their distribution appear to compensate in at least two ways: (1) through an increase in egg size (developmental energy reserve) with latitude which supports a higher development rate and shorter hatching times at higher latitudes (Kooijman 2000), and (2) a corresponding shift in the period of spawning to later in the season when temperatures are approaching their seasonal peak. Both patterns have been reported (see Minami & Tanaka 1992). In these species one limit to their distribution may be the point at which increases in egg size reach a maximum (possibly constrained by energy availability to the adult, or by the trade-off between egg size and number), thereby limiting the capacity to compensate for the effects of further decreases in temperature on development times.

Tropical flatfish species, in contrast, appear to be limited only by cold waters, these limits occurring at both the southern and northern limits of their ranges. The limiting factors in their distribution are likely, therefore, to mirror those experienced by temperate flatfishes occupying habitats near the northern limit of their range. However, for tropical flatfishes the limits to their distribution are also likely to be influenced by the higher average temperatures and the relative paucity of food in the habitats they occupy, both of which would tend to limit adult body size and egg size, thereby limiting their scope for adaptive responses.

It follows from the above that, for tropical species, egg size (which is typically small) will remain relatively constant over the protracted spawning period that is typical at those latitudes. Furthermore, because of their small egg, the high temperatures they experience, and the corresponding short development period they exhibit, spawning locations should be closer to the nursery areas. Walsh (1994a) reported a strong relationship between latitude, egg size, the season of spawning and the duration of the transport phase towards the nursery areas – all of which are consistent with the above expectations.

6.3 Average recruitment levels

In the standard analysis of recruitment processes, as applied to most marine fish populations including the flatfishes, total egg production at spawning (total reproductive effort) is considered to be directly proportional to the total biomass of the parent stock. While there are good theoretical reasons to expect a decoupling at large stock sizes (Beverton & Holt 1957), experience has shown the biomass of the parent stock to be largely independent of recruitment over a wide range of parent stock sizes (Rothschild 1988; Hilborn & Walters 1992). This implies that either (1) density-dependent factors sufficient in magnitude to dampen and possibly even offset any relationship between spawning stock biomass or total egg production operate during the egg and larval stages or (2) the assumption of a causal relationship between spawning stock biomass, total egg production and recruitment is flawed.

While density-dependent forces clearly operate at the juvenile stage (Leggett 1977; Van der Veer 1986), and these appear to be particularly important in flatfishes (Beverton & Iles 1992a, b), evidence in support of the existence of density-dependent regulation during the egg and larval stages is virtually non-existent, notwithstanding the frequent assumption of its importance (Iles 1994; Iles & Beverton 1998). The existence of density-dependent mortality during the egg stage is highly unlikely because developing embryos rely exclusively on stored energy reserves. Density-dependent predation is a possibility, but is unlikely to be a significant factor. Densities during the egg stage are relatively low even in the most concentrated egg patches (typically <10 eggs m^{-3}; Harding *et al.* 1978; Van der Land 1991; Cameron *et al.* 1992). However, predation on eggs might be selective, as in the case of predation by sprat and herring on plaice eggs whereby selection for later developmental stages was observed (Ellis & Nash 1997).

The potential for density-dependent mortality factors to act during the pelagic larval stage at levels sufficient to obscure an otherwise robust relationship between stock biomass (or total egg production) and recruitment, while real, is also remote. Larval densities during the pelagic life phase are known to be dramatically lower than those of eggs (<1 individual m^{-2}; Van der Veer 1985; Hovenkamp 1991; Modin & Pihl 1994; Van der Veer *et al.* 2000). Moreover, while starvation and the impact of food limitation on the susceptibility of larvae to predation may influence survival (Hovenkamp 1991; Leggett & DeBlois 1994), there is no evidence that this effect has a density-dependent basis (Van der Veer *et al.* 2000).

In contrast, the transition from a pelagic larval existence to the sessile benthic existence of juveniles that is characteristic of flatfishes results in a concentration of individuals. In plaice, demersal densities are a factor of 10–100 higher that those in the plankton (Van der Veer 1985). The high juvenile densities of >1 individual m^{-2} that result on the nursery areas (Van der Veer & Witte 1999) has been shown to produce both density-dependent growth and density-dependent mortality in coastal species (Van der Veer & Bergman 1987; Rijnsdorp & Van Leeuwen 1992; Modin & Pihl 1994; Nash 1998; Van der Veer et al. 2000). Intense density-dependent mortality at this stage could create a 'bottleneck' sufficiently intense to obscure any underlying relationship between spawner biomass and recruitment. A similar density-dependent bottleneck might also operate in offshore, deeper water nurseries. However, it is unclear (but considered doubtful) whether juvenile densities are sufficiently high in these habitats to produce a similarly strong effect (Rogers 1994).

Marshall et al. (1999) have shown that, for Atlantic cod (*Gadus morhua*) in the Barents Sea, total spawner biomass is not a viable surrogate for total egg production or total reproductive effort. For this population complex, at least, the underlying assumption of a relationship between spawner biomass and recruitment is thus flawed. Marshall et al. (1999) found a strong positive correlation between recruitment and both liver weights and total liver lipid energy. Simulation studies have demonstrated that this variation in lipid reserves is consistent with annual variation in the abundance of capelin (*Mallotus villosus*), a key prey of cod. The resulting implication is that the nutritional state of the spawning stock is more important than the numbers of spawners in determining year-class strength (level of recruitment). This finding of maternal effects (Solemdal 1997) appears to hold for other species. For example, Marshall et al. (1999) reported a positive relationship between recruitment in South African anchovy (*Engraulis encrasicolus*) and the quantity of oil, expressed as a percentage of meal production, in the commercial anchovy catch. A similar relationship may apply in flatfishes. At the moment there are no data available with which to test this possibility.

For a large number of marine fish species the correlation between the abundance of immatures at any life stage and recruitment is well established by the early juvenile period (e.g. Leggett & DeBlois 1994). Given this reality, the work of Marshall et al. (1999) suggests that the abundance and quality of reproductive products, and of the resulting larvae, may be more important than the spawner biomass in determining recruitment. Indeed, for Barents Sea cod, Marshall et al. (1999) demonstrated that spawner biomass was not significantly different at the extremes of the capelin (food) biomass, but that recruitment was independent of the former and varied dramatically with the latter, suggesting that the long-term average recruitment experienced by a stock or population will be determined by the carrying capacity of the system it occupies. Furthermore, variance about the long-term average will reflect the impact of annual, or longer-term, variations about that long-term mean carrying capacity on the abundance and the condition of the reproductive stock.

The term 'carrying capacity' is used to indicate the general productive capacity of the ecosystem occupied. In this sense it equates to the parameter K in the logistic growth equation:

$dN/dt = r * N * (1 - N/K)$

where N is the actual population size, r is the instantaneous population growth rate and K represents the maximum sustainable population size. Both specify the asymptotic population biomass that can be supported by a given ecosystem in its equilibrium state (MacCall 1990; Kashiwai 1995). From a theoretical perspective, the carrying capacity of a region (e.g. nursery area, larval retention zone) is reached when the *per capita* population growth rate equates to zero (i.e. at which mortality-induced biomass losses are in equilibrium with increases in biomass due to growth). For juvenile flatfishes on the nursery areas, mortality is determined primarily by two processes: (1) direct density-dependent mortality (starvation-induced), and (2) indirect density-dependent mortality resulting from selective predation operating differentially against weaker (nutritionally challenged) individuals in the population (Van der Veer *et al*. 1994b, 1997).

The strongest evidence of density-dependent regulation of recruitment in temperate zone flatfishes, both coastal (plaice and common sole (*Solea solea*)) and offshore (dab (*Limanda limanda*), scaldfish (*Arnoglossus laterna*) and solenette (*Buglossidium luteum*)) is the strong and direct relationship between the size of the nursery area and the output of juveniles (level of recruitment) (Rijnsdorp *et al*. 1992; Gibson 1994; Van der Veer *et al*. 2000). The theory underlying the operation of this nursery size constraint is given by Beverton (1995). It is worth noting that the nursery-size hypothesis applies at the level of individual species. When two or more species share a nursery area (e.g. plaice and common sole) the total biomass of the two can, and often does, exceed the biomass that would be supported if only one occupied the nursery. This observation implies that the carrying capacity of the system is distinct from the carrying capacity of the system for a particular species (see MacCall 1990). Frank & Shackell (2001) have recently demonstrated that carrying capacity in marine systems (in their case expressed as the number or density of species on a given bank area) is related to ecosystem area. Recruitment in flatfish stocks has also been found to be proportional to nursery zone area (Rijnsdorp *et al*. 1992; Van der Veer *et al*. 2000).

To conclude, long-term average recruitment level achieved by flatfishes in any given year appears to be governed by two distinct processes: (1) the effect of environment (food availability to the adults) on adult condition at the time of reproduction (which in turn regulates egg production and (possibly) directly influences the survival of the resulting eggs and larvae) and (2) the forces of density-dependent mortality operating on juveniles on the nursery sites. The degree to which the former dominates recruitment processes in species with pelagic juveniles (for which density-dependence may be less severe or even non-existent) is presently unknown. For flatfishes, the decrease in system productivity toward the equator is likely to be expressed in both processes. This decrease is consistent with Pauly's (1994) hypothesis regarding the lower overall production of benthic fish species in tropical waters, with the observed decline in carrying capacity of Atlantic cod with increasing temperature (Myers *et al*. 2001) and with the lower overall densities of flatfishes that is characteristic for tropical nurseries (Van der Veer *et al*. 1994a, 1997).

6.4 Recruitment variability

Marine fish populations are generally highly variable in numbers through time (see, for example, Rothschild 1988), a consequence of variability in recruitment. It is worth recalling, however, Heath's (1992) admonition that this focus on temporal variability in recruitment and abundance masks the reality of the underlying stability of marine populations. It is now well documented that even slight variations in daily mortality rates operating over the egg and larval stages are capable of generating variations in recruitment orders of magnitude greater than those actually observed (Heath 1992). In fact, recruitment to most fish populations is remarkably stable when viewed from this perspective.

In flatfishes, inter-annual variability in recruitment is typically smaller than that experienced by other teleosts (see, for example, May 1984; Iles & Beverton 2000). This low variability implies a dynamic link between recruitment and the specific life cycle characteristics of this group. The life cycle of flatfishes (pelagic egg and larval stages, demersal juvenile and adult stages) and their general biology (body size, growth rate, fecundity, longevity) do not differ dramatically from those of other teleosts, the one notable exception being their dramatic physical metamorphosis and the accompanying transition from a three-dimensional pelagic lifestyle to a demersal two-dimensional existence. This life stage transition occurs after a pelagic stage, the duration of which is comparatively short among teleosts, as indicated by the accelerated development of the mechanosensory system in flatfish larvae relative to those of other species having pelagic larvae (Fuiman 1997).

In flatfishes that inhabit continental shelf habitats this transition to benthic living occurs within well-defined shallow-water nursery areas. The restricted scale of these nurseries typically results in juvenile densities that may exceed 1 individual m^{-2} (see for example Zijlstra 1972; Modin & Pihl 1994). These nurseries can occur in estuaries, shallow seas and ocean banks (for overview see Gibson 1994, 1997). The distribution of juveniles within these nurseries is characterised by strong depth gradients, the highest densities occurring in shallow waters (see, for example, Kuipers 1973; Fonds 1978; Berghahn 1983).

In contrast, the juvenile nursery areas of flatfishes that inhabit deeper offshore waters tend to be larger. As a consequence, juvenile densities are much lower and rarely exceed 0.01 individual m^{-2} (Bolle et al. 1994; Baltus & Van der Veer 1995; Norcross et al. 1995). In both shallow- and deep-water species, annual recruitment success seems well established (reliably predicted) at the early juvenile stage (Van der Veer et al. 2000).

It is now well established that the concentrating effects imposed by these stock-specific geographic nursery areas and the high larval densities that result induce strong density-dependent mortality that acts to dampen the annual variability in the numbers of settling metamorphs, and hence in the numbers of recruits to the adult population (Beverton & Iles 1992b). This dampening implies that, for flatfishes, the primary source of recruitment variation is likely to be found in biological or physical processes that influence survival during the relatively short pelagic larval stage and possibly the early post-settlement demersal stage.

With the exception of fish species characterised by highly localised distributions, latitudinal variation in life history traits is well documented (e.g. Schaffer & Elson 1975; Leggett & Carscadden 1978; Houde 1989). These variations are adaptive, therefore it is likely that individual populations within the species range will vary in their susceptibility to variability to specific environmental controls. The net effect of these population-specific adaptations, however, is to maximise population stability (minimise recruitment variability) in the face of the particular features of the environment occupied (Leggett & Carscadden 1978). Miller et al. (1991) were among the first to apply this reasoning to flatfishes and to formalise it in what later has been referred to as the 'species range hypothesis' (Leggett & Frank 1997). This hypothesis explicitly states that recruitment variability in flatfishes should increase polewards from the centre of distribution of individual species, with the greatest variability occurring near the northern limits of ranges. However, in developing their arguments Miller *et al.* (1991) restricted their reasoning to the benthic juvenile stages. Given that year-class strength in these and other marine fishes is primarily determined during the pre-juvenile stage (see Leggett & DeBlois 1994, and above), their conclusions would be relevant only with respect to factors operating during the fine-scale regulation of recruitment variation that takes place after year-class strength has largely been determined. This fact may explain the general weakness of the outcome of their predictions when tested for multiple species and over broad geographic areas (Leggett & Frank 1997; Philippart *et al.* 1998). Their hypotheses may, however, provide useful insight if applied to the pelagic (egg and larval) stage of the life history during which recruitment levels are largely determined.

6.4.1 Processes influencing recruitment variability

For the purposes of this discussion, the questions of principal interest are (1) what are the key processes influencing recruitment variability and (2) why is recruitment variability in flatfishes low relative to that experienced by other fish species?

Most marine fishes that have pelagic early life stages require enormous quantities of eggs to produce a small number of surviving offspring (Fig. 6.1). In these fishes egg and larval mortality rates are very high, often exceeding 90% per day (for review see Heath 1992; Bunn *et al.* 2000). Even modest changes in this daily rate, expressed over the duration of the egg and larval period, can induce order of magnitude differences in the number of individuals exiting the larval stage (see for example Hjort 1914, 1926; Gulland 1965; Rothschild 1988; Houde 1987, 1989; Chambers & Leggett 1992; Heath 1992; Leggett & DeBlois 1994). Inter-annual

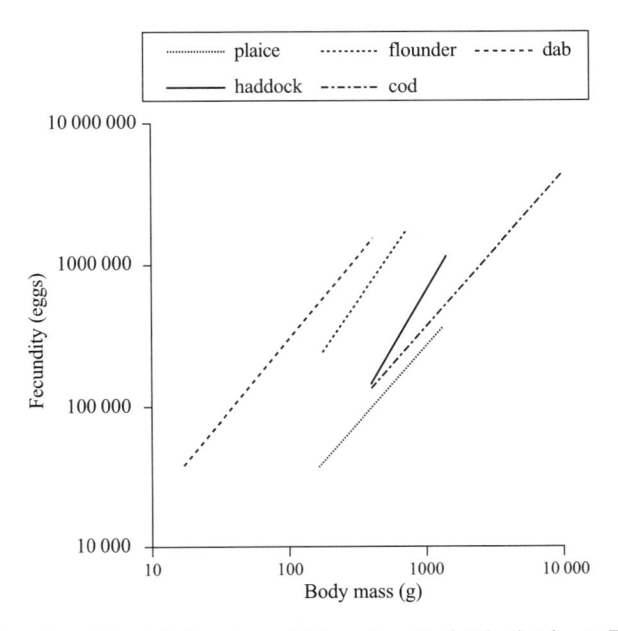

Fig. 6.1 Fecundity (eggs) in relation to body wet mass (g) for various North Atlantic teleosts. Data for plaice, flounder (*Platichthys flesus*) and dab after Roff (1982); for haddock (*Melanogrammus aeglefinus*) after Hodder (1963); for Atlantic cod after Oosthuizen & Daan (1974).

variability in recruitment is believed to be generated by daily or seasonal differences in these rates (Leggett & DeBlois 1994).

If, as previously discussed, the findings of Marshall *et al.* (1991) are generally applicable, this indicates that in contrast to the findings of Rijnsdorp (1994) variability in recruitment is at least partly caused by processes operating in the adult phase. The findings of Marshall *et al.* (1991) imply that the level of recruitment achieved by flatfishes in any given year is likely to be governed in an important way by two distinct processes. The first of these processes is the effect of environment (food availability to the adults) on adult condition at the time of reproduction. It would regulate both total egg production and the size (quality) of individual eggs and is likely to directly influence the survival of both eggs and the resulting larvae. Changes in survival would, in turn, influence the numbers of metamorphs recruiting to the juvenile nursery areas. The second major process is the inter-annual variability in the intensity of the density-dependent mortality operating on the juvenile nursery sites (the latter being influenced by both the number of metamorphs entering the nursery and the productivity of the nursery areas during the juvenile stage).

The degree to which the former process dominates recruitment in species that have pelagic juveniles (for which density dependence may be less severe or even non-existent) is presently unknown. However, in flatfishes, for which strong density-dependence during the juvenile stage is well documented (for overview see Beverton & Iles 1992b), the effect of density-dependent factors operating during this stage will be to dampen the inter-annual variability in the abundance of metamorphs entering the nursery system (i.e. to act as a pipeline or 'bottleneck' of fixed capacity) and to produce a lower annual variation in recruitment to the adult stock.

In flatfishes, the decrease in ecosystem productivity that occurs toward the equator (Gross *et al.* 1988) is likely to influence both the adult and juvenile stages, leading to lower overall production and lower inter-annual variability in recruitment. This finding is consistent with Pauly's (1994) hypothesis regarding the lower overall production of benthic fish species in tropical waters, and with the lower overall densities characteristic of flatfishes inhabiting tropical nurseries (Van der Veer *et al.* 1994a, 1995). Individual stocks and populations of marine fish species that have broad geographic distributions are known to evolve discrete life history strategies in response to the particular physical and biological characteristics of the ecosystems they occupy (e.g. Schaffer & Elson 1975; Leggett & Carscadden 1978; Houde 1989). The distinctly different life history characteristics of these low-latitude flatfishes suggest an adaptation to these larger-scale ecosystem features.

There is strong evidence that imbedded within these regulatory processes is another, founded on latitudinal differences in the duration of the larval period and the resulting duration and length of larval drift. The most dramatically variable marine abiotic gradient over latitude is temperature, the annual mean of which decreases toward the poles while the variation about this mean increases polewards. Because of the increasing seasonality of temperature with latitude, the duration of the spawning period declines from near continuous at the equator to highly contracted near the poles (for overview see Fig. 4 in Minami & Tanaka 1992). Thus, the number of 'reproductive trials' available to a stock or species declines with latitude. This decline results in a gradual transition from k selected life histories at low latitudes to r selected life histories nearer the poles. Given the 'big bang' (sensu Cole 1954) strategy pursued by northern species, recruitment variability at higher latitudes should be inherently greater, as found in American plaice (*Hippoglossoides platessoides*) (Walsh 1994b). In all fishes, time to hatching and the rate of larval development are known to be strongly and inversely related to temperature. Within a species, the effect of small gradients in temperature on development toward the poles can be offset by larger egg size, which facilitates accelerated development (Pauly & Pullin 1988; Kooijman 2000; Van der Veer *et al*. 2003). However, in general, the duration of the egg and larval (pelagic) stage exhibits a strong positive relationship with latitude (Minami & Tanaka 1992).

In flatfishes, successful recruitment requires that the developing eggs and larvae reach the nursery grounds within the appropriate time/space window. Given that development time increases towards the poles, the duration of the period of egg and larval transport also increases polewards (e.g. Walsh 1994a). In principle, one result will be an extended and expanded dispersion of eggs and larvae and a heightened probability that individual larvae may settle in suboptimal nursery habitats. Given this variability, it is likely that a relationship also exists between the size of the nursery area and recruitment stability – the larger the area the greater the dampening effect on variance in recruitment. Inter-annual variability in the duration and patterns of drift could, therefore, be equal in importance to, or even more important than, the impact of food availability and predator intensity during drift in determining recruitment success (Hollowed *et al*. 1987; Nielsen *et al*. 1998; Van der Veer *et al*. 1998). Several investigators have explored the potential or real impact of increased stage duration on recruitment (Ware 1975; Shepherd & Cushing 1990; Pepin 1991; Chambers & Leggett 1992; Houde 1994). Their analyses have, however, been principally focused on the effect of differences in growth rate on duration and its influence on predator-induced mortality at the egg and larval stage. In flatfishes, the combined effects of the concentration of spawning times and increasing drift times polewards on variability in the probability of larvae reaching known nursery areas and, ultimately, on recruitment has been formalised as the 'pelagic stage duration' hypothesis (Van der Veer *et al*. 2000). The work of Pepin & Myers (1991) showing recruitment variability in marine fishes to be positively correlated with stage duration, and of Rijnsdorp *et al*. (1992) and Walsh (1994b) (who respectively reported higher variance in recruitment in northern stocks of common sole and American plaice in the North Atlantic), is consistent with the hypothesis.

The stage duration hypothesis introduces a new and useful dimension into the analysis of recruitment variation at higher latitudes. Until its formulation, the generally held view was that variations in the food supply and the intensity of predation were the dominant direct regulators of recruitment (see Leggett & DeBlois 1994 and Dower *et al*. 2000 for reviews). Under this hypothesis the importance of the effect of food availability and predation intensity is subjugated to the effects of dispersion resulting from extended drift times and to the size of the receiving nursery zones. For example, with respect to feeding, even in relatively small larvae, such as those of plaice (approx. 6–15 mm), the time to irreversible starvation at 6°C can be more than 5 weeks (Hovenkamp 1991). Turning to predation Rijnsdorp & Jaworski (1990) have shown that within a given geographic area, such as the central North Sea, egg mortality is inversely related to egg size. If this pattern is general, the increase in average egg size polewards could lead to lower average egg mortality at higher latitudes. The general decrease in average temperature towards the poles would make this trend become even more pronounced. Similar trends could also occur with respect to larvae, an expression of the 'bigger is better' hypothesis (see Leggett & DeBlois 1994 for review).

Notwithstanding the stage duration hypothesis, however, it is reasonable to assume that individual species will have evolved adaptive responses to systematic changes in food availability and predation levels with latitude. From an evolutionary perspective, for example, the observed pattern of increasing larval sizes polewards may reflect an adaptation to the character of the food regime. That is, larger size is an energetic cushion in the face of increasing variability, but higher overall primary and secondary productivity at higher latitudes. Smaller size, on the other hand, is an adaptation to low but relatively constant and uniform productivity at lower latitudes (Gross *et al.* 1988; Kooijman 2000). From a recruitment perspective, the factor of interest is the inter-annual variability in mortality. There is no evidence that predation intensity exhibits a regular pattern with latitude.

An important extension of the species range hypotheses, when applied to the egg and larval stages of fishes, is the prediction that recruitment variability will increase with latitude both *within* and *between* species in parallel with the pattern of increasing variability in dispersal and transport times. However, the other possibility is that the egg size/development rate relationship, and the relationship (spatial) between spawning site and nursery ground, and possibly the relationship between water temperature and spawning time (a signal for when to spawn), represents an evolutionary optimisation to the average condition (temperature, current speed) likely to be experienced over a time interval relevant in evolutionary terms. If these relationships hold, then recruitment variability should be proportional to the extent to which conditions (temperature, currents) deviate from this average.

6.4.2 Recruitment variability in flatfishes relative to other marine fish species

In flatfishes as a group, recruitment variability is generally low, relative to that observed in other species (Fig. 6.2). There is no evidence to suggest that the levels of variance induced at the egg and larval stages of flatfishes differ meaningfully from those in other species. Egg size

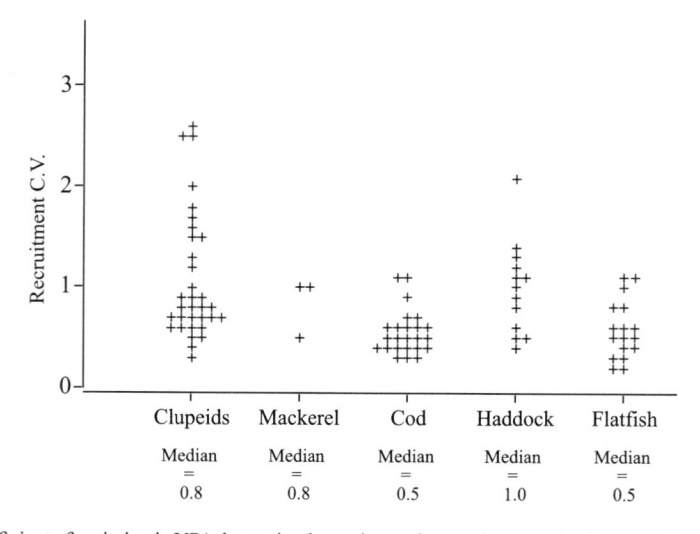

Fig. 6.2 Coefficient of variation in VPA determined recruitment by species grouping from the North Atlantic Ocean and North Pacific Ocean. After Bailey (1994), based on data from Myers *et al.* (1990) and from Hollowed (personal communication).

in flatfishes ranges between 0.50 and 4.25 mm and does not differ from that in other teleosts (see Chapter 4). Moreover, newly hatched larvae of flatfishes reflect the range of sizes of the eggs from which they hatch (Van der Veer et al. 2003) and do not differ from the range of sizes seen in other species (Russell 1976). Finally, their morphology and their pelagic existence mirror that of other teleosts throughout the larval period. The factor distinguishing the flatfishes from other teleosts is the dramatic morphological change associated with the metamorphosis to the juvenile form characterised by eye migration, asymmetrical pigmentation and a 90° rotation in posture (Norman 1934; Fuiman 1997; Osse & Van den Boogaart 1997). These changes are virtually coincident with the transition to a benthic, two-dimensional lifestyle, a transition that also occurs in many non-flatfish species.

This suggests that, as a group, the role of factors that operate to damp or limit the translation of year-class variability from the egg and larval stages to the later stages is more important in this group. In this context, Beverton (1984) hypothesised that one of the prime characteristics determining the overall stability (lack of variability) of a population was the degree of spatial concentration that occurs in the juvenile phase. The 'concentration hypothesis' as it has come to be known (Beverton 1995; Iles & Beverton 2000) argues that species which experience high density concentrations in specific nursery habitats during their early life history may have a high probability of 'saturating' the carrying capacity of those habitats, leading to density-dependent growth and size-selective, numerical or functional predatory responses, leading to density-dependent mortality. These factors have all been shown to operate in nursery areas (Van der Veer 1986; Bergman et al. 1988; Rijnsdorp & Van Leeuwen 1992; Van der Veer et al. 2000).

Beverton's (1995) hypothesis does not explicitly account for the reality that, for the flatfishes as a group, this juvenile stage concentration operates in two dimensions. In a detailed comparison of fish species in west England, which were classified according to their use of space (benthic, proximo-benthic and pelagic), Henderson & Holmes (1991) found that those species which occupied a two-dimensional (benthic) environment had the lowest variability in year-class strength. Van der Veer et al. (2000) have suggested that it is this two-dimensional concentration, which also results in density-dependent regulating processes, that is the important determinant of the reduced recruitment variability characterised by flatfishes (and other species occupying habitats that are similarly dimensionally constrained during the juvenile phase). One such species group, the cods (Gadidae), are known to experience strong density-dependent mortality during the demersal juvenile stage (Sundby et al. 1989). The observed pattern of juvenile distributions in lemon sole (*Microstomus kitt*) in the North Sea also supports this view: lemon sole do not concentrate spatially but do concentrate two-dimensionally, and they exhibit low recruitment variability (Rae 1970). It is concluded that the low recruitment variability characteristic of the flatfishes as a group is directly related to the life history characteristics of the group and, specifically, to their adoption of a size-constrained (nursery zone) two-dimensional (demersal) juvenile stage. It is hypothesised that recruitment variability in flatfishes and in other teleosts should also, therefore, be related to the duration of the pelagic versus the demersal early life stages in both a relative and absolute sense. This hypothesis resembles aspects of the member-vagrant hypothesis postulated by Sinclair (1988) and as a next step it would be worthwhile to apply the member-vagrant hypothesis to recruitment in flatfishes.

References

Bailey, K.M. (1994) Predation on juvenile flatfish and recruitment variability. *Netherlands Journal of Sea Research*, **32**, 175–189.

Baltus, C.A.M. & Van der Veer, H.W. (1995) Nursery grounds of the solenette *Buglossidium luteum* (Risso, 1810) and the scaldfish *Arnoglossus laterna* (Walbaum, 1792) in the southern North Sea. *Netherlands Journal of Sea Research*, **34**, 81–88.

Berghahn, R. (1983) Untersuchungen an Plattfischen und Nordseegarnelen (*Crangon crangon*) im Eulitoral des Wattenmeeres nach dem Übergang zum Bodenleben. *Helgoländer Meeresuntersuchungen*, **36**, 163–181.

Bergman, M.J.N., Van der Veer, H.W. & Zijlstra, J.J. (1988) Plaice nurseries: effects on recruitment. *Journal of Fish Biology*, **33** (Suppl. A), 201–218.

Beverton, R.J.H. (1984) Dynamics of single species. In: *Exploitation of Marine Communities* (ed. R.M. May). pp. 13–58. Springer Verlag, Berlin.

Beverton, R.J.H. (1995) Spatial limitation of population size: the concentration hypothesis. *Netherlands Journal of Sea Research*, **34**, 1–6.

Beverton, R.J.H. & Holt, S.J. (1957) On the dynamics of exploited fish populations. *Fisheries Investigations London, Series 2, Marine Fisheries* **19**, 1–533.

Beverton, R.J.H. & Iles, T.C. (1992a) Mortality rates of 0-group plaice (*Pleuronectes platessa* L.), dab (*Limanda limanda* L.) and turbot (*Scophthalmus maximus* L.) in European waters II. Comparison of mortality rates and construction of life table for 0-group plaice. *Netherlands Journal of Sea Research*, **29**, 49–59.

Beverton, R.J.H. & Iles, T.C. (1992b) Mortality rates of 0-group plaice (*Pleuronectes platessa* L.), dab (*Limanda limanda* L.) and turbot (*Scophthalmus maximus* L.) in European waters III. Density-dependence of mortality rates of 0-group plaice and some demographic implications. *Netherlands Journal of Sea Research*, **29**, 61–79.

Bolle, L.J., Dapper, R., Witte, J.IJ. & Van der Veer, H.W. (1994) Nursery grounds of dab (*Limanda limanda* L.) in the southern North Sea. *Netherlands Journal of Sea Research*, **32**, 299–307.

Bunn, N.A., Fox, C.J. & Webb, T. (2000) A literature review of studies on fish egg mortality: implications for the estimation of spawning stock biomass by the annual egg production method. *Scientific Series Technical Report CEFAS, Lowestoft*, **111**, 1–37.

Cameron, P., Berg, J., Dethlefsen, V. & Von Westernhagen, H. (1992) Developmental defects in pelagic embryos of several flatfish species in the southern North Sea. *Netherlands Journal of Sea Research*, **29**, 239–256.

Chambers, R.C. (1997) Environmental influences on egg and propagule sizes in marine fishes. In: *Early Life History and Recruitment in Fish Populations* (eds R.C. Chambers & E.A. Trippel). pp. 63–102. Chapman & Hall, London.

Chambers, R.C. & Leggett, W.C. (1992) Possible causes and consequences of variation in age and size at metamorphosis in flatfish (Pleuronectiformes): an analysis at the individual, population, and species levels. *Netherlands Journal of Sea Research*, **29**, 7–24.

Cole, L.C. (1954) The population consequences of life history phenomena. *Quarterly Review of Biology*, **29**, 103–137.

Conover, D.O. (1992) Seasonality and the scheduling of life history at different latitudes. *Journal of Fish Biology*, **41** (Suppl. B), 161–178.

Cushing, D.H. (1995) *Recruitment. Population and Regulation in the Sea*. Cambridge University Press, Cambridge.

Dower, J.F., Leggett, W.C. & Frank, K.T. (2000) Improving fisheries oceanography in the future. In: *Fisheries Oceanography: An Integrative Approach to Fisheries Ecology and Management* (eds P.J. Harrison & T.R. Parsons). pp. 263–281. Blackwell Scientific, New York.

Ellis, T. & Nash, R.D.M. (1997) Predation by sprat and herring on pelagic fish eggs in a plaice spawning area in the Irish Sea. *Journal of Fish Biology,* **50**, 1195–1202.

Fonds, M. (1978) The seasonal distribution of some fish species in the western Dutch Wadden Sea. In: *Fishes and Fisheries of the Wadden Sea* (eds N. Dankers, W.J. Wolff & J.J. Zijlstra). pp. 42–77. Balkema Press, Rotterdam.

Fox, C.J., Planque, B.P. & Darby, C.D. (2000) Synchrony in the recruitment time-series of plaice (*Pleuronectes platessa* L.) around the United Kingdom and the influence of sea temperature. *Journal of Sea Research,* **44**, 159–168.

Frank, K.T. & Shakell, N.L. (2001) Area-dependent patterns of finfish activity in a large marine ecosystem. *Canadian Journal of Fisheries and Aquatic Sciences,* **58**, 1703–1711.

Fuiman, L.A. (1997) What can flatfish ontogenies tell us about pelagic and benthic lifestyles? *Journal of Sea Research,* **37**, 257–267.

Gibson, R.N. (1994) Impact of habitat quality and quantity on the recruitment of juvenile flatfishes. *Netherlands Journal of Sea Research,* **32**, 191–206.

Gibson, R.N. (1997) Behaviour and the distribution of flatfishes. *Journal of Sea Research,* **37**, 241–256.

Greer Walker, M., Harden Jones, F.R. & Arnold, G.P. (1978) The movements of plaice (*Pleuronectes platessa* L.) tracked in the open sea. *Journal du Conseil International pour l'Exploration de la Mer*, **38**, 58–86.

Gross, M.R., Coleman, R.M. & McDowall, R.M. (1988) Aquatic productivity and the evolution of diadromous fish migration. *Science,* **239**, 1291–1293.

Gulland, J.A. (1965) Survival of the youngest stages of fish and its relation to year-class strength. *Publications of the International Commission for the Northwest Atlantic Fisheries,* **6**, 365–371.

Harden Jones, F.R. (1968) *Fish Migration.* Arnold, London.

Harding, D., Nichols, J.H. & Tungate, D.S. (1978) The spawning of the plaice (*Pleuronectes platessa* L.) in the Southern North Sea and the English Channel. *Rapports et Procès-Verbaux des Réunions du Conseil International pour l'Exploration de la Mer,* **172**, 102–113.

Heath, M.R. (1992) Field investigations of the early life stages of marine fish. *Advances in Marine Biology,* **28**, 2–133.

Henderson, P. & Holmes, R.H.A. (1991) On the population dynamics of dab, sole and flounder within Bridgewater Bay in the Lower Severn estuary, England. *Netherlands Journal of Sea Research,* **27**, 337–344.

Hilborn R. & Walters, C.J. (1992) *Quantitative Fisheries Stock Assessment: Choice, Dynamics and Uncertainty.* Chapman & Hall, New York.

Hjort, J. (1914) Fluctuations in the great fisheries of Northern Europe viewed in the light of biological research. *Rapports et Procès-Verbaux des Réunions du Conseil International pour l'Exploration de la Mer,* **20**, 1–228.

Hjort, J. (1926) Fluctuations in the year classes of important food fishes. *Journal du Conseil International pour l'Exploration de la Mer,* **1**, 1–38.

Hodder, V.M. (1963) Fecundity of Grand Bank haddock. *Journal of the Fisheries Research Board of Canada,* **20**, 1465–1487.

Hollowed, A.B., Bailey, K.M. & Wooster, W.S. (1987) Patterns in recruitment of marine fishes in the Northeast Pacific Ocean. *Biological Oceanography,* **5**, 99–131.

Houde, E.D. (1987) Fish early life dynamics and recruitment variability. *American Fisheries Society Symposium,* **2**, 17–29.

Houde, E.D. (1989) Comparative growth, mortality, and energetics of marine fish larvae: temperature and implied latitudinal effects. *Fishery Bulletin*, **87**, 471–495.

Houde, E.D. (1994) Differences between marine and freshwater fish larvae: implications for recruitment. *ICES Journal of Marine Science,* **51**, 91–97.

Hovenkamp, F. (1991) *On the growth of larval plaice in the North Sea.* PhD thesis, State University Groningen, Groningen.

Iles, T.C. (1994) A review of stock-recruitment relationships with reference to flatfish populations. *Netherlands Journal of Sea Research,* **32**, 399–420.

Iles, T.C. & Beverton, R.J.H. (1998) Stock, recruitment and moderating processes in flatfish. *Journal of Sea Research*, **39**, 41–55.

Iles, T.C. & Beverton, R.J.H. (2000) The concentration hypothesis: the statistical evidence. *ICES Journal of Marine Science*, **57**, 216–227.

Kashiwai, M. (1995) History of carrying capacity concept and an index of ecosystem productivity (Review). *Bulletin of the Hokkaido National Fisheries Research Institute*, **59**, 81–100.

Kooijman, S.A.L.M. (2000) *Dynamic Energy and Mass Budgets in Biological Systems.* Cambridge University Press, Cambridge.

Kuipers, B.R. (1973) On the tidal migration of young plaice (*Pleuronectes platessa*) in the Wadden Sea. *Netherlands Journal of Sea Research,* **6**, 376–388.

Leggett, W.C. (1977) Density dependence, density independence and recruitment in the American shad (*Alosa sapidissima*) population of the Connecticut River. In: *Proceedings of the Conference on Assessing the Effects of Power-plant-induced Mortality on Fish Populations* (ed. W. Van Winkle). pp. 3–17. Pergamon Press, New York.

Leggett, W.C. & Carscadden, J.E. (1978) Latitudinal variation in reproductive characteristics of American shad (*Alosa sapidissima*): evidence for population specific life history strategies in fish. *Journal of the Fisheries Research Board of Canada*, **35**, 1469–1478.

Leggett, W.C. & DeBlois, E.M. (1994) Recruitment in marine fishes: is it regulated by starvation and regulation in the egg and larval stages? *Netherlands Journal of Sea Research,* **32**, 119–134.

Leggett, W.C. & Frank, K.T. (1997) A comparative analysis of recruitment variability in North Atlantic flatfishes – testing the species range hypothesis. *Journal of Sea Research,* **37**, 281–299.

MacCall, A.D. (1990) *Dynamic Geography of Marine Fish Populations.* University of Washington Press, Seattle, WA.

Marshall, C.T., Yaragina, N.A., Lambert, Y. & Kjesbu, O.S. (1999) Total lipid energy as a proxy for total egg production by fish stocks. *Nature,* **402**, 288–290.

May, R.M. (1984) *Exploitation of Marine Communities.* Springer Verlag, Berlin.

Miller, J.M., Burke, J.S. & Fitzhugh, G.R. (1991) Early life history patterns of Atlantic North American flatfish: likely and (unlikely) factors controlling recruitment. *Netherlands Journal of Sea Research,* **27**, 261–275.

Minami, T. & Tanaka, M. (1992) Life history cycles in flatfish from the northwestern Pacific, with particular reference to their early life histories. *Netherlands Journal of Sea Research,* **29**, 35–48.

Modin, J. & Pihl, L. (1994) Differences in growth and mortality of juvenile plaice, *Pleuronectes platessa* L., following normal and extremely high settlement. *Netherlands Journal of Sea Research,* **32**, 331–341.

Myers, R.A., Blanchard, W. & Thompson, K.R. (1990) Summary of North Atlantic fish recruitment 1942–87. *Canadian Technical Report of Fisheries and Aquatic Sciences,* **1743**, 1–97.

Myers, R.A., Mertz, G. & Bridson, J. (1997) Spatial scales of interannual recruitment variations of marine, anadromous, and freshwater fish. *Canadian Journal of Fisheries and Aquatic Sciences*, **54**, 1400–1407.

Myers, R.A., MacKenzie, B.R., Bowen, K.G. & Barrowman, N.J. (2001) What is the carrying capacity for fish in the ocean? A meta-analysis of population dynamics of North Atlantic cod. *Canadian Journal of Fisheries and Aquatic Sciences*, **58**, 1464–1476.

Nash, R.D.M. (1998) Exploring the population dynamics of Irish Sea plaice, *Pleuronectes platessa* L. through the use of Paulik diagrams. *Journal of Sea Research*, **40**, 1–18.

Nielsen, E., Bagge, O. & MacKenzie, B.R. (1998) Wind-induced transport of plaice (*Pleuronectes platessa*): early life-history stages in the Skagerrak-Kattegat. *Journal of Sea Research*, **39**, 11–28.

Norcross, B.L., Holladay, B.A. & Müter, F.J. (1995) Nursery area characteristics of pleuronectids in coastal Alaska, USA. *Netherlands Journal of Sea Research*, **34**, 161–175.

Norman, J.R. (1934) *A Systematic Monograph of the Flatfishes* (*Heterosomata*), Vol. I, *Psettodidae, Bothidae, Pleuronectidae*. British Museum (National History), London.

Oosthuizen, E. & Daan, N. (1974) Egg fecundity and maturity of North Sea cod, *Gadus morhua*. *Netherlands Journal of Sea Research*, **8**, 378–397.

Osse, J.W.M. & Van den Boogaart, J.G.M. (1997) Size of flatfish larvae at transformation, functional demands and historical constraints. *Journal of Sea Research*, **37**, 229–239.

Pauly, D. (1994) A framework for latitudinal comparisons of flatfish recruitment. *Netherlands Journal of Sea Research*, **32**, 107–118.

Pauly, D. & Pullin, R.S.V. (1988) Hatching time in spherical, pelagic, marine eggs in response to temperature and egg size. *Environmental Biology of Fishes*, **22**, 261–271.

Pepin, P. (1991) Effect of temperature and size on development, mortality, and survival rates of the pelagic early life history stages of marine fish. *Canadian Journal of Fisheries and Aquatic Sciences*, **48**, 503–518.

Pepin, P. &. Myers, R.A. (1991) Significance of egg and larval size to recruitment variability of temperate marine fish. *Canadian Journal of Fisheries and Aquatic Sciences*, **48**, 1820–1828.

Philippart, C.J.M., Henderson, P.A., Johannessen, T. Rijnsdorp, A.D. & Rogers, S.I. (1998) Latitudinal variation in fish recruits in Northwest Europe. *Journal of Sea Research*, **39**, 69–77.

Rae, B.B. (1970) The distribution of flatfishes in Scottish and adjacent waters. *Marine Research*, **2**, 1–39.

Rickman, S.J., Dulvy, N.K., Jennings, S. & Reynolds, J.D. (2000) Recruitment variation related to fecundity in marine fishes. *Canadian Journal of Fisheries and Aquatic Sciences*, **57**, 116–124.

Rijnsdorp, A.D. (1994) Population-regulating processes during the adult phase in flatfish. *Netherlands Journal of Sea Research*, **32**, 207–223.

Rijnsdorp, A.D. & Jaworski, A. (1990) Size-selective mortality in plaice and cod eggs: a new method in the study of egg mortality. *Journal du Conseil International pour l'Exploration de la Mer*, **47**, 256–263.

Rijnsdorp, A.D. & Van Leeuwen, P.I. (1992) Density-dependent and density-independent changes in somatic growth of female plaice *Pleuronectes platessa* between 1930 and 1985 as revealed by back-calculation of otoliths. *Marine Ecology Progress Series*, **88**, 19–32.

Rijnsdorp, A.D., Van Beek, F.A., Flatman, S., *et al.* (1992) Recruitment in sole stocks, *Solea solea* (L.) in the northeast Atlantic. *Netherlands Journal of Sea Research*, **29**, 173–192.

Roff, D.A. (1982) Reproductive strategies in flatfish: a first synthesis. *Canadian Journal of Fisheries and Aquatic Sciences*, **39**, 1686–1698.

Rogers, S.I. (1994) Population density and growth rate of juvenile sole *Solea solea* (L.). *Netherlands Journal of Sea Research*, **32**, 353–360.

Rothschild, B.J. (1988) *Dynamics of Marine Fish Populations*. Harvard University Press, Cambridge, MA.

Rothschild, B.J. & DiNardo, G.T. (1987) Comparison of recruitment variability and life history data among marine and anadromous fishes. *American Fisheries Society Symposium*, **1**, 531–546.

Russell, F.S. (1976) *The Eggs and Planktonic Stages of British Marine Fishes*. Academic Press, London.

Schaffer, W.M. & Elson, P.F. (1975) The adaptive significance of variations in life history among local populations of Atlantic salmon in North America. *Ecology*, **56**, 577–590.

Shepherd, J.G. & Cushing, D.H. (1990) Regulation in fish populations: myth or mirage? *Philosophical Transactions of the Royal Society London Series B*, **330**, 151–164.

Sinclair, M. (1988) *Marine Populations: An Essay on Population Regulation and Speciation*. Washington Sea Grant Program, Seattle.

Solemdal, P. (1997) Maternal effects – a link between the past and the future. *Journal of Sea Research*, **37**, 213–227.

Sundby, S., Bjørke, H., Soldal, A.V. & Olsen, S. (1989) Mortality rates during the early life stages and year-class strength of northeast Arctic cod (*Gadus morhua* L.). *Rapports et Procès-Verbaux des Réunions du Conseil International pour l'Exploration de la Mer*, **191**, 351–358.

Van der Land, M.A. (1991) Distribution of flatfish eggs in the 1989 egg surveys in the southeastern North Sea, and mortality of plaice and sole eggs. *Netherlands Journal of Sea Research*, **27**, 277–286.

Van der Veer, H.W. (1985) Impact of coelenterate predation on larval plaice *Pleuronectes platessa* and flounder *Platichthys flesus* stock in the western Wadden Sea. *Marine Ecology Progress Series*, **25**, 229–238.

Van der Veer, H.W. (1986) Immigration, settlement and density-dependent mortality of a larval and early post-larval 0-group plaice (*Pleuronectes platessa*) population in the western Dutch Wadden Sea. *Marine Ecology Progress Series*, **29**, 223–236.

Van der Veer, H.W. & Bergman, M.J.N. (1987) Predation by crustaceans on a newly settled 0-group plaice *Pleuronectes platessa* population in the western Wadden Sea. *Marine Ecology Progress Series*, **35**, 203–215.

Van der Veer, H.W. & Witte, J.IJ. (1999) Year-class strength of plaice *Pleuronectes platessa* L in the Southern Bight of the North Sea: a validation and analysis of the inverse relationship with winter seawater temperature. *Marine Ecology Progress Series*, **184**, 245–257.

Van der Veer, H.W., Aliaume, C., Miller, J.M., Adriaans, E.J., Witte, J.IJ. & Zerbi, A. (1994a) Ecological observations on juvenile flatfish in a tropical coastal system, Puerto Rico. *Netherlands Journal of Sea Research*, **32**, 453–460.

Van der Veer, H.W., Berghahn, R. & Rijnsdorp, A.D. (1994b) Impact of juvenile growth on recruitment in flatfish. *Netherlands Journal of Sea Research*, **32**, 153–173.

Van der Veer, H.W., Adriaans, E.J., Bolle, L.J., *et al.* (1995) Ecological observations on juvenile flatfish in a tropical estuary: Arquipelago dos Bijagos, Guinea-Bissau. *Netherlands Journal of Sea Research*, **34**, 221–228.

Van der Veer, H.W., Ellis, T., Miller, J.M., Pihl, L. & Rijnsdorp, A.D. (1997) Size-selective predation on juvenile North Sea flatfish and possible implications for recruitment. In: *Early Life History and Recruitment in Fish Populations* (eds R.C. Chambers & E.A. Trippel). pp. 279–303. Chapman & Hall, London.

Van der Veer, H.W., Ruardij, P., Van den Berg, A.J. & Ridderinkhof, H. (1998) Impact of interannual variability in hydrodynamic circulation on egg and larval transport of plaice *Pleuronectes platessa* L. in the southern North Sea. *Journal of Sea Research*, **39**, 29–40.

Van der Veer, H.W., Berghahn, R., Miller, J.M. & Rijnsdorp, A.D. (2000) Recruitment in flatfish, with special emphasis on North Atlantic species: progress made by the Flatfish Symposia. *ICES Journal of Marine Science*, **57**, 202–215.

Van der Veer, H.W., Kooijman, S.A.L.M. & Van der Meer, J. (2003) Body size scaling relationships in flatfish as predicted by Dynamic Energy Budgets (DEB theory): implications for recruitment. *Journal of Sea Research,* **50**, 255–270.

Walsh, S.J. (1994a) Life history traits and spawning characteristics in populations of long rough dab (American plaice) *Hippoglossoides platessoides* (Fabricius) in the North Atlantic. *Netherlands Journal of Sea Research,* **32**, 241–254.

Walsh, S.J. (1994b) Recruitment variability in populations of long rough dab (American plaice) *Hippoglossoides platessoides* (Fabricius) in the North Atlantic. *Netherlands Journal of Sea Research,* **32**, 421–431.

Ware, D.M. (1975) Relation between egg size, growth, and natural mortality of larval fish. *Journal of the Fisheries Research Board of Canada,* **32**, 2503–2512.

Wheeler, A. (1978) *Key to the Fishes of Northern Europe.* Frederick Warne, London.

Zijlstra, J.J. (1972) On the importance of the Wadden Sea as a nursery area in relation to the conservation of the southern North Sea fishery resources. *Symposia of the Zoological Society of London,* **29**, 233–258.

Chapter 7
Age and growth

Richard D.M. Nash and Audrey J. Geffen

7.1 Introduction

Flatfishes are very accessible in the wild and hardy in the laboratory, thus many of the early studies of fish growth used flatfishes, especially plaice (*Pleuronectes platessa*) in the Atlantic and a number of species in the Pacific. As the science of fisheries developed, so did the need to quantify the population structure and growth characteristics of the different flatfish species. In fact, the importance of ageing fishes and determining their growth rate was realised early in the last century (Allen 1916).

Much of the early information on the ageing and growth of flatfishes (primarily plaice) is referenced in Wimpenny (1953), Graham (1956) and Beverton & Holt (1957). Prior to the 1950s researchers had gained a fairly good understanding of the methods. A clear pattern of summer and winter growth was recognised in the otoliths, which were first used in the late 1800s. Other bony structures such as opercular bones, the pectoral girdle and the concave faces of the vertebrae exhibited seasonal growth patterns (Cunningham 1905) but these were not as distinctive as those on the otoliths. The observation that a pair of rings may not delimit 1 year's growth led to early verification studies based on marginal increment analyses. Experimental work on plaice and flounder (*Platichthys flesus*) showed that the seasonal pattern on both otoliths and scales was primarily driven by seasonal changes in water temperature rather than by variations in food availability. The use of otoliths for age estimation of flatfishes was not universal. Species differences slowly became apparent and methodological refinements followed.

Direct measurements of the growth of flatfishes were afforded by series of tagging and transplantation experiments, and laboratory or enclosure experiments (Johnstone *et al.* 1921). In all cases it was apparent that there was considerable variability in individual growth rates and that growth rates varied between areas. The widespread sexual dimorphism in growth with females growing faster and reaching larger sizes than males was also recognised (e.g. Johnstone *et al.* 1921; Bigelow & Schroeder 1953; Bagenal 1955).

The effects of gear selectivity and ontogenetic behavioural changes of flatfishes on the accurate estimation of age structure and growth rates were recognised, especially with the offshore movement of larger juvenile plaice from the nursery grounds and a general offshore movement with size and age. The possibility that fishing pressure could make major changes to the age structure and growth of commercially exploited flatfish populations was mentioned by Jones (1958), citing the prevalence of Rosa Lee's phenomenon in plaice.

The transplantation experiments in the late 1800s and early 1900s were the first comprehensive studies on the manipulation of plaice growth rates. Transplantations from low to high productivity areas (Jutland coast to the shallow Limfjord (Anonymous 1909) or from the English coastal region to the Dogger Bank) resulted in an enhanced growth rate. However, anthropogenic effects were not well studied and only a few studies demonstrated the influence of contaminants on growth rates (e.g. Dilling *et al.* 1926).

Beginning in the 1950s more studies focused on growth during the juvenile stages, especially on nursery grounds. These studies were motivated by the drive to understand recruitment and the increasing interest in ecology. Also, the development of modern mariculture depended primarily on experimental work with flatfishes. These studies confirmed previous findings and provided new information on growth during the larval stages. The rapid advances in methodology and general understanding of subcellular biology from the early 1970s onward allowed very rapid advances to be made into new approaches to the study of age and growth in fishes.

7.2 Age estimation

7.2.1 Larvae and juveniles

Age estimation studies of larval flatfishes are confined to rather few species, despite the value of determining the age of larvae for recruitment studies. The otoliths of juvenile flatfishes have a number of characteristics that provide valuable information. There are usually one or two increments close to the core of the otolith that indicate hatching, or other events soon after hatching. The otolith is nearly spherical during larval development, but becomes more hemispherical close to metamorphosis.

The majority of validation studies have confirmed that primary increments are formed daily in the otoliths of larval and juvenile flatfish. Experimental studies have supported the use of primary increment counts to estimate age for wild larvae of plaice (Hovenkamp 1990), common sole (*Solea solea*) (Amara *et al.* 1994), flounder (Bos 1999), and the greenback flounder (*Rhombosolea tapirina*) and longsnout flounder *Ammotretis rostratus* in Australia (Jenkins 1987). However, winter flounder (*Pseudopleuronectes americanus*) (Casas 1998), turbot (*Scophthalmus maximus*) (Geffen 1982) and summer flounder (*Paralichthys dentatus*) (Szedlmayer & Able 1992) have exhibited non-daily increment formation during the larval stage. Specific developmental events can be recorded as distinct otolith features and this enables the estimation of individual age at different stages during early life history. Such developmental events include hatching (e.g. in plaice; Karakiri & von Westernhagen 1989), and mouth opening (e.g. in common sole; Lagardère & Troadec 1997) or first feeding (e.g. in California flounder (*Paralichthys californicus*); Kramer 1991).

During metamorphosis the shape of the otolith changes dramatically. Accessory growth centres are formed at points on the surface of the otolith, and these tend to shape the growing otolith into the flattened rectangular shape characteristic of the adults. The formation of the growth centres is clearly associated with metamorphosis but the exact timing seems to differ among species. For example, in thickback sole (*Microchirus variegatus*) accessory growth centres form before the migrating eye has crossed the midplane (dorsal edge) but in common

Table 7.1 Validated otolith age estimates for larval and juvenile flatfishes

Species	Comments	References
Starry flounder (*Platichthys stellatus*)	Disrupted increments during metamorphosis	Campana 1984
Plaice	Larvae	Karakiri & von Westernhagen 1989
	During metamorphosis	Modin *et al.* 1996
	Juveniles	Al-Hossaini & Pitcher 1988
Winter flounder	Larvae	Casas 1998
	Juveniles	Sogard 1991
Common sole	Low contrast increments during metamorphosis	Lagardère & Troadec 1997
Dover sole	Larvae	Butler *et al.* 1996
Greenback flounder and long-snout flounder	Disrupted increments during metamorphosis	Jenkins 1987
Summer flounder	Larvae	Szedlmayer & Able 1992
Fringed flounder (*Etropus crossotus*)	Juveniles	Reichert *et al.* 2000

sole the growth centres do not form until after the eye has crossed the midplane (Amara *et al.* 1998). In plaice, the first of the accessory growth centres forms at the end of stage 4 (Ryland 1966) when the body is already flattened and the eye has completed migration (Modin *et al.* 1996). In Dover sole (*Microstomus pacificus*) the growth centres form when the eye migration begins (Markle *et al.* 1992). In windowpane (*Scophthalmus aquosus*) the formation of the accessory growth centres begins after the migrating eye has crossed the midplane and continues until the end of metamorphosis (Neuman *et al.* 2001). The otoliths of juvenile greenback flounder, however, do not seem to form accessory growth centres, even after settlement (May & Jenkins 1992). Disruptions in increment formation have been associated with metamorphosis in some species (Campana 1984; Jenkins 1987; Lagardère & Troadec 1997), and age estimations covering this period may have higher associated errors. During metamorphosis and for a short period afterwards the otoliths are asymmetrical, both in shape and in size, and this asymmetry may also affect age estimation (Sogard 1991). Increments formed within the accessory growth centres may not represent daily growth, and it is not clear whether the otolith surface in the areas between growth centres continues to accrete daily increments. However, once the individual growth centres expand and come into contact with one another, new material is once again accreted over the whole surface. In most species examined, the post-metamorphic, post-growth centre, otolith increments are formed daily and thus age estimates derived from these counts are considered valid (Table 7.1). Counts of the primary increments in the post-metamorphic area of juvenile flatfish otoliths can give valuable information about the timing and patterns of settlement. Separate settlement cohorts of different ages were identified in plaice (Al-Hossaini *et al.* 1989) and common sole (Amara & Lagardère 1995).

7.2.2 Adults

Age estimation in flatfish is primarily accomplished using otoliths. In fact, the work that is most commonly cited as the first example of age estimation using otoliths is a study of plaice (Reibisch 1899). In the majority of species examined the otoliths display clear and unambiguous increments, a factor that led to the early acceptance of their use as an accurate method of age estimation. Notable exceptions are yellowtail flounder (*Limanda ferruginea*) and summer flounder, which are aged using scales (http://www.nefsc.noaa.gov/fbi/speciest-bl.html; Penttila & Dery 1988). Difficulties in age estimation using otoliths sometimes occur in species such as plaice (Nash *et al.* 1992), American plaice (*Hippoglossoides platessoides*) and winter flounder (Penttila & Dery 1988) because the first annulus is missing or indistinct. Some warm-water or short-lived species also lack an easily interpreted otolith annulus pattern for age estimation (Reichert 1998), e.g. wide-eyed flounder (*Bothas podas*) (Nash *et al.* 1991) and some tropical cynoglossids (Terwilliger & Munroe 1999).

7.3 Growth of larvae

The potential value of flatfishes in the aquaculture industry has stimulated detailed studies of larval growth, and the factors influencing growth rates, for many species. In contrast, the studies of larval flatfish growth in nature are relatively few. Prior to the 1980s most studies of larval growth consisted of studying changes in mean size or in length frequency distributions over time. Because the growth estimations were crude, there was little attempt to relate growth rates to physical or biological variables. Shelbourne (1957) was probably the first to estimate the effect of food supply on larval growth, by comparing the size of plaice larvae in 'good and bad plankton patches'. Larval growth rates, determined from laboratory and field observations, are often slow during the yolk-sac stage, but rapid from first feeding until metamorphosis.

7.3.1 Variation in growth

Individual variation in growth rates is a significant feature of larval and juvenile flatfishes (Mollander & Mollander-Swedmark 1957; Shelbourne *et al.* 1963; Chambers *et al.* 1988; Bertram *et al.* 1997; Benoit & Pepin 1999a). Within sibling groups there is usually little variation in size at hatching, and variability in yolk-sac utilisation is also limited. Thus, most of the variation in growth rate is probably attributable to differences in food acquisition and metabolism. Differential growth rates in settling and juvenile flatfishes may be key factors to understanding mortality patterns in natural populations (Chambers & Leggett 1992; Fitzhugh *et al.* 1996; Amara *et al.* 1997). These same growth features of flatfish populations cause problems in commercial aquaculture and significant effort has been invested in reducing individual variation in growth (Bengtson 1999; Klokseth & Øiestad 1999; Gavlik *et al.* 2002). In aquaculture systems the production of uniform cohorts of juveniles can improve feeding regimes and reduce the need for handling and grading, and reduce harassment within groups (Bengtson 1999; Burke *et al.* 1999; Dou *et al.* 2000).

7.3.2 Factors affecting larval growth

The size at hatching for flatfish varies considerably between species, and this variation generates very different patterns of feeding and larval growth rates. Larvae at hatching range from 2–3 mm in turbot to 15 mm in Atlantic halibut (*Hippoglossus hippoglossus*). Even between closely related Pleuronectidae, newly hatched plaice larvae are five times larger than flounder. Accompanying these differences are differences in the amount of yolk-sac at hatching, yolk utilisation efficiencies and differences in mouth size. There are also differences in the development of the mouth at hatching. Some species can feed immediately but others hatch without functional mouths. Yolk-sac utilisation and growth on an endogenous food supply extends from less than 1 day in tropical cynoglossids, to months in the case of Atlantic halibut. Yolk utilisation efficiencies can be high, especially in cold-water species (Houde & Zastrow 1993). Flatfish larvae continue to increase in length until shortly before the end of the yolk-sac period (Ehrlich & Blaxter 1976; Howell 1980; Fukuhara 1990).

Four factors that can affect larval growth rates have been highlighted, namely food, temperature, density and maternal effects (see Table 7.2). In general, there is a positive relationship between food levels and growth rate but there are instances where growth does not appear to be affected by prey concentrations. Food quality is also an important factor for some species, but some species appear to be very resilient to food quality and this probably distinguishes some warm-water species from cold-water species. Temperature also has a profound effect on growth rates over both temporal and spatial scales. Tracking changes in the thermal history of larvae has been undertaken using Sr/Ca ratios in the otoliths but this method is not always reliable (Toole *et al.* 1993). The density of individuals can cause the development of size variation in larvae, presumably through changes in growth rate. Often the differences in growth rate are caused by differences in food acquisition. Around metamorphosis various behaviours and/or cannibalism come into play. The last factor is maternal effects. The term 'maternal effects' is used to group together a set of influences that are not strictly genetic, but relate to the effects of maternal condition and age on the amount and quality of yolk and the size of larvae at hatching. In general, larger eggs produce larger larvae, often with more yolk reserves and better feeding success and survival. Most flatfish are serial spawners (Chapter 4), and the interval between ovulations can be as short as 24 hours in species such as common sole (Child *et al.* 1991), Senegalese sole (*Solea senegalensis*) (Dinis *et al.* 1999), the New Zealand turbot (*Colistium nudipinnis*) (Tait & Hickman 2001) or as long as 3 days in turbot (Howell & Scott 1989; McEvoy & McEvoy 1992), Atlantic halibut (Olsen *et al.* 1999) and plaice (Nash *et al.* 2000). Smaller females seem to produce smaller eggs, and egg size often decreases over successive spawnings.

7.4 Growth during metamorphosis

During metamorphosis flatfish larvae often spend the majority of their time in the water column, feeding on planktonic prey (Jenkins 1987; Grover 1998; Fernandez-Diaz *et al.* 2001). The developmental changes associated with metamorphosis may take priority over somatic growth, especially growth in length (Osse & Van den Boogaart 1997). These same developmental changes may also temporarily impair the ability of larvae to capture prey and thus reduce food consumption (Wyatt 1972; Keefe & Able 1993) and growth. Although relatively few studies address this question specifically, declines in growth during metamorphosis have been measured in both laboratory and field studies (Table 7.3).

The relationship between larval growth rate, size and metamorphosis has received considerable attention in both laboratory and field studies. In several species it is clear that

Table 7.2 Factors that affect larval flatfish growth

Factor	Response	Comment	Species	Location	Reference
Food	Growth rate affected by prey concentration	Growth rates of many flatfish larvae seem to respond quickly to changes in food availability	Various	Laboratory	Wyatt 1972; Houde & Schekter 1980; Bisbal & Bengtson 1995; Rabe & Brown 2000
		Demonstration of growth compensation with food levels	Various	Field	Shelbourne 1957; Wyatt 1972; Lyczkowski & Richardson 1979; Grover 1998
	Growth rate not affected by prey concentration		Summer flounder	Laboratory	Bisbal & Bengtson 1995
			Witch (*Glyptocephalus cynoglossus*)	Laboratory	Rabe & Brown 2001
			Flathead sole (*Hippoglossoides elassodon*)	Field	Haldorson et al. 1989
	Growth rates affected by food quality	Some species are resilient to changes in nutritional quality (highly unsaturated fatty acid (HUFAs) composition) of the food	Plaice, Japanese flounder (*Paralichthys olivaceus*), southern flounder (*P. lethostigma*), Atlantic halibut	Laboratory	Dickey-Collas & Geffen 1992; Alam et al. 2001; Denson & Smith 1997; Hamre et al. 2002
		Some species dependent on specific fatty acids, more vulnerable to poor feeding conditions	Japanese flounder, turbot, summer flounder, yellowtail flounder, Atlantic halibut	Laboratory	Estevez et al. 1997, 1999; Baker et al. 1998; Copeman et al. 2002; Olsen et al. 1999
Temperature	Temperature has a significant effect on growth rate	Warm-water species	Lined sole (*Archirus bilineatus*), southern flounder, turbot, summer flounder	Laboratory	Houde 1974; Burke et al. 1999; Gibson & Johnston 1995; Johns et al. 1981
		Colder-water species	Yellowtail flounder, greenback flounder, winter flounder, Atlantic halibut	Laboratory	Howell 1980; Benoit & Pepin 1999a; Hart et al. 1996; Keller & Klein-MacPhee 2000; Bidwell & Howell 2001; Pittman et al. 1989
		Seasonal and inter-annual variations in temperature	Common sole, plaice	Field	Amara et al. 1994; Hovenkamp 1989; Hovenkamp & Witte 1991
		Over latitudinal gradients differences in growth rate of a species may also be due to genetic or feeding conditions	Various	Field	Miller et al. 1991; Minami & Tanaka 1992; Chambers et al. 1995

Table 7.2 (Continued.)

Factor	Response	Comment	Species	Location	Reference
Density	Development of size variations within populations	Assumed to be the result of behavioural differences in prey detection and capture, leading to differential food acquisition	Various	Laboratory	Wyatt 1972; Rabe & Brown 2000, 2001; Mollander & Mollander-Swedmark 1957
		Harassing or aggressive behaviours and cannibalism are frequently observed during metamorphosis and settlement	Various	Laboratory	Houde 1977; Takahashi 1994; Daniels et al. 1996; Dou et al. 2000; King et al. 2000
Maternal effects	Size of eggs varies which influences the size at hatching and growth rate of larvae	There is some influence of egg size on egg development times, and eggs that hatch earlier produce smaller larvae with larger yolk-sacs. Growth rates of individual larvae are often affected by their size at hatching and amount of yolk reserves	Yellowtail flounder, plaice	Laboratory	Benoit & Pepin 1999b; Fox et al. 2003
		Maternal effects have been shown to influence growth and survival past the larval period	Winter flounder, American plaice, yellowtail flounder	Laboratory	Chambers & Leggett 1992; Walsh 1994; Benoit & Pepin 1999b

Table 7.3 Evidence of decrease in growth rate associated with metamorphosis

Species	Comments	Reference
Common sole	Field (Bay of Biscay)	Boulhic et al. 1992
Dover sole	Laboratory and field, based on otolith ageing	Butler et al. 1996
Plaice	Laboratory, confirmed with protein metabolism, RNA/DNA	Christensen & Korsgaard 1999
Senegalese sole	Laboratory, confirmed with biochemical measures	Fernandez-Diaz et al. 2001
Starry flounder	Laboratory, confirmed with otolith pattern	Campana 1984
Windowpane	Laboratory and field, based on otolith ageing	Neuman et al. 2001
Winter flounder	Laboratory, latency period defined	Bertram et al. 1997
Witch	Laboratory	Bidwell & Howell 2001

metamorphosis is a size-related event, yet in others it seems to depend more on larval growth rate rather than absolute size. In some species there is a sharp size threshold for metamorphosis resulting in a fairly synchronised settling and uniform post-settlement size distribution (Fernandez-Diaz et al. 2001). In other species the size threshold for metamorphosis or settling is broader (Gavlik et al. 2002). Within a species, environmental conditions that affect larval growth rate may also affect growth during metamorphosis and the pattern of metamorphosis. Burke et al. (1999) reported that higher temperatures increased larval growth rate and resulted in more synchronised settlement in summer flounder. Poor feeding conditions resulted in slower larval growth followed by smaller size at metamorphosis in the Senegalese sole, and metamorphosis was also less synchronised (Fernandez-Diaz et al. 2001).

In general, the variation in size at metamorphosis is greater than the variation in age at metamorphosis, at least for most of the experimental data on flatfish species reviewed by Chambers and Leggett (1987). The same is true of common sole (Boulhic et al. 1992; Amara & Lagardère 1995) and arrowtooth flounder (*Reinhardtius stomias*) (Bouwens et al. 1999) larvae collected in the wild. Higher temperatures leading to increased larval growth rates result in metamorphosis at larger sizes in most species examined (Benoit & Pepin 1999a), although in some species, such as Japanese flounder (Seikai et al. 1986), length at metamorphosis increased for slower growing larvae. Growth rate and size at metamorphosis are often uncoupled in plaice, and larvae may grow quickly and metamorphose at a small size or grow slowly and metamorphose at a larger size (Hovenkamp & Witte 1991).

In the wild, flatfishes may be particularly vulnerable during metamorphosis, although one study showed that the escape response of winter flounder was not worse during metamorphosis (Williams & Brown 1992). The most dramatic example of growth disruption associated with metamorphosis is in Dover sole. In this species metamorphosis may extend to 1 year in duration, during which time there is no apparent growth in length or weight (Markle et al. 1992; Butler et al. 1996). Kramer (1991) used size-at-age data to confirm that the growth rates of California flounder were lowest immediately after metamorphosis. These patterns lead to wide size distributions in post-settlement fish, presumably because those individuals that complete metamorphosis first resumed feeding and growth first, and often at a higher rate on the new food. Information about growth during the period of metamorphosis is vital for models of settlement and mortality. The growth of plaice decreases around the time of metamorphosis and settlement in many experimental studies. This is usually attributed to poor feeding either because of the type of prey offered, or to changes in behaviour (Neave 1985), or to the inability to process visual information and thus feed effectively. In the case of turbot visual acuity reaches its maximum after metamorphosis, whereas in plaice the eye is fully developed before metamorphosis (Neave 1984). Individuals in metamorphic stages often have the lowest indication of food consumption in field studies (Grover 1998; Lagardère *et al.* 1999). There is no evidence that flatfishes can control their growth rate directly in order to manipulate settling in response to favourable conditions such as substratum (Gibson & Batty 1990). However, Markle *et al.* (1992) discuss the possibility of delayed metamorphosis in Dover sole. Reduction in growth during metamorphosis is a concern in aquaculture where focus is usually on obtaining maximum growth rates. However, it may be advantageous to manipulate growth and development so as to produce more uniform cohorts after settlement (Gavlik *et al.* 2002).

7.5 Growth on the nursery grounds

Flatfish nursery grounds vary between species, with different characteristics such as depth range, salinity and substratum. In principle flatfish nursery grounds are the areas occupied, starting at or shortly after settlement, for a time through the juvenile phase. Metamorphosis and the learning of new behaviours associated with a benthic mode of life may increase the vulnerability of flatfishes at the start of the nursery ground stage. After metamorphosis juvenile flatfishes generally grow rapidly relative to the rest of their lifespan. Nursery grounds generally provide a partial refuge from predation and a highly productive area that promotes growth. The duration of the nursery ground phase varies between species.

Very rarely do nursery grounds provide the ideal situation with ample prey and little to no competition for food, resulting in maximal growth of the juveniles. Variability in numbers of juvenile flatfishes settling on the nursery grounds, the amount and quality of available prey, and environmental conditions such as temperature all contribute to variation in growth rate of juvenile flatfishes on the nursery grounds. One of the major problems associated with measuring the variability in growth rate on the nursery grounds relates to continued arrival of new fishes while those already present are growing. Using mean size from samples tends to underestimate the growth in this case. To combat this problem, the primary otolith increments have been used to estimate age and hence growth rate and these in turn revealed the presence of sub-cohorts and different growth rates associated with the sub-cohorts (e.g. Al-Hossaini *et al.* 1989; Hovenkamp 1991; Karakiri *et al.* 1991; May & Jenkins 1992; Sogard & Able 1992; Dau 1994; Modin & Pihl 1994). At the end of the nursery ground phase the growth rates estimated from mean size also tend to be inaccurate as larger fish begin emigrating to deeper water. Direct growth measurements can be made using tagging techniques (Nash *et al.* 1992, 1994). Both otolith and tagging studies have revealed large individual variability of growth rates on nursery grounds.

7.5.1 Growth models/growth experiments

The growth rates of post-metamorphic flatfishes have been studied intensively for aquaculture purposes, and especially for ecological studies seeking to test hypotheses about density dependence and food limitations in nursery areas. In many species the plasticity of growth rates, showing both compensatory and depensatory growth patterns, has inhibited the development of clear models of juvenile growth. However, experimental work on growth and metabolism has established a variety of models that often serve to highlight patterns that may occur in the wild.

Laboratory studies designed to examine the relative effects of temperature and food availability suggest that growth rate is very sensitive to fluctuations in feeding conditions (Malloy *et al.* 1996). When food is not limiting, juvenile flatfishes may continue to grow at a wide range of temperatures (Fonds *et al.* 1995). This is likely to be the result of adaptation to variable shallow nursery ground conditions, but is also observed in species with offshore nursery grounds (Hallaraker *et al.* 1995). Food consumption and growth efficiencies were size-related in plaice and flounder, and the response of these variables to temperature was also size-dependent such that smaller fish grew faster at higher temperatures than did larger fish (Fonds *et al.* 1992). In Atlantic halibut growth rate declined as juveniles became larger (Hallaraker *et al.* 1995).

7.5.2 *Maximum achievable growth and evidence for deviations from maximum growth*

The search to define maximum achievable growth on nursery grounds has been fuelled historically by the desire to be able to link juvenile growth with recruitment success (Van der Veer *et al.* 1994; Chambers *et al.* 1995). However, Rogers (1994) did not find a link between growth in juvenile common sole and subsequent recruitment. In this case, other factors – such as an algal bloom altering the nursery ground feeding conditions – may have been responsible for low growth rates that did not translate into survival to recruitment.

Density-dependent effects on growth have been suggested for a number of species: English sole (*Parophrys vetula*) (Peterman & Bradford 1987), plaice (Steele & Edwards 1970; Zilstra *et al.* 1982; Poxton *et al.* 1983; Modin & Pihl 1994) and four spotted megrim (*Lepidorhombus boscii*) (Landa 1999). A re-examination of some of the plaice data suggested that the variation in growth was similar to that predicted by a model based on water temperature and maximum feeding rates and differences could be explained by settling dates (Bergman *et al.* 1988; Van der Veer *et al.* 1990).

More recently, comparative growth studies have been initiated to help distinguish between good and poor settlement areas, based on the assumption that good sites will produce good growth. The maximum growth/optimal feeding conditions (MG/OFC) hypothesis assumes optimal food conditions where there is no competition for food and hence no density-dependent effects on growth (Van der Veer & Witte 1993). Under these circumstances, the maximum growth is determined solely by the ambient water temperature. Conflicting evidence has been presented for whether food is ever a limiting factor on nursery grounds and whether density-dependent effects on growth are visible. Van der Veer *et al.* (2001) demonstrated that growth of flounder and common sole in the Dutch Wadden Sea was dependent on the prevailing water temperatures and was not food limited. Neither dab (*Limanda limanda*) nor common sole in the Kattegat had growth rates that were food limited (Pihl 1989). Van der Veer *et al.* (1990) analysed a number of different 0-group plaice populations and concluded that differences in sizes of fish at the end of the nursery ground phase (August) reflected differences in thermal regime and growing season but did not support the suggestion of density-dependent effects due to food limitation. Studies on winter flounder (Sogard *et al.* 2001) and southern flounder, (Kamermans *et al.* 1995) also failed to find clear evidence of food-limited growth rates. In contrast Berghahn *et al.* (1995), working with plaice in the Wadden Sea, demonstrated variability in growth rate with food quality and quantity and Van der Veer & Witte (1993) showed a positive correlation between growth of plaice and food abundance. They also showed that where food abundance was at a similar level the presence of *Arenicola,* the preferred prey, resulted in higher growth rates. Juvenile southern flounder grew faster in areas of historically lower abundance (Guindon & Miller 1995). On winter flounder nursery grounds, more prey was available over coarser sediments, resulting in higher growth rates (Sogard 1992). Other factors such as periodic or sustained disruptions to normal behaviour patterns (Moore & Moore 1976; Gibson 1994; Geffen & Nash 1995), habitat quality (Able *et al.* 1999), day length (Poxton *et al.* 1983), salinity (Gutt 1985; Malloy & Targett 1991) and dissolved oxygen (Phelan *et al.* 2000) have also been shown to affect growth rates. While this does not argue for or against density-dependent effects it does point out that a number of other factors besides temperature will affect growth rates.

The reason for the rather inconclusive evidence for density-dependent growth effects on nursery grounds may be that it is only rarely, due to over-exploitation, that settlement on the nursery grounds is high enough to invoke competition for food. Prime nursery grounds should have relatively high 'carrying capacities' through relatively high productivity. How often nursery grounds such as the Wadden Sea reach their carrying capacity is unknown.

One further factor complicating growth rates on nursery grounds is the effect of latitude. A species may not respond to the same set of physiological variables in the same way throughout its range. In Atlantic halibut, the optimal temperature for growth was lower for high latitude fish compared with low latitude fish, compensating for the shorter growing season (Jonassen *et al.* 2000). In turbot the growth performance of the high latitude population was superior to the lower latitude populations (Imsland *et al.* 2000).

7.5.3 Growth compensation/depensation

Compensatory growth can occur when there is a negative relationship between age and growth rate, i.e. the growth slows with size. In this case the smaller fishes exhibit increased growth rates and catch up in size with the larger fishes (Ricker 1975). Under growth depensation variance of size distribution increases with time due to differential growth rates between individuals (Magnuson 1962). While there are numerous studies that demonstrate individual variability in growth rate of juvenile flatfishes on nursery grounds there are no definitive field studies that show that this is the result of behavioural hierarchies. Laboratory-based studies show the formation and maintenance of size variation in groups of juvenile flatfishes (Purdom 1974; Hallaraker *et al.* 1995; Carter *et al.* 1996). However, Bertram *et al.* (1993) found negative correlations between larval and juvenile growth rates in winter flounder, which challenges the notion that size variation is maintained over the early life history stages.

The common pattern on nursery grounds is an increase in mean length, an associated increase in standard deviation of length and a reduction in coefficient of variation with time (Van der Veer *et al.* 1994). There are spatial variations within large nursery grounds and between species but the pattern is remarkably similar between species and over geographic zones (temperate to tropical). Variability in size during the months after metamorphosis or settlement may be partially due to variations in settling date. The variation in size toward the end of the summer could be partially due to the length of time an individual had been on the nursery ground. However, Fitzhugh *et al.* (1996) determined that variability in growth rate rather than birth date of southern flounder was the primary cause of the variability in length.

The variability in growth rates between individuals may be genetic but is more likely to be due to differences in food consumption and temperature regimes. Food quantity and quality and temperature vary over the nursery grounds (Van der Veer *et al.* 1994) and that can lead to variation in growth rates. However, some species can attain similar growth rates over wide geographical areas (Bolle *et al.* 1994).

7.6 Growth of adults

Growth in adult fishes is governed by different constraints than in the juvenile phase because reproduction is a significant competitor for energy. As in juvenile populations, food levels, thermal regimes, habitat size and quality, genetics and anthropogenic effects should all affect adult growth rates and in addition have an influence on the age structure of a fish population. Although there is ample information on the age and growth of specific populations, few laboratory or field experiments have examined the factors that influence the growth of adult flatfishes.

The method used to measure growth rates depends on the level of resolution required. Changes over years are usually measured as population changes in length-at-age, and are reliant on good age estimation techniques. Seasonal changes in growth rate and individual growth rates are measured using different techniques, including tagging, back-calculation and biochemical methods. Population growth rates are usually represented as von Bertalanffy, Gompertz or other growth curves (see Ricker 1975). In adult populations back-calculation of length-at-age, to estimate the growth trajectory of an individual, is also used (see for example Rijnsdorp & Van Leeuwen 1992; Millner & Whiting 1996). Length-at-age data can be problematic, especially in exploited populations, as gear selectivity or avoidance can bias the estimate of mean length-at-age and thus lead to incorrect growth curves. Fishery-derived data can similarly be biased due to the selectivity of the fishery as a whole.

In flatfish populations in general there is a sexual dimorphism with females growing larger than males (Terwilliger & Munroe 1999; see also Chapter 4). Allied to the greater maximum size, females generally live longer (e.g. Bowering 1989; Chen *et al.* 1992; Vassilopoulou & Ondrias 1999). Notable exceptions to the females being larger than males are stone flounder *Platichthys bicoloratus* (Dou 1995) and the wide-eyed flounder (Nash *et al.* 1991).

7.6.1 Factors affecting adult growth rates

One of the main factors that has been assumed to affect growth is density, mediated through variation in food supply (Beverton & Holt 1957), although studies of marine fishes have often failed to show density-dependent growth (Rijnsdorp 1994). This apparent lack of density-dependent growth may be due to high levels of exploitation resulting in population densities that are below a threshold for these effects to come into play. Globally, many species exhibit increased growth rates with increasing levels of exploitation. Exploitation may reduce population size sufficiently to increase food availability. However, Rijnsdorp (1994) also points out that other factors could be involved in the changes in growth rate. These potential factors were categorised as direct, indirect (on the food availability) and artificial effects. Changes in growth rate could be due directly to changes in temperature, oxygen or food availability. Indirect effects on food availability could result from (1) natural changes in the ecosystem, (2) fishery-induced changes giving a relaxation of density-dependent growth, (3) shifts in the composition and abundance of food or shifts in community or inter-specific food competitors or (4) changes in food abundance due to eutrophication or pollution. Artificial changes in apparent growth rate were considered sampling artefacts from changes in fishing gear or in age estimation techniques. Examples of the factors and effects on growth rate in flatfishes are summarised in Table 7.4.

7.6.2 Trade-off between growth and reproduction

The difference in the growth rates of maturing and non-maturing female plaice supports the general concept that growth and reproduction are activities that compete for limited resources (Rijnsdorp 1993b; see also Chapter 4). Recruit spawners have to build up energy reserves but do not have to recover from depleted body condition from the previous year's spawning. Reproduction will thus have its maximum effect on somatic growth after the first reproductive season. Annual differences in growth rate can affect the length and age at first maturity, as in plaice (Rijnsdorp 1989) and American plaice (Morgan & Colbourne 1999).

As growth, maturation and egg production are closely related processes it is difficult to disentangle the influence of age at first maturity, age or size-related variability in somatic growth and fecundity/egg size relationships. There is ample evidence that juvenile growth is a major factor in the transition to maturity. Maturation is a growth-dependent process and individual growth rate influences the allocation of available energy between egg production and somatic growth in adults. Rijnsdorp (1990) proposed that during the growing season adult fishes set a fixed energy reserve to be used for reproduction, and any surplus energy is shifted to somatic growth. Deteriorating feeding conditions means a shut-off of somatic growth first, then an effect on the size-specific reproductive output.

7.7 Longevity

Flatfishes follow the general trend among fishes with longevity greater in larger sized species and those with deeper water distributions (Fig. 7.1). This general pattern is likely to reflect interactions between temperature, food availability and energetics. Maximum ages range from 60 years in the Dover sole (Munk 2001) to 1.5 years in the fringed flounder (Reichert 1998). In contrast the solenette (*Buglossidium luteum*) is thought to live up to 10 years (Nottage & Perkins 1983), the blackcheek tonguefish (*Symphurus plagiusa*) only about 5 years (Terwilliger & Munroe 1999) and *Tarphops oligolepis* as little as 2 years (Minami & Tanaka 1992). Maximum ages can vary widely between populations within species, especially those that have wide distributions. For example, American plaice populations

Table 7.4 Factors that can affect adult growth rates

Factor	Response	Comment	Species	Location	Reference
Food	Increased growth rate with increased food availability	Due to spatial variability in food quality and abundance	Dab, common sole and plaice	Southern and central North Sea	Rijnsdorp & Van Beek 1991; Henderson 1998
		Related to beam trawl effort	Common sole	North Sea	De Veen 1976
Temperature	Growth rates vary with temperature	Complex effect of temperature and length of growing season	Dab	North Sea	Henderson 1998
		No correlation between temperature and growth	Plaice	North Sea	Rijnsdorp & Van Leeuwen 1996
		Changes in growth due to unspecified oceanographic conditions	English sole	NE Pacific	Sampson & Al Jufaily 1999
Habitat	The quality and quantity of available habitat affects the growth rate	Sediment characteristics affect growth	English sole	NE Pacific	Sampson & Al Jufaily 1999
	Salinity: reduced growth rates at lower salinity	Due to higher energy expenditure associated with a higher branchial sodium, potassium and ATPase activity	*Paralichthys orbignyanus*	Brazilian waters	Sampaio & Bianchini 2002
Stock/genetics	Stock or genetic effect on growth rate	Geographical differences in growth	American plaice	Atlantic and Mediterranean	Vassilopoulou & Ondrias 1999
		Potential for broodstock selection	Common sole	Atlantic and Mediterranean	Exadactylos et al. 1998
		Effects of fishing pressure on the genetic variability	Plaice	Atlantic	Rijnsdorp 1993a
Anthropogenic effects	Positive effects on growth rate	Eutrophication and beam trawling effects	Plaice	North Sea	Rijnsdorp & Van Leeuwen 1996
	Negative effects on growth rate		Common sole	Bristol Channel	Horwood 1993
		Chemical pollutants	English sole	Laboratory	Johnson et al. 1998

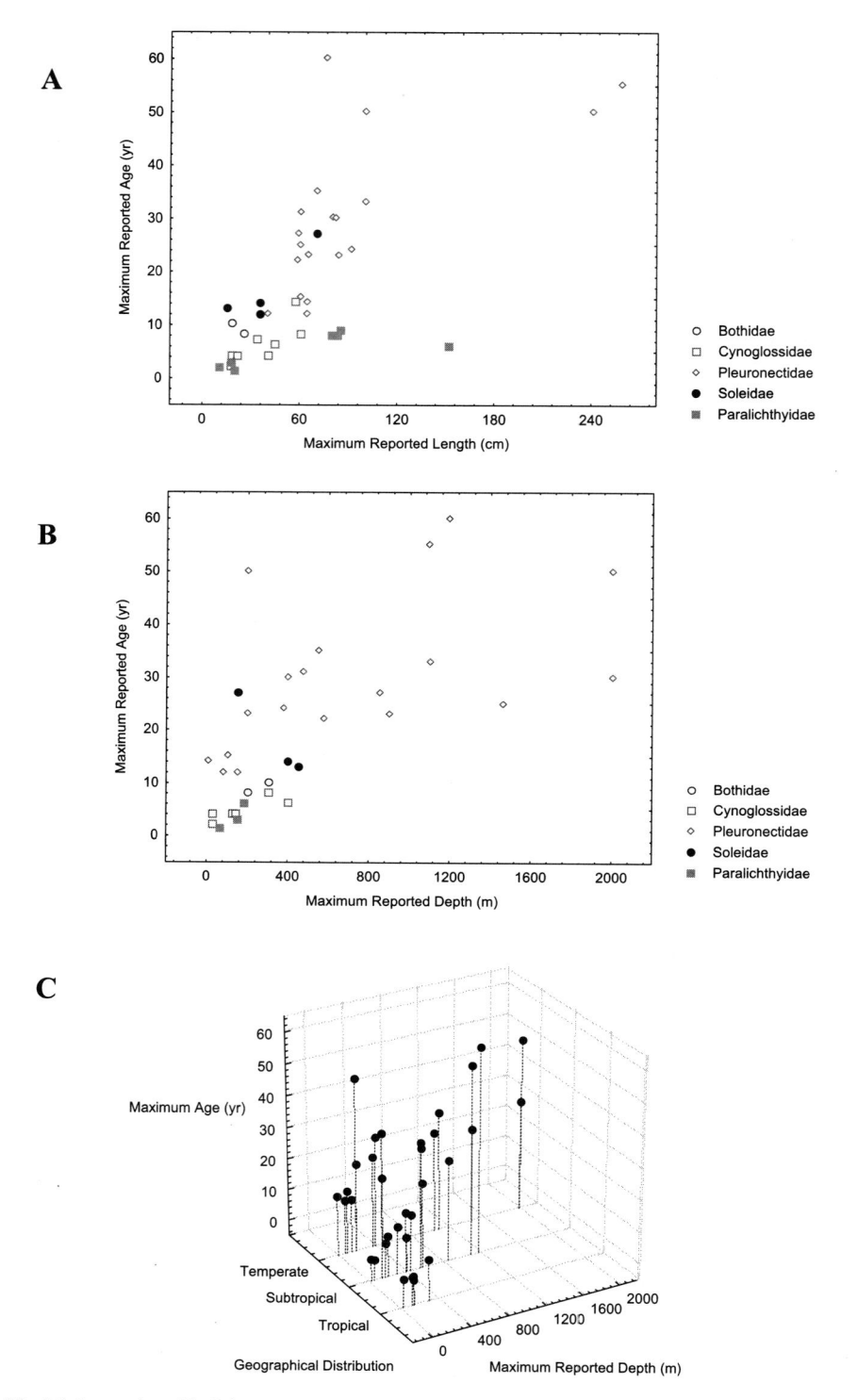

Fig. 7.1 Longevity of flatfishes. A. Maximum ages and lengths reported for species from five flatfish families. B. Maximum ages and depth distributions reported for species from five flatfish families. C. Pattern of longevity in relation to depth and latitudinal distribution. Data for all graphs from Minami & Tanaka (1992), Terwilliger & Munroe (1999), FishBase (Froese & Pauly 2002) and Munk (2001).

show strong latitudinal trends in longevity (Walsh 1994), as do megrim (*Lepidorhombus whiffiagonis*) (Vassilopoulou & Ondrias 1999), witch (Albert *et al.* 1998) and several North Pacific species (Munk 2001).

Differences in inter- and intra-specific longevity have been ascribed to latitudinal differences (see Fig. 7.1), as well as to the effects of temperature, food availability and life history strategies. The pattern of commercial exploitation is probably the most significant factor that causes local differences in longevity both between species and between populations (Rijnsdorp 1993a; Albert *et al.* 1998). In the early 1900s there were frequent reports of plaice of 20–30 years old, as estimated from otolith readings. Although one 40-year-old female common sole was captured in 1999 (Anonymous 1999), maximum ages in the catches of most exploited flatfish species are often in decline (Millner & Whiting 1996).

References

Able, K.W., Manderson, J.P. & Studholme, A.L. (1999) Habitat quality for shallow water fishes in an urban estuary: the effects of man-made structures on growth. *Marine Ecology Progress Series*, **187**, 227–235.

Alam, M.S., Teshima, S., Ishikawa, M. & Koshio, S. (2001) Effects of ursodeoxycholic acid on growth and digestive enzyme activities of Japanese flounder *Paralichthys olivaceus* (Temminck & Schlegel). *Aquaculture Research*, **32**, 235–243.

Albert, O.T., Eliassen, J.E. & Hoeines, A. (1998) Flatfishes of Norwegian coasts and fjords. *Journal of Sea Research*, **40**, 153–171.

Al-Hossaini, M. & Pitcher, T.J. (1988) The relation between daily rings, body growth and environmental factors in plaice, *Pleuronectes platessa* L., juvenile otoliths. *Journal of Fish Biology*, **33**, 409–418.

Al-Hossaini, M., Liu, Q. & Pitcher, T. (1989) Otolith microstructure indicating growth and mortality among plaice, *Pleuronectes platessa* L., post-larval sub-cohorts. *Journal of Fish Biology*, **35** (Suppl. A), 81–90.

Allen, E.J. (1916) The age of fishes and the rate at which they grow. *Journal of the Marine Biological Association of the United Kingdom*, **11**, 399–424.

Amara, R. & Lagardère, F. (1995) Size and age at onset of metamorphosis in sole (*Solea solea* (L)) of the Gulf of Gascogne. *ICES Journal of Marine Science*, **52**, 247–256.

Amara, R., Desaunay, Y. & Lagardère, F. (1994) Seasonal variation in growth of larval sole *Solea solea* (L.) and consequences on the success of larval immigration. *Netherlands Journal of Sea Research*, **32**, 287–298.

Amara, R., Galois, R. & Lagardère, F. (1997) Nutritional condition and growth rate of sole, *Solea solea* (L.) larvae during metamorphosis: a spatial and temporal analysis. *Journal of Fish Biology*, **51** (Suppl. A), 397.

Amara, R., Poulard, J.C., Lagardère, F. & Desaunay, Y. (1998) Comparison between the life cycles of two Soleidae, the common sole, *Solea solea*, and the thickback sole, *Microchirus variegatus*, in the Bay of Biscay (France). *Environmental Biology of Fishes*, **53**, 193–209.

Anonymous (1909) Marking and transplantation experiments with plaice, and some notes on the natural history of that fish. *Proceedings of the Suffolk Institute of Archaeology and Natural History*, 1–13.

Anonymous (1999) Not all sole die young! *Fishing News*, No. 4461, 20 August 1999, 3pp.

Bagenal, T.B. (1955) The growth rate of the long rough dab *Hippoglossoides platessoides* (Fabr.). *Journal of the Marine Biological Association of the United Kingdom*, **34**, 297–311.

Baker, E.P., Alves, D. & Bengtson, D.A. (1998) Effects of rotifer and *Artemia* fatty-acid enrichment on survival, growth and pigmentation of summer flounder *Paralichthys dentatus* larvae. *Journal of the World Aquaculture Society*, **29**, 494–498.

Bengtson, D.A. (1999) Aquaculture of summer flounder (*Paralichthys dentatus*): status of knowledge, current research and future research priorities. *Aquaculture*, **176**, 39–49.

Benoit, H.P. & Pepin, P. (1999a) Individual variability in growth rate and the timing of metamorphosis in yellowtail flounder *Pleuronectes ferrugineus*. *Marine Ecology Progress Series*, **184**, 231–244.

Benoit, H.P. & Pepin, P. (1999b) Interaction of rearing temperature and maternal influence on egg development rates and larval size at hatch in yellowtail flounder (*Pleuronectes ferrugineus*). *Canadian Journal of Fisheries and Aquatic Sciences*, **56**, 785–794.

Berghahn, R., Ludemann, K. & Ruth, M., (1995) Differences in individual growth of newly settled 0-group plaice (*Pleuronectes platessa* L) in the intertidal of neighbouring Wadden Sea areas. *Netherlands Journal of Sea Research*, **34**, 131–138.

Bergman, M.J.N., Van der Veer, H.W. & Zijlstra, J.J. (1988) Plaice nurseries: effects on recruitment. *Journal of Fish Biology*, **33**, 201–218.

Bertram, D.F., Chambers, R.C. & Leggett, W.C. (1993) Negative correlations between larval and juvenile growth rates in winter flounder: implications of compensatory growth for variation in size-at-age. *Marine Ecology Progress Series*, **96**, 209–215.

Bertram, D.F., Miller, T.J. & Leggett, W.C. (1997) Individual variation in growth and development during the early life stages of winter flounder, *Pleuronectes americanus*. *Fishery Bulletin*, **95**, 1–10.

Beverton, R.J.H. & Holt, S.J. (1957) On the dynamics of exploited fish populations. *Fishery Investigations, Series II, Marine Fisheries, Great Britain Ministry of Agriculture, Fisheries and Food*, **19**, 1–533.

Bidwell, D.A. & Howell, W.H. (2001) The effect of temperature on first feeding, growth, and survival of larval witch flounder *Glyptocephalus cynoglossus*. *Journal of the World Aquaculture Society*, **32**, 373–384.

Bigelow, H.B. & Schroeder, W.C. (1953) Fishes of the Gulf of Maine. *Fishery Bulletin of the Fish and Wildlife Service*, **53**, 1–577.

Bisbal, G.A. & Bengtson, D.A. (1995) Effects of delayed feeding on survival and growth of summer flounder *Paralichthys dentatus* larvae. *Marine Ecology Progress Series*, **121**, 301–306.

Bolle, L.J., Dapper, R., Witte, J.IJ & Van der Veer, H.W. (1994) Nursery grounds of dab (*Limanda limanda* L) in the southern North Sea. *Netherlands Journal of Sea Research*, **32**, 299–307.

Bos, A.R. (1999) Tidal transport of flounder larvae (*Pleuronectes flesus*) in the Elbe River, Germany. *Archive of Fishery and Marine Research*, **47**, 47–60.

Boulhic, M., Galois, R., Koutsikopoulos, C., Lagardère, F. & Person-Le Ruyet, J. (1992) Nutritional status, growth and survival of the pelagic stages of the Dover sole *Solea solea* (L), in the Bay of Biscay. *Annales de l'Institut Oceanographique*, **68**, 117–139.

Bouwens, K.A., Smith, R.L., Paul, A.J. & Rugen, W. (1999) Length at and timing of hatching and settlement for arrowtooth flounders in the Gulf of Alaska. *Alaska Fishery Research Bulletin*, **6**, 41–48.

Bowering, W.R. (1989) Witch flounder distribution of southern Newfoundland, and changes in age, growth, and sexual maturity patterns with commercial exploitation. *Transactions of the American Fisheries Society*, **118**, 659–669.

Burke, J.S., Seikai, T., Tanaka, Y. & Tanaka, M. (1999) Experimental intensive culture of summer flounder, *Paralichthys dentatus*. *Aquaculture*, **176**, 135–144.

Butler, J.L., Dahlin, K.A. & Moser, H.G. (1996) Growth and duration of the planktonic phase and a stage based population matrix of Dover sole, *Microstomus pacificus*. *Bulletin of Marine Science*, **58**, 29–43.

Campana, S.E. (1984) Microstructural growth patterns in the otoliths of larval and juvenile starry flounder, *Platichthys stellatus*. *Canadian Journal of Zoology*, **62**, 1507–1512.

Carter, C.G., Purser, G.J., Houlihan, D.F. & Thomas, P. (1996) The effect of decreased ration on feeding hierarchies in groups of greenback flounder (*Rhombosolea tapirina*: Teleostei). *Journal of the Marine Biological Association of the United Kingdom*, **76**, 505–516.

Casas, M.C. (1998) Increment formation in otoliths of slow-growing winter flounder (*Pleuronectes americanus*) larvae in cold water. *Canadian Journal of Fisheries and Aquatic Sciences*, **55**, 162–169.

Chambers, R. & Leggett, W.C. (1987) Size and age at metamorphosis in marine fishes: an analysis of laboratory-reared winter flounder (*Pseudopleuronectes americanus*) with a review of variation in other species. *Canadian Journal of Fisheries and Aquatic Sciences*, **44**, 1936–1947.

Chambers, R.C. & Leggett, W.C. (1992) Possible causes and consequences of variation in age and size at metamorphosis in flatfishes (Pleuronectiformes) – an analysis at the individual, population, and species levels. *Netherlands Journal of Sea Research*, **29**, 7–24.

Chambers, R., Leggett, W.C. & Brown, J.A. (1988) Variation in and among early life history traits of laboratory-reared winter flounder (*Pseudopleuronectes americanus*). *Marine Ecology Progress Series*, **47**, 1–15.

Chambers, R.C., Rose, K.A. & Tyler, J.A. (1995) Recruitment and recruitment processes of winter flounder, *Pleuronectes americanus*, at different latitudes: implications of an individual-based simulation model. *Netherlands Journal of Sea Research*, **34**, 19–43.

Chen, D.G., Liu, C.G. & Dou, S.Z. (1992) The biology of flatfish (Pleuronectinae) in the coastal waters of China. *Netherlands Journal of Sea Research*, **29**, 25–33.

Child, A.R., Howell, B.R. & Houghton, R.G. (1991) Daily periodicity and timing of the spawning of sole, *Solea solea* (L), in the Thames estuary. *ICES Journal of Marine Science*, **48**, 317–323.

Christensen, M.N. & Korsgaard, B. (1999) Protein metabolism, growth and pigmentation patterns during metamorphosis of plaice (*Pleuronectes platessa*) larvae. *Journal of Experimental Marine Biology and Ecology*, **237**, 225–241.

Copeman, L.A., Parrish, C.C., Brown, J.A. & Harel, M. (2002) Effects of docosahexaenoic, eicosapentaenoic, and arachidonic acids on the early growth, survival, lipid composition and pigmentation of yellowtail flounder (*Limanda ferruginea*): a live food enrichment experiment. *Aquaculture*, **210**, 285–304.

Cunningham, J.T. (1905) Zones of growth in the skeletal structures of Gadidae and Pleuronectidae. *Annual Report of the Fishery Board for Scotland*, **23**, 125–140.

Daniels, H.V., Berlinsky, D.L., Hodson, R.G. & Sullivan, C.V. (1996) Effects of stocking density, salinity, and light intensity on growth and survival of southern flounder *Paralichthys lethostigma* larvae. *Journal of the World Aquaculture Society*, **27**, 153–159.

Dau, K. (1994) *Population dynamics of 0-group plaice,* Pleuronectes platessa *L., on a Manx nursery ground, including the study of sub-cohorts by use of otolith microstructure*. Diploma thesis, University of Geneva.

Denson, M.R. & Smith, T.I.J. (1997) Diet and light intensity effects on survival, growth and pigmentation of southern flounder *Paralichthys lethostigma*. *Journal of the World Aquaculture Society*, **28**, 366–373.

De Veen, J.F. (1976) On changes in some biological parameters in the North Sea sole (*Solea solea* L.). *Journal du Conseil International pour l'Exploration de la Mer*, **37**, 60–90.

Dickey-Collas, M. & Geffen, A.J. (1992) Importance of the fatty-acids 20:5 ω 3 and 22:6 ω 3 in the diet of plaice (*Pleuronectes platessa*) larvae. *Marine Biology*, **113**, 463–468.

Dilling, W.J., Healey, C.W. & Smith, W.C. (1926) Experiments on the effects of lead on the growth of plaice (*Pleuronectes platessa*). *Annals of Applied Biology*, **13**, 168–176.

Dinis, M.T., Ribeiro, L., Soares, F. & Sarasquete, C. (1999) A review of the cultivation potential of *Solea senegalensis* in Spain and Portugal. *Aquaculture*, **176**, 27–38.

Dou, S.Z. (1995) Life history cycles of flatfish species in the Bohai Sea, China. *Netherlands Journal of Sea Research*, **34**, 195–210.

Dou, S.Z., Seikai, T. & Tsukamoto, K. (2000) Cannibalism in Japanese flounder juveniles, *Paralichthys olivaceus*, reared under controlled conditions. *Aquaculture*, **182**, 149–159.

Ehrlich, K.F. & Blaxter, J.H.S. (1976) Morphological and histological changes during the growth and starvation of herring and plaice larvae. *Marine Biology*, **35**, 105–118.

Estevez, A., Ishikawa, M. & Kanazawa, A. (1997) Effects of arachidonic acid on pigmentation and fatty acid composition of Japanese flounder, *Paralichthys olivaceus* (Temminck an Schlegel). *Aquaculture Research*, **28**, 279–289.

Estevez, A., McEvoy, L.A., Bell, J.G. & Sargent, J.R. (1999) Growth, survival, lipid composition and pigmentation of turbot (*Scophthalmus maximus*) larvae fed live-prey enriched in arachidonic and eicosapentaenoic acids. *Aquaculture*, **180**, 321–343.

Exadactylos, A., Geffen, A. & Thorpe, J. (1998) Population structure of the Dover sole, *Solea solea* L., in a background of high gene flow. *Journal of Sea Research*, **40**, 117–129.

Fernandez-Diaz, C., Yufera, M., Canavate, J.P., Moyano, F.J., Alarcon, F.J. & Diaz, M. (2001) Growth and physiological changes during metamorphosis of Senegal sole reared in the laboratory. *Journal of Fish Biology*, **58**, 1086–1097.

Fitzhugh, G.R., Crowder, L.B. & Monaghan, J.P., Jr (1996) Mechanisms contributing to variable growth in juvenile southern flounder (*Paralichthys lethostigma*). *Canadian Journal of Fisheries and Aquatic Sciences*, **53**, 1964–1973.

Fonds, M., Cronie, R., Vethaak, A.D. & Van der Puyl, P. (1992) Metabolism, food consumption and growth of plaice (*Pleuronectes platessa*) and flounder (*Platichthys flesus*) in relation to fish size and temperature. *Netherlands Journal of Sea Research*, **29**, 127–143.

Fonds, M., Tanaka, M. & Van der Veer, H.W. (1995) Feeding and growth of juvenile Japanese flounder *Paralichthys olivaceus* in relation to temperature and food supply. *Netherlands Journal of Sea Research*, **34**, 111–118.

Fox, C.J., Geffen, A.J., Blyth, R. & Nash, R.D.M. (2003) An evaluation of the temperature dependent development rates of plaice (*Pleuronectes platessa* L.) eggs from the Irish Sea. *Journal of Plankton Research*, **25**, 1319–1329.

Froese, R. & Pauly, D. (2002) FishBase, www.fishbase.org.

Fukuhara, O. (1990) Effects of temperature on yolk utilization, initial growth, and behaviour of unfed marine fish-larvae. *Marine Biology*, **106**, 169–174.

Gavlik, S., Albino, M. & Specker, J.L. (2002) Metamorphosis in summer flounder: manipulation of thyroid status to synchronize settling behavior, growth, and development. *Aquaculture*, **203**, 359–373.

Geffen, A.J. (1982) Otolith ring deposition in relation to growth rate in herring (*Clupea harengus*) and turbot (*Scophthalmus maximus*) larvae. *Marine Biology*, **71**, 317–326.

Geffen, A.J. & Nash, R.D.M. (1995) Periodicity of otolith check formation in juvenile plaice *Pleuronectes platessa* L. In: *Recent Developments in Fish Otolith Research* (eds D.H. Secor, J.M. Dean & S.E. Campana). pp. 65–76. Belle W. Baruch Library in Marine Science, No. 19. University of South Carolina Press, Columbia, SC.

Gibson, R.N. (1994) Impact of habitat quality and quantity on the recruitment of juvenile flatfishes. *Netherlands Journal of Sea Research*, **32**, 191–206.

Gibson, R.N. & Batty, R.S. (1990) Lack of substratum effect on the growth and metamorphosis of larval plaice *Pleuronectes platessa*. *Marine Ecology Progress Series*, **66**, 219–223.

Gibson, S. & Johnston, I.A. (1995) Temperature and development in larvae of the turbot *Scophthalmus maximus*. *Marine Biology*, **124**, 17–25.

Graham, M. (1956) Plaice. In: *Sea Fisheries: Their Investigation in the United Kingdom* (ed. M. Graham). pp. 332–371. Arnold, London.

Grover, J.J. (1998) Feeding habits of pelagic summer flounder, *Paralichthys dentatus*, larvae in oceanic and estuarine habitats. *Fishery Bulletin*, **96**, 248–257.

Guindon, K.Y. & Miller, J.M. (1995) Growth potential of juvenile southern flounder, *Paralichthys lethostigma*, in low salinity nursery areas of Pamlico Sound, North Carolina, USA. *Netherlands Journal of Sea Research*, **34**, 89–100.

Gutt, J. (1985) The growth of juvenile flounders (*Platichthys flesus* L.) at salinities of 0, 5, 15 and 35‰. *Journal of Applied Ichthyology*, **1**, 17–21.

Haldorson, L., Paul, A.J., Sterritt, D. & Watts, J. (1989) Annual and seasonal variation in growth of larval walleye pollock and flathead sole in a southeastern Alaskan bay. *Rapports et Procès-Verbaux des Réunions Conseil International pour l'Exploration de la Mer*, **191**, 220–225.

Hallaraker, H., Folkvord, A. & Stefansson, S.O. (1995) Growth of juvenile halibut (*Hippoglossus hippoglossus*) related to temperature, day length and feeding regime. *Netherlands Journal of Sea Research*, **34**, 139–147.

Hamre, K., Opstad, I., Espe, M., Solbakken, J., Hemre, G.I. & Pittman, K. (2002) Nutrient composition and metamorphosis success of Atlantic halibut (*Hippoglossus hippoglossus*, L.) larvae fed natural zooplankton or *Artemia*. *Aquaculture Nutrition*, **8**, 139–148.

Hart, P.R., Hutchinson, W.G. & Purser, G.J. (1996) Effects of photoperiod, temperature and salinity on hatchery-reared larvae of the greenback flounder (*Rhombosolea tapirina* Gunther, 1862). *Aquaculture*, **144**, 303–311.

Henderson, P.A. (1998) On the variation in dab *Limanda limanda* recruitment: a zoogeographic study. *Journal of Sea Research*, **40**, 131–142.

Horwood, J.W. (1993) The Bristol Channel sole (*Solea sole* (L.)): a fisheries case study. *Advances in Marine Biology*, **29**, 215–367.

Houde, E. (1974) Effects of temperature and delayed feeding on growth and survival of larvae of three species of subtropical marine fishes. *Marine Biology*, **26**, 271–285.

Houde, E.D. (1977) Food concentration and stocking density effects on survival and growth of laboratory reared larvae of bay anchovy, *Anchoa mitchilli* and lined sole *Achirus lineatus*. *Marine Biology*, **43**, 333–341.

Houde, E.D. & Schekter, R.C. (1980) Feeding by marine fish larvae: developmental and functional responses. *Environmental Biology of Fishes*, **5**, 315–334.

Houde, E.D. & Zastrow, C.E. (1993) Ecosystem specific and taxon specific dynamic and energetics properties of larval fish assemblages. *Bulletin of Marine Science*, **53**, 290–335.

Hovenkamp, F. (1989) Within-season variation in growth of larval plaice (*Pleuronectes platessa* L.). *Rapports et Procès-Verbaux des Réunions Conseil International pour l'Exploration de la Mer*, **191**, 248–257.

Hovenkamp, F. (1990) Growth differences in larval plaice *Pleuronectes platessa* in the Southern Bight of the North Sea as indicated by otolith increments and RNA/DNA ratios. *Marine Ecology Progress Series*, **58**, 205–215.

Hovenkamp, F. (1991) Immigration of larval plaice (*Pleuronectes platessa* L.) into the western Wadden Sea: a question of timing. *Netherlands Journal of Sea Research*, **27**, 287–296.

Hovenkamp, F. & Witte, J.I. (1991) Growth, otolith growth and RNA/DNA ratios of larval plaice *Pleuronectes platessa* in the North Sea 1987 to 1989. *Marine Ecology Progress Series*, **70**, 105–116.

Howell, B.R. & Scott, A.P. (1989) Ovulation cycles and post-ovulatory deterioration of eggs of the turbot (*Scophthalmus maximus* L.). *Rapports et Procès-Verbaux des Réunions Conseil International pour l'Exploration de la Mer*, **191**, 21–26.

Howell, W.H. (1980) Temperature effects on growth and yolk utilization in yellowtail flounder, *Limanda ferruginea*, yolk-sac larvae. *Fishery Bulletin*, **78**, 731–739.

Imsland, A.K., Foss, A., Nævdal, G., *et al.* (2000) Countergradient variation in growth and food conversion efficiency of juvenile turbot. *Journal of Fish Biology*, **57**, 1213–1226.

Jenkins, G.P. (1987) Age and growth of co-occurring larvae of two flounder species *Rhombosolea tapirina* and *Ammotretis rostratus*. *Marine Biology*, **95**, 157–166.

Johns, D.M., Howell, W.H. & Klein-MacPhee, G. (1981) Yolk utilization and growth to yolk-sac absorption in summer flounder (*Paralichthys dentatus*) larvae at constant and cyclic temperatures. *Marine Biology*, **63**, 301–308.

Johnson, L.L., Landahl, J.T., Kubin, L.A., *et al.* (1998) Assessing the effects of anthropogenic stressors on Puget Sound flatfish populations. *Journal of Sea Research*, **39**, 125–137.

Johnstone, J., Birtwistle, W. & Smith, W.C. (1921) The plaice fisheries of the Irish Sea. *Report of the Lancashire Sea Fishery Laboratory, 1921*, **30**, 39–179.

Jonassen, T.M., Imsland, A.K., Fitzgerald, R., *et al.* (2000) Geographic variation in growth and food conversion efficiency of juvenile Atlantic halibut related to latitude. *Journal of Fish Biology*, **56**, 279–294.

Jones, R. (1958) Lee's phenomenon of 'apparent change in growth rate' with particular reference to haddock and plaice. *International Commission for North Atlantic Fisheries, Special Publication*, **1**, 229–242.

Kamermans, P., Guindon, K.Y. & Miller, J.M. (1995) Importance of food availability for growth of juvenile southern flounder (*Paralichthys lethostigma*) in the Pamlico River estuary, North Carolina, USA. *Netherlands Journal of Sea Research*, **34**, 101–109.

Karakiri, M. & von Westernhagen, H. (1989) Daily growth patterns in otoliths of larval and juvenile plaice (*Pleuronectes platessa* L.): influence of temperature, salinity, and light conditions. *Rapports et Procès-Verbaux des Réunions Conseil International pour l'Exploration de la Mer*, **191**, 376–382.

Karakiri, M., Berghahn, R. & Van der Veer, H.W. (1991) Variations in settlement and growth of 0-group plaice (*Pleuronectes platessa* L.) in the Dutch Wadden Sea as determined by otolith microstructure analysis. *Netherlands Journal of Sea Research*, **27**, 345–351.

Keefe, M. & Able, K.W. (1993) Patterns of metamorphosis in summer flounder, *Paralichthys dentatus*. *Journal of Fish Biology*, **42**, 713–728.

Keller, A.A. & Klein-MacPhee, G. (2000) Impact of elevated temperature on the growth, survival, and trophic dynamics of winter flounder larvae: a mesocosm study. *Canadian Journal of Fisheries and Aquatic Sciences*, **57**, 2382–2392.

King, N.J., Howell, W.H., Huber, M. & Bengtson, D.A. (2000) Effects of larval stocking density on laboratory-scale and commercial-scale production of summer flounder *Paralichthys dentatus*. *Journal of the World Aquaculture Society*, **31**, 436–445.

Klokseth, V. & Øiestad, V. (1999) Forced settlement of metamorphosing halibut (*Hippoglossus hippoglossus* L.) in shallow raceways: growth pattern, survival, and behaviour. *Aquaculture*, **176**, 117–133.

Kramer, S.H. (1991) Growth, mortality, and movements of juvenile California halibut *Paralichthys californicus* in shallow coastal and bay habitats of San Diego County, California. *Fishery Bulletin*, **89**, 195–207.

Lagardère, F. & Troadec, H. (1997) Age estimation in common sole *Solea solea* larvae: validation of daily increments and evaluation of a pattern recognition technique. *Marine Ecology Progress Series*, **155**, 223–237.

Lagardère, F., Amara, R. & Joassard, L. (1999) Vertical distribution and feeding activity of metamorphosing sole, *Solea solea*, before immigration to the Bay of Vilaine nursery (northern Bay of Biscay, France). *Environmental Biology of Fishes*, **56**, 213–228.

Landa, J. (1999) Density-dependent growth of four spot megrim (*Lepidorhombus boscii*) in the northern Spanish shelf. *Fisheries Research*, **40**, 267–276.

Lyczkowski, L.J. & Richardson, S.L. (1979) Winter-spring abundance of larval English sole, *Parophrys vetulus*, between the Columbia River and Cape Blanco, Oregon during 1972–1975 with notes on occurrences of three other pleuronectids. *Estuarine and Coastal Marine Science*, **8**, 455–476.

McEvoy, L.A. & McEvoy, J. (1992) Multiple spawning in several commercial fish species and its consequences for fisheries management, cultivation and experimentation. *Journal of Fish Biology*, **41**, 125–136.

Magnuson, J.J. (1962) An analysis of aggressive behavior, growth and competition for food and space in medaka (*Oryzias latipes* (Pisces, Cyprinodontidae)). *Canadian Journal of Zoology*, **40**, 313–384.

Malloy, K.D. & Targett, T.E. (1991) Feeding, growth, and survival of juvenile summer flounder (*Paralichthys dentatus*): experimental analysis of the effects of temperature and salinity. *Marine Ecology Progress Series*, **72**, 213–223.

Malloy, K.D., Yamashita, Y., Yamada, H. & Targett, T.E. (1996) Spatial and temporal patterns of juvenile stone flounder *Kareius bicoloratus* growth rates during and after settlement. *Marine Ecology Progress Series*, **131**, 49–59.

Markle, D.F., Harris, P.M. & Toole, C.L. (1992) Metamorphosis and an overview of early-life-history stages in Dover sole *Microstomus pacificus*. *Fishery Bulletin*, **90**, 285–301.

May, H.M.A. & Jenkins, G.P. (1992) Patterns of settlement and growth of juvenile flounder *Rhombosolea tapirina* determined from otolith microstructure. *Marine Ecology Progress Series*, **79**, 203–214.

Miller, J.M., Burke, J.S. & Fitzhugh, G.R. (1991) Early life history patterns of Atlantic North-American flatfish: likely (and unlikely) factors controlling recruitment. *Netherlands Journal of Sea Research*, **27**, 261–275.

Millner, R.S. & Whiting, C.L. (1996) Long-term changes in growth and population abundance of sole in the North Sea from 1940 to the present. *ICES Journal of Marine Science*, **53**, 1185–1195.

Minami, T. & Tanaka, M. (1992) Life history cycles in flatfish from the northwestern Pacific, with particular reference to their early life histories. *Netherlands Journal of Sea Research*, **29**, 35–48.

Modin, J. & Pihl, L. (1994) Differences in growth and mortality of juvenile plaice, *Pleuronectes platessa* L., following normal and extremely high settlement. *Netherlands Journal of Sea Resesearch*, **32**, 331–341.

Modin, J., Fagerholm, B., Gunnarsson, B. & Pihl, L. (1996) Changes in otolith microstructure at metamorphosis of plaice, *Pleuronectes platessa* L. *ICES Journal of Marine Science*, **53**, 745–748.

Mollander, A.R. & Mollander-Swedmark, M. (1957) Experimental investigations on variation in plaice (*Pleuronectes platessa*, L.). *Reports of the Institute of Marine Research, Lysekil*, **7**, 3–45.

Moore, J.W. & Moore, I.A. (1976) The basis of food selection in flounders, *Platichthys flesus* (L.) in the Severn estuary. *Journal of Fish Biology*, **9**, 139–156.

Morgan, M.J. & Colbourne, E.B. (1999) Variation in maturity-at-age and size in three populations of American plaice. *ICES Journal of Marine Science*, **56**, 673–688.

Munk, K.M. (2001) Maximum ages of groundfishes in waters off Alaska and British Columbia and considerations of age determination. *Alaska Fishery Research Bulletin*, **8**, 12–21.

Nash, R.D.M., Geffen, A.J. & Santos, R.S. (1991) The wide-eyed flounder *Bothus podas* Delaroche a singular flatfish in varied shallow-water habitats of the Azores. *Netherlands Journal of Sea Research*, **27**, 367–373.

Nash, R.D.M., Geffen, A.J. & Hughes, G. (1992) Winter growth of juvenile plaice on the Port Erin Bay (Isle of Man) nursery ground. *Journal of Fish Biology*, **41**, 209–215.

Nash, R.D.M., Geffen, A.J. & Hughes, G. (1994) Individual growth of juvenile plaice (*Pleuronectes platessa* L.) on a small Irish Sea nursery ground (Port Erin Bay, Isle of Man, UK). *Netherlands Journal of Sea Research*, **32**, 369–378.

Nash, R.D.M., Witthames, P.R., Pawson, M. & Alesworth, E. (2000) Regional variability in the dynamics of reproduction and growth of Irish Sea plaice, *Pleuronectes platessa* L. *Journal of Sea Research*, **44**, 55–64.

Neave, D.A. (1984) The development of visual acuity in larval plaice (*Pleuronectes platessa* L.) and turbot (*Scophthalmus maximus* L.). *Journal of Experimental Marine Biology and Ecology*, **78**, 167–175.

Neave, D.A. (1985) The dorsal light reactions of larval and metamorphosing flatfish. *Journal of Fish Biology*, **26**, 629–640.

Neuman, M.J., Witting, D.A. & Able, K.W. (2001) Relationships between otolith microstructure, otolith growth, somatic growth and ontogenetic transitions in two cohorts of windowpane. *Journal of Fish Biology*, **58**, 967–984.

Nottage, A.S. & Perkins, E.J. (1983) The biology of solenette, *Buglossidium luteum* (Risso), in the Solway Firth. *Journal of Fish Biology*, **22**, 21–27.

Olsen, Y., Evjemo, J.O. & Olsen, A. (1999) Status of the cultivation technology for production of Atlantic halibut (*Hippoglossus hippoglossus*) juveniles in Norway/Europe. *Aquaculture*, **176**, 3–13.

Osse, J.W.M. & Van den Boogaart, J.G.M. (1997) Size of flatfish larvae at transformation, functional demands and historical constraints. *Journal of Sea Research*, **37**, 229–239.

Penttila, J. & Dery, L.M. (1988) Age determination methods for northwest Atlantic species. *NOAA Technical Report* NMFS 72.

Peterman, R.M. & Bradford, M.J. (1987) Density-dependent growth of age 1 English sole (*Parophrys vetulus*) in Oregon and Washington coastal waters. *Canadian Journal of Fisheries and Aquatic Sciences*, **44**, 48–53.

Phelan, B.A., Goldberg, R., Bejda, A.J., *et al.* (2000) Estuarine and habitat-related differences in growth rates of young-of-the-year winter flounder (*Pseudopleuronectes americanus*) and tautog (*Tautoga onitis*) in three northeastern estuaries. *Journal of Experimental Marine Biology and Ecology*, **247**, 1–28.

Pihl, L. (1989) Abundance, biomass and production of juvenile flatfish in southeastern Kattegat. *Netherlands Journal of Sea Research*, **24**, 69–81.

Pittman, K., Skiftesvik, A.B. & Harboe, T. (1989) Effect of temperature on growth rates and organogenesis in the larvae of halibut (*Hippoglossus hippoglossus* L.). *Rapports et Procès-Verbaux des Réunions Conseil International pour l'Exploration de la Mer*, **191**, 421–430.

Poxton, M.G., Eleftheriou, A. & McIntyre, A.D. (1983) The food and growth of 0-group flatfish on nursery grounds in the Clyde Sea area. *Estuarine and Coastal Shelf Science*, **17**, 319–337.

Purdom, C.E. (1974) Variation in fish. In: *Sea Fisheries Research* (ed. F.R. Harden-Jones). pp. 347–355. Elek Science, London.

Rabe, J. & Brown, J.A. (2000) A pulse feeding strategy for rearing larval fish: an experiment with yellowtail flounder. *Aquaculture*, **191**, 289–302.

Rabe, J. & Brown, J.A. (2001) The behavior, growth, and survival of witch flounder (*Glyptocephalus cynoglossus*) larvae in relation to prey availability: adaptations to an extended larval period. *Fishery Bulletin*, **99**, 465–474.

Reibisch, J. (1899) Ueber die Einzahl bei *Pleuronectes platessa* und die Altersbestimmung dieser Form aus den Otolithen. *Wissenschaftliche Meeresuntersuchungen (Kiel)*, **4**, 233–248.

Reichert, M.J.M. (1998) *Etropus crossotus*, an annual flatfish species; age and growth of the fringed flounder in South Carolina. *Journal of Sea Research*, **40**, 323–332.

Reichert, M.J.M., Dean, J.M., Feller, R.J. & Grego, J.M. (2000) Somatic growth and otolith growth in juveniles of a small subtropical flatfish, the fringed flounder, *Etropus crossotus*. *Journal of Experimental Marine Biology and Ecology*, **254**, 169–188.

Ricker, W.E. (1975) Computation and interpretation of biological statistics of fish populations. *Bulletin of the Fisheries Research Board of Canada*, **191**, 1–382.

Rijnsdorp, A.D. (1989) Maturation of male and female North Sea plaice (*Pleuronectes platessa* L.). *Journal du Conseil International pour l'Exploration de la Mer*, **46**, 35–51.

Rijnsdorp, A.D. (1990) The mechanism of energy allocation over reproduction and somatic growth in female North Sea plaice, *Pleuronectes platessa* L. *Netherlands Journal of Sea Research* **25**, 279–290.

Rijnsdorp, A.D. (1993a) Fisheries as a large-scale experiment on life-history evolution: disentangling phenotypic and genetic effects in changes in maturation and reproduction of North Sea plaice, *Pleuronectes platessa* L. *Oecologia*, **96**, 391–401.

Rijnsdorp, A.D. (1993b) Relationship between juvenile growth and the onset of sexual maturity of female North Sea plaice, *Pleuronectes platessa*. *Canadian Journal of Fisheries and Aquatic Sciences*, **50**, 1617–1631.

Rijnsdorp, A.D. (1994) Population regulating processes during the adult phase in flatfish. *Netherlands Journal of Sea Research*, **32**, 207–223.

Rijnsdorp, A.D. & Van Beek, F.A. (1991) Changes in growth of plaice *Pleuronectes platessa* L. and sole *Solea solea* (L.) in the North Sea. *Netherlands Journal of Sea Research*, **27**, 441–457.

Rijnsdorp, A.D. & Van Leeuwen, P.I. (1992) Density-dependent and independent changes in somatic growth of female North Sea plaice *Pleuronectes platessa* between 1930 and 1985 as revealed by back-calculation of otoliths. *Marine Ecology Progress Series*, **88**, 19–32.

Rijnsdorp, A.D. & Van Leeuwen, P.I. (1996) Changes in growth of North Sea plaice since 1950 in relation to density, eutrophication, beam-trawl effort, and temperature. *ICES Journal of Marine Science*, **53**, 1199–1213.

Rogers, S.I. (1994) Population density and growth rate of juvenile sole *Solea solea* (L). *Netherlands Journal of Sea Research*, **32**, 353–360.

Ryland, J.S. (1966) Observations on the development of larvae of plaice, *Pleuronectes platessa* L., in aquaria. *Journal du Conseil International pour l'Exploration de la Mer*, **30**, 177–195.

Sampaio, L.A. & Bianchini, A. (2002) Salinity effects on osmoregulation and growth of the euryhaline flounder *Paralichthys orbignyanus*. *Journal of Experimental Marine Biology and Ecology*, **269**, 187–196.

Sampson, D.B. & Al Jufaily, S.M. (1999) Geographic variation in the maturity and growth schedules of English sole along the US west coast. *Journal of Fish Biology*, **54**, 1–17.

Seikai, T., Tanangonan, J.B. & Tanaka, M. (1986) Temperature influence on larval growth and metamorphosis of the Japanese flounder *Palalichthys olivaceus* in the laboratory. *Bulletin of the Japanese Society of Scientific Fisheries*, **52**, 977–982.

Shelbourne, J.E. (1957) The feeding and condition of plaice larvae in good and bad plankton patches. *Journal of the Marine Biological Association of the United Kingdom*, **36**, 539–552.

Shelbourne, J.E., Riley, J.D. & Thacker, G.T. (1963) Marine fish culture in Britain. I. Plaice rearing in closed circulation at Lowestoft, 1957–1960. *Journal du Conseil International pour l'Exploration de la Mer*, **28**, 50–69.

Sogard, S.M. (1991) Interpretation of otolith microstructure in juvenile winter flounder (*Pseudopleuronectes americanus*): ontogenic development, daily increment validation, and somatic growth relationships. *Canadian Journal of Fisheries and Aquatic Sciences*, **48**, 1862–1871.

Sogard, S.M. (1992) Variability in growth rates of juvenile fishes in different estuarine habitats. *Marine Ecology Progress Series*, **85**, 35–53.

Sogard, S.M. & Able, K.W. (1992) Growth variation of newly settled winter flounder (*Pseudopleuronectes americanus*) in New Jersey estuaries as determined by otolith microstructure. *Netherlands Journal of Sea Research*, **29**, 163–172.

Sogard, S.M., Able, K.W. & Hagan, S.M. (2001) Long-term assessment of settlement and growth of juvenile winter flounder (*Pseudopleuronectes americanus*) in New Jersey estuaries. *Journal of Sea Research*, **45**, 189–204.

Steele, J.H. & Edwards, R.R.C. (1970) The ecology of 0-group plaice and common dabs in Loch Ewe. IV. Dynamics of the plaice and dab population. *Journal of Experimental Marine Biology and Ecology*, **4**, 174–187.

Szedlmayer, S.T. & Able, K.W. (1992) Validation studies of daily increment formation for larval and juvenile summer flounder, *Paralichthys dentatus*. *Canadian Journal of Fisheries and Aquatic Sciences*, **49**, 1856–1862.

Tait, M.J. & Hickman, R.W. (2001) Reproduction, gamete supply and larval rearing of New Zealand turbot *Colistium nudipinnis* (Waite 1910) and brill *Colistium guntheri* (Hutton 1873): a potential new aquaculture species. *Aquaculture Research*, **32**, 717–725.

Takahashi, Y. (1994) Influence of stocking density and food at late-phase of larval period on hypermelanosis on the blind body side in juvenile Japanese flounder. *Nippon Suisan Gakkaishi*, **60**, 593–598.

Terwilliger, M.R. & Munroe, T.A. (1999) Age, growth, longevity, and mortality of blackcheek tonguefish, *Symphurus plagiusa* (Cynoglossidae: Pleuronectiformes), in Chesapeake Bay, Virginia. *Fishery Bulletin*, **97**, 340–361.

Toole, C.L., Markle, D.F. & Harris, P.M. (1993) Relationships between otolith microstructure, microchemistry, and early life history events in Dover sole, *Microstomus pacificus*. *Fishery Bulletin*, **91**, 732–753.

Van der Veer, H.W. & Witte, J.IJ. (1993) The 'maximum growth/optimal food condition' hypothesis: a test for 0-group plaice *Pleuronectes platessa* in the Dutch Wadden Sea. *Marine Ecology Progress Series*, **101**, 81–90.

Van der Veer, H.W., Pihl, L. & Bergman, M.J.N. (1990) Recruitment mechanisms in North Sea plaice *Pleuronectes platessa*. *Marine Ecology Progress Series*, **64**, 1–12.

Van der Veer, H.W., Berghahn, R. & Rijnsdorp, A.D. (1994) Impact of juvenile growth on recruitment in flatfish. *Netherlands Journal of Sea Research*, **32**, 153–173.

Van der Veer, H.W., Dapper, R. & Witte, J.IJ. (2001) The nursery function of the intertidal areas in the western Wadden Sea for 0-group sole *Solea solea* (L.). *Journal of Sea Research*, **45**, 271–279.

Vassilopoulou, V. & Ondrias, I. (1999) Age and growth of the four-spotted megrim (*Lepidorhombus boscii*) in eastern Mediterranean waters. *Journal of the Marine Biological Association of the United Kingdom*, **79**, 171–178.

Walsh, S.J. (1994) Life history traits and spawning characteristics in populations of long rough dab (American plaice) *Hippoglossoides platessoides* (Fabricius) in the North-Atlantic. *Netherlands Journal of Sea Research*, **32**, 241–254.

Williams, P.J. & Brown, J.A. (1992) Development changes in the escape response of larval winter flounder *Pleuronectes americanus* from hatch through metamorphosis. *Marine Ecology Progress Series*, **88**, 185–193.

Wimpenny, R.S. (1953) *The Plaice.* Edward Arnold, London.

Wyatt, T. (1972) Some effects of food density on the growth and behaviour of plaice larvae. *Marine Biology*, **14**, 210–216.

Zijlstra, J.J., Dapper, R. & Witte, J.IJ. (1982) Settlement, growth and mortality of post-larval plaice (*Pleuronectes platessa* L.) in the western Wadden Sea. *Netherlands Journal of Sea Research*, **15**, 250–272.

Chapter 8
Ecology of juvenile and adult stages of flatfishes: distribution and dynamics of habitat associations

Kenneth W. Able, Melissa J. Neuman and Håkan Wennhage

8.1 Introduction

The juvenile and adult stages of flatfishes are of past and continuing interest, largely because they are the focus of harvests by commercial and recreational fishermen from temperate to tropical waters of the world's oceans and estuaries. As such, they have received much attention in the published literature, especially because of the recent Flatfish Symposia (see Preface), and as a result, many of the other chapters in this book. In fact, these complementary chapters such as those on reproduction (Chapter 4), trophic ecology (Chapter 9), age and growth (Chapter 7) and behaviour (Chapter 10) provide some boundaries for our own comments on the ecology of the juvenile and adult stages.

Our purpose is to review and synthesise the available information on the distribution and dynamics of habitat use for flatfishes including the roles that ontogeny, and short (tidal, diel, lunar, episodic) and long (annual, decadal) temporal scales have on habitat use and habitat quality, and thus identification of nurseries. We further define the limits of our treatment with a number of definitions.

8.2 Definitions

Ecology: The distribution and abundance of flatfishes in space and time.
Habitat: Often defined as simply the place where an animal lives (Krebs 1994) yet this is often very difficult to characterise more fully, partly because of the complex life history of flatfishes in which the different life history stages occupy different habitats. As an example, just for juveniles and adults these may include settlement habitats, nursery habitats, feeding habitats and spawning habitats. Furthermore, attempts to determine habitats are confounded by the high mobility of primarily benthic fishes such as flatfishes. We recognise this difficulty by focusing, whenever possible, on the dynamics of flatfish habitat associations. Wherever possible we make a distinction between habitat 'use' and habitat 'selection' (Kramer *et al.* 1997), assuming that distinction is available from individual papers.
Nursery: A habitat is a nursery of a species if its contribution per area to the production of recruits to the adult population is greater, on average, than production from all other juvenile

habitats (Beck *et al.* 2001). Furthermore, nursery habitats support greater adult recruitment from any contribution of four factors: (1) density, (2) growth and (3) survival of juveniles, and (4) movement to adult habitats.

Juveniles: The portion of the life history that, for most species, includes from settlement and/or the completion of metamorphosis to sexual maturity.

Adults: The portion of the life history from sexual maturity to senescence.

8.3 Distribution and ontogeny

Differences in ontogenetic state (e.g. juvenile vs adult for the purposes of this chapter) interact with seasonal fluctuations in abiotic (e.g. temperature, salinity, dissolved oxygen, substratum, depth) and biotic (e.g. predation, competition, food availability, parasites and disease) factors to produce differential distribution and habitat use at a variety of spatial and temporal scales (Gibson *et al.* 1996). Variability in one or a multitude of these factors may generate different levels of recruitment variation both within and among species. Therefore, given that many flatfishes are economically important, and given the recent emphasis placed on identifying and conserving essential fish habitat (NOAA 1996; Able 1999; Beck *et al.* 2001), understanding the effects of abiotic and biotic factors on the distribution dynamics of flatfishes during ontogeny is critical for those who need to make informed management decisions. Clearly, the degree to which we can address the above is limited by the scope of the published literature. Of the 15 recognised family level taxa of flatfishes (Chapter 2), only six are well represented in the literature (Paralichthyidae, Pleuronectidae, Bothidae, Scophthalmidae, Soleidae, Cynoglossidae), in large part because these families contain species of economic importance.

Many flatfish studies have aimed to identify which abiotic factors significantly influence the distribution of juveniles and adults belonging to a particular species, or species complex. Investigation of approximately 50 species from the Atlantic, 35 from the Pacific, two species from the Indian Ocean and one species from the Arctic Ocean, in addition to approximately 10 species from the Sea of Japan, China, Okhotsk and Bering seas, and four species from the Caribbean Sea, have concluded that a combination of variables – most often including depth, temperature, salinity and substratum type – are the best environmental predictors of habitat utilisation within a study area (e.g. Powell & Schwartz 1977; Kramer 1991; Jager *et al.* 1993; Szedlmayer & Able 1993; Vetter *et al.* 1994; Norcross *et al.* 1995; Allen & Baltz 1997; Albert *et al.* 1998; Rogers *et al.* 1998; Steves *et al.* 1999; Walsh *et al.* 1999; Sullivan *et al.* 2000; Stoner & Abookire 2002; Diaz de Astarloa & Fabré 2003; Stoner & Ottmar 2003). These variables, particularly temperature and salinity, may vary dramatically across seasons in temperate regions and may exhibit gradients at a variety of spatial scales. Other factors may vary over a range of spatial scales (e.g. depth and substratum type), but are more stable across seasons. Examination of these abiotic variables is common, perhaps because they are inexpensive to measure and fairly easy to quantify, and thus, the ubiquitous citation of these factors being correlated with flatfish abundance may be a methodological artifact. A caution noted in many studies is that while multivariate analyses may indicate that certain abiotic factors, or combinations of them, are correlated with patterns of distribution and abundance, they may not be the causal mechanisms responsible for producing these patterns. Instead, abiotic variables may be proxies for biological attributes of the habitat, such as reduced risk of predation or increased food availability and it is these that are truly responsible for controlling patterns of flatfish distribution. If so, flatfishes could use the gradients in abiotic factors to locate areas where the biotic factors are likely to match their habitat requirements. Biotic factors have been shown to affect the distribution of approximately 20 species of flatfishes from the Atlantic, 12 species from the Pacific and 2 species from the Indian Ocean, as well as approximately 14 species from the Sea of China and Japan.

A logical starting point for examining changes in flatfish distribution is settlement, given that the start of the juvenile stage coincides with settlement and/or the completion of metamorphosis for most species (see Toole *et al.* 1997 for some exceptions). Settlement patterns are governed by larval supply, but can be modified by larval behaviour and habitat selection during the settlement period. These changes can be expected to be particularly important over long temporal periods (see Sogard *et al.* 2001) or large spatial scales where habitat selection has a limited ability to operate. Pihl *et al.* (2000) examined the spatial variation in the abundance of newly settled plaice, *Pleuronectes platessa*, in the Swedish Skagerrak archipelago by sampling 32 isolated nursery grounds. Plaice settlement densities exhibited significant variation at several spatial scales from hundreds of metres to 200 km. Some of the variation could be explained by differences in habitat characteristics, but these patterns largely seemed to be obscured by differences in larval supply. In California flounder, *Paralichthys californicus*, a large proportion of the larvae settle on open coasts and subsequently migrate to more protected areas assumed to be nursery habitats (Allen & Herbinson 1990; Kramer 1991).

Species where the nurseries only constitute a limited part of the distribution range and/or species where the adults aggregate on spawning grounds must, by definition, have ontogenetic changes in distribution patterns. Furthermore, it appears that ontogenetic shifts in distribution during the juvenile and adult stages of flatfishes are a common phenomenon. In a North Carolina, USA estuary, four of seven species studied (summer flounder, *Paralichthys dentatus*; bay whiff, *Citharichthys spilopterus*; blackcheek tonguefish, *Symphurus plagiusa*; and hogchoker, *Trinectes maculatus*) showed ontogenetic changes in distribution (Walsh *et al.* 1999), and in a Louisiana, USA estuary all of the four most common flatfish species (offshore tonguefish, *Symphurus civitatium*; bay whiff; fringed flounder, *Etropus crossotus*; and blackcheek tonguefish) showed similar changes (Allen & Baltz 1997). Both abiotic and biotic factors were considered to be important in the ontogenetic changes of juvenile flatfishes on the coast of Scotland (Gibson *et al.* 2002). It has been suggested that the nurseries of all flatfish species in Japanese waters seem to be located at the shallow end of the adult's distribution range (Minami & Tanaka 1992). In some species the use of shallow habitats by small juveniles during early ontogeny is followed by movement of larger fish to deeper waters. One of the earliest records of the positive relationship between size-related ontogenetic shifts and depth was found in plaice (Heinke 1913), and the pattern was later termed 'Heinke's law'. This relationship has since been confirmed and demonstrated in a number of other flatfish species (e.g. McCracken 1963; Gibson 1973; Lockwood 1974; Dorel *et al.* 1991; Zimmerman & Goddard 1996). These shallow habitats for other species are often found in several types of coastal habitats such as open beaches (Riley *et al.* 1981; Pihl 1989), sheltered tidal flats (Kuipers 1977; Burke *et al.* 1991), and estuaries (Potter *et al.* 1990; Cabral & Costa 1999) including salt-marsh creeks (Rountree & Able 1992). These examples also show that shallow habitats are used by members of several different families of flatfishes (Paralichthyidae, Pleuronectidae, Scophthalmidae, Soleidae) and that recently settled flatfishes utilise shallow habitats in many parts of the world.

There are, however, exceptions to this pattern of settlement in shallow waters and subsequent dispersal into deeper habitats. In yellowtail flounder (*Limanda ferruginea*) on the Grand Banks of Newfoundland, there was no apparent ontogenetic change in distribution relative to depth and temperature and this was stable among year-classes and over the season (Walsh 1992). Witch (*Glyptocephalus cynoglossus*) along the Canadian east coast showed a negative relationship between fish size and depth during the summer (Powles & Kohler 1970). Despite the exceptions, there is mounting evidence suggesting that shallow-water habitats may enhance survival of early juvenile stages of a large number of flatfish species (reviewed in Gibson 1994). Lacking, however, are studies that provide direct information on the processes that cause individuals of many species to choose greater depth with increasing size. Most of the information on ontogenetic changes in habitat use comes from descriptive studies, and the associations found with different biotic and abiotic factors are consequently correlative in nature. The abiotic factors commonly manifest themselves in easily distinguishable gradients amenable for correlative studies. There is, however, little evidence to support the proposition that the optimum for these variables changes with ontogeny during the juvenile and adult stages of flatfishes (but see Fonds *et al.* 1992). A number of abiotic factors, including different combinations of sediment type, turbidity, salinity, depth and temperature, were correlated with species-specific ontogenetic state of the flatfish assemblage in a Louisiana estuary (Allen & Baltz 1997).

Among biotic factors, prey distribution has been suggested to cause ontogenetic changes in habitat use of summer flounder and southern flounder, *Paralichthys lethostigma* (Burke 1995). These two species settled in the same portion of estuaries in North Carolina, USA, and initially relied on the same food types. As the two species developed, their diets became different and the distribution of the species subsequently diverged, reflecting spatial differences in abundance of the major prey species within the estuary. Other studies have suggested that the size spectrum of the prey is important in habitat choice (e.g. Lockwood 1984; Castillo-Rivera *et al.* 2000). Predation could also be an important part of ontogenetic changes in habitat use. Brown shrimp (*Crangon crangon*) is the major predator on newly settled plaice until plaice reach a size refuge at a length of approximately 30 mm (Van der Veer & Bergman 1987). On tidal flats, plaice initially remain in intertidal pools (Berghahn 1983), minimising the spatial overlap with larger predatory shrimps. On open coasts the early benthic stage of plaice concentrates at depths <1 m (Gibson 1973; Lockwood 1974) reducing the overlap with the shrimps, which are primarily found in deeper waters (Gibson *et al.* 1996, 2002). By the time the spatial overlap between plaice and this predator starts to increase in both areas, a major portion of the plaice population have reached a size where their susceptibility to shrimp predation is low. The influence of brown shrimp is also evident on the small-scale distribution of flounder, *Platichthys flesus* (Modin & Pihl 1996). Similar predator/prey interactions have been identified in Japanese waters for *Crangon affinis* on stone flounder, *Platichthys bicoloratus* (Yamashita *et al.* 1996). Other studies of predation risk suggest patterns of size refugia for winter flounder (*Pseudopleuronectes americanus*) relative to predation by fishes (Manderson *et al.* 1999).

8.3.1 Early juvenile habitat associations

Over 50% of the flatfish papers examined report that juveniles use shallow nearshore areas as habitats. However, because these shallow habitats are logistically easier and less costly to study, and with the recent emphasis on the preservation and conservation of coastal habitats, they have been the emphasis of most of the literature. As methodological advancements are made (e.g. in situ observations with submersibles and/or mounted video camera systems), allowing investigators to broaden the geographic range over which habitat utilisation can be studied, the present-day view of the importance of nearshore nursery habitats to young flatfishes may change (e.g. Sullivan *et al.* 2000).

Early juvenile habitat characteristics vary among and within flatfish species, and it is no surprise that not all newly settled flatfishes utilise shallow, warm estuaries as nurseries. In some cases these types of nearshore nursery habitats are not available; in others, even if these habitats are available, they may not be used by young flatfishes. An example is provided by juvenile pleuronectid flounders occurring in coastal Alaskan waters (Norcross *et al.* 1995; Aboookire & Norcross 1998). The bays and nearshore habitats off Alaska are considerably deeper (>10 m), at similar distances from shore, than those in other parts of the world. The juvenile nursery areas for these species occur between approximately 10 and 80 m depth, where water temperatures are cold (4–11°C) and salinities vary little (30–32 ppt). Newly settled Pacific halibut (*Hippoglossus stenolepis*) and rock sole (*Lepidopsetta bilineata*), arrive in these relatively deep, cold habitats following an inshore migration from even greater depths. The factors controlling such movements remain uncertain. However, even at these depths, fish can experience as much as a 5°C increase in temperature by migrating inshore.

Another example from different populations of a western Atlantic bothid, windowpane (*Scophthalmus aquosus*), suggests that there may be geographical and cohort-specific variation in juvenile habitats. Juveniles occurring on Georges Bank, approximately 50–75 km from the Massachusetts, USA coast, presumably do not make the extensive migrations from these relatively deep waters (27–39 m) to nearshore nursery areas (Morse & Able 1995). However, the pattern exhibited by Middle Atlantic Bight windowpane is different in that the spring-spawned cohort of windowpane initially settle in relatively shallow water (<20 m), and then enter nearby estuaries (Neuman & Able 2003). Autumn-spawned windowpane also initially settle in shallow water on the inner continental shelf, but unlike spring-spawned fish, they do not appear to use estuaries during their first few months of life. Factors controlling the passive and active movements of windowpane to their nursery areas on Georges Bank may be quite different from those in the Middle Atlantic Bight, given that the spatial occurrence of newly settled and larger juveniles are highly correlated on Georges Bank, but differ considerably in the Middle Atlantic Bight, especially with respect to the spring cohort (Morse & Able 1995). In addition, this example illustrates that different controlling mechanisms may be responsible for dissimilar patterns of habitat use between distinct cohorts of the same species, an observation that has rarely been noted in the flatfish literature.

Spawning of some species occurs considerably farther offshore (continental shelf and beyond), almost guaranteeing that juvenile stages will not encounter shallow, nearshore habitats. Two species (yellowtail flounder and Gulf Stream flounder, *Citharichthys arctifrons*) occurring in a region of the northwest Atlantic referred to as the New York Bight and one species occurring in the northeast Pacific (Dover sole, *Microstomus pacificus*) utilise habitats

>40 m depth (Toole *et al.* 1997; Steves *et al.* 1999). Two of these species (Gulf Stream flounder and Dover sole) engage in cross-shelf migrations from initial settlement areas to juvenile habitats. Ekman transport has been suggested as the mechanism by which Dover sole are delivered to the juvenile habitat (Toole *et al.* 1997), but mechanisms of transport for the Gulf Stream flounder are unknown (Steves *et al.* 1999). Unlike these species, yellowtail juveniles exhibit temporal and spatial stability within the nursery area. This stability is thought to occur because early life stages of yellowtail flounder track a cold cell of remnant winter-bottom water (approx. 6°C), a dominant physical feature that occurs annually on the continental shelf of the New York Bight (Steves *et al.* 1999; Sullivan *et al.* 2000).

For a small subset of species, which are resident in estuaries, settlement into juvenile habitat is dynamic. Evidence suggests that for the hogchoker, a western Atlantic soleid, summer spawning occurs in the polyhaline portions of estuaries and that settling larvae follow (or may be transported by the salt wedge transport system) up the salinity gradient to the salt/freshwater interface (0–8 ppt) in a short period of time (Dovel *et al.* 1969). For this species, congregation in the juvenile habitat may be more dependent on the salinity, and perhaps turbidity, conditions. Preliminary evidence suggests that some populations of the South African blackhand sole, *Solea bleekeri*, may follow a similar pattern of habitat use (Cyrus 1991). As the preceding examples demonstrate, recently settled flatfishes tend to occupy distinct habitats during discrete time intervals and these spatial and temporal boundaries may be explained, in part, by abiotic characteristics, most commonly depth, salinity and nature of the substratum (Powell & Schwartz 1977; Jager *et al.* 1993; Le Clus *et al.* 1994; Norcross *et al.* 1995; Abookire & Norcross 1998) or extent of river plumes in coastal regions (Le Pape *et al.* 2003a, b).

On the other hand, a growing body of literature has focused on the role that biotic factors play in governing habitat dynamics. Studies from around the world (MacDonald & Green 1986; Le Clus *et al.* 1994; Tokranov & Maksimenkov 1994; Toepfer & Fleeger 1995a, b; Aarnio *et al.* 1996; Gibson *et al.* 1998) demonstrate that prey availability, predator presence/absence, and species-specific morphology and behaviour may be driving differential patterns of settlement among juvenile flatfishes more than the physical properties of the habitat itself. The majority of these studies suggest that dietary partitioning, along with the staggered, seasonal ingress of different species into the juvenile habitat, may serve to limit predatory and competitive interactions. A series of examples follows.

Spatial, temporal and dietary overlap among the juveniles of five flatfish species (plaice; dab, *Limanda limanda*; turbot, *Scophthalmus maximus*; brill, *S. rhombus*; and common sole, *Solea solea*) was examined on a sandy beach along the Belgian west coast (Beyst *et al.* 1999). Even though this habitat is physically dynamic and characterised by low spatial heterogeneity, the species exhibited little dietary overlap and were segregated spatially and temporally during their time of residency. In another example, juvenile summer flounder and southern flounder tended to be spatially segregated within North Carolina, USA estuaries (Powell & Schwartz 1977; Burke 1995). Substratum and salinity were the two most important abiotic factors governing their distributions, with higher southern flounder abundances over clay/silt or organic-rich bottom in areas of low salinity and higher summer flounder abundances over sand bottom in areas of moderate to high salinity. Thus, spatial segregation may have resulted from feeding differences between the species that were related to their habitat preferences. Observations made in southern New Jersey, USA estuaries indicate that age-0 summer flounder make extensive use of salt-marsh creeks. Their movements into and out of the creeks may be aided by tidal stream transport, and residency and movement within the creeks was correlated with specific habitat parameters including temperature, salinity, dissolved oxygen and food availability (Rountree & Able 1992; Szedlmayer & Able 1993). The summer movements of flatfishes and benthic crustaceans in Gullmarsfjord Bay, Sweden were attributed to a combination of foraging activities, predator avoidance and selection of suitable environmental conditions (Gibson et al. 1998).

These studies are some of the few that provide qualitative information regarding the potential impact that predators and competitors may have on the movements of juvenile flatfishes. In related efforts, attempts to calculate mortality estimates due to predation by benthic crustaceans, fishes and birds on juvenile flatfishes in European, South African, US and Japanese waters have been made (Whitfield & Blaber 1979; Van der Veer & Bergman 1987; Tanaka et al. 1989; Beverton & Iles 1992a, b; Witting 1995; Gibson et al. 1998; Leopold et al. 1998; Manderson et al. 1999). While these studies provide important estimates of predation rates imposed by predators on juvenile flatfishes, they are limited in terms of the study duration, the spatial scale examined, and the assumptions made regarding prey and predator population sizes. With these limitations, it has been very difficult to predict whether mortality due to predation contributes significantly to habitat use and to inter-annual variation in recruitment. The European literature on flatfishes suggests that mortality imposed by benthic crustaceans only dampens inter-annual variability in year-class strength (Van der Veer & Bergman 1987; Pihl 1990; Van der Veer et al. 1997).

8.3.2 Late juvenile and adult habitat associations

Late juvenile and adult flatfishes tend to be distributed over broader spatial and temporal scales than early juveniles. This difference is probably because the known life histories of many species are such that early juveniles congregate in nearshore, protected areas, or other nursery areas, that are spatially limited. Unfortunately, the available information on late juvenile and adult distributions (except perhaps with regard to spawning distributions) is less detailed than those for early juveniles. Of the abiotic factors thought to affect the distribution of late juvenile and adult flatfishes, temperature, depth and substratum type are cited most regularly (Rijnsdorp et al. 1992; Díaz de Astarloa & Munroe 1998; Rogers et al. 1998; Yamashita et al. 2001). Salinity may play a less important role, except perhaps with respect to species that inhabit estuaries throughout their lives, such as certain populations of hogchoker and blackhand sole (Dovel et al. 1969; Cyrus 1991). It has been proposed that turbidity affects the density of at least two species of flatfishes that are known to form aggregations on patches of homogenous substrata in the Sea of Japan (Gomelyuk & Shchetkov 1999). Biotic factors, especially prey availability, also influence the timing of movements (Tokranov & Maksimenkov 1994) and the formation of mixed patchily distributed conspecific and groups of adult flatfishes (Gomelyuk & Shchetkov 1999).

Dispersal of flatfishes from juvenile to adult habitats, whether seasonal or size/age-related, has been attributed to changes in dietary requirements, reduced risk of predation by larger predators and an assortment of abiotic variables, most commonly the avoidance of decreasing temperatures in shallow water with the onset of winter in temperate climates (see Seasonal cycles, p. 172). Correlations between dissolved oxygen levels and abundance of some flatfishes have been identified. A negative relationship between oxygen concentrations and the biomass of plaice and dab was found in Scandinavian waters (southeast Kattegat) where hypoxic (<3 mg l^{-1}) conditions occur during the autumn months (Petersen & Pihl 1995). Other studies have shown that some flatfishes may be particularly well adapted to withstanding low dissolved oxygen conditions. For example, the spawning adults of Dover sole occupy a depth range from 600 to 1000 m where there is an oxygen minimum zone (dissolved oxygen levels 0.7 mg l^{-1}) (Vetter *et al.* 1994).

While the previous examples illustrate that the information available on factors controlling the distribution of late juvenile and adult flatfishes is often descriptive and somewhat speculative, the literature on factors that affect timing and location of spawning in adult flatfishes is relatively plentiful. This difference occurs because abundances can be particularly high during discrete time intervals and in specific locations. In addition, it is relatively easy to quantify the reproductive state of field-captured adults through a variety of techniques (e.g. gonadosomatic indices), enabling investigators to identify where and when spawning adults occur (Wilk *et al.* 1990). In instances where data are available the spatial boundaries of spawning in some flatfishes may be related to depth (Zimmermann & Goddard 1996; Tok & Biryukov 1998), salinity and hydrodynamics (Stoner *et al.* 1999).

8.3.3 Dynamics of habitat associations

The estuaries and marine habitats of flatfishes can be highly dynamic environments where biotic and abiotic factors change over a multitude of timescales. The distribution of mobile animals, such as flatfishes, would therefore be expected to change over time in response to the changing environment. Some changes are cyclic (e.g. seasonal) and the animals may have adapted their habitat choices through behaviour or life cycle characteristics relative to the predictability of the dynamic environment. Other changes may be episodic (e.g. hypoxia) and a direct response to the changing environment would be expected unless the magnitude and extent of the change causes mortality. Changes in habitat characteristics could also be long-term (e.g. climatic) leading to gradual changes in distribution that could be mediated by an evolutionary response by the population. There are, however, most certainly constraints on the dynamics of habitat associations that could lead to discrepancies between the spatial pattern in occurrence of flatfishes in relation to habitat quality. For example, flatfishes might have limited capabilities to locate the habitat with the highest habitat quality at all times. The cost of movement/migration may also exceed the benefits of changing habitat.

8.3.3.1 Tidal, diel and seasonal cycles

Tidal and diel cycles. Changes in distribution with the tidal and diel cycles have primarily been studied in flatfishes utilising shallow areas as feeding areas and little information is available for deeper waters. The tidal and diel cycles are normally out of phase, which may add to the complexity of the temporal changes in distribution. However, the difference in frequency does makes it possible to distinguish between the tidal and diel component of the dynamics in habitat associations through repeated sampling (see Gibson *et al.* 1996). The spatial scale of movement would be expected to be small because of the high frequency of tidal and diel cycles. Ultrasonic tagging has shown that juvenile summer flounder (Szedlmayer & Able 1993) and adult flounder (Wirjoatmodjo & Pitcher 1984) move on the order of 10^2–10^3 m with each tide. In several examples, tidal migrations, which may be species-specific (Gibson 1973), have been attributed to increased feeding potential in the intertidal zone (Tyler 1971; Berghahn 1987; Raffaelli *et al.* 1990) and to predator avoidance (Gibson 1973; Ansell & Gibson 1990). The influence of tides can also extend into the subtidal zone. Juvenile summer flounder utilise salt-marsh creeks as feeding grounds during high tide (Rountree & Able 1992) but these areas are characterised by substantial fluctuations in environmental conditions. Movements out of the marsh creeks at low tide may therefore partly result from 'behavioural homeostasis', whereby the fish move to remain at nearly constant physical conditions over the tidal cycle (Szedlmayer & Able 1993).

Diel movements occur both in combination with tidal movements (Burrows *et al.* 1994) and alone where the tidal amplitude is negligible (Bregneballe 1961; Gibson *et al.* 1998). In the latter, flatfishes move inshore during the evening and return to deeper waters at dawn, and the movements and feeding rhythms of potential predators are sometimes reported to match the diel cycles. Gadoids can be important predators on flatfishes in tidal (Ellis & Gibson 1995) and non-tidal areas (Pihl 1982; Arnott & Pihl 2000), and these gadoids are most abundant in the deeper parts of the juvenile habitat during the night (Gibson *et al.* 1996, 1998). Predation by diurnal predators such as birds can also be of significant magnitude (e.g. Summers 1979; Leopold *et al.* 1998), but it has not been established if this predation risk affects the diel movements of flatfishes or of their fish predators.

For one of the most extensively studied flatfishes, plaice, tidal and diel movements have been studied in several areas within the species range. On the extensive mud flats of the Wadden Sea, newly settled plaice remain in the intertidal zone during the entire tidal cycle and tidal migration patterns gradually develop during the course of approximately 1 month (Berghahn 1983; Van der Veer & Bergman 1986). This change in behaviour corresponds to increasing fluctuations in temperature and oxygen saturation that may force them to leave the tidal pools at low tide. Once the tidal movements have developed, diel comparisons show that the tidal flats are used to the same extent during the day and night (Berghahn 1986). On open coasts with a tidal regime, plaice primarily seem to settle in the subtidal zone and initiate onshore and offshore movements with each tidal cycle (Gibson 1973; Lockwood 1974; Wennhage, personal observations). The distribution in these subtidal areas is modified by a superimposed diel movement pattern with onshore tidal movements being more pronounced during the night (Burrows *et al.* 1994). On the Swedish west coast, where the tidal amplitude is low (<0.3 m) the diel movements resemble those on open tidal coasts (Gibson *et al.* 1998).

Seasonal cycles. Descriptive studies of flatfish populations have commonly found changes in the distribution patterns over the season. Seasonal changes in habitat use may, however, be hard to separate from those caused by ontogenetic changes in habitat use. This is especially true for the juvenile stage, as a large proportion of the flatfish species have their life history adjusted to produce offspring at a certain time of the year. We have therefore restricted our comments to a few examples where the seasonal movements persist for a considerable part of the life cycle. A number of flatfish species use estuarine habitats seasonally. Flounder use estuaries as juvenile habitat and the adult fish also use the estuaries as feeding grounds during the summer (Bregneballe 1961). Summer flounder show a similar use of estuaries (Able & Kaiser 1994). Plaice use shallow coastal areas as juvenile habitat during their first summer and a proportion of these fishes return as I and II age-group juveniles the following summers

(Kuipers 1977). Winter flounder may display a latitudinal change in seasonal movements along the east coast of North America (McCracken 1963). During the summer this species leaves the shore zone in areas where the water temperature rises above 15°C to reside in colder waters offshore. In the winter the northern populations move to deeper water when the temperature gradient is reversed (but see Howe & Coates 1975 and Phelan et al. 2000 for different interpretations).

In most flatfish species in temperate regions, seasonal dispersal of late juveniles from nearshore habitats into deeper, oceanic waters is common during the autumn months when water temperatures inshore begin to drop. This temperature decline may serve as a cue to juveniles of some species (e.g. smallmouth flounder, *Etropus microstomus*, Able & Fahay 1998; summer flounder, Szedlmayer & Able 1993) that it is time to make seasonal offshore migrations to waters that exhibit more stable temperatures during the winter months. In polar and tropical regions, one might expect water temperature to play less of a role in explaining seasonal movements of flatfishes, but so few studies have been conducted that it remains uncertain how abiotic factors affect flatfish distributions in these regions. Movements of late juveniles and adults in tropical regions may be mediated by seasonal changes in rainfall and the subsequent effect this may have on salinity, rather than by changes in temperature (Manickchand-Heileman 1994; Sánchez-Gil et al. 1994).

Bathymetric shifts to deeper water are not limited to those flatfishes that utilise nearshore areas as juveniles. For example, species that settle in relatively deep, oceanic habitats, such as arrowtooth flounder (*Reinhardtius stomias*) and Dover sole (Hunter et al. 1990; Rickey 1995), undergo seasonal migrations to even deeper habitats, but the reasons for these migrations are not well understood. Field surveys have suggested that late juveniles and adults undergo seasonal offshore migrations to particular depth zones and substratum characteristics (Ford 1965; Vetter et al. 1994; Zimmermann & Goddard 1996; Tok & Biryukov 1998). It has been hypothesised that some flatfishes are fairly immobile, may remain buried in sediments, and grow very little during winter months in temperate regions (Able & Fahay 1998; Hales & Able 2001; Neuman et al. 2001). Following this season of inactivity, adults may migrate inshore to shallower depths during the warmer months to presumably take advantage of warmer waters and enhanced feeding opportunities due to increased productivity (Bregneballe 1961; Tokranov & Maksimenkov 1994).

Habitat associations change most dramatically, for both juveniles and adults, when they make seasonal or reproductive migrations. A prior, extensive review of the behaviour of flatfishes, including that for juveniles and adults, has covered the topic of migration and the possible cues (Gibson 1997). One of the most intriguing findings in recent decades is the ability of adults to use vertical migrations and thus tidal stream transport during migration. To date, this has only been verified for plaice (Arnold & Metcalfe 1995) and common sole (de Veen 1978; Greer Walker et al. 1980). Tidal stream transport might also occur in the juveniles or adults of other species. For example, summer flounder migrate long distances (Able & Kaiser 1994; Packer & Hoff 1999), are known to move with tidal currents in a selective manner (Szedlmayer & Able 1993) and are known to move off the bottom regularly (Olla et al. 1972). Other species that might be likely to exhibit tidal stream transport include California flounder (Kramer 1991) and Japanese flounder, *Paralichthys olivaceus* (Tanaka et al. 1989).

During the adult stage most migrations are associated with spawning. As an example, tag/recapture studies have clarified migratory pathways for summer flounder and these help to explain the difference in the timing of spawning in different parts of the species range (Burke et al. 2000). However, one intriguing tag/recapture study in northern New Jersey waters (Phelan 1992) noted the apparent presence of non-migratory adult winter flounder on the continental shelf during the winter when they are expected to be in the estuary in preparation for spawning there (Bigelow & Schroeder 1953). Alternatively, spawning may occur on the continental shelf during the winter (Howe et al. 1976) or the location of reproduction varies geographically. It would be interesting to confirm if there are non-migratory, spawning winter flounder on the shelf in the winter and to determine whether the same phenomenon occurs in other flatfishes as well.

Long-term changes. There are few examples available, beyond those of fishing impacts, but observations over a 7-year period in Narragansett Bay, USA, suggested that a reduction in the abundance of winter flounder was related to climatic changes, i.e. a trend for increasing temperatures, while that for a co-occurring species, windowpane, did not vary (Jeffries & Johnson 1974). In another example, the increasing occurrence and abundance of green macroalgae in recent decades (Pihl et al. 1999) may have influenced vulnerability to predation of plaice (Wennhage 2002).

Episodic events. Extremes in abiotic factors such as temperature, salinity and oxygen concentration would be expected to affect the dynamics of flatfish habitat use. Extremely low water temperature during the winter can lead to thermal death. During the severe winter of 1963 in the southern North Sea, large numbers of dead adult common sole were caught by research and fishing vessels (Woodhead 1964). Other flatfish species found dead, but in lower numbers, were dabs, plaice, brill and turbot. It was assumed that most of the mortality was caused by the low temperature per se, even though a skin disease was also observed in a number of the species. In other examples, newly settled plaice that remain in tidal pools during calm conditions and sunshine can, however, be subjected to lethal temperatures and detrimental UV radiation (Berghahn et al. 1993). Also, oxygen concentrations can drop quickly at high rates of respiration or decomposition in estuarine and marine ecosystems. This drop could be especially prominent in stratified water bodies or where water circulation is impaired for other reasons (e.g. vegetated areas). The Kattegat area between Sweden and Denmark is shallow (average depth 23 m) and has a pronounced halocline at 15–20 m depth, leaving a thin layer of poorly mixed bottom water. During the autumn when oxygen concentrations are at their lowest levels, the biomass of flatfishes can be reduced drastically (Pihl 1989; Baden et al. 1990; Petersen & Pihl 1995). Migration of fishes was considered to be the main cause of this change in distribution, as fishermen reported unusually large catches of flatfishes in gill nets at or above the halocline. A similar migratory response of adult summer flounder, away from low dissolved oxygen habitats on the continental shelf and onto ocean beaches and adjacent estuaries, has been reported in the New York Bight (Freeman & Turner 1977; Swanson & Sindermann 1979). In plaice juvenile habitats, proliferation of ephemeral macroalgae can concentrate individuals in the remaining unvegetated parts (Wennhage & Pihl 1994). Cage experiments have shown that growth rate is reduced in the presence of these macroalgae and that the oxygen levels under floating algae at night were lower than in the unvegetated areas (Wennhage, unpublished observations). Other studies have demonstrated that habitat suitability for winter flounder, as demonstrated by variation in growth of caged small juveniles, indicates that multiple environmental variables (temperature, salinity) can change rapidly (Manderson et al. 2002).

Typically, we know little about how biotic interactions (e.g. competition, predation, etc.) influence the dynamics of habitat associations. Perhaps one of the more intriguing possibilities is that social behaviour can play an important role. In one example, habitat use for the Caribbean eyed flounder, *Bothus ocellatus*, varied on a diel cycle with sexually dimorphic habitat use (Konstantinou & Shen 1995). Both sexes maintained overlapping territories during the day, when reproduction was occurring and dispersed to 'retirement' sites at night that were tens of metres away. Similarities in social organisation and habitat use have been suggested for other bothids (Moyer *et al.* 1985; Carvalho *et al.* 2003). In another example, two pleuronectids (marbled flounder, *Pseudopleuronectes yokohamae* and littlemouth flounder, *P. herzensteini*) formed distinct groups composed of conspecific or congeneric animals in homogenous habitats (Gomelyuk & Shchetkow 1999). It was suggested that these flatfishes form and maintain groups through visual contact.

8.4 Future emphasis

Enhanced understanding of the distribution of flatfishes and the dynamics of their habitat associations can come from a variety of approaches including improved data gathering and modelling (e.g. Boisclair 2001; Stoner *et al.* 2001), manipulative experiments in the field and numerous technological advances, especially when they can be used in tandem. Some techniques, such as the use of otoliths to back-calculate the life history and ecology, are continuing to improve with the added ability to detect and interpret extreme events (e.g. Berghahn 2000) or natal homing with analyses of microchemistry (e.g. Thorrold *et al.* 2001). Others have used field manipulation experiments to assess habitat-specific measures of growth to address habitat quality such as growth (Sogard 1992; Duffy-Anderson & Able 1999; Phelan *et al.* 2000), feeding success (Duffy-Anderson & Able 2001) and predation, e.g. tethering (Haywood & Pendrey 1996, but see Kneib & Scheele 2000).

Tagging techniques have been enhanced in a number of ways, especially by the miniaturisation of the tags themselves (e.g. coded wire tags, Wallin *et al.* 1997; ultrasonic tags, Boehlert 1997), to environmental data collection with archival tags that dramatically enhances the interpretation of habitat use and movements (Metcalfe & Arnold 1997; Kasai *et al.* 2000). Other approaches, not yet used on flatfishes, include the use of data storage and satellite transmission of that data (Block *et al.* 1998; Lutcavage *et al.* 2000). Among the most potentially insightful approaches are those that address the habitat and behaviour of flatfishes with in situ techniques such as underwater video (e.g. Gibson *et al.* 1998) in relatively shallow waters, and ROVs (e.g. Norcross & Müter 1999), submersibles (e.g. Sullivan *et al.* 2000) or unmanned lander platforms (e.g. Priede & Bagley 2000) in deeper waters.

Despite these possibilities, for juvenile and adult flatfishes we still lack the most basic data on distribution and abundance of many species, especially for species in polar and tropical waters. Thus, one approach that still needs considerable attention is the description of pattern (Underwood *et al.* 2000) of habitat use in order to improve our ability to ask better questions with the improved techniques available. As a further example, studies reporting ontogenetic changes in habitats are often inferred but not always tested. Accordingly, the reported changes could be caused by differences in size-selective mortality between habitats or along environmental gradients, where mortality may either be natural or caused by fishing, either directly or indirectly (e.g. habitat degradation). Studies that provide estimates of swimming capabilities and mortality rates relative to the magnitude of ontogenetic change, and the spatial and temporal scales at which changes in the distribution occur, would help to clarify the role that ontogeny plays in mediating flatfish distributions. As with most advances in science, conceptual changes in the way we view and evaluate progress can come from a variety of sources including governmental initiatives (e.g. Essential Fish Habitat in the US; NOAA 1996) or conceptual redirection of older paradigms (Beck et al. 2001) and these can apply to juvenile and adult flatfishes as well.

In summary, our review and the conclusions we have drawn may be somewhat biased by the fact that abiotic and biotic factors controlling patterns of flatfish distribution have only been examined for a small subset of the total number of flatfish species known (approximately 20%). Furthermore, the majority of these species represent only six of the 15 recognised flatfish families (see Chapter 2). Even with these limitations, we can still draw some general conclusions about the factors affecting distribution patterns of juvenile and adult flatfishes. Spatial and temporal boundaries on both broad and finer scales are, in part, defined by water temperature, salinity, depth and substratum characteristics. At finer scales, biotic variables such as food availability and predator and competitor avoidance may play important roles. In our view, and those of others (Moyle & Cech 1996; Helfman et al. 1997), these results are not unique to flatfishes. The one exception may be with respect to the significant role that substratum type plays in mediating flatfish distributions (Gibson 1994; see also Chapter 10).

In conclusion, there is convincing evidence that the distribution patterns of many flatfish species change with ontogeny. The change in habitat utilisation could be a response to more than one factor and these factors are likely to vary through ontogeny due to the change in habitat requirements over the life cycle. Experimental studies are therefore needed to evaluate how the effects of single factors and combinations of factors affect habitat use with ontogeny. Tagging experiments are needed to establish the relative importance of migrations and differential mortality in producing these ontogenetic patterns, as well as seasonal, episodic and short-term changes in distribution.

Acknowledgements

This chapter was largely prepared while the senior author was on sabbatical at the National Oceanic and Atmospheric Administration's National Ocean Service, Center for Coastal Fisheries and Habitat Research, Beaufort, NC, USA. This manuscript is Rutgers University Institute of Marine and Coastal Sciences Publication Number 2004–04.

References

Aarnio, K., Bonsdorff, E. & Rosenback, N. (1996) Food and feeding habits of juvenile flounder *Platichthys flesus* (L.), and turbot *Scophthalmus maximus* (L.) in the Imsland Archipelago, northern Baltic Sea. *Journal of Sea Research*, **36**, 311–320.

Able, K.W. (1999) Measures of juvenile fish habitat quality: examples from a national estuarine research reserve. In: *Fish Habitat: Essential Fish Habitat and Rehabilitation. American Fisheries Society Symposium,* **22**, 134–147.

Able, K.W. & Fahay, M.P. (1998) *The First Year in the Life of Estuarine Fishes in the Middle Atlantic Bight.* Rutgers University Press, New Brunswick, NJ.

Able, K.W. & Kaiser, S.C. (1994) Synthesis of summer flounder habitat parameters. *NOAA Coastal Ocean Program Decision Analysis Series No. 1.* NOAA Coastal Ocean Office, Silver Spring, MD.

Abookire, A.A. & Norcross, B.L. (1998) Depth and substrate as determinants of distribution of juvenile flathead sole (*Hippoglossoides elassodon*) and rock sole (*Pleuronectes bilineatus*) distribution in Kachemak Bay, Alaska. *Journal of Sea Research,* **39**, 113–123.

Albert, O.T., Eliassen, J.E. & Hoines, A. (1998) Flatfishes of Norwegian coasts and fjords. *Journal of Sea Research,* **40**, 153–171.

Allen, M.J. & Herbinson, K.T (1990) Settlement of juvenile California halibut, *Paralichthys californicus*, along the coasts of Los Angeles, Orange, and San Diego counties in 1989. *CalCOFI Report,* **31**, 84–92.

Allen, R.L. & Baltz, D.M. (1997) Distribution and microhabitat use by flatfishes in a Louisiana estuary. *Environmental Biology of Fishes,* **50**, 85–103.

Ansell, A.D. & Gibson, R.N. (1990) Patterns of feeding and movement of juvenile flatfishes on an open sandy beach. In: *Trophic Relationships in the Marine Environment* (eds M. Barnes & R.N. Gibson). pp. 191–207. Aberdeen University Press, Aberdeen, Scotland.

Arnold, G.P. & Metcalfe, J.D. (1995) Seasonal migrations of plaice (*Pleuronectes platessa*) through the Dover Strait. *Marine Biology,* **127**, 151–160.

Arnott, S.A. & Pihl, L. (2000) Selection of prey size and prey species by 1-group cod *Gadus morhua*: effects of satiation level and prey handling times. *Marine Ecology Progress Series,* **198**, 225–238.

Baden, S.P., Loo, L.-O., Pihl, L. & Rosenberg, R. (1990) Effects of eutrophication on benthic communities including fish: Swedish west coast. *Ambio,* **19**, 113–122.

Beck, M.W., Heck K.L., Jr., Able, K.W., *et al.* (2001) The identification, conservation, and management of estuarine and marine nurseries for fish and invertebrates. *Bioscience,* **51**, 633–641.

Berghahn, R. (1983) Untersuchungen an Plattfischen und Nordseegarnelen (*Crangon crangon*) im Eulitoral des Wattenmeeres nach dem Übergang zum Bodenleben. *Helgoländer Wissenschaftliche Meeresuntersuchungen,* **36**, 163–181.

Berghahn, R. (1986) Determining abundance, distribution, and mortality of 0-group plaice (*Pleuronectes platessa* L.) in the Wadden Sea. *Journal of Applied Ichthyology,* **2**, 11–22.

Berghahn, R. (1987) Effects of tidal migration on growth of 0-group plaice (*Pleuronectes platessa* L.) in the North Frisian Wadden Sea. *Meeresforschung,* **31**, 209–226.

Berghahn, R. (2000) Response to extreme conditions in coastal areas: biological tags in flatfish otoliths. *Marine Ecology Progress Series,* **192**, 277–285.

Berghahn, R., Bullock, A.M. & Karakiri, M. (1993) Effects of solar radiation on the population dynamics of juvenile flatfish in the shallows of the Wadden Sea. *Journal of Fish Biology,* **42**, 329–345.

Beverton, R.J.H. & Iles, T.C. (1992a) Mortality rates of 0-group plaice (*Pleuronectes platessa* L.), dab (*Limanda limanda* L.), and turbot (*Scophthalmus maximus* L.) in European waters II. Comparison of mortality rates and construction of life tables for 0-group plaice. *Netherlands Journal of Sea Research,* **29**, 49–59.

Beverton, R.J.H. & Iles, T.C. (1992b) Mortality rates of 0-group plaice (*Pleuronectes platessa* L.), dab (*Limanda limanda* L.) and turbot (*Scophthalmus maximus* L.) in European waters. III. Density-dependence of mortality rates of 0-group plaice and some demographic implications. *Netherlands Journal of Sea Research,* **29**, 61–79.

Beyst, B., Cattrijsse, A. & Mees, J. (1999) Feeding ecology of juvenile flatfishes of the surf zone of a sandy beach. *Journal of Fish Biology,* **55**, 1171–1186.

Bigelow, H.B. & Schroeder, W.C. (1953) Fishes of the Gulf of Maine. *Fishery Bulletin,* **74**, 1–576.

Block, B.A., Dewar, H., Farwell, C. & Prince, E.D. (1998) A new satellite technology for tracking the movements of the Atlantic bluefin tuna. *Proceedings of the National Academy of Sciences of the United States of America,* **95**, 9384–9389.

Boehlert, G.W. (1997) Application of acoustic and archival tags to assess estuarine, nearshore, and offshore habitat utilization and movement by salmonids. *NOAA Technical Memorandum NOAA-TM-NMFS-SWFSC-236.*

Boisclair, D. (2001) Fish habitat modeling: from conceptual framework to functional tools. *Canadian Journal of Fisheries and Aquatic Sciences,* **58**, 1–9.

Bregneballe, F. (1961) Plaice and flounder as consumers of the microscopic bottom fauna. *Meddelelser fra Danmarks Fiskeri-og Havsundersogelser,* **3**, 133–182.

Burke, J.S. (1995) Role of feeding and prey distribution of summer and southern flounder in selection of estuarine nursery habitats. *Journal of Fish Biology,* **47**, 355–366.

Burke, J.S., Miller, J.M. & Hoss, D.E. (1991) Immigration and settlement pattern of *Paralichthys dentatus* and *P. lethostigma* in an estuarine nursery ground, North Carolina, USA. *Netherlands Journal of Sea Research,* **27**, 393–405.

Burke, J.S., Monaghan, J.P., Jr & Yokoyama, S. (2000) Efforts to understand stock structure of summer flounder (*Paralichthys dentatus*) in North Carolina, USA. *Journal of Sea Research,* **44**, 111–122.

Burrows, M.T., Gibson, R.N., Robb, L. & Comely, C.A. (1994) Temporal patterns of movement in juvenile flatfishes and their predators – underwater television observations. *Journal of Experimental Marine Biology and Ecology,* **177**, 251–268.

Cabral, H. & Costa, M.J. (1999) Differential use of nursery areas within the Tagus estuary by sympatric soles, *Solea solea* and *Solea senegalensis. Environmental Biology of Fishes,* **56**, 389–397.

Carvalho, N., Alfonso, P. & Santos, R.S. (2003) The haremic mating system and mate choice in the wide-eyed flounder *Bothus podas. Environmental Biology of Fishes,* **66**, 249–258.

Castillo-Rivera, M., Kobelkowsky, A. & Chavez, A.M. (2000) Feeding biology of the flatfish *Citharichthys spilopterus* (Bothidae) in a tropical estuary of Mexico. *Journal of Applied Ichthyology,* **16**, 73–78.

Cyrus, D.P. (1991) The biology of *Solea bleekeri* (Teleostei) in Lake St Lucia on the southeast coast of Africa. *Netherlands Journal of Sea Research,* **27**, 209–216.

de Veen, J.F. (1978) On selective tidal transport in the migration of North Sea plaice (*Pleuronectes platessa*) and other flatfish species. *Netherlands Journal of Sea Research,* **12**, 115–147.

Díaz de Astarloa, J.M. & Munroe, T.A. (1998) Systematics, distribution and ecology of commercially important paralichthyid flounders occurring in Argentinean-Uruguayan waters (*Paralichthys*, Paralichthyidae): an overview. *Journal of Sea Research,* **39**, 1–9.

Diaz de Astarloa, J.M. & Fabré, N.N. (2003) Abundance of three flatfish species (Pleuronectiformes, Paralichthyidae) off northern Argentina and Uruguay in relation to environmental factors. *Archive of Fishery and Marine Research,* **50**, 123–140.

Dorel, D., Koutsikopoulos, C., Desaunay, Y. & Marchand, J. (1991) Seasonal distribution of young sole (*Solea solea* (L.)) in the nursery ground of the Bay of Vilaine (northern Bay of Biscay). *Netherlands Journal of Sea Research,* **27**, 297–306.

Dovel, W.L., Mihursky, J.A. & McErlean, A.J. (1969) Life history aspects of the hogchoker, *Trinectes maculatus*, in the Patuxent River Estuary, Maryland. *Chesapeake Science,* **10**, 104–119.

Duffy-Anderson, J.T. & Able, K.W. (1999) Effects of municipal piers on the growth of juvenile fish in the Hudson River estuary: a study across the pier edge. *Marine Biology,* **133**, 409–418.

Duffy-Anderson, J.T. & Able, K.W. (2001) An assessment of the feeding success of young-of-the-year winter flounder (*Pseudopleuronectes americanus*) near a municipal pier in the Hudson River estuary, U.S.A. *Estuaries,* **24**, 430–440.

Ellis, T. & Gibson, R.N. (1995) Size-selective predation of 0-group flatfishes on a Scottish coastal nursery ground. *Marine Ecology Progress Series,* **127**, 27–37.

Fonds, M., Cronie, R., Vethaak, A.D. & Van der Puyl, P. (1992) Metabolism, food consumption and growth of plaice (*Pleuronectes platessa*) and flounder (*Platichthys flesus*) in relation to fish size and temperature. *Netherlands Journal of Sea Research,* **29**, 127–143.

Ford, R.F. (1965) *Distribution, population dynamics and behavior of a bothid flatfish,* Citharichthys stigmaeus. PhD thesis, University of California, San Diego, CA.

Freeman, B.L. & Turner, S.C. (1977) The effects of anoxic water on the summer flounder (*Paralichthys dentatus*), a bottom dwelling fish. In: Oxygen depletion and associated environmental disturbances in the Middle Atlantic Bight in 1976. *NOAA Technical Series Report No. 3,* pp. 451–462. Northeast Fisheries Center, NMFS/NOAA, US Department of Commerce, Sandy Hook, NJ.

Gibson, R.N. (1973) The intertidal movements and distribution of young fish on a sandy beach with special reference to the plaice (*Pleuronectes platessa* L.). *Journal of Experimental Marine Biology and Ecology,* **12**, 79–102.

Gibson, R.N. (1994) Impact of habitat quality and quantity on the recruitment of juvenile flatfishes. *Netherlands Journal of Sea Research,* **32**, 191–206.

Gibson, R.N. (1997) Behaviour and the distribution of flatfishes. *Journal of Sea Research,* **37**, 241–256.

Gibson, R.N., Robb, L., Burrows, M.T. & Ansell, A.D. (1996) Tidal, diel and longer term changes in the distribution of fishes on a Scottish sandy beach. *Marine Ecology Progress Series,* **130**, 1–17.

Gibson, R.N., Pihl, L., Burrows, M.T., Modin, J., Wennhage, H. & Nickell, L.A. (1998) Diel movements of juvenile plaice *Pleuronectes platessa* in relation to predators, competitors, food availability and abiotic factors on a microtidal nursery ground. *Marine Ecology Progress Series,* **165**, 145–159.

Gibson, R.N., Robb, L., Wennhage, H. & Burrows, M.T. (2002) Ontogenetic changes in depth distribution of juvenile flatfishes in relation to predation risk and temperature on a shallow-water nursery ground. *Marine Ecology Progress Series* **229**, 233–244.

Gomelyuk, V.E. & Shchetkov, S.Y. (1999) Small-scale spatial structure of two flatfish species in Peter the Great Bay, Sea of Japan. *Journal of the Marine Biological Association of the United Kingdom,* **79**, 509–520.

Greer Walker, M., Harden Jones, F.R. & Arnold, G.P. (1980) The movements of plaice (*Pleuronectes platessa* L.) tracked in the open sea. *Journal du Conseil International pour l'Exploration de la Mer,* **38**, 58–86.

Hales, L.S., Jr & Able, K.W. (2001) Winter mortality, growth, and behavior of young-of-the-year of four coastal marine fishes in New Jersey (USA) waters. *Marine Biology,* **139**, 45–54.

Haywood, M.D.E. & Pendrey, R.C. (1996) A new design for a submersible chronographic tethering device to record predation in different habitats. *Marine Ecology Progress Series,* **143**, 307–312.

Heinke, F. (1913) Investigations on the plaice. I. The plaice fishery and protective regulations. *Rapports et Procès-verbaux des Réunions Conseil International pour l'Exploration de la Mer,* **17**, 1–153.

Helfman, G.S., Collette, B.B. & Facey, D.E. (1997) *The Diversity of Fishes.* Blackwell Science, Malden, MA.

Howe, A.B. & Coates, P.G. (1975) Winter flounder movements, growth, and mortality off Massachusetts. *Transactions of the American Fisheries Society,* **104**, 13–29.

Howe, A.B., Coates, P.G. & Pierce, D.E. (1976) Winter flounder estuarine year-class abundance, mortality and recruitment. *Transactions of the American Fisheries Society,* **105**, 647–657.

Hunter, J.R., Butler, J.L., Kimbrell, C. & Lynn, E.A. (1990) Bathymetric patterns in size, age, sexual maturity, water content, and caloric density of Dover sole, *Microstomus pacificus*. *CalCOFI Report,* **31**, 132–144.

Jager, Z., Kleef, H.L. & Tydeman, P. (1993) The distribution of 0-group flatfish in relation to abiotic factors on the tidal flats in the brackish Dollard (Ems estuary, Wadden Sea). *Journal of Fish Biology,* **43**, 31–43.

Jeffries, H.P. & Johnson, W.C. (1974) Seasonal distributions of bottom fishes in the Narragansett Bay area: seven-year variations in the abundance of winter flounder (*Pseudopleuronectes americanus*). *Journal of the Fisheries Research Board of Canada,* **31**, 1057–1066.

Kasai, A., Sakamoto, W., Mitsunaga, Y. & Yamamoto, S. (2000) Behaviour of immature yellowtail (*Seriola quinqueradiata*) observed by electronic data-recording tags. *Fisheries Oceanography,* **9**, 259–270.

Kneib, R.T. & Scheele, C.E.H. (2000) Does tethering of mobile prey measure relative predation potential? An empirical test using mummichogs and grass shrimp. *Marine Ecology Progress Series,* **198**, 181–190.

Konstantinou, H. & Shen, D.C. (1995) The social and reproductive behavior of the eyed flounder, *Bothus ocellatus*, with notes on the spawning of *Bothus lunatus* and *Bothus ellipticus*. *Environmental Biology of Fishes,* **44**, 311–324.

Kramer, D.L., Rangeley, R.W. & Chapman, L.J. (1997) Habitat selection: patterns of spatial distribution from behavioural decisions. In: *Behavioural Ecology of Teleost Fishes* (ed. J.J. Godin). pp. 37–80. Oxford University Press, New York.

Kramer, S.H. (1991) Growth, mortality, and movements of juvenile California halibut *Paralichthys californicus* in shallow coastal and bay habitats of San Diego County, California. *Fishery Bulletin,* **89**, 195–207.

Krebs, C.J. (1994) *Ecology: The Experimental Analyses of Distribution and Abundance.* Harper Collins College Publishers, New York.

Kuipers, B.R. (1977) On the ecology of juvenile plaice on a tidal flat in the Wadden Sea. *Netherlands Journal of Sea Research,* **11**, 56–91.

Le Clus, F., Hennig, H.F-K.O., Melo, Y.C. & Boyd, A.J. (1994) Impact of the extent and locality of mud patches on the density and geographic distribution of juvenile Agulhas sole *Austroglossus pectoralis* (Soleidae). *South African Journal of Marine Science,* **14**, 19–36.

Leopold, M.F., Van Damme, C.J.G. & Van der Veer, H.W. (1998) Diet of cormorants and the impact of cormorant predation on juvenile flatfish in the Dutch Wadden Sea. *Journal of Sea Research,* **40**, 93–107.

Le Pape, O., Chauvet, F., Mahévas, S., Lazure, P., Guérault, D. & Désaunay, Y. (2003a) Quantitative description of habitat suitability for the juvenile common sole (*Solea solea*, L.) in the Bay of Biscay (France) and the contribution of different habitats to the adult population. *Journal of Sea Research,* **50**, 139–149.

Le Pape, O., Chauvet, F., Desaunay, Y. & Guerault, D. (2003b) Relationship between interannual variations of the river plume and the extent of nursery grounds for the common sole (*Solea solea* L.) in Vilaine Bay: effects of recruitment variability. *Journal of Sea Research,* **50**, 177–185.

Lockwood, S.J. (1974) The settlement, distribution and movements of 0-group plaice *Pleuronectes platessa* (L.) in Filey Bay, Yorkshire. *Journal of Fish Biology,* **6**, 465–477.

Lockwood, S.J. (1984) The daily food intake of 0-group plaice (*Pleuronectes platessa* L.) under natural conditions: changes with size and season. *Journal du Conseil International pour l' Exploration de la Mer,* **41**, 181–193.

Lutcavage, M.E., Brill, R.W., Skomal, G.B., Chase, B.C., Goldstein, J.L. & Tutein, J. (2000) Tracking adult North Atlantic bluefin tuna (*Thunnus thynnus*) in the northwestern Atlantic using ultrasonic telemetry. *Marine Biology,* **137**, 347–358.

McCracken, F.D. (1963) Seasonal movements of the winter flounder, *Pseudopleuronectes americanus* (Walbaum), on the Atlantic coast. *Journal of the Fisheries Research Board of Canada,* **20**, 551–586.

MacDonald, J.S. & Green, R.H. (1986) Food resource utilization by five species of benthic feeding fish in Passamaquoddy Bay, New Brunswick. *Canadian Journal of Fisheries and Aquatic Sciences,* **43**, 1534–1546.

Manderson, J.P., Phelan, B.A., Bejda, A.J., Stehlik, L.L. & Stoner, A.W. (1999) Predation by striped searobin (*Prionotus evolans*, Triglidae) on young-of-the-year winter flounder (*Pseudopleuronectes americanus*, Walbaum): examining prey size selection and prey choice using field observations and laboratory experiments. *Journal of Experimental Marine Biology and Ecology,* **242**, 211–231.

Manderson, J.P., Phelan, B.A., Meise, C., *et al.* (2002) Spatial dynamics of habitat suitability for growth of newly-settled winter flounder in an estuarine nursery. *Marine Ecology Progress Series,* **228**, 227–239.

Manickchand-Heileman, S.C. (1994) Distribution and abundance of flatfish on the South American continental shelf from Suriname to Colombia. *Netherlands Journal of Sea Research,* **32**, 441–452.

Metcalfe, J.D. & Arnold, G.P. (1997) Tracking fish with electronic tags. *Nature,* **387**, 665–666.

Minami, T. & Tanaka, M. (1992) Life history cycles in flatfish from the northwestern Pacific, with particular reference to their early life histories. *Netherlands Journal of Sea Research,* **29**, 35–48.

Modin, J. & Pihl, L. (1996) Small-scale distribution of juvenile plaice and flounder in relation to predatory shrimp in a shallow Swedish bay. *Journal of Fish Biology,* **49**, 1070–1085.

Morse, W.W. & Able, K.W. (1995) Distribution and life history of windowpane, *Scophthalmus aquosus*, off the northeastern United States. *Fishery Bulletin,* **93**, 674–692.

Moyer, J.T., Yogo, Y., Zaiser, M.J. & Tsukahara, H. (1985) Spawning behavior and social organization of the flounder *Crossorhombus kobensis* (Bothidae) at Miyake-jima, Japan. *Japanese Journal of Ichthyology,* **32**, 363–356.

Moyle, P.B. & Cech, J.J., Jr (1996) *Fishes: An Introduction to Ichthyology*, 3rd edn. Prentice-Hall, Upper Saddle River, NJ.

Neuman, M.J. & Able, K.W. (2003) Inter-cohort differences in spatial and temporal settlement patterns of young-of-the-year windowpane, *Scophthalmus aquosus*, in southern New Jersey. *Estuarine Coastal and Shelf Science,* **56**, 527–538.

Neuman, M.J., Witting, D.A. & Able, K.W. (2001) Relationships between otolith microstructure, otolith growth, somatic growth and ontogenetic transitions in two cohorts of windowpane. *Journal of Fish Biology,* **58**, 967–984.

NOAA (National Oceanic and Atmospheric Administration) (1996) Magnuson-Stevens Fishery Conservation and Management Act amended through 11 October 1996. *National Marine Fisheries Service, National Oceanic and Atmospheric Administration Technical Memorandum NMFS-F/SPO-23.* US Department of Commerce, Washington, DC.

Norcross, B.L. & Müter, F.J. (1999) The use of an ROV in the study of juvenile flatfishes. *Fisheries Research,* **39**, 241–251.

Norcross, B.L., Holladay, B.A. & Müter, F.-J. (1995) Nursery area characteristics of pleuronectids in coastal Alaska, USA. *Netherlands Journal of Sea Research,* **34**, 161–175.

Olla, B.L., Samet, C.E & Studholme, A.L. (1972) Activity and feeding behavior of the summer flounder (*Paralichthys dentatus*) under controlled laboratory conditions. *Fishery Bulletin,* **70**, 1127–1136.

Packer, D.B. & Hoff, T. (1999) Life history, habitat parameters and essential habitat of Mid-Atlantic summer flounder. In: Fish habitat: essential fish habitat and rehabilitation. *American Fisheries Society Symposium* **22**, 76–92.

Petersen, J.K. & Pihl, L. (1995) Responses to hypoxia of plaice, *Pleuronectes platessa*, and dab, *Limanda limanda*, in the south-east Kattegat: distribution and growth. *Environmental Biology of Fishes,* **43**, 311–321.

Phelan, B.A. (1992) Winter flounder movements in the inner New York Bight. *Transactions of the American Fisheries Society*, **121**, 777–784.

Phelan, B.A., Goldberg, R., Bejda, A.J., *et al.* (2000) Estuarine and habitat-related differences in growth rates of young-of-the-year winter flounder (*Pseudopleuronectes americanus*) and tautog (*Tautoga onitis*) in three northeastern US estuaries. *Journal of Experimental Marine Biology and Ecology*, **247**, 1–28.

Pihl, L. (1982) Food-intake of young cod and flounder in a shallow bay on the Swedish west coast. *Netherlands Journal of Sea Research*, **15**, 419–432.

Pihl, L. (1989) Abundance, biomass and production of juvenile flatfish in southeastern Kattegat. *Netherlands Journal of Sea Research*, **24**, 69–81.

Pihl, L. (1990) Year-class strength regulation in plaice (*Pleuronectes platessa* L.) on the Swedish west coast. *Hydrobiologia*, **195**, 79–88.

Pihl, L., Svenson, A., Koksnes, P.-O. & Wennhage, H. (1999) Distribution of green algal mats throughout shallow soft bottoms of the Swedish Skagerrak archipelago. *Journal of Sea Research*, **44**, 65–80.

Pihl, L., Modin, J. & Wennhage, H. (2000) Spatial distribution patterns of newly settled plaice (*Pleuronectes platessa* L.) along the Swedish Skagerrak archipelago. *Journal of Sea Research*, **44**, 65–80.

Potter, I.C., Beckley, L.E., Whitfield, A.K. & Lenanton, R.C.J. (1990) Comparisons between the roles played by estuaries in the life cycle of fishes in temperate Western Australia and Southern Africa. *Environmental Biology of Fishes*, **28**, 143–178.

Powell, A.B. & Schwartz, F.J. (1977) Distribution of paralichthid flounders (Bothidae: *Paralichthys*) in North Carolina estuaries. *Chesapeake Science*, **18**, 334–339.

Powles, P.M. & Kohler, A.C. (1970) Depth distributions of various stages of witch flounder (*Glyptocephalus cynoglossus*) of Nova Scotia and in the Gulf of St. Lawrence. *Journal of the Fisheries Research Board of Canada*, **27**, 2053–2062.

Priede, I.G. & Bagley, P.M. (2000) In situ studies on deep-sea demersal fishes using autonomous unmanned lander platforms. *Oceanography and Marine Biology: An Annual Review*, **38**, 357–392.

Raffaelli, D., Richner, H., Summers, R. & Northcott, S. (1990) Tidal migrations in the flounder (*Platichthys flesus*). *Marine Behaviour and Physiology*, **16**, 249–260.

Rickey, M.H. (1995) Maturity, spawning, and seasonal movement of arrowtooth flounder, *Atheresthes stomias*, off Washington. *Fishery Bulletin*, **93**, 127–138.

Rijnsdorp, A.D., Van Beek, F.A., Flatman, S., *et al.* (1992) Recruitment of sole stocks, *Solea solea* (L.), in the northeast Atlantic. *Netherlands Journal of Sea Research*, **29**, 173–192.

Riley, J.D., Symonds, D.J. & Woolner, L. (1981) On the factors influencing the distribution of 0-group demersal fish in coastal waters. *Rapports et Procès-verbaux des Réunions Conseil International pour l'Exploration de la Mer*, **178**, 223–228.

Rogers, S.I., Rijnsdorp, A.D., Damm, U. & Vanhee, W. (1998) Demersal fish populations in the coastal waters of the UK and continental NW Europe from beam trawl survey data collected from 1990 to 1995. *Journal of Sea Research*, **39**, 79–102.

Rountree, R.A. & Able, K.W. (1992) Foraging habits, growth, and temporal patterns of salt-marsh creek habitat use by young-of-the-year summer flounder in New Jersey. *Transactions of the American Fisheries Society*, **121**, 765–776.

Sánchez-Gil, P., Arreguín-Sanchez, F. & García-Abad, M.C. (1994) Ecological strategies and recruitment of *Syacium gunteri* (Pisces: Bothidae) in the southern Gulf of Mexico shelf. *Netherlands Journal of Sea Research*, **32**, 433–439.

Sogard, S. (1992) Variability in growth rates of juvenile fishes in different estuarine habitats. *Marine Ecology Progress Series*, **85**, 35–53.

Sogard, S.M., Able, K.W. & Hagan, S.M. (2001) Long-term assessment of settlement and growth of juvenile winter flounder (*Pseudopleuronectes americanus*) in New Jersey estuaries. *Journal of Sea Research,* **45**, 189–204.

Steves, B.P., Cowen, R.K. & Malchoff, M.H. (1999) Settlement and nursery habitats for demersal fishes on the continental shelf of the New York Bight. *Fishery Bulletin,* **98**, 167–188.

Stoner, A.W. & Abookire, A.A. (2002) Sediment preferences and size-specific distribution of young-of-the-year Pacific halibut in an Alaska nursery. *Journal of Fish Biology,* **61**, 540–559.

Stoner, A.W. & Ottmar, M.L. (2003) Relationships between size-specific sediment preferences and burial capabilities in juveniles of two Alaska flatfishes. *Journal of Experimental Marine Biology and Ecology,* **282**, 85–101.

Stoner, A.W., Bejda, A.J., Manderson, J.P., Phelan, B.A., Stehlik, L.L. & Pessutti, J.P. (1999) Behavior of winter flounder, *Pseudopleuronectes americanus*, during the reproductive season: laboratory and field observations on spawning, feeding, and locomotion. *Fishery Bulletin,* **97**, 999–1016.

Stoner, A.W., Manderson, J.P. & Pessutti, J.P. (2001) Spatially explicit analysis of estuarine habitat for juvenile winter flounder: combining generalized additive models and geographic information system. *Marine Ecology Progress Series,* **213**, 253–271.

Sullivan, M.C., Cowen, R.K., Able, K.W. & Fahay, M.P. (2000) Spatial scaling of recruitment in four continental shelf fishes. *Marine Ecology Progress Series,* **207**, 141–154.

Summers, R.W. (1979) Life-cycle and population ecology of the flounder *Platichthys flesus* (L) in the Ythan Estuary, Scotland. *Journal of Natural History,* **13**, 703–723.

Swanson, R.L. & Sindermann, C.J.E. (1979) Oxygen depletion and associated benthic mortalities in the New York Bight, 1976. *NOAA Professional Paper 11*. US Department of Commerce.

Szedlmayer, S.T. & Able, K.W. (1993) Ultrasonic telemetry of age-0 summer flounder, *Paralichthys dentatus*, movements in a southern New Jersey estuary. *Copeia,* **1993**, 728–736.

Tanaka, M., Goto, T., Tomiyama, M. & Sudo, H. (1989) Immigration, settlement and mortality of flounder (*Paralichthys olivaceus*) larvae and juveniles in a nursery ground, Shijiki Bay, Japan. *Netherlands Journal of Sea Research,* **24**, 57–67.

Thorrold, S.R., Latkoczy, C., Swart, P.K. & Jones, C.M. (2001) Natal homing in a marine fish metapopulation. *Science,* **291**, 297–299.

Toepfer, C. & Fleeger, J.W. (1995a) Effects of marsh-edge habitat variables on feeding success by juvenile bay whiff, *Citharichthys spilopterus* (Teleostei: Bothidae). *Southwestern Naturalist,* **40**, 297–300.

Toepfer, C. & Fleeger, J.W. (1995b) Diet of juvenile fishes, *Citharichthys spilopterus*, *Symphurus plagiusa*, and *Gobionellus boleosoma*. *Bulletin of Marine Science,* **56**, 238–249.

Tok, K.S. & Biryukov, I.A. (1998) Distribution and some features of the biology of *Reinhardtius hippoglossoides matsuurae* (Pleuronectidae) and *Sebastolulbus macrochir* (Scorpaenidae) from the eastern coast of Sakhalin. *Journal of Ichthyology,* **38**, 143–146.

Tokranov, A.M. & Maksimenkov, V.V. (1994) Feeding of the starry flounder, *Platichthys stellatus*, in the Bol'shaya River estuary (western Kamchatka). *Journal of Ichthyology,* **34**, 76–83.

Toole, G.L., Markle, D.F. & Donohoe, C.J. (1997) Settlement timing, distribution, and abundance of Dover sole (*Microstomus pacificus*) on an outer continental shelf nursery area. *Canadian Journal of Fisheries and Aquatic Sciences,* **54**, 531–542.

Tyler, A.V. (1971) Surges of winter flounder, *Pseudopleuronectes americanus*, into the intertidal zone. *Journal of the Fisheries Research Board of Canada,* **28**, 1727–1732.

Underwood, A.J., Chapman, M.G. & Connell, S.D. (2000) Observations in ecology: you can't make progress on processes without understanding the patterns. *Journal of Experimental Marine Biology and Ecology,* **250**, 97–115.

Van der Veer, H.W. & Bergman, M.J.N. (1986) Development of tidally related behaviour of a newly settled 0-group plaice (*Pleuronectes platessa*) population in the western Wadden Sea. *Marine Ecology Progress Series,* **31**, 121–129.

Van der Veer, H.W. & Bergman, M.J.N. (1987) Predation by crustaceans on a newly settled 0-group plaice *Pleuronectes platessa* population in the western Wadden Sea. *Marine Ecology Progress Series,* **35**, 203–215.

Van der Veer, H.W., Ellis, T., Miller, J.M., Pihl, L. & Rijnsdorp, A.D. (1997) Size-selective predation on juvenile North Sea flatfish and possible implications for recruitment. In: *Early Life History and Recruitment in Fish Populations* (eds R.C. Chambers & E.A. Trippel). pp. 279–303. Chapman & Hall, London.

Vetter, R.D., Lynn, E.A. & Coasta, A.S. (1994) Depth zonation and metabolic adaptation in Dover sole, *Microstomus pacificus*, and other deep-living flatfishes: factors that affect the sole. *Marine Biology,* **120**, 145–159.

Wallin, J.E., Ransier, J.M., Fox, S. & McMichael, R.H., Jr (1997) Short-term retention of coded wire and internal anchor tags in juvenile common snook, *Centropomus undecimalis*. *Fishery Bulletin,* **95**, 873–878.

Walsh, H.J., Peters, D.S. & Cyrus, D.P. (1999) Habitat utilization by small flatfishes in a North Carolina estuary. *Estuaries,* **22**, 803–813.

Walsh, S.J. (1992) Factors influencing distribution of juvenile yellowtail flounder, *L. ferruginea*, on the Grand Banks of Newfoundland. *Netherlands Journal of Sea Research,* **29**, 193–203.

Wennhage, H. (2002) Vulnerability of newly settled plaice (*Pleuronectes platessa* L) to predation: effects of habitat structure and predator functional response. *Journal of Experimental Marine Biology and Ecology,* **269**, 129–145.

Wennhage, H. & Pihl, L. (1994) Substratum selection by juvenile plaice (*Pleuronectes platessa* L.): impact of benthic microalgae and filamentous macroalgae. *Netherlands Journal of Sea Research,* **32**, 343–351.

Whitfield, A.K. & Blaber, S.J.M. (1979) Feeding ecology of piscivorous birds at Lake St Lucia. Part 3: Swimming birds. *Ostrich,* **50**, 10–20.

Wilk, S.J., Morse, W.W. & Stehlik, L.L. (1990) Annual cycles of gonad-somatic indices as indicators of spawning activity for selected species of finfish collected from the New York Bight. *Fishery Bulletin,* **88**, 775–786.

Wirjoatmodjo, S. & Pitcher, T.J. (1984) Flounders follow the tides to feed: evidence from ultrasonic tracking in an estuary. *Estuarine Coastal and Shelf Science,* **19**, 231–241.

Witting, D.A. (1995) *Settlement of winter flounder,* Pleuronectes americanus, *in a southern New Jersey estuary: spatial and temporal dynamics and the effect of decapod predation*. PhD dissertation, Rutgers, State University of New Jersey, USA.

Woodhead, P.M.J. (1964) The death of North Sea fish during the winter of 1962/63, particularly with reference to the sole *Solea vulgaris*. *Helgoländer Wissenschaftliche Meeresuntersuchungen,* **10**, 283–300.

Yamashita, Y., Yamada, H., Malloy, K.D., Targett, T.E. & Tsuruta, Y. (1996) Sand shrimp predation on settling and newly-settled stone flounder and its relationship to optimal nursery habitat selection in Sendai Bay, Japan. In: *Survival Strategies in Early Life Stages of Marine Resources* (eds Y. Watanabe, Y. Yamashita & Y. Ooozeki). pp. 271–283. A.A. Balkema, Rotterdam.

Yamashita, Y., Tanaka, M. & Miller, J.M. (2001) Ecophysiology of juvenile flatfish in nursery grounds. *Journal of Sea Research*, **45**, 205–218.

Zimmermann, M. & Goddard, P. (1996) Biology and distribution of arrowtooth, *Atheresthes stomias*, and Kamchatka, *A. evermanni*, flounders in Alaskan waters. *Fishery Bulletin,* **94**, 358–370.

Chapter 9
The trophic ecology of flatfishes

Jason S. Link, Michael J. Fogarty and Richard W. Langton

9.1 Introduction

Flatfishes are an economically and ecologically important component of continental shelf, deep ocean, small sea, riverine and estuarine ecosystems worldwide (Chapter 3). Landings of flatfishes account for >10^6 t yr^{-1} for the past several decades (Garcia & Newton 1997; FAO 1998). US landings of flatfishes during 2002 alone were valued at nearly $102 million or 85 million Euros (NMFS 2003).

Flatfishes serve as a major energy pathway for conversion of benthic production into a form suitable for consumption by higher predators and humans. As such, flatfishes are critical components of benthic communities. The ecological role of flatfishes provides a tangible 'ecosystem service' for human benefit in a benthic environment that is otherwise inherently difficult to evaluate in an economic sense. Any understanding of marine benthic ecosystems needs to incorporate trophic dynamics of flatfishes.

Because of the benthic affinity of flatfishes, in addition to effects from directed fisheries, these species are highly susceptible to perturbations to the ocean bottom (reviewed in Jennings & Kaiser 1998; Auster & Langton 1999; Kaiser & de Groot 2000). Several researchers have shown that fishing (i.e. trawling, dredging, etc.) impacts on the ocean bottom can alter critical habitat, trophic dynamics, and ultimately survivability (particularly of juveniles) in several ecosystems. However, some researchers have demonstrated that, at least at short timescales, the effects of fishing may in fact be positive by providing energy subsidies in the form of discards or resuspended sediment and associated benthos (e.g. Kaiser & Ramsay 1997). A full understanding of the impacts of fishing on benthic habitats and communities also requires a detailed knowledge of flatfish ecology.

Another consideration is that some flatfishes are important piscivores. Many of the fishes eaten by piscivorous flatfishes are commercially valuable. The magnitude of flatfish predatory removals may be similar to fisheries for economically valuable fishes (e.g. Livingston 1993; Overholtz *et al.* 2000). Thus, the potential for direct competition between fisheries and flatfishes merits examination. Additionally, some flatfish piscivores, particularly halibuts, are the largest teleosts in many ecosystems (e.g. Rodriguez-Marin *et al.* 1995; Orlov 1997; Yang 1997). How impacts on these apex predators influence an ecosystem is not entirely clear.

This chapter reviews fundamental issues in flatfish trophic ecology and attempts to provide a global perspective, emphasising work within the past 20 years without excluding 'classic' references. In addition to cataloguing the qualitative and quantitative relationships between flatfishes and their prey, predators and competitors, the magnitude and significance of flatfish trophic dynamics in several example ecosystems are examined.

9.2 Major flatfish feeding groups

Flatfishes are important predators in benthic communities. Evidence from ecosystems around the world suggests that flatfishes primarily consume two general prey types: flatfishes either eat polychaetes and small benthic crustaceans or larger, wider-gaped flatfishes eat almost entirely fishes and squids (Table 9.1). Other common but less dominant prey items are: harpacticoid copepods, bivalves, echinoderms, oligochaetes, insect larvae (chironomids), decapods, mysids, euphausids and similar shrimps. Although these observations generally confirm the feeding categorisations first proposed by Yazdani (1969) and de Groot (1969, 1971), they are subtly different, particularly with respect to Pleuronectiformes taxonomy. The classification in this chapter represents the realisation of feeding behaviour and morphology (i.e. mouth shape, gape width, body shape, swimming ability, etc.) as seen in the diet. That the flatfish taxonomic group is so functionally similar across a broad range of depth, latitude, salinity, temperature and productivity gradients is intriguing. Equally intriguing is the broad representation of flatfish families across the different feeding groups, representing both adaptive radiation and ontogenetic changes in diet.

9.2.1 Polychaete and crustacean eaters

Annelid-feeding species are very common among flatfishes. In collections as early as 1915–16, Linton (1921) noted the high percentage of annelids and amphipods in the diet of winter flounder. Other studies for winter flounder (Langton & Bowman 1981; Carlson et al. 1997) and, for example, yellowtail flounder (Langton 1983; Bowman & Michaels 1984; Collie 1987a, b), English sole, rex sole, Dover sole (Kravitz et al. 1977; Pearcy & Hancock 1978), yellowfin sole (Tokranov 1990), Alaska plaice (Zhang 1988; Tokranov 1990), banded-fin flounder (Tokhranov & Maksimenkov 1995a), dab (Beare & Moore 1997), witch (Langton & Bowman 1981; Bowman & Michaels 1984), lemon sole (Piet et al. 1998) and flounder (Summers 1980; Piet et al. 1998) demonstrate the preponderance of polychaetes and amphipods (particularly gammarids) in the diet of flatfishes (Table 9.1). For most of these species, polychaetes and amphipods usually comprise 40–70% of the diet (Fig. 9.1). In some cases polychaetes alone can constitute >90% of the diet (e.g. Pearcy & Hancock 1978). Amphipods and similar small crustaceans appear to augment the polychaete portion (and vice versa) of the diet for flatfishes during periods of the year or at locations where polychaetes are less available/abundant.

Polychaetes and gammarid amphipods are important in the diet of numerous flatfishes. It is unlikely that flatfish predation alone regulates these prey populations, although flatfish predation can remove a large fraction of benthic production (Evans 1983; Collie 1987b; Gee 1987; Hostens & Mees 1999). However, other factors that result in declines in the abundance of these prey could strongly influence the population dynamics of several commercially valuable flatfishes (e.g. Collie 1987b). These small, benthic organisms are not usually considered in a fisheries context, yet perhaps should be in the broader ecosystem context for fisheries management that has been recently prescribed (Larkin 1996; Jennings & Kaiser 1998; NMFS 1999; NRC 1999; Link 2002).

Table 9.1 Major prey items of representative flatfish species from around the world

Species	Common name	Location	Major prey	Reference
Polychaete and small crustacean feeders				
Bothidae				
Arnoglossus laterna	Scaldfish	W. Mediterranean	Juveniles – amphipods, harpacticoids, other crustaceans	de Morais & Bodiou 1984
Arnoglossus thori	Thor's scaldfish	W. Mediterranean	Juveniles – amphipods, harpacticoids, other crustaceans	de Morais & Bodiou 1984
Bothus podas	Wide-eyed flounder	Azores	Polychaetes, molluscs, shrimps, amphipods, other crustaceans	Nash et al. 1991
Hippoglossina macrops	Bigeye flounder	Chile	Red shrimp, other crustaceans	Arancibia & Melendez 1987
Syacium gunteri	Shoal flounder	Gulf of Mexico	Penaeids, fish, other crustaceans	Garcia-Abad et al. 1992
Citharidae				
Citharus linguatula	Atlantic spotted flounder	Morocco	Mysids, euphasiids, shrimps, squid, fish	Belghyti et al. 1993
		E. Spain	Crustaceans (mysids, decapods), molluscs, fish	Redon et al. 1994
Cynoglossidae				
Cynoglossus joyneri	Red tonguesole	Bohai Sea (China)	*Crangon*, mysids, polychaetes, squid	Dou et al. 1992
Paralichthyidae				
Citharichthys arctifrons	Gulfstream flounder	NW Atlantic	Polychaetes, amphipods	Langton & Bowman 1981
		NW Atlantic	Polychaetes, gammarids, other crustaceans	Link et al. 2002
Citharichthys sordidus	Pacific sanddab	Oregon, USA	Euphasiids, shrimps, amphipods	Pearcy & Hancock 1978
Paralichthys dentatus	Summer flounder	North Carolina Estuaries, USA	Juveniles – polychaetes	Burke 1995
Paralichthys lethostigma	Southern flounder	North Carolina Estuaries, USA	Juveniles – mysids, amphipods, copepods	Burke 1995
Paralichthys microps	Small-eyed flounder	Concepcion Bay, Chile	Smaller sizes crustaceans (euphasiids, mysids)	Gonzalez & Chong 1997
Paralichthys olivaceus	Japanese flounder	NE Japan	Smaller sizes mysids	Yamada et al. 1998
Pleuronectidae				
Reinhardtius stomias	Arrowtooth flounder	Gulf of Alaska	Smaller sizes shrimps	Yang 1993, 1995, 1997
Cleisthenes herzensteini	Sōhachi	Bohai Sea (China)	*Crangon*, mysids, squid	Dou et al. 1992
Eopsetta grigorjewi	Shotted halibut	Japan Sea	Euphasiids, crabs, fish	Tominaga & Nashida 1991
Glyptocephalus cynoglossus	Witch	NW Atlantic	Polychaetes, echinoderms	Bowman & Michaels 1984
		NW Atlantic	Polychaetes	Langton & Bowman 1981
		NW Atlantic	Polychaetes, crustaceans	Link & Almeida 2000, Link et al. 2002

Table 9.1 (Continued.)

Species	Common name	Location	Major prey	Reference
		Passamaquoddy Bay, NB, Canada	Polychaetes, amphipods	Macdonald & Green 1986
		Flemish Cap (NW Atlantic)	Polychaetes	Rodriguez-Marin et al. 1994
Glyptocephalus zachirus	Rex sole	Oregon, USA	Amphipods, polychaetes	Kravitz et al. 1977
		Oregon, USA	Smaller sizes amphipods, crustaceans; larger sizes polychaetes	Pearcy & Hancock 1978
Hippoglossoides dubius	Flathead flounder	Funka Bay, Japan	Smaller sizes gammarids, cumaceans	Yokoyama 1995
Hippoglossoides platessoides	American plaice	Barents Sea	Polychaetes (palps), ophiuroids, mysids, amphipods, other crustaceans	Berestoskiy 1995
		NW Atlantic	Small sizes polychaetes, crustaceans	Link et al. 2002
		Passamaquoddy Bay, NB, Canada	Polychaetes, molluscs, amphipods	Macdonald & Green 1986
		Sable Island Bank, Canada	Amphipods, cumaceans, other crustaceans	Martell & McClelland 1992
		North Sea	Crustaceans, fish, echinoderms	Piet et al. 1998
Hippoglossus stenolepis	Pacific halibut	Gulf of Alaska	Smaller sizes shrimps	Yang 1993, 1995, 1997
Lepidopsetta bilineata	Rock sole	Bering Sea	Polychaetes, bivalves, fish	Tokranov 1990
		E. Bering Sea	Polychaetes, echinoderms	Zhang 1988
Limanda aspera	Yellowfin sole	Bering Sea	Polychaetes, bivalves	Tokranov 1990
		E. Bering Sea	Amphipods, echinoderms, polychaetes	Zhang 1988
Limanda ferruginea	Yellowtail flounder	NW Atlantic	Polychaetes, amphipods	Bowman & Michaels 1984
		Georges Bank, NW Atlantic	Amphipods, polychaetes	Collie 1987a, b
		Grand Bank	Gammarids, polychaetes, fish (sand lance), mysids	Gonzalez et al. 1998
		NW Atlantic	Polychaetes, amphipods	Langton & Bowman 1981
		NE USA	Polychaetes, small crustaceans	Langton 1983
		Massachusetts, USA	Amphipods, polychaetes, cumaceans	Libey & Cole 1979
		NW Atlantic	Polychaetes, amphipods, other crustaceans	Link & Almeida 2000, Link et al. 2002
		Sable Island Bank, Canada	Polychaetes, amphipods, other crustaceans	Martell & McClelland 1992
		New York Bight, USA	Polychaetes, amphipods	Steimle & Terranova 1991
Limanda limanda	Dab	Kames Bay, UK	Amphipods	Beare & Moore 1997
		North Sea	Crustaceans, echinoderms	Piet et al. 1998
Pleuronectes pinnifasciatus	Far-Eastern smooth flounder	Bol'shaya R., Kamchatka, Russia	Amphipods, polychaetes, other crustaceans	Tokranov & Maksimenkov 1995a

Species	Common name	Location	Diet	Reference
Lyopsetta exilis	Slender sole	Oregon, USA	Euphausiids, shrimps, amphipods	Pearcy & Hancock 1978
Microstomus kitt	Lemon sole	North Sea	Polychaetes, crustaceans	Piet *et al.* 1998
Microstomus pacificus	Dover sole	Oregon, USA	Polychaetes, small crustaceans, bivalves, ophiuroids depending on depth	Pearcy & Hancock 1978
Parophrys vetula	English sole	Oregon, USA	Polychaetes, amphipods	Kravitz *et al.* 1977
Platichthys flesus	Flounder	Baltic Sea	Oligochaetes, amphipods, chironomids; smaller sizes harpacticoids	Aarnio *et al.* 1996
		River Frome, England	Juveniles – chironomids, other aquatic insects, molluscs, crustaceans	Beaumont & Mann 1984
		SW English Estuaries, UK	Smaller sizes harpacticoid copepods	Gee 1987
		Netherlands Estuary	Amphipods, mysids, molluscs	Hostens & Mees 1999
		North Sea	Crustaceans, polychaetes	Piet *et al.* 1998
		Ythan Estuary (Scotland)	Amphipods, polychaetes	Summers 1980
		Elbe Estuary, Germany	Copepods, oligochaetes	Thiel *et al.* 1997
		River Dee, North Wales, UK	Juveniles – chironomids, oligochaetes, copepods	Weatherly 1989
		Aber Estuary, UK	Chironomids	Williams & Williams 1998
Platichthys stellatus	Starry flounder	Bol'shaya R., Kamchatka, Russia	Small sizes crustaceans, chironomids	Tokranov & Maksimenkov 1994, 1995b
Pseudopleuronectes herzensteini	Littlemouth flounder	Japan Sea	Polychaetes	Tominaga & Nashida 1991
Pleuronectes platessa	Plaice	Gullmar Fjord, Sweden	Juveniles – polychaetes, crustaceans, molluscs	Evans 1983
		SW English Estuaries, UK	Smaller sizes harpacticoid copepods	Gee 1987
		Netherlands Estuary	Mysids, molluscs	Hostens & Mees 1999
		North Sea	Polychaetes, echinoderms, crustaceans	Piet *et al.* 1998
Pleuronectes quadrituberculatus	Alaska plaice	Bering Sea	Polychaetes, bivalves	Tokranov 1990
		E. Bering Sea	Polychaetes, amphipods	Zhang 1988, Zhang *et al.* 1998
Pseudopleuronectes americanus	Winter flounder	Long Island Sound, USA	Amphipods, polychaetes	Carlson *et al.* 1997
		NW Atlantic	Polychaetes, bivalves, amphipods	Langton & Bowman 1981
		NW Atlantic	Polychaetes, cnidarians, amphipods, other crustaceans	Link & Almeida 2000, Link *et al.* 2002
		Woods Hole, USA	Polychaetes, amphipods	Linton 1921
		Passamaquoddy Bay, NB, Canada	Polychaetes, amphipods	Macdonald & Green 1986
Reinhardtius hippoglossoides	Greenland halibut	Sable Island Bank, Canada	Polychaetes, amphipods	Martell & McClelland 1992
		Flemish Cap (NW Atlantic)	Smaller sizes decapods, amphipods	Rodriguez-Marin *et al.* 1994, 1995
Verasper variegatus	Spotted halibut	Bohai Sea (China)	*Crangon*, squid	Dou *et al.* 1992

Table 9.1 (*Continued.*)

Species	Common name	Location	Major prey	Reference
Rhombosoleidae				
Peltoretis flavilatus	Southern lemon sole	Wellington Harbour, New Zealand	Polychaetes, crabs	Livingston 1987
Peltorhampus scapha	New Zealand sole	Wellington Harbour, New Zealand	Polychaetes, cnidarians	Livingston 1987
Rhombosolea leporina	Yellowbelly flounder	Wellington Harbour, New Zealand	Crabs, polychaetes	Livingston 1987
Scophthalmidae				
Scophthalmus aquosus	Windowpane	NW Atlantic	Mysids, pandalids, *Crangon*, fish	Langton & Bowman 1981
		NW Atlantic	Mysids, fish (sand lance), polychaetes, amphipods, *Crangon*	Link & Almeida 2000, Link *et al.* 2002
Scophthalmus maximus	Turbot	New York Bight, USA	Mysids, fish (sand lance), polychaetes, shrimps	Steimle & Terranova 1991
Soleidae		Baltic Sea	Mysids, fish, amphipods; smaller sizes amphipods	Aarnio *et al.* 1997
Buglossidium luteum	Solenette	W. Mediterranean	Juveniles – harpacticoid copepods	de Morais & Bodiou 1984
		North Sea	Polychaetes, molluscs, amphipods	Piet *et al.* 1998
Dicologlossa cuneata	Wedge sole	Morocco	Benthic infauna, epifauna	Belghyti *et al.* 1993
		Netherlands Estuary	Amphipods	Hostens & Mees 1999
Solea solea	Common sole	North Sea	Polychaetes	Piet *et al.* 1998
Echinoderm feeders				
Pleuronectidae				
Hippoglossoides platessoides	American plaice	NW Atlantic	Echinoderms, decapods	Bowman & Michaels 1984
		NW Atlantic	Echinoderms (echinoids, ophiuroids), molluscs	Langton & Bowman 1981
		NW Atlantic	Echinoderms	Link & Almeida 2000, Link *et al.* 2002
		Gulf of Maine	Ophiuroids	Packer *et al.* 1994
		Flemish Cap (NW Atlantic)	Echinoderms	Rodriguez-Marin *et al.* 1994
Lepidopsetta bilineata	Rock sole	Oregon, USA	Ophiuroids, molluscs, polychaetes	Kravitz *et al.* 1977
Limanda limanda	Dab	North Sea	Ophiuroids, clams	Kaiser & Ramsay 1997
		German Bight	Ophiuroids, molluscs	Temming & Hammer 1994
Microstomus pacificus	Dover sole	Oregon, USA	Ophiuroids, polychaetes	Gabriel & Pearcy 1981
Platichthys stellatus	Starry flounder	NE Bering Sea & SE Chukchi Sea	Brittle stars, clams	Jewett & Feder 1980

Rhombosoleidae				
Rhombosolea plebeia	New Zealand flounder	Wellington Harbour, New Zealand	Ophiuroids, polychaetes	Livingston 1987
Siphon tip and other benthos feeders				
Achiridae				
Trinectes maculatus	Hogchoker	Chesapeake Bay, USA	Bivalve (tellinid) siphons, polychaetes, amphipods, other crustaceans	Derrick & Kennedy 1997
Paralichthyidae				
Paralichthys californicus	California flounder	California	Bivalve siphons	Peterson & Quammen 1982
Pleuronectidae				
Pleuronichthys guttulatus	Diamond turbot	California	Bivalve siphons	Peterson & Quammen 1982
Limanda limanda	Dab	Netherlands Estuary	Molluscs	Hostens & Mees 1999
Platichthys flesus	Flounder	Wadden Sea	Bivalves, siphon tips, polychaetes	de Vlas 1979
Pleuronectes platessa	Plaice	Kames Bay, UK	Siphon tips; larger sizes polychaetes	Beare & Moore 1997
		Wadden Sea	Bivalves, siphon tips, polychaetes	de Vlas 1979
Pseudopleuronectes americanus	Winter flounder	Newfoundland, NW Atlantic	Anthozoa, fish eggs, algae, sea urchins	Keats 1990
		New York Bight, USA	Anemone, polychaetes, some molluscs	Steimle & Terranova 1991
Soleidae				
Solea bleekeri	Blackhand sole	Lake St Lucia, South Africa	Siphon tips, amphipod, polychaetes	Cyrus 1991
Fish feeders				
Bothidae				
Arnoglossus laterna	Scaldfish	North Sea	Fish, crustaceans	Piet *et al.* 1998
Arnoglossus scapha	Mahue or megrim	Wellington Harbour, New Zealand	Fish (anchovies), decapods	Livingston 1987
Pseudorhombus arsius	Largetooth flounder	Kuwait	Fish, shrimp, other crustaceans	Euzen 1987
Pseudorhombus pentophthalmus	Fivespot flounder	Japan Sea	Fish, euphasiids, mysids	Tominaga & Nashida 1991
Paralichthyidae				
Citharichthys sordidus	Pacific sanddab	Oregon, USA	Fish (northern anchovy), euphasiids	Kravitz *et al.* 1977
Lepidorhombus whiffiagonis	Megrim	Celtic Sea	Fish	du Buit 1992
Paralichthys dentatus	Summer flounder	NW Atlantic	Fish, squid	Langton & Bowman 1981
		NW Atlantic	Fish, squid	Link & Almeida 2000, Link *et al.* 2002
		North Carolina Estuaries, USA	Fish, crustaceans (mysids, shrimps)	Powell & Schwartz 1979
		New York Bight, USA	Fish, mysids, amphipods	Steimle & Terranova 1991
Paralichthys lethostigma	Southern flounder	North Carolina Estuaries, USA	Fish, crustaceans (mysids, shrimps)	Powell & Schwartz 1979
Paralichthys microps	Small-eyed flounder	Concepcion Bay, Chile	Fish	Gonzalez & Chong 1997

Table 9.1 (*Continued.*)

Species	Common name	Location	Major prey	Reference
Paralichthys oblongus	American fourspot flounder	NW Atlantic	Fish, decapods, squid	Bowman & Michaels 1984
		NW Atlantic	Fish, squid, pandalids, other crustaceans	Langton & Bowman 1981
		NW Atlantic	Fish, squid	Link & Almeida 2000, Link *et al.* 2002
		New York Bight, USA	Fish, mysids, *Crangon*, crabs, other decapods	Steimle & Terranova 1991
Paralichthys olivaceus	Japanese flounder	Japan	Fish (sardines, anchovy), crustaceans (shrimps, *Crangon*)	Nashida & Tominaga 1987
		Japan	Fish, crustaceans (mysids)	Nashida *et al.* 1984
		Japan Sea	Fish, crabs	Tominaga & Nashida 1991
		NE Japan	Fish	Yamada *et al.* 1998
Pleuronectidae				
Reinhardtius evermanni	Kamchatka flounder	Kamchatka, Russia	Fish, squid	Orlov 1997, 1999
Reinhardtius stomias	Arrowtooth flounder	Bering Sea	Fish (Alaska pollock, herring)	Livingston 1993
		Gulf of Alaska	Fish (Alaska pollock, capelin, herring)	Yang 1993, 1995, 1997
Eopsetta jordani	Petrale sole	Oregon, USA	Fish, decapod crustaceans	Kravitz *et al.* 1977
Hippoglossoides dubius	Flathead flounder	Funka Bay, Japan	Fish	Yokoyama 1995
Hippoglossoides elassodon	Flathead sole	Bering Sea	Fish (Alaska pollock, herring)	Livingston 1993
Hippoglossoides platessoides	American plaice	Grand Bank	Fish (sand lance), mysids	Gonzalez *et al.* 1998
Hippoglossus hippoglossus	Atlantic halibut	NW Atlantic	Fish, squid, decapods	Link & Almeida 2000, Link *et al.* 2002
Hippoglossus stenolepis	Pacific halibut	Kamchatka, Russia	Fish, squid, decapods	Orlov 1997, 1999
		Gulf of Alaska	Fish (Alaska pollock)	Yang 1993, 1995, 1997
Platichthys bicoloratus	Stone flounder	Bohai Sea (China)	Sand lance, squid, *Crangon*	Dou *et al.* 1992
Platichthys stellatus	Starry flounder	Bol'shaya R., Kamchatka, Russia	Large sizes fish; medium sizes bivalves, fish offal	Tokranov & Maksimenkov 1994, 1995b
Limanda aspera	Yellowfin sole	Bering Sea	Fish (walleye pollock, herring)	Livingston 1993
Lepidopsetta bilineata	Rock sole	Bering Sea	Fish (walleye pollock, herring)	Livingston 1993
Pleuronectes quadrituberculatus	Alaska plaice	Bering Sea	Fish (walleye pollock, herring)	Livingston 1993
Reinhardtius hippoglossoides	Greenland halibut	Kamchatka, Russia	Fish, squid	Orlov 1997, 1999
		Bering Sea	Fish (walleye pollock, herring)	Livingston 1993
		Flemish Cap (NW Atlantic)	Fish, fish offal, squid	Rodriguez-Marin *et al.* 1994, 1995
Scophthalmidae				
Scophthalmus maximus	Turbot	North Sea	Fish, crustaceans	Piet *et al.* 1998
Scophthalmus rhombus	Brill	North Sea	Fish, crustaceans	Piet *et al.* 1998

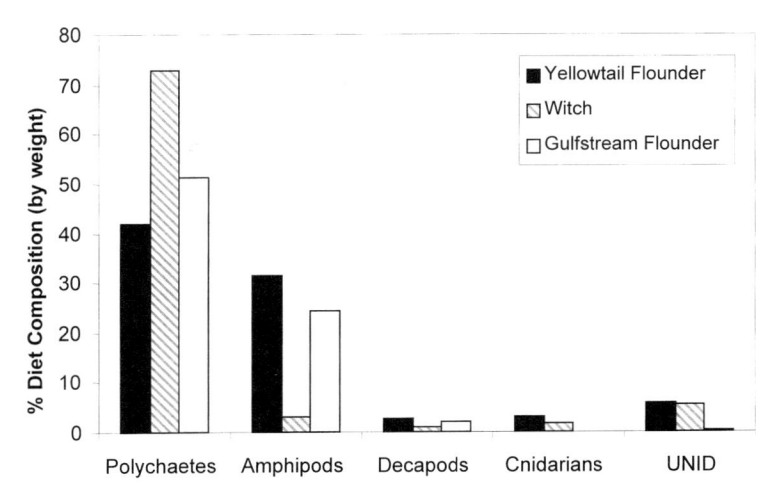

Fig. 9.1 Example diet composition (by weight) of three polychaete-small crustacean feeding flatfishes (yellowtail flounder, witch and Gulfstream flounder) from the northwest Atlantic. Data adapted from Table 1 in Langton & Bowman (1981).

A simple calculation illustrates the importance of these prey items and the role of flatfishes in converting benthic production into fish flesh. Several reports (Langton & Bowman 1981; Langton 1983; Bowman & Michaels 1984) have shown that the average weight of yellowtail flounder stomach contents ranges between 0.5 and 1.0 g, and approximately 40–50% of the diet consists of polychaetes. If the food in a flounder stomach weighing 0.5 g consists of 40% polychaetes, an average flounder would eat 0.2 g of polychaetes. If it is assumed that an average polychaete weighs 25 mg then this average flounder would have eaten 8 polychaete worms (a similar estimate derived from Bowman & Michaels (1984) provides 7–9 polychaetes per yellowtail stomach). Assuming a conservative digestion rate of 2 days for all food items of this average flounder (sensu Overholtz *et al.* 1999, 2000), then this flounder will need to eat 0.5 g every other day, or will only feed 182 days each year, ultimately consuming 1456 polychaete worms each year. The average yellowtail flounder is caught at age 3 and by the time it enters the market will have converted over 4300 worms into fish flesh.

9.2.2 Piscivores

Many flatfishes consume large amounts of other fishes. Piscivorous flatfishes have a larger mouth gape, and similar morphological adaptations, than their non-piscivorous counterparts (Yazdani 1969; de Groot 1971; Podoskina 1993; Aarnio *et al.* 1996). Additionally, many of the piscivorous flatfishes are large; for example, halibuts can reach lengths over 2 m. These characteristics determine the degree of piscivory exhibited by a particular flatfish. Examples of piscivorous flatfishes include Pacific and Atlantic halibut, summer flounder, American fourspot flounder, arrowtooth flounder, Japanese flounder, flathead sole, turbot and Kamchatka flounder (Table 9.1).

The majority of fishes eaten by flatfishes are forage species. Species such as Alaska pollock (*Theragra chalcogramma*), Pacific and Atlantic herring (*Clupea pallasii*, *C. harengus*) other clupeids, small scombrids, hakes (Merlucciidae), sand lance (Ammodytidae), capelin (*Mallotus villosus*), engraulids, cottids, myctophids, stichaeids, zoarcids and other flatfishes

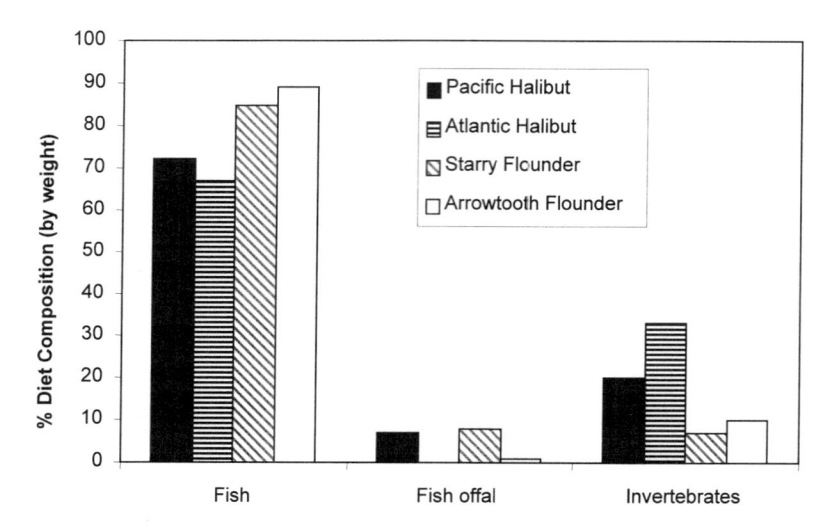

Fig. 9.2 Example of piscivorous flatfishes, with a high amount of fish (by weight) in the diet. Data for Pacific halibut adapted from Table 4.1 in Yang (1993). Data for Atlantic halibut adapted from Fig. 15A in Link & Almeida (2000). Data for starry flounder adapted from the largest size class in Fig. 2 in Tokranov & Maksimenkov (1994). Data for arrowtooth flounder adapted from Table 3.1 in Yang (1993).

are the most common fish prey. Squids are also common prey in piscivorous flatfishes. For many piscivorous flatfishes, fishes constitute 40–80% of the diet (Fig. 9.2). In some instances fishes can make up >90% of the diet, particularly for larger flatfishes (e.g. Yang 1993, 1995; Tokranov & Maksimenkov 1994) (Fig. 9.2). Fish eggs are an additional component of the diet of some flatfishes (e.g. Maurer & Bowman 1975; Bowman & Michaels 1984; Rodriguez-Marin et al. 1994), but usually account for <0.1% of the diet. Juveniles of commercially valuable species can also be major prey for piscivorous flatfishes. For example, small Atlantic cod (*Gadus morhua*) form 18.9% by volume of the diet of 55–80 cm Greenland halibut (Rodriguez-Marin et al. 1994). Whether or not predatory removal of juvenile stages of commercially valuable species affects recruitment of these prey to the fishery is unclear.

The magnitude of piscivorous flatfish feeding on forage fishes or juvenile fish populations, particularly those that are commercially valuable, is a very relevant question. Livingston (1993) provides an excellent example of an evaluation of predatory removal by various flatfishes on Alaska pollock. In 1985, more than 176 000 metric tons of Alaska pollock were removed by flatfishes in the eastern Bering Sea. Although the biomass removed was small compared with other predators or the fishery, flatfish predation contributed to a high mortality rate of juvenile pollock which notably dampened the 1985 year-class. Bax (1991, 1998) and Overholtz et al. (1991, 1999, 2000) confirm that predatory removal of fish by other fishes (including but not limited to flatfishes) is often the largest source of fish mortality in several ecosystems. In addition to potential effects on recruitment, when a flatfish and a fishery remove the same size of a particular prey fish, there is the potential for strong competition between the predator and the fishery (e.g. Livingston 1993; Overholtz et al. 2000). The merits of predator control to protect fisheries have been debated at length (Larkin 1996; Bax 1998; Yodzis 2000). The conclusion is that if the magnitude and size classes of removal are similar between the predators and fisheries, then fisheries managers should at least be aware of the potential conflict and set management objectives accordingly.

There is also a potential for strong predator–prey interactions between some flatfishes and other targeted resources such as squids, scallops, other bivalves, pandalids, paeneids and other shrimps (Table 9.1). Flatfish predation could limit the population size available for harvest of these molluscs or shrimps. Fisheries management should at least be cognisant of the qualitative implications of these predator–prey relationships.

9.2.3 Specialists

There are few exceptions to the general paradigm of polychaete-crustacean or fish-eating flatfishes, yet among the flatfishes there are some specialist feeders (Table 9.1). The specialist feeding flatfishes eat primarily echinoderms or bivalve siphons.

One non-typical, specialist feeder is American plaice, which eats almost exclusively echinoids (sand dollars), ophiuroids (brittle stars) and other echinoderms (Langton & Bowman 1981; Bowman & Michaels 1984; Packer *et al.* 1994; Berestovskiy 1995). In some instances >90% of the diet consists of echinoderms (e.g. Langton & Bowman 1981) (Fig. 9.3). There are only a few other flatfish species (e.g. starry flounder, Dover sole, dab) that specialise on ophiuroids or echinoids (e.g. Jewett & Feder 1980; Gabriel & Pearcy 1981; Temming & Hammer 1994; Kaiser & Ramsay 1997). These flatfishes also eat a larger amount of bivalves than other flatfishes. Most conspecifics, and often these same species at different locations, consume a more typical flatfish diet of polychaetes, benthic crustaceans or small fishes (Beare & Moore 1997; Gonzalez *et al.* 1998). Why certain flatfishes specialise on prey items that one would intuitively think are difficult to digest and low in energy content remains an interesting question. This is particularly true given that the same species often do not eat echinoderms at all locales.

Another specialist feeding strategy among flatfishes is the consumption of bivalve siphons. For example, hogchoker (Derrick & Kennedy 1997), plaice (Edwards & Steele 1968; de Vlas 1979; Ansell & Gibson 1990; Beare & Moore 1997), diamond turbot, California

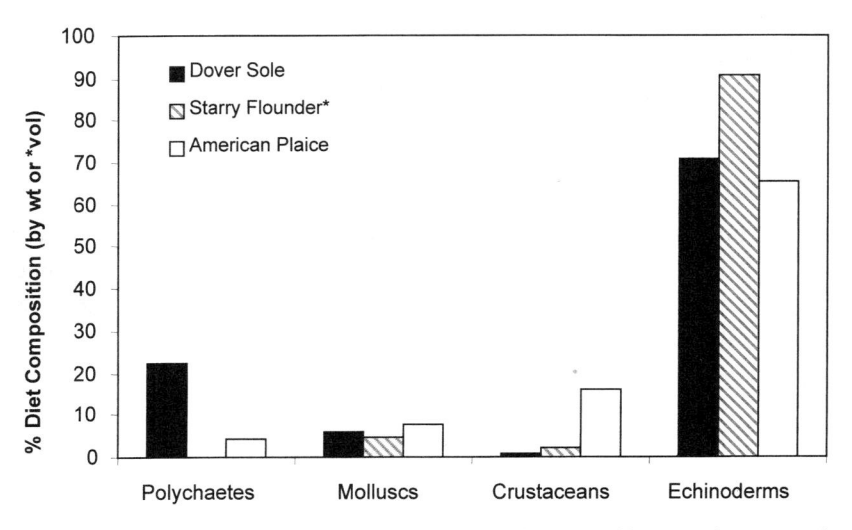

Fig. 9.3 Example of echinoderm specialists, showing percentage diet composition. Data for Dover sole adapted from Table 1, station S29, in Gabriel & Pearcy (1981). Data for starry flounder (by volume*) adapted from Table 1, Norton Sound, in Jewett & Feder (1980). These data are percentage diet composition by volume, not weight. Data for American plaice adapted from Table 1 in Langton & Bowman (1981).

flounder (Peterson & Quammen 1982), blackhand sole (Cyrus 1991) and flounder (de Vlas 1979) all exhibit this feeding behaviour. When present, siphons can form up to ca. 80% of the diet (Ansell & Gibson 1990; Cyrus 1991). It is highly probable that extended polychaete tails and clam siphon tips are detected by the same foraging strategy. However, feeding on siphon tips appears to be a transient pattern (practised primarily by juveniles) instead of a feeding behaviour that persists into adulthood (cf. Kuipers 1977). Feeding on siphon tips is interesting from the perspective of trophic ecology because it is one of the few examples of sublethal predation.

9.2.4 Other considerations

9.2.4.1 Anthropogenically produced food

Any chapter on flatfish trophic ecology would be incomplete without discussing food that flatfishes derive from scavenging, particularly after perturbations to the ocean bottom. Kaiser & Ramsay (1997) demonstrate that dab changes its diet and increases the amount of food ingested in areas that have been disturbed by bottom trawling. The implications are that populations of dab have increased in the North Sea from additional food made available due to trawling, primarily in the form of benthic organisms that have been displaced from the ocean bottom (Greenstreet & Hall 1996; Kaiser & Ramsay 1997). Species such as Greenland halibut (Rodriguez-Marin *et al.* 1995), starry flounder (Tokranov & Maksimenkov 1994, 1995b) and Pacific halibut (Yang 1993) similarly feed on offal and discards from fish-processing ships and factories. Winter flounder (Carlson *et al.* 1997) also demonstrate patterns of opportunistic feeding after dumping, sewage disposal and dredging in Long Island Sound. This phenomenon is probably more widespread than is currently suspected (cf. Kaiser & de Groot 2000).

9.2.4.2 Seasonality

In addition to short-term changes in diet, seasonal differences in diet have been observed for many flatfishes (e.g. Libey & Cole 1979; Bowman & Michaels 1984; Nashida & Tominaga 1987; Tokranov 1990; Williams & Williams 1998). As one might expect, these differences in diet reflect the seasonal cycles in benthic community composition and production. Other studies have similarly documented long-term changes in diet, particularly among piscivorous flatfishes, as a response to changes in available prey over the period of several years to decades (Tokranov 1990; Wainright *et al.* 1993; Link *et al.* 2002). However, most of these flatfishes exhibit shifts in diet among specific prey species, not drastic changes among broad prey groups. An analysis from Georges Bank confirms the generally static guild structure of these benthivores (Garrison & Link 2000b).

9.2.4.3 Ontogeny

All flatfishes eat polychaetes and small benthic crustaceans at some point in their life history. Many of the piscivores or specialists consume polychaetes and meiofauna at smaller sizes but grow out of this feeding mode (e.g. Bowman & Michaels 1984; Gee 1987; Tokranov & Maksimenkov 1994; Berestovskiy 1995; Rodriguez-Marin *et al.* 1995; Yang 1995; Yokoyama 1995;

Aarnio *et al.* 1996; Gonzalez & Chong 1997; Yamada *et al.* 1998; Garrison & Link 2000a, b). These patterns represent a classical ontogenetic shift in diet (Werner & Gilliam 1984). However, several flatfishes remain principally worm-benthic invertebrate eaters throughout their life history, particularly the major polychaete feeders described above. This constancy of diet reflects the differences in oral morphology among the flatfish species. The differential (relatively minimal) shift with ontogeny in relative mouth gape, and hence diet, observed for some of these flatfishes is unusual among fishes (e.g. Podoskina 1993).

9.2.4.4 Spatial factors

A final consideration of factors determining flatfish diet is spatial differences in prey availability. Accounting for ontogeny, most flatfishes feed within one of the major feeding guilds described above. Yet the particular prey items in the benthic community vary widely, as seen in the diverse diet composition of flatfishes at different locales in river-estuary ecosystems (e.g. Summers 1980; Derrick & Kennedy 1997; Thiel *et al.* 1997), exposed and protected bays (e.g. Yamada *et al.* 1998), different tidal zones (e.g. Wells *et al.* 1973; Ansell & Gibson 1990), and different regions and depths on continental shelves (e.g. Pearcy & Hancock 1978; Gabriel & Pearcy 1981; Langton & Bowman 1981; Methven 1999). Many of these studies document that functionally similar and preferred prey are generally selected but that differences in the availability of these prey result in differing diets.

In summary, within the constraints imposed by their morphology, flatfishes feed opportunistically and eat what is most readily abundant at a particular time and place.

9.3 Flatfish predators

Studies of flatfish predation document a suite of predators that change with the life history stage of flatfishes. For example, predation on flatfish eggs by herring has been documented in the North Sea (Daan 1976; Daan *et al.* 1984) and Irish Sea (Ellis & Nash 1997). Coelenterates have been identified as potential predators of larval fish. Larval flatfishes have been positively identified in the stomach contents of the ctenophore (*Pleurobrachia pileus*) and the jellyfish (*Aurelia aurita*) from the Wadden Sea (Van der Veer 1985). The data suggest that the spatial and temporal overlap between the larval fish and coelenterates could cause the observed annual decline in larval flatfish abundance in the Wadden Sea. Interestingly, Van der Veer (1985) noted that plaice larvae might find a refuge both in size and time when compared with other larvae, as most of the larval plaice settle to the benthos prior to the annual population explosion of coelenterates, whereas dab larvae remain available to these invertebrate predators.

Once flatfishes settle on the seafloor they congregate in nursery areas (Gibson 1994) and are consequently vulnerable to predation. Predation at this life history stage can be substantial, being particularly acute immediately after settlement (Van der Veer & Bergman 1987), to the extent that predation on this life stage may influence year-class strength (Bailey 1994). The sand shrimp (*Crangon crangon*) and shore crab (*Carcinus maenas*) are reported to be especially voracious predators of newly settled flatfishes (Van der Veer & Bergman 1987; Ansell *et al.* 1999). Additionally, younger individuals of species such as Atlantic cod, which forage over the sand flats at night, can also consume a large number of flatfishes (Ellis & Gibson 1995). Like larval predation, there is a refuge in space and time, as ~20–30 mm plaice and winter flounder have some immunity to attack by *Crangon* (Van der Veer & Bergman 1987; Witting & Able 1995).

Other studies (e.g. Minami 1986; Ellis & Gibson 1995; Gibson & Robb 1996) have concentrated on predators of 0-group flatfishes. Major juvenile flatfish predators include gadids, eels and other flatfishes. Predatory fishes generally prey on mobile epibenthic species and only consume flatfishes when available. Predation is also size-specific for juvenile flatfishes. Ellis & Gibson (1995) noted that dab predation by Atlantic cod was restricted to cod >58 mm in length. Despite a large population of smaller cod no dab were identified in the stomach contents of these smaller cod. Again, the impact of predation on young growing fishes has a strong temporal component because the flatfish prey species escape these predators by growing out of vulnerable sizes. However, this does not totally eliminate predation since other non-fish predators such as birds maintain predatory pressure on juvenile flatfishes (e.g. cormorants; Leopold *et al.* 1998). The cumulative impact of predation on year-class strength is hard to assess. Evidence from juvenile flatfishes in European waters indicates that predation at this life history stage may be the controlling factor that mediates annual differences in year-class size (sensu Van der Veer & Bergman 1987). Yet it is likely that predation acts more as a moderating factor in flatfish population maintenance rather than as an absolute control (Gibson 1994).

Adult flatfishes are also not immune from predation and have been observed to play a minor but identifiable role in the diet of other fishes. For example, studies from the northwest Atlantic and Kamchatkan shelf have identified flatfishes in the stomachs of piscivorous fishes such as cods, red hake (*Urophycis chuss*), white hake (*U. tenuis*), spotted codling (*U. regia*), sculpins (Cottidae), sea raven (*Hemitripterus americanus*), other cottids, bluefish (*Pomatomus saltatrix*) and American angler (*Lophius americanus*) (Langton & Bowman 1980; Tokranov 1992; Buckel *et al.* 1999; Link & Almeida 2000). Flatfishes have been identified from the faecal material of grey seals (*Halichoerus grypus*) (e.g. Hammond *et al.* 1994). Sharks are also an important predator of flatfishes (e.g. Stilwell & Kohler 1993; Gelsleichter *et al.* 1999). The role of flatfishes in the trophic dynamics of marine ecosystems is therefore broad but probably less significant than other, more common forage fishes. The most important predators of many flatfishes, of course, are humans.

9.4 Flatfish competitors

There are four requirements that must be fulfilled to demonstrate competition: spatio-temporal overlap, similarity of resource utilisation (i.e. diet), limiting resources and notable population impacts from the interaction. These are not trivial issues in productive, diverse, highly connected food webs that are embedded in a host of biological and physical interactions that are rarely, if ever, fully elucidated (sensu Link 1999). Given the caveats and difficulties of directly evaluating competition in the field, it is true that organisms that eat the same food at least have the potential for competition.

All species that consume benthic invertebrates or forage fishes have the potential to compete with flatfishes. Among the flatfishes, polychaete-benthic crustacean feeders have high dietary overlaps (e.g. Langton & Bowman 1981; Zhang 1988; Piet *et al.* 1998; Zhang *et al.* 1998; Garrison & Link 2000a, b). Piscivorous flatfishes similarly exhibit high dietary overlap (e.g. Yang 1993; Piet *et al.* 1998; Orlov 1999; Garrison & Link 2000a, b). Inter- and intraspecific competition among the flatfishes is mitigated primarily by morphological constraints and ontogenetic shifts in diet (Tominaga & Nashida 1991; Yokoyama 1995; Piet *et al.* 1998).

Shared resource utilisation not only exists among the flatfishes but also between flatfishes and other benthic-oriented species. The major competitors of flatfishes include skates, zoarcids, cottids, gadids, similar demersal fishes and decapod crustaceans (e.g. Tyler 1972; Langton 1982; Evans 1983; MacDonald 1983; Vinogradov 1984; MacDonald & Green 1986; Arancibia & Melendez 1987; Tominaga & Nashida 1991; Yokoyama 1995; Hostens & Mees 1999; Orlov 1999; Garrison & Link 2000a, b). The extent of competition between flatfishes and benthic invertebrates is generally less well understood. By definition, members of the same feeding guild have a strong potential for competition. However, species that utilise similar resources do not necessarily influence other intra-guild populations, and changes in intra-guild populations are not necessarily caused by competition.

An early demonstration of the potential role of competition in structuring a flatfish community was made in experimental studies in Scotland; this work started in 1886 and extended for over a decade (Garstang 1900). In two bays, comparison of areas open and closed to fishing showed that plaice and lemon sole abundance decreased in open areas relative to protected areas. However, in the open areas, dab (American plaice) numbers increased during the experiment. Garstang was 'inclined to attribute the increase to the advantage conferred on the dabs by the reduced numbers of their competitors, the plaice and lemon soles.'

The evidence is ambiguous as to whether competition is a strong factor influencing populations of flatfishes. There are instances where competition is minimised by subtle differences in preference for prey such as polychaetes, cnidarians, molluscs, amphipods, different decapods, etc., by highly productive prey populations (i.e. effectively unlimited resources), by dietary switching among particular prey items and by low spatio-temporal overlap (Evans 1983; Zhang 1988; Tominaga & Nashida 1991; Burke 1995; Yokoyama 1995; Armstrong 1997; Zhang *et al.* 1998). Conversely, there are instances where competition is strongly suspected, as indicated by high dietary overlap, concurrently opposite trends in abundance of competitor populations, and/or expulsion of competitive inferiors to suboptimal habitat (Persson 1981; Beaumont & Mann 1984; Piet *et al.* 1998). Further work is required for a better elucidation of competition among flatfishes in particular and marine fishes in general.

9.5 Flatfish trophic dynamics: a case study of Georges Bank

Georges Bank is a highly productive marine ecosystem located off the northeastern United States (Fogarty & Murawski 1998). Dramatic declines in key components of the flatfish assemblage in this area have occurred with important implications for the economics of the region and for overall fish community structure. These changes highlight the need to understand both the role of flatfishes as predators and the potential role of competition in structuring marine systems, and thus serve as a useful case study of flatfish community dynamics.

9.5.1 Shifts in abundance and species composition

Major changes in the abundance of several flatfish species on Georges Bank can be inferred from historical landings data and, in more recent times, from estimates of abundance derived from research vessel surveys and from fishery-dependent sources (Fogarty *et al.* 1987; German 1987; Mayo *et al.* 1992; Fogarty & Murawski 1998). Collectively, these changes in abundance hold important implications for the structure of the Georges Bank ecosystem. In particular, depletion of once dominant flatfish species may have resulted in important changes in the abundance and productivity of their competitors and prey.

The Atlantic halibut supported the earliest major flatfish fishery in the Gulf of Maine and adjacent waters (Bigelow & Schroeder 1953). Anecdotal reports dating to the colonial period (mid to late 1700s) indicate that this species was very abundant in the Gulf of Maine and concentrations were later identified on offshore banks and adjacent areas. Commercial exploitation of halibut was initiated around 1820 in the Gulf of Maine and approximately 1830 on Georges Bank (Goode 1887). Local depletions of halibut were recorded within two to three decades of the inception of fishing for this species and by 1850 the directed fishery for halibut in the Gulf of Maine and on Georges Bank was in decline. Complete records of landings in this region commence in the 1890s and show a continued decline to extremely low levels.

The development of the winter flounder fishery on Georges Bank can be traced to the early twentieth century. The winter flounder fishery on the bank intensified with the development of otter trawling. Declines in landings of winter flounder on Georges Bank, however, were noted by the 1930s under increasing exploitation (Hennemuth & Rockwell 1987). The fishery for yellowtail flounder developed in response to the decline in the winter flounder fishery. By the 1950s, however, landings of this species had also undergone decline (Royce *et al.* 1959). Increased fishing pressure was then placed on other flatfish species, including American plaice, witch and windowpane, in an attempt to develop substitute products as the abundance of winter and yellowtail flounder declined.

Estimates of relative abundance derived from research vessel surveys document further changes in the flatfish assemblage on Georges Bank from the early 1960s to present (Fig. 9.4). Sharp declines in yellowtail flounder abundance occurred under increasing exploitation by distant water fleets in the 1960s and subsequently under increased effort by the domestic

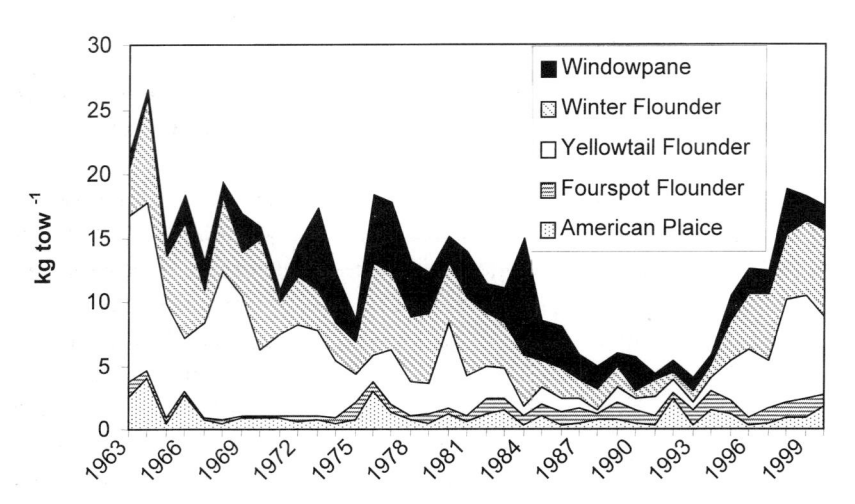

Fig. 9.4 Trends in relative abundance of flatfish species on Georges Bank derived from standardised research vessel surveys for American plaice, American fourspot flounder, yellowtail flounder, winter flounder and windowpane.

fleet. Winter flounder initially increased in abundance during the 1960s, concurrent with the decline in yellowtail flounder, but had decreased in abundance by the late 1980s under increased exploitation. American plaice also declined rapidly under increased exploitation. However, recent (early 1990s) area closures on Georges Bank have led to some recovery of yellowtail and winter flounder (Murawski et al. 2000).

9.5.2 Potential competitive interactions

Competition has been advanced as a hypothesis to explain patterns of co-variation in abundance of flatfishes and other fish taxa on Georges Bank. Overholtz & Tyler (1986) modelled abundance of fish communities on Georges Bank and noted that changes in abundance of yellowtail flounder and windowpane, sculpins and skates were consistent with the hypothesis of competition with haddock (*Melanogrammus aeglefinus*) during the period of high haddock abundance in the 1960s and 1970s. Fogarty & Brodziak (1994) examined changes in abundance of major species groups including flatfishes, gadids, other finfishes, skates and pelagic fishes in a multivariate time series analysis. Statistically significant inverse relationships at biologically appropriate lags were found for several groups including flatfishes and skates. Collie & DeLong (1999) developed multi-species production models for aggregate biomass of four species groups including flatfishes, gadids, elasmobranchs and small pelagic fishes on Georges Bank. Statistically significant inverse relationships between flatfishes and elasmobranchs and between flatfishes and gadids were reported. However, the resulting models for flatfish dynamics predicted extinction of flatfishes even in the absence of exploitation and were deemed unrealistic.

Royce et al. (1959) considered competitive interactions between yellowtail flounder and haddock on Georges Bank and adjacent areas as a potentially important factor responsible for the dynamics of these species. Pitt (1970) reached a similar conclusion for Newfoundland flatfish populations. Evidence for diet overlap between flatfishes and other fish species on Georges Bank has been examined to assess the potential for competition. Grosslein et al. (1980) reported dietary overlap values of approximately 60% between winter flounder and haddock and between winter flounder and yellowtail flounder over all size classes during 1969–72. Dietary overlap of approximately 45% between yellowtail flounder and little skate (*Leucoraja erinacea*) was reported. Estimated diet overlap between windowpane and red hake was about 40%.

Nelson (1993) also reported high diet overlap estimates for little skate and yellowtail flounder. For smaller size classes of little skate (<39 cm TL) and yellowtail (>26 cm TL), estimated diet overlap ranged from 72% to 92% in spring samples and somewhat lower values for winter estimates. Overlap was low between larger size classes of little skate (>40 cm TL) and all sizes of yellowtail. Larger skates exhibit higher levels of piscivory and therefore lower levels of diet overlap with yellowtail. Nelson further examined growth rates of yellowtail flounder and abundance of little skate and found no relationship. On this basis, Nelson concluded that, despite high dietary overlap, yellowtail flounder and little skate were not competitors.

Garrison & Link (2000a, b) and Garrison (2000) examined feeding guild structure of fishes on Georges Bank and the northeast continental shelf based on diet similarity estimates. Flatfish species constitute a dominant component of several identified guilds, most notably the benthivore assemblage. Garrison & Link note that subtle differences in prey preferences, feeding morphology, and in spatial distribution may alleviate competitive interactions among flatfishes in this guild. The potential for competition between flatfishes, skates and other groundfishes was identified as an important area for future research.

9.5.3 Predation by flatfishes

Larger Atlantic halibut are known to feed on Atlantic cod, tusk (*Brosme brosme*), haddock, ocean perch (*Sebastes marinus*), sculpins, grenadiers (Macrouridae), silver hake (*Merluccius bilinearis*), Atlantic herring, sand lance, capelin, skates, wolf-fish (*Anarhichas lupus*), Atlantic mackerel (*Scomber scombrus*) and other flatfishes (Bigelow & Schroeder 1953; Kohler 1967; Maurer & Bowman 1975; Nickerson 1978; Reid *et al.* 1999; Link *et al.* 2002). The reported abundance of this species prior to extensive exploitation and its large size and presumed energetic requirements suggest the potential for high removals of fishes and macroinvertebrates through predation. A release in predation pressure on fish and squid populations with the sharp reduction in the abundance of halibut could have had significant effects on these prey taxa. The observation of the importance of other flatfish species in the diet of halibut is of particular interest in the present context.

The decline in the abundance of yellowtail and winter flounder has presumably resulted in a sharp decrease in the consumption of benthic invertebrates by these species. As noted above, this decreased consumption may have increased the availability of these prey items to other fishes. Coupled with the changes in abundance, apparent changes in mean stomach content weight over time of several flatfish species have been reported (Link *et al.* 2002). Despite relatively consistent diet compositions of major prey groups, the mean stomach content weight for eight flatfish species on the northeast continental shelf peaked in the early 1980s and subsequently declined, returning approximately to levels observed earlier (Fig. 9.5). The factors underlying the apparent changes over time have not been identified. These observations, coupled with information on flatfish abundance levels, suggest that consumption by several flatfish was substantially higher in the mid-1980s than in the succeeding period.

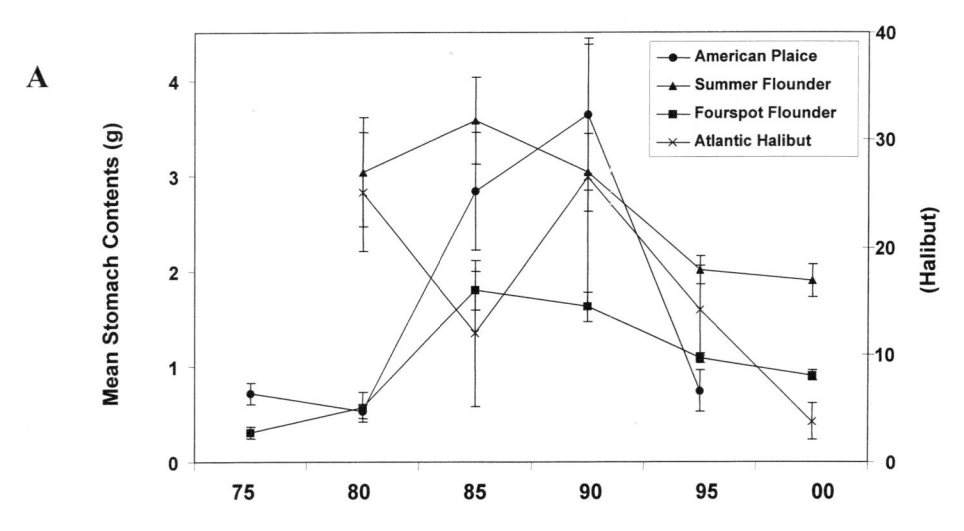

Fig. 9.5 Mean stomach contents (g) of the flatfish species in the northwest Atlantic across the 5-year blocks of the time series. A. The major piscivores/echinoderm feeders.

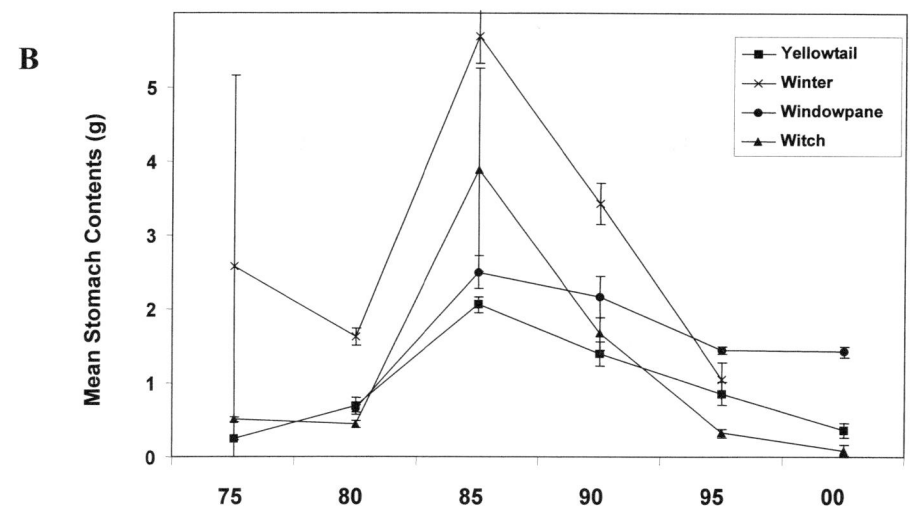

Fig. 9.5 *contined.* B. The major polychaete/amphipod feeders. Data adapted from Link *et al.* (2002).

9.5.4 Have changes in flatfish populations influenced the Georges Bank ecosystem?

Fishery-induced changes in the flatfish community on Georges Bank have been profound, with major reductions in biomass of key species. These changes have potentially important cascading effects in the system. Predation pressure on other fishes by Atlantic halibut has been dramatically reduced by the depletion of this species. The full implications of this reduction cannot be determined because of the lack of adequate baseline information on halibut and prey populations at the inception of the fishery.

It can also be inferred that predation on benthic invertebrates, particularly polychaetes and decapod crustaceans, by species such as yellowtail and winter flounder is markedly reduced relative to historical levels. A central question in understanding subsequent changes in the system is whether other species were able to exploit additional prey resources (or space) with the decline of yellowtail and winter flounder as well as other groundfish species, resulting in population increases for skates or other flatfishes. As noted earlier, competition is inherently difficult to demonstrate in systems in which controlled and replicated experiments are not possible. Given this concern, there is evidence of spatial, temporal and dietary overlap of flatfishes with other fish species on Georges Bank that indicate a strong *potential* for competitive interactions. The observed changes in abundance are consistent with a competitive mechanism operating at a system-wide level.

9.6 Summary and conclusions

Flatfishes generally feed on worms and small crustaceans, fishes and squids, or echinoderms. Morphology, ontogeny, and spatio-temporal availability of prey determine the diet. In some cases, predation by flatfishes can have significant impacts on their prey populations. The magnitude of flatfish predation, and the influences thereof, on various prey populations is a topic that merits much further examination. In particular, the potential competition between fisheries and flatfishes for fish prey is an area that needs to be fully elucidated in a fisheries management context. The influences of flatfish predation on non-targeted but ecologically valuable species (e.g. polychaetes, amphipods) are also usually undetermined.

Flatfishes are eaten by many different species, including cnidarians, crustaceans, other fishes, birds and mammals. The degree of predation varies across life history and may influence annual variations in year-class strength. Flatfishes are notable prey in most ecosystems, but less so than other forage fishes. The cumulative impact of predation on year-class strength is difficult to assess but should be incorporated into flatfish population models, particularly as an explicit mortality term.

Flatfishes have high spatial, temporal and dietary overlaps with many other species. Opposite trends in abundance of possible flatfish competitor populations suggest that competitive processes do occur. Although difficult to ascertain, the potential for competition among flatfishes and between flatfishes and other species is strong. Better assessments of in situ competition and population-level impacts are fruitful areas for further research.

Effects on benthic communities as a result of harvesting activities and their implications for benthivorous flatfishes deserve further attention. Reductions in biomass and species diversity in benthic organisms in exploited areas have been documented on Georges Bank (Collie *et al.* 1997, 2000). However, the potential implications of these changes on flatfish populations have not been assessed. Time series of biomass and species composition of the benthos are not available for Georges Bank and changes in benthic production cannot currently be related to changes in fish production. Link *et al.* (2002) do note that, in contrast to piscivorous flatfishes, there have been no apparent changes in diet composition of benthivorous flatfishes that might reflect changes in prey availability or abundance. The establishment of marine reserves on Georges Bank (Fogarty & Murawski 1998) and elsewhere offers the possibility to examine these questions experimentally.

Acknowledgements

We thank members of the Food Web Dynamics Program, past and present, for their dedicated effort at understanding fish trophic ecology in the northwest Atlantic and providing us with the basis for much of our work. We thank M. Woodruff and J. Riley for their assistance with the literature search. We thank R. Gibson for the invitation to contribute to this book. We thank anonymous reviewers for their constructive comments on early drafts. J.L. asserts *Soli Deo Gloria*.

References

Aarnio, K., Bonsdorff, E. & Rosenback, N. (1996) Food and feeding habits of juvenile flounder *Platichthys flesus* (L.), and turbot *Scophthalmus maximus* L. in the Aland Archipelago, northern Baltic Sea. *Journal of Sea Research*, **36**, 311–320.

Ansell, A.D. & Gibson, R.N. (1990) Patterns of feeding and movement of juvenile flatfishes on an open sandy beach. In: *Trophic Relationships in the Marine Environment* (eds M. Barnes & R.N. Gibson). Proceedings of the 24th European Marine Biology Symposium, pp. 191–207. Aberdeen University Press, Aberdeen, Scotland.

Ansell, A.D., Comely, C.A. & Robb, L. (1999) Distribution, movements and diet of macrocrustaceans on a Scottish sandy beach with particular reference to predation on juvenile fishes. *Marine Ecology Progress Series*, **176**, 115–130.

Arancibia, H.F. & Melendez, R.C. (1987) Alimentacion de peces concurrentes en la pesqueria de *Pleuroncodes monodon* Milne Edwards. *Investagacion Pesquera Santiago (Chile)*, **34**, 113–128.

Armstrong, M.P. (1997) Seasonal and ontogenetic changes in distribution and abundance of smooth flounder, *Pleuronectes putnami*, and winter flounder, *Pleuronectes americanus*, along estuarine depth and salinity gradients. *Fishery Bulletin*, **95**, 414–430.

Auster, P.J. & Langton, R.W. (1999) The effects of fishing on fish habitat. In: Fish habitat: essential fish habitat and rehabilitation (ed. L. Benaka). *American Fisheries Society Symposium*, **22**, 150–187.

Bailey, K.M. (1994) Predation on juvenile flatfish and recruitment variability. *Netherlands Journal of Sea Research*, **32**, 175–189.

Bax, N.J. (1991) A comparison of the fish biomass flow to fish, fisheries, and mammals on six marine ecosystems. *ICES Marine Science Symposia*, **193**, 217–224.

Bax, N.J. (1998) The significance and prediction of predation in marine fisheries. *ICES Journal of Marine Science*, **55**, 997–1030.

Beare, D.J. & Moore, P.G. (1997) The contribution of Amphipoda to the diet of certain inshore fish species in Kames Bay, Millport. *Journal of the Marine Biological Association of the United Kingdom*, **77**, 907–910.

Beaumont, W.R.C. & Mann, R.H.K. (1984) The age, growth and diet of a freshwater population of the flounder, *Platichthys flesus* (L.), in southern England. *Journal of Fish Biology*, **25**, 607–616.

Belghyti, D., Aguesse, P. & Gabrion, C. (1993) Feeding ethology of *Citharus linguatula* and *Dicologoglossa cuneata* of the Moroccan Atlantic area. *Vie et Milieu*, **43**, 95–108.

Berestovskiy, Y.G. (1995) Feeding habitats and strategy of American Plaice, *Hippoglossoides platessoides limandoides*, in the Barents and Norwegian Seas. *Journal of Ichthyology*, **35**, 40–54.

Bigelow, H.B. & Schroeder, W.C. (1953) Fishes of the Gulf of Maine. *Fishery Bulletin*, **53**, 1–577.

Bowman, R.E. & Michaels, W.L. (1984) Food of seventeen species of northwest Atlantic fish. *NOAA Technical Memorandum*, NMFS-F/NEC-28.

Buckel, J.A., Fogarty, M.J. & Conover, D.O. (1999) Foraging habits of bluefish, *Pomatomus saltatrix*, on the U.S. east coast continental shelf. *Fishery Bulletin*, **97**, 758–775.

Burke, J.S. (1995) Role of feeding and prey distribution of summer and southern flounder in selection of estuarine nursery habitats. *Journal of Fish Biology*, **47**, 355–366.

Carlson, J.K., Randall, T.A. & Mroczka, M.E. (1997) Feeding habits of winter flounder (*Pleuronectes americanus*) in a habitat exposed to anthropogenic disturbance. *Journal of Northwest Atlantic Fisheries Science*, **21**, 65–73.

Collie, J.S. (1987a) Food selection by yellowtail flounder (*Limanda ferruginea*) on Georges Bank. *Canadian Journal of Fisheries and Aquatic Sciences*, **44**, 357–367.

Collie, J.S. (1987b) Food consumption by yellowtail flounder in relation to production of its benthic prey. *Marine Ecology Progress Series*, **36**, 205–213.

Collie, J.S. & DeLong, A.K. (1999) Multispecies interactions in the Georges Bank fish community. In: Ecosystem Considerations in Fisheries Management, Proceedings of the 16th Lowell Wakefield Fisheries Symposium. *Alaska Sea Grant College Report No. 99–01*, pp. 187–210. University of Alaska, Anchorage, AK.

Collie, J.S., Escanero, G.A. & Valentine, P.C. (1997) Effects of bottom fishing on the benthic megafauna of Georges Bank. *Marine Ecology Progress Series*, **155**, 159–172.

Collie, J.S., Escanero, G.A. & Valentine, P.C. (2000) Photographic evaluation of the impacts of bottom fishing on benthic epifauna. *ICES Journal of Marine Science*, **57**, 987–1001.

Cyrus, D.P. (1991) The biology of *Solea bleekeri* (Teleostei) in Lake St. Lucia on the southeast coast of Africa. *Netherlands Journal of Sea Research*, **27**, 209–216.

Daan, N. (1976) Some preliminary investigations into predation on fish eggs and larvae in the southern North Sea. *International Council for the Exploration of the Sea,* CM/L:15, 1–5.

Daan, N., Rijnsdorp, A.D. & Overbeeke, G.R. (1984) Predation of plaice and cod eggs by North Sea herring. *Council Meeting of the International Council for the Exploration of the Sea,* CM/L:13, 1–16.

de Groot, S.J. (1969) Digestive system and sensorial factors in relation to the feeding behavior of flatfish (Pleuronectiformes). *Journal du Conseil International pour l'Exploration de la Mer*, **32**, 385–395.

de Groot, S.J. (1971) On the interrelationships between morphology of the alimentary tract, food and feeding behaviour in flatfishes (Pisces: Pleuronectiformes). *Netherlands Journal of Sea Research*, **5**, 121–196.

de Morais, T.L. & Bodiou, J.Y. (1984) Predation on meiofauna by juvenile fish in a western Mediterranean flatfish nursery ground. *Marine Biology*, **82**, 209–215.

Derrick, P.A. & Kennedy, V.S. (1997) Prey selection by the hogchoker, *Trinectes maculatus* (Pisces: Soleidae), along summer salinity gradients in Chesapeake Bay, USA. *Marine Biology*, **129**, 699–711.

de Vlas, J. (1979) Annual food intake by plaice and flounder in a tidal flat area in the Dutch Wadden Sea, with special reference to consumption of regenerating parts of macrobenthic prey. *Netherlands Journal of Sea Research*, **13**, 117–153.

Dou, S., Yang, J. & Chen, D. (1992) Food habits of stone flounder, spotted flounder, high-eyed flounder and red tongue sole in the Bohai Sea. *Journal of Fisheries of China*, **16**, 162–166.

du Buit, M.H. (1992) Alimentation de la cardine, *Lepidorhombus whiffiagonis* en Mer Celtique. *Cahiers de Biologie Marine*, **33**, 501–514.

Edwards, R. & Steele, J.H. (1968) The ecology of 0-group plaice and common dabs at Lochewe. I. Population and food. *Journal of Experimental Marine Biology and Ecology*, **2**, 215–238.

Ellis, T. & Gibson, R.N. (1995) Size-selective predation of 0-group flatfishes on a Scottish coastal nursery ground. *Marine Ecology Progress Series*, **127**, 27–37.

Ellis, T. & Nash, R.D.M. (1997) Predation by sprat and herring on pelagic fish eggs in a plaice spawning area in the Irish Sea. *Journal of Fish Biology*, **50**, 1195–1202.

Euzen, O. (1987) Food habits and diet composition of some fish of Kuwait. *Kuwait Bulletin of Marine Science*, **1987**, 65–85.

Evans, S. (1983) Production, predation and food niche segregation in a marine shallow soft-bottom community. *Marine Ecology Progress Series*, **10**, 147–157.

FAO (1998) *The State of World Fisheries and Aquaculture 1998.* FAO, Rome.

Fogarty, M.J. & Brodziak, J. (1994) Multivariate time series analysis of fish community dynamics on Georges Bank. *International Council for the Exploration of the Sea,* CM/P: 3.

Fogarty, M.J. & Murawski, S.A. (1998) Large-scale disturbance and the structure of marine systems: fishery impacts on Georges Bank. *Ecological Applications*, **8** (Suppl.), S6–S22.

Fogarty, M.J., Sissenwine, M.P. & Grosslein, M.D. (1987) Fish population dynamics. In: *Georges Bank* (ed. R. Backus). pp. 494–521. MIT Press, Cambridge, MA.

Gabriel, W.L. & Pearcy, W.G. (1981) Feeding selectivity of Dover sole, *Microstomus pacificus*, off Oregon. *Fishery Bulletin*, **79**, 749–763.

Garcia, S. & Newton, C. (1997) Current situation, trends, and prospects in world capture fisheries. In: Global trends: fisheries management (eds D. Pikitch, D. Huppert & M.P. Sissenwine). *American Fisheries Society Symposium*, **20**, 3–27.

Garcia-Abad, M.C., Yanez-Aranciba, A., Sanchez-Gil, P. & Garcia, M.P. (1992) Distribucion, reproducion y alimentacion de *Syacium gunteri* Ginsburg (Pisces:Bothidae), en el Golfo de Mexico. *Revista de Biologia Tropical*, **39**, 27–34.

Garrison, L.P. (2000) Spatial and dietary overlap in the Georges Bank groundfish community. *Canadian Journal of Fisheries and Aquatic Sciences*, **57**, 1679–1691.

Garrison, L.P. & Link, J.S. (2000a) Dietary guild structure of the fish community in the northeast United States continental shelf ecosystem. *Marine Ecology Progress Series*, **202**, 231–240.

Garrison, L.P. & Link, J.S. (2000b) Fishing effects on spatial distribution and trophic guild structure in the Georges Bank fish community. *ICES Journal of Marine Science*, **57**, 723–730.

Garstang, W. (1900) The impoverishment of the sea – critical summary of the experimental and statistical evidence bearing upon the alleged depletion of the trawling grounds. *Journal of the Marine Biological Association of the United Kingdom*, **6**, 1–69.

Gee, J.M. (1987) Impact of epibenthic predation on estuarine intertidal harpacticoid copepod populations. *Marine Biology*, **96**, 497–510.

Glesleichter, J., Musick, J.A. & Nichols, S. (1999) Food habits of the smooth dogfish, *Mustelus canis*, dusky shark, *Carcharinus obscurus*, Atlantic sharpnose shark, *Rhizoprionodon terraenovae*, and the sand tiger, *Carcharias taurus*, from the northwest Atlantic Ocean. *Environmental Biology of Fishes*, **54**, 205–217.

German, A.W. (1987) History of the early fisheries: 1720–1930. In: *Georges Bank* (ed. R. Backus). pp. 406–424. MIT Press, Cambridge, MA.

Gibson, R.N. (1994) Impact of habitat quality and quantity on the recruitment of juvenile flatfishes. *Netherlands Journal of Sea Research*, **32**, 191–206.

Gibson, R.N. & Robb, L. (1996) Piscine predation on juvenile fishes on a Scottish sandy beach. *Journal of Fish Biology*, **49**, 120–138.

Gonzalez, C., Costas, G., Bruno, I. & Paz, X. (1998) Feeding chronology of American plaice (*Hippoglossoides platessoides*) and yellowtail flounder (*Limanda ferruginea*) in the Grand Bank (NAFO Division 3N). *NAFO Scientific Council Research Document*, *1998*, No. 98/42.

Gonzalez, P. & Chong, J. (1997) Alimentacion del lenguado de ojos chicos *Paralichthys microps* (Guenther, 1881) (Pleuronectiformes, Paralichthyidae) en Bahia de Concepcion (VIII region, Chile). *Gayana Zoologia*, **61**, 7–13.

Goode, G.B. (1887) *The Fisheries and Fishing Industries of the United States, Section II. A Geographical Review of the Fisheries Industries and Fishing Communities for the Year 1880*. US Government Printing Office, Washington, DC.

Greenstreet, S.P.R. & Hall, S.J. (1996) Fishing and groundfish assemblage structure in the northwestern North Sea: an analysis of long-term and spatial trends. *Journal of Animal Ecology*, **65**, 577–598.

Grosslein, M.D., Langton, R.W. & Sissenwine, M.P. (1980) Recent fluctuations in pelagic fish stocks of the Northwest Atlantic, Georges Bank region, in relation to species interactions. *Rapports et Procès-Verbaux des Réunions Conseil Permanent International pour l'Exploration de la Mer*, **177**, 374–404.

Hammond, P.S., Hall, A.J. & Prime, J.H. (1994) The diet of grey seals in the Inner and Outer Hebrides. *Journal of Applied Ecology*, **31**, 737–746.

Hennemuth, R.C. & Rockwell, S. (1987) History of fisheries conservation and management. In: *Georges Bank* (ed. R. Backus). pp. 430–446. MIT Press, Cambridge, MA.

Hostens, K. & Mees, J. (1999) The mysid-feeding guild of demersal fishes in the brackish zone of the Westerschelde estuary. *Journal of Fish Biology*, **55**, 704–719.

Jennings, S. & Kaiser, M.J. (1998) The effects of fishing on marine ecosystems. *Advances in Marine Biology*, **34**, 201–352.

Jewett, S.C. & Feder, H.M. (1980) Autumn food of adult starry flounders, *Platichthys stellatus*, from the northeastern Bering Sea and the southeastern Chukchi Sea. *Journal du Conseil pour l' Exploration de la Mer*, **39**, 7–14.

Kaiser, M.J. & de Groot, S.J. (2000) *Effects of Fishing on Non-Target Species and Habitats: Biological, Conservation, and Socio-Economic Issues*. Blackwell Science, Oxford.

Kaiser, M.J. & Ramsay, K. (1997) Opportunistic feeding by dabs within areas of trawl disturbance: possible implications for increased survival. *Marine Ecology Progress Series*, **152**, 307–310.

Keats, D.W. (1990) Food of winter flounder *Pseudopleuronectes americanus* in a sea urchin dominated community in eastern Newfoundland. *Marine Ecology Progress Series*, **60**, 13–22.

Kohler, A.C. (1967) Size at maturity, spawning season, and food of Atlantic halibut. *Journal of the Fisheries Research Board of Canada*, **24**, 53–66.

Kravitz, M.J., Pearcy, W.G. & Guin, M.P. (1977) Food of five co-occurring flatfishes on Oregon's continental shelf. *Fishery Bulletin*, **74**, 984–990.

Kuipers, B.R. (1977) On the ecology of juvenile plaice on a tidal flat in the Wadden Sea. *Netherlands Journal of Sea Research*, **11**, 56–91.

Langton, R.W. (1982) Diet overlap between Atlantic cod, *Gadus morhua*, silver hake, *Merluccius bilinearis*, and fifteen other Northwest Atlantic finfish. *Fishery Bulletin*, **80**, 745–759.

Langton, R.W. (1983) Food habits of yellowtail flounder, *Limanda ferruginea* (Storer), from off the northeastern United States. *Fishery Bulletin*, **81**, 15–22.

Langton, R.W. & Bowman, R.E. (1980) Food of fifteen northwest Atlantic gadiform fishes. *NOAA Technical Report*, NMFS SSRF B740.

Langton, R.W. & Bowman, R.E. (1981) Food of eight northwest Atlantic pleuronectiform fishes. *NOAA Technical Report*, NMFS-SSRF749.

Larkin, P.A. (1996) Concepts and issues in marine ecosystem management. *Reviews in Fish Biology and Fisheries*, **6**, 139–164.

Leopold, M.F., Van Damme, C.J.G. & Van der Veer, H.W. (1998) Diet of cormorants and the impact of cormorant predation on juvenile flatfish in the Dutch Wadden Sea. *Journal of Sea Research*, **40**, 93–107.

Libey, G.S. & Cole, C.F. (1979) Food habits of yellowtail flounder, *Limanda ferruginea* (Storer). *Journal of Fish Biology*, **15**, 371–374.

Link, J.S. (1999) (Re)Constructing food webs and managing fisheries. In: Ecosystem considerations in fisheries management. Proceedings of the 16th Lowell Wakefield Fisheries Symposium. *Alaska Sea Grant College Report No. 99–01*, pp. 571–588. University of Alaska, Anchorage, AK.

Link, J.S. (2002) Ecological considerations in fisheries management: when does it matter? *Fisheries*, **27**, 10–17.

Link, J.S. & Almeida, F.P. (2000) An overview and history of the food web dynamics program of the Northeast Fisheries Science Center, Woods Hole, Massachusetts. *NOAA Technical Memorandum*, NMFS-NEFSC-159.

Link, J.S., Bolles, K. & Milliken, C.G. (2002) The feeding of flatfish in the northeast United States continental shelf ecosystem. *Journal of Northwest Atlantic Fisheries Science*, **30**, 1–17.

Linton, E. (1921) Food of young winter flounders. Appendix IV. *US Report of Commissioner of Fisheries for 1921*. B.F. Doc. 907.

Livingston, M.E. (1987) Food resource use among five flatfish species (Pleuronectiformes) in Wellington Harbour, New Zealand. *New Zealand Journal of Marine and Freshwater Research*, **21**, 281–293.

Livingston, P.A. (1993) Importance of predation by groundfish, marine mammals and birds on walleye pollock *Theragra chalcogramma* and Pacific herring *Clupea pallasi* in the eastern Bering Sea. *Marine Ecology Progress Series*, **102**, 205–215.

MacDonald, J.S. (1983) Laboratory observations of feeding behaviour of the ocean pout (*Macrozoarces americanus*) and winter flounder (*Pseudopleuronectes americanus*) with reference to niche overlap of natural populations. *Canadian Journal of Zoology,* **61**, 539–546.

MacDonald, J.S. & Green, R.H. (1986) Food resource utilization by five species of benthic feeding fish in Passamaquoddy Bay, New Brunswick. *Canadian Journal of Fisheries and Aquatic Sciences,* **43**, 1534–1546.

Martell, D.J. & McClelland, G. (1992) Prey spectra of pleuronectids (*Hippoglossoides platessoides, Pleuronectes ferrugineus, Pleuronectes americanus*) from Sable Island bank. *Canadian Technical Report of Fisheries and Aquatic Sciences,* No. 1895.

Maurer, R.O. & Bowman, R.E. (1975) Food habits of marine fishes of the northwest Atlantic – data report. *US NMFS Northeast Fisheries Science Center, Woods Hole Laboratory Reference Document,* 75–3.

Mayo, R.K., Fogarty, M.J. & Serchuk, F.M. (1992) Aggregate fish biomass and production on Georges Bank. *Journal of Northwest Atlantic Fisheries Science,* **14**, 59–78.

Methven, D.A. (1999) Annotated bibliography of demersal fish feeding with emphasis on selected studies from the Scotian Shelf and Grand Banks of the northwestern Atlantic. *Canadian Technical Report of Fisheries and Aquatic Sciences,* No. 2267.

Minami, T. (1986) Predations on the young flatfishes in the Japan Sea. *Bulletin of the Japan Sea Regional Fisheries Research Laboratory,* **36**, 39–47.

Murawski, S.A., Brown, R., Lai, H.-L., Rago, P.J. & Hendrickson, L. (2000) Large-scale closed areas as a fishery-management tool in temperate marine systems: the Georges Bank experience. *Bulletin of Marine Science,* **66**, 775–798.

Nash, R.D.M., Geffen, A.J. & Santos, R.S. (1991) The wide-eyed flounder, *Bothus podas* Delaroche, a singular flatfish in varied shallow-water habitats of the Azores. *Netherlands Journal of Sea Research,* **27**, 367–373.

Nashida, K. & Tominaga, O. (1987) Studies on groundfish communities in the coastal waters of northern Niigata Prefecture II. Seasonal changes of feeding habits and daily rations of young flounder, *Paralichthys olivaceus. Bulletin of the Japan Sea Regional Fisheries Research Laboratory,* **37**, 39–56.

Nashida, K., Tominaga, O., Miyajima, H. & Ito, M. (1984) Studies on groundfish communities in the coastal waters of northern Niigata Prefecture I. Estimation of daily food consumption of young flounder, *Paralichthys olivaceus. Bulletin of the Japan Sea Regional Fisheries Research Laboratory,* **34**, 1–17.

Nelson, G.A. (1993) *The potential impacts of skate abundance upon the invertebrate resources and growth of yellowtail flounder* (Pleuronectes ferruginea) *on Georges Bank*. PhD dissertation, University of Massachusetts, Amherst, MA.

Nickerson, J.T.R. (1978) The Atlantic halibut and its utilization. *Marine Fisheries Review,* **40**, 21–25.

NMFS (1999) *Ecosystem-based Fishery Management*. A report to Congress by the Ecosystems Principles Advisory Panel. US DOC, Silver Spring, MD.

NMFS (2003) *Fisheries of the United States, 2000.* US DOC, Silver Spring, MD.

NRC (1999) *Sustaining Marine Fisheries*. National Academy Press, Washington, DC.

Orlov, A.M. (1997) Role of fishes in predator diets of the Pacific slope of the northern Kuril Islands and southeastern Kamchatka. In: International Symposium on the Role of Forage Fishes in Marine Ecosystems. Proceedings of the 14th Lowell Wakefield Fisheries Symposium. *Alaska Sea Grant College Report No. 97–01,* pp. 209–229. University of Alaska, Anchorage, AK.

Orlov, A.M. (1999) Some aspects of trophic relations among Pacific predatory fishes off the northern Kuril Islands and southeastern Kamchatka. In: *Fish Feeding Ecology and Digestion: GUTSHOP '98* (eds D. MacKinlay & D. Houlihan). pp. 41–52. American Fisheries Society Physiology Section, Bethesda, MD.

Overholtz, W.J. & Tyler, A.V. (1986) An exploratory simulation model of competition and predation in a demersal fish assemblage on Georges Bank. *Transactions of the American Fisheries Society*, **115**, 805–817.

Overholtz, W.J., Murawski, S.A. & Foster, K.L. (1991) Impact of predatory fish, marine mammals, and seabirds on the pelagic fish ecosystem of the northeastern USA. *ICES Marine Science Symposium*, **193**, 198–208.

Overholtz, W.J., Link, J.S. & Suslowicz, L.E. (1999) Consumption and harvest of pelagic fishes in the Gulf of Maine-Georges Bank ecosystem: implications for fishery management. In: Ecosystem Approaches for Fisheries Management, Proceedings of the 16th Lowell Wakefield Fisheries Symposium. *Alaska Sea Grant College Report No. 99–01*, pp. 163–186. University of Alaska, Anchorage, AK.

Overholtz, W.J., Link, J.S. & Suslowicz, L.E. (2000) The impact and implications of fish predation on pelagic fish and squid on the eastern USA shelf. *ICES Journal of Marine Science*, **57**, 1147–1159.

Packer, D.B., Watling, L. & Langton, R.W. (1994) The population structure of the brittle star *Ophiura sarsi* Luetken in the Gulf of Maine and its trophic relationship to American plaice (*Hippoglossoides platessoides* Fabricius). *Journal of Experimental Marine Biology and Ecology*, **179**, 207–222.

Pearcy, W.G. & Hancock, D. (1978) Feeding habits of Dover sole, *Microstomus pacificus*; rex sole, *Glyptocephalus zachirus*; slender sole, *Lyopsetta exilis*; and Pacific sanddab, *Citharichthys sordidus*, in a region of diverse sediments and bathymetry off Oregon. *Fishery Bulletin*, **76**, 641–651.

Persson, L.E. (1981) Were macrobenthic changes induced by thinning out of flatfish stocks in the Baltic proper? *Ophelia*, **20**, 137–152.

Peterson, C.H. & Quammen, M.L. (1982) Siphon nipping: its importance to small fishes and its impact on growth of the bivalve *Protothaca staminea* (Conrad). *Journal of Experimental Marine Biology and Ecology*, **63**, 249–268.

Piet, G.J., Pfisterer, A.B. & Rijnsdop, A.D. (1998) On factors structuring the flatfish assemblage in the southern North Sea. *Netherlands Journal of Sea Research*, **40**, 143–152.

Pitt, T.K. (1970) Distribution, abundance, and spawning of yellowtail flounder (*Limanda ferruginea*) in the Newfoundland area of the northwest Atlantic. *Journal of the Fisheries Research Board of Canada*, **27**, 1161–1171.

Podoskina, T.A. (1993) Morphology of supporting elements in jaw-pharyngeal apparatus of some pleuronectiform fishes in relation to feeding specializations. *Journal of Ichthyology*, **33**, 122–128.

Powell, A.B. & Schwartz, F.J. (1979) Food of *Paralichthys dentatus* and *P. lethostigma* (Pisces: Bothidae) in North Carolina estuaries. *Estuaries*, **2**, 276–279.

Redon, M.J., Morte, M.S. & Sanz-Brau, A. (1994) Feeding habits of the spotted flounder *Citharus linguatula* off the eastern coast of Spain. *Marine Biology*, **120**, 197–201.

Reid, R.F., Almeida, F. & Zetlin, C. (1999) Essential fish habitat source document: fishery independent surveys, data sources, and methods. *NOAA Technical Memorandum*, NMFS-NE-122.

Rodriguez-Marin, E., Punzon, A., Paz, J. & Olaso, I. (1994) Feeding of most abundant fish species in Flemish Cap in summer 1993. *NAFO Scientific Council Research Document*, 1994 No. 94–35.

Rodriguez-Marin, E., Punzon, A. & Paz, J. (1995) Feeding patterns of Greenland halibut (*Reinhardtius hippoglossoides*) in Flemish Pass (Northwest Atlantic). *NAFO Scientific Council Studies*, **23**, 43–54.

Royce, W.R., Butler, R.J. & Premetz, E.D. (1959) Decline of the yellowtail flounder (*Limanda ferruginea*) off southern New England. *US Fisheries and Wildlife Service Fish Bulletin*, **146**, 1–98.

Steimle, F.W. & Terranova, R. (1991) Trophodynamics of select demersal fishes in the New York Bight. *NOAA Technical Memorandum*, NMFS-F/NEC-84.

Stillwell, C.E. & Kohler, N.E. (1993) Food habits of the sandbar shark *Carcharhinus plumbeus* off the U.S. northeast coast, with estimates of daily ration. *Fishery Bulletin*, **91**, 138–150.

Summers, R.W. (1980) The diet and feeding behaviour of the flounder *Platichthys flesus* (L.) in the Ythan Estuary, Aberdeenshire, Scotland. *Estuarine and Coastal Marine Science*, **11**, 217–232.

Temming, A. & Hammer, C. (1994) Sex specific food consumption of dab (*Limanda limanda* L.) based on a 24h fishery. *Archive of Fishery and Marine Research*, **42**, 123–136.

Thiel, R., Nellen, W., Sepulveda, A., Oesmann, S., Tebbe, K. & Drenkelfort, T. (1997) Spatial gradients of food consumption and production of dominant fish species in the Elbe estuary, Germany. *International Council for the Exploration of the Sea* CM/S:04.

Tokranov, A.M. (1990) Feeding of the yellowfin sole, *Limanda aspera*, in the southwestern part of the Bering Sea. *Journal of Ichthyology*, **30**, 59–67.

Tokranov, A.M. (1992) Features of feeding of benthic predatory fishes of the West Kamchatka Shelf. *Journal of Ichthyology*, **32**, 45–55.

Tokranov, A.M. & Maksimenkov, V.V. (1994) Feeding of the starry flounder, *Platichthys sellatus*, in the Bol'shaya River Estuary (Western Kamchatka). *Journal of Ichthyology*, **34**, 76–83.

Tokranov, A.M. & Maksimenkov, V.V. (1995a) Some features of biology of the banded flounder, *Liopsetta pinnifasciata* (Pleuronectidae), in the Bol'shaya River Estuary (Western Kamchatka). *Journal of Ichthyology*, **35**, 22–28.

Tokranov, A.M. & Maksimenkov, V.V. (1995b) Feeding habits of predatory fishes in the Bol'shaya River Estuary (West Kamchatka). *Journal of Ichthyology*, **35**, 102–112.

Tominaga, O. & Nashida, K. (1991) Interspecific relationship between brown sole *Pleuronectes herzensteini* and other demersal fishes in the coastal waters of northern Niigata prefecture. *Bulletin of the Japan Sea National Fisheries Research Institute*, **41**, 11–26.

Tyler, A.V. (1972) Food resource division among northern, marine, demersal fishes. *Journal of the Fisheries Research Board of Canada*, **29**, 997–1003.

Van der Veer, H.W. (1985) Impact of coelenterate predation on larval plaice *Pleuronectes platessa* and flounder *Platichthys flesus* stock in the western Wadden Sea. *Marine Ecology Progress Series*, **25**, 229–238.

Van der Veer, H.W. & Bergman, M.J.N. (1987) Predation by crustaceans on newly settled 0-group plaice *Pleuronectes platessa* population in the western Wadden Sea. *Marine Ecology Progress Series*, **35**, 203–215.

Vinogradov, V.I. (1984) Food of silver hake, red hake and other fishes of Georges Bank and adjacent waters, 1968–1974. *NAFO Scientific Council Studies*, **7**, 87–94.

Wainright, S.C., Fogarty, M.J., Greenfield, R.C. & Fry, B. (1993) Long-term changes in the Georges Bank food web: trends in stable isotopic compositions of fish scales. *Marine Biology*, **115**, 481–493.

Weatherly, N.S. (1989) The diet and growth of 0-group flounder, *Platichthys flesus* (L.), in the River Dee, North Wales. *Hydrobiologia*, **178**, 193–198.

Wells, B., Steele, D.H. & Tyler, A.V. (1973) Intertidal feeding of winter flounders (*Pseudopleuronectes americanus*) in the Bay of Fundy. *Journal of the Fisheries Research Board of Canada*, **30**, 1374–1378.

Werner, E.E. & Gilliam, J.F. (1984) The ontogenetic niche and species interactions in size-structured populations. *Annual Review of Ecology and Systematics*, **15**, 393–425.

Williams, D.D. & Williams, N.E. (1998) Seasonal variation, export dynamics and consumption of freshwater invertebrates in an estuarine environment. *Estuarine, Coastal and Shelf Science*, **46**, 393–410.

Witting, D.A. & Able, K.W. (1995) Predation by sevenspine bay shrimp *Crangon septemspinosa* on winter flounder *Pleuronectes americanus* during settlement: laboratory observations. *Marine Ecology Progress Series*, **123**, 23–31.

Yamada, H., Sato, K., Nagahora, S., Kumagai, A. & Yamashita, Y. (1998) Feeding habits of the Japanese flounder *Paralichthys olivaceus* in Pacific coastal waters of Tohoku District, Northeastern Japan. *Nippon Suisan Gakkaishi*, **64**, 249–258.

Yang, M.S. (1993) Food habits of the commercially important groundfishes in the Gulf of Alaska in 1990. *NOAA Technical Memorandum*, NMFS-AFSC-22.

Yang, M.S. (1995) Food habits and diet overlap of arrowtooth flounder (*Atheresthes stomias*) and Pacific halibut (*Hippoglossus stenolepis*) in the Gulf of Alaska. In: Proceedings of the International Symposium on North Pacific Flatfish. *Alaska Sea Grant College Program Report No. 95–04.* pp. 205–223. University of Alaska, Fairbanks, AK.

Yang, M.S. (1997) Trophic role of Atka mackerel (*Pleurogrammus monopterygius*) in the Aleutian Islands. In: International Symposium on the Role of Forage Fishes in Marine Ecosystems, Proceedings of the 14th Lowell Wakefield Fisheries Symposium Series. *Alaska Sea Grant College Program Report No. 97–01*, pp. 277–279. University of Alaska, Fairbanks, AK.

Yazdani, G.M. (1969) Adaptation in the jaws of flatfish (Pleuronectiformes). *Journal of Zoology, London*, **159**, 181–222.

Yodzis, P. (2000) Must top predators be culled for the sake of fisheries? *Trends in Ecology and Evolution*, **16**, 78–84.

Yokoyama, S.I. (1995) Feeding habits of red flounder *Hippoglossoides dubius* in Funka Bay and its offshore waters, Hokkaido, Japan. In: Proceedings of the International Symposium on North Pacific Flatfish. *Alaska Sea Grant College Program Report No. 95–04.* pp. 247–259. University of Alaska, Fairbanks, AK.

Zhang, C.I. (1988) Food habits and ecological interactions of Alaska plaice, *Pleuronectes quadrituberculatus*, with other flatfish species in the eastern Bering Sea. *Bulletin of the Korean Fisheries Society*, **21**, 150–160.

Zhang, C.I., Wilderbuer, T.K. & Walters, G.E. (1998) Biological characteristics and fishery assessment of Alaska plaice, *Pleuronectes quadrituberculatus*, in the eastern Bering Sea. *Marine Fisheries Review*, **60**, 16–27.

Chapter 10
The behaviour of flatfishes

Robin N. Gibson

10.1 Introduction

Their seemingly sedentary lifestyle and remarkable camouflage abilities make flatfishes relatively inconspicuous members of the benthic fish fauna. It may be for these reasons that their behaviour has been little studied compared with other highly motile, brightly coloured and pelagic groups that show complex reproductive, agonistic and schooling behaviour. Nevertheless, flatfishes do have a characteristic behavioural repertoire that reflects their bottom-living habit but also enables them to take advantage of life in the water column. This chapter provides an overview of the principal aspects of flatfish behaviour during the benthic stage of their life history. Aspects of larval behaviour are described in Chapter 5. This chapter builds on the earlier reviews of de Groot (1969a) and Gibson (1997) and includes summaries of locomotion and colour change but concentrates on the key activities of spawning, feeding and reactions to predators that enable fishes to reproduce, grow and survive. These activities take place at a range of spatial scales from localised foraging to long-distance migrations and the chapter also includes a brief account of movement patterns that have evolved to make the most effective use of the environment, a topic that is covered in greater detail in Chapter 8. Finally, the role of behaviour in the capture and culture of flatfishes is discussed and its importance in attempts to augment and conserve wild flatfish populations through stock enhancement.

10.2 Locomotion and related behaviour

10.2.1 Locomotion

Metamorphosed flatfishes are negatively buoyant and spend most of their time on the bottom but they can also swim effectively in the water column. When moving in contact with the bottom, locomotion is effected by antero-posterior movements of the dorsal and anal fin rays that lever the fish over the substratum (Orcutt 1950; Kruuk 1963; Olla *et al.* 1969; Holmes & Gibson 1983). In this way the fish can creep slowly forwards or backwards and turn in either direction. Progression in this mode usually consists of short movements interspersed with pauses. When active, the head is often raised in the 'alert posture' (Fig. 10.1) and supported by the anterior rays of the median fins and by the blind side pectoral and pelvic fins (Olla *et*

Fig. 10.1 Behaviour of the lemon sole (*Microstomus kitt*). Top: Drawings illustrating the range of eye movements. Bottom: The 'head up' alert posture. Reproduced with permission from Steven (1930).

al. 1969; Stickney *et al.* 1973). The rays of the blind side pectoral fin are often shorter and less branched than those of the eyed side and in this form may provide more effective support (Marsh 1977). In the groups that do not feed predominantly by sight and hence do not scan their environment visually, the pectoral fins are small or absent (Soleidae, Achiropsettidae, Cynoglossidae; Nelson 1994). Faster movements over the seabed are caused by rapid propulsive strokes of the tail and undulations of the body (anguilliform locomotion; Lindsey 1978) and are aided by the pectorals (Orcutt 1950). By keeping close to the bottom flatfishes may make use of the 'ground effect' which reduces the energy costs of swimming (Videler 1993; Webb & Gerstner 2000).

In order to swim in the water column a fish has first to leave the bottom. In slow departures the fish first orientates itself in the direction it intends to travel (Stickney *et al.* 1973), slowly raises its head and then propels itself upwards using the body and tail. If the start is rapid, lift-off is opposed by water viscosity and a suction pressure (Stefan adhesion) develops under the fish making it more difficult to leave the seabed. Studies on winter flounder (*Pseudopleuronectes americanus*) and hogchoker (*Trinectes maculatus*) have shown that this adhesive force is reduced by producing a jet of water from the blind side opercular valve (Brainerd *et al.* 1997). Fast starts in the speckled sanddab (*Citharichthys stigmaeus*) take the form of a 'U-start' that results in vertical displacement of the centre of mass but the posterior portion of the body remains in contact with the substratum until the fish is 'waterborne'. The bottom prevents recoil and prolonged contact results in the attainment of greater final speeds. Resting on the bottom in a U-shape, as some fishes often do (Fig.10.1), would reduce the time taken to reach maximum speed. Fast starts in the water column, in contrast, are produced by smaller amplitude body movements and the centre of mass is propelled forwards (Webb 1981).

Once in the water column, fishes swim in more or less straight lines using rhythmic undulations of the whole body and can glide without such body movements for considerable distances (Olla *et al.* 1972; Stickney *et al.* 1973; Kawabe *et al.* 2003). This 'swim and glide' behaviour can result in significant energy savings (Weihs 1973). Because of their 90° rotation in swimming position compared with other fishes, flatfishes control their vertical direction by changing the angle of the body and the median and caudal fins. Horizontal direction is determined by using the raised pectoral fin of the eyed side as a rudder. Directional change is therefore efficient in the vertical plane but less so horizontally, especially at high speeds (Stickney *et al.* 1973). When swimming in midwater the body is held in the same position as on the bottom, i.e. with the anatomical dorso-ventral axis parallel to the seabed. However, some species (e.g. Greenland halibut *Reinhardtius hippoglossoides*, de Groot 1970; summer flounder *Paralichthys dentatus*, Stickney *et al.* 1973) have been seen to swim with their dorso-ventral axis perpendicular to the bottom. This orientation may provide greater lateral manoeuvrability for these two piscivorous species (Stickney *et al.* 1973).

Swimming speeds of flatfishes have rarely been measured but maximum speeds in the range of 1–6 body lengths s^{-1} have been recorded in the laboratory for four North Atlantic species (Blaxter & Dickson 1959). Similar speeds have been observed for small winter flounder (Beamish 1966) and turbot (*Scophthalmus maximus*) (Gibson & Johnston 1995). In large aquaria the 'preferred' swimming speeds of summer flounder (Olla *et al.* 1972) and Japanese flounder (*Paralichthys olivaceus*) (Kawabe *et al.* 2003) are much slower. At the other end of the scale, small brill (*Scophthalmus rhombus*) have been observed stalking prey at speeds of <1 mm s^{-1} (Holmes & Gibson 1983).

10.2.2 Rheotaxis and station holding

Flatfishes on the seabed may be subjected to a range of current speeds, some of which may be sufficient to displace them. On sediment bottoms the risk of displacement and/or the energetic costs of holding station against the current can be reduced by burying (see below). Even when buried, however, fish can still be exposed to current forces by erosion of the sediment around them (Arnold 1969). At least one species, the plaice (*Pleuronectes platessa*), has evolved behavioural responses to counteract this problem. In still water and at low current speeds it orientates itself randomly to the water flow but as the current speed increases to more than about 8 cm s^{-1} it shows a rheotactic response and heads into the current. Streamlining is most effective in this position. Further increases in current speed result in burying or clamping the body down onto the sediment surface. At even higher speeds the fish shows a posterior fin beating response and an arched back posture. Eventually the fish has to swim against the current or be displaced (Arnold 1969). Heading into the current reduces drag but at the same time causes a lifting force to be exerted on the body. The fin beating response, which pumps water from under the body, and the arched back posture reduce the pressure difference between the upper and lower body surfaces and counteract the lift (Arnold & Weihs 1978; Webb 1989).

10.2.3 Burying

The ability to bury themselves in the sediment is a characteristic of most flatfishes and may have functions other than the avoidance of currents (see below). Semi-quantitative descriptions of this behaviour in a range of species (Orcutt 1950; Kruuk 1963; Arnold 1969; Olla *et al.* 1972; Kawabe *et al.* 2003) show that it consists principally of vigorous beats of the head against the sediment accompanied by a wave of muscular contraction that travels with decreasing amplitude down the body. The head movements waft the sediment from beneath the body so that it falls back onto the upper surface. Vertical movements of the marginal fin rays have the same effect. The vigour and amplitude of the head and fin movements determine the extent to which the fish is buried, varying from complete burial to a slight covering of sediment on the body margins. The duration of a bout of burying behaviour seems to vary with the species but lasts only a few seconds or less (Kruuk 1963; Kawabe *et al.* 2003).

10.3 Colour change

The remarkable ability of flatfishes to change the colour and pattern of their eyed side is a well-known characteristic of the group and has stimulated research on the topic since the end of the nineteenth century. The early works by Sumner (1911) and Mast (1914) clearly established this ability and demonstrated that colour change was mediated by visual stimuli received from the substratum by the eyes. Subsequent studies on common sole, *Solea solea*, indicated that the upper part of the retina needs to be activated for the response to occur (de Groot *et al.* 1969). Changes in colour and pattern (Fig. 10.2) are caused by the relative expansion and contraction of three types of chromatophores (Healey 1999). Changes can be rapid (within seconds to minutes; Ramachandran *et al.* 1996) or slow, and the time taken respectively reflects the nervous and hormonal control of the process (Healey 1999). A third much slower process (weeks; Ellis *et al.* 1997) involves a quantitative change in the number of chromatophores and the amount of pigment they contain (Sumner 1940). The changes, which are reversible, consist of a modification of a permanent pattern of spots and flecks (Norman 1934; Healey 1999). The ability to match their background means that flatfishes are well camouflaged against both their predators and their prey and are capable of modifying this camouflage to suit local conditions.

10.4 Reproduction

The behaviour involved in reproduction can be separated into three components: the response to internal and external stimuli that bring the sexes into reproductive condition, a movement to the spawning grounds and spawning itself. The first two components are dealt with elsewhere (Chapters 8 and 15) and only spawning behaviour is described here.

10.4.1 Spawning behaviour

The first description of flatfish spawning was by Butler (1897) for the common sole and subsequently expanded by Baynes *et al.* (1994). Here the male approaches a female and works his way under her body. The pair then swim off the bottom with their heads and anterior portion of the body closely pressed closely together and the genital pores adjacent. No involvement of fish other than the pair was observed and spawning always took place at night. The presumed reproductive behaviour of the finless sole, *Pardachirus marmoratus*, consisted of a male and a female resting close together or swimming in pairs with the larger female leading. The

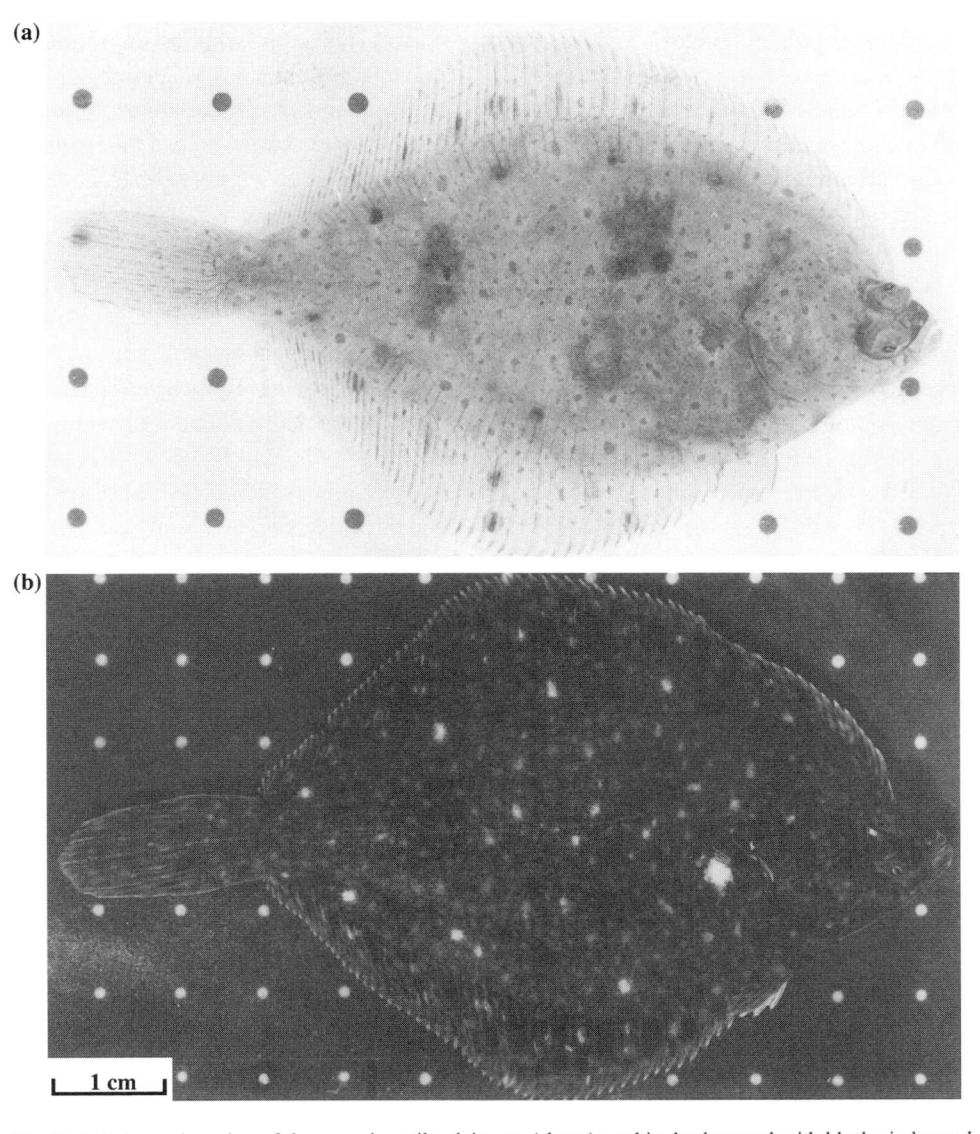

Fig. 10.2 Colour adaptation of the same juvenile plaice on (above) a white background with black circles and (below) a black background with white circles. Reproduced with permission from Healey (1999).

male then came to rest on the back of the female with their genital openings in close contact (Clark & George 1979).

There is only one detailed description of spawning in pleuronectids. In the winter flounder one or more males follow a female and eventually make contact with her. The female may then avoid the males or a pair may swim in rapid tight circles with the female's body in a vertical position. In most of the observed spawnings, several males were involved with a single female and pair spawnings accounted for only about one-fifth of the total. Very occasionally the female would spawn again a short distance from the first event. Despite the fact that several males were often involved in a single spawning no agonistic behaviour was ever observed between them (Stoner *et al.* 1999). All spawning sequences were recorded just before sunset or at night. Plaice also spawn at night (Nicholls 1989) but details of the behaviour involved are scant (Forster 1953).

Observations of several small bothid species spawning in their natural habitat represent the most detailed accounts of flatfish reproductive behaviour. Males defended territories against other males and all spawned during the afternoon or late evening near sunset. Females occupied home ranges within or overlapping the territories of the males. In the largescale flounder, *Engyprosopon grandisquama*, males defended territories against other males and agonistic behaviour consisted of 'rearing up' and 'parallel swimming' (Manabe *et al.* 2000). In this species most males spawned with one or two smaller females within their territories and generally pair formation was assorted by body size. Courtship and spawning consisted of approaches by the male, contact, following and a spawning rise (Fig. 10.3). Similar haremic mating systems and social organisation have been described in Kobe flounder *Crossorhombus kobensis* (Moyer *et al.* 1985), eyed flounder *Bothus ocellatus*, plate fish *B. lunatus* and *B. ellipticus* (Konstantinou & Shen 1995) and in wide-eyed flounder *Bothus podas* (Carvalho *et al.* 2003). In Kobe flounder, females indicate readiness to spawn by rapidly raising and lowering their heads. This behaviour attracts males who contact the female and then circle around flagging with the pectoral fin. The pair then rise to spawn with the male on top. Spawning of the three *Bothus* species is essentially similar except that the male is positioned below the female during the spawning rise. After the last spawning of the day the males of eyed flounder move to 'retirement sites' in shallower water whereas the females 'retire' to deeper water. They re-occupy their former territories the following morning. All of these bothids are sexually dimorphic in various characters.

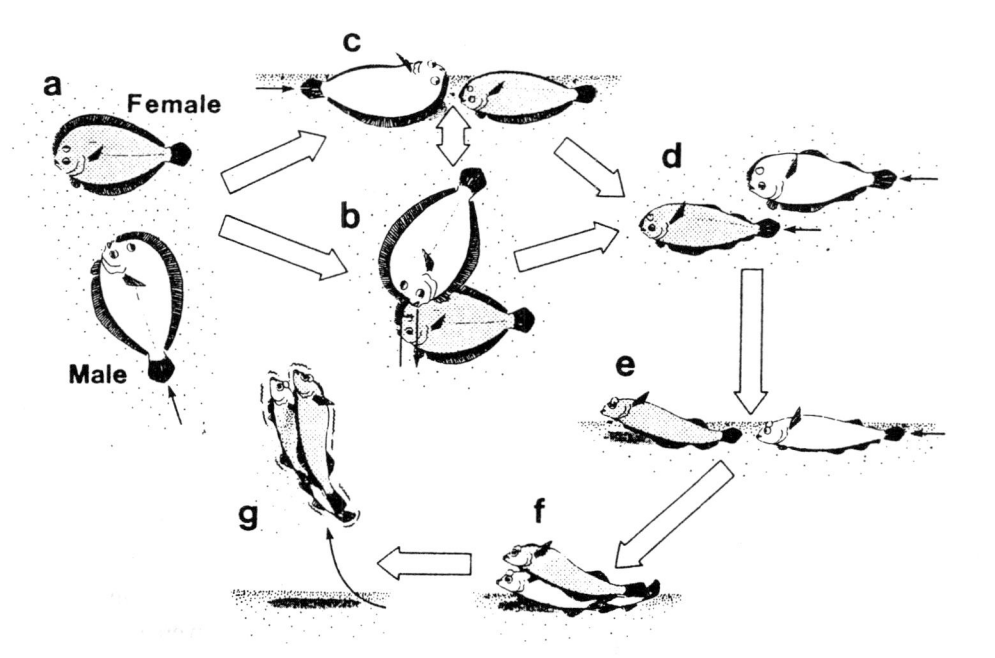

Fig. 10.3 Spawning behaviour in the largescale flounder (*Engyprosopon grandisquama*). (a) Male approaches female; (b) male touches eyed side of female; (c) male approaches front of female; (d) male follows female; (e) female raises head; (f) male moves under female; (g) spawning rise and release of gametes. Reproduced with permission from Manabe *et al.* (2000).

10.5 Feeding

10.5.1 Flatfish feeding types

Behaviourally, flatfishes fall into three basic feeding types (de Groot 1969b, 1971; Yazdani 1969; see also Chapter 9):

(1) Visual feeders feeding on free-swimming prey
(2) Predominantly visual feeders taking principally benthic, slow-moving and sedentary prey
(3) Non-visual nocturnal feeders eating benthic sedentary prey.

Although diet can change with development, the adult body shape, morphology of the digestive tract and jaws (de Groot 1969b, 1971; Yazdani 1969; Livingston 1987), and the structure of the brain (Evans 1937) and sense organs (Livingston 1987) all correlate closely with feeding type. Feeding behaviour also differs between the types.

10.5.2 Feeding behaviour

Feeding behaviour is best described with reference to the 'predation cycle' which comprises four basic elements: searching, encountering, capturing and ingesting a prey item (Fig. 10.4). The way in which each of these elements is performed depends on the species and which feeding type it represents (Holmes & Gibson 1983).

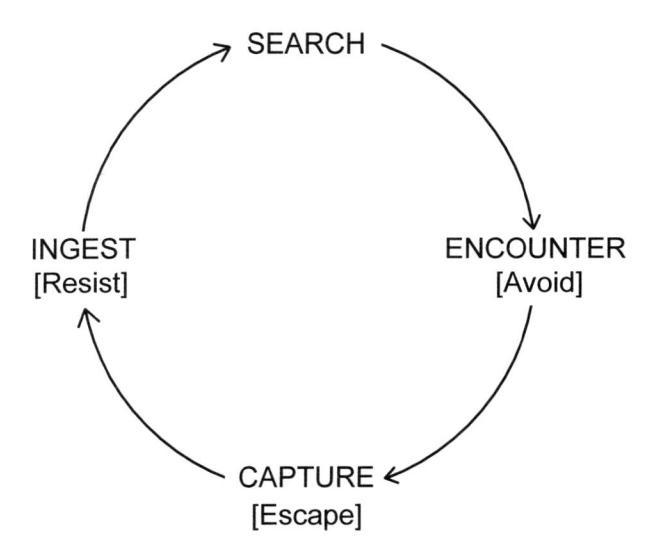

Fig. 10.4 The predation cycle. For a predator (capital letters) the cycle consists of four basic stages: searching for, encountering, capturing and ingesting the prey. In response, the prey (lower case letters in brackets) attempts to avoid the predator, escape if attacked and resist ingestion if captured. A successful predator completes the cycle; the prey successfully avoids predation if it breaks the cycle.

10.5.2.1 Search

Many species that are predominantly visual feeders scan the immediate environment with their very mobile (Fig. 10.1) and independently moveable eyes. Fields of view are extensive and give almost 360° coverage in all planes. Vision is binocular for about 40–50° anteriorly and above but only about 10° posteriorly (Scheuring 1921; Fujimoto *et al.* 1992). To increase their field of view, all flatfishes can protrude their eyes by contracting the muscular walls of the recessus orbitalis, which forces fluid into the orbit and protrudes the eye. Retraction is caused by contraction of the eye muscles (Cole & Johnston 1901). Adoption of the alert head-up posture (Steven 1930; Olla *et al.* 1969, 1972; Stickney *et al.* 1973; Macdonald 1983) further elevates the eyes (Fig. 10.1). This raised head posture is not seen in species that mostly rely on non-visual cues for prey detection. Such species have smaller eyes placed further back on the head than those of visual feeders whose eyes are situated closer to the anterior margin of the head (Livingston 1987). Their searching behaviour, as exemplified by the common sole, is also simpler (Holmes & Gibson 1983) and consists of slow creeping movements over the bottom and the head is frequently patted on the bottom or moved from side to side (Steven 1930; Holmes & Gibson 1983; Batty & Hoyt 1995). In some species, the anterior dorsal fin rays may also be used as detectors in the search process (Livingston 1987). The 'coughing' behaviour observed in rock sole (*Lepidopsetta bilineata*) and starry flounder (*Platichthys stellatus*) may be an olfactory searching behaviour analogous to sniffing in air-breathing animals (Nevitt 1991).

Active searching consists of swimming short distances, usually in a slow discontinuous way (Kruuk 1963; Olla *et al.* 1972; Gibson 1980; Holmes & Gibson 1983; Hill *et al.* 2000). Alternatively, ambush tactics may be employed (e.g. flowery flounder *Bothus mancus*, Hiatt & Strasburg 1960; topknot *Zeugopterus punctatus* and Eckström's topknot *Phrynorhombus regius*, Holmes & Gibson 1983; southern flounder *Paralichthys lethostigma*, Minello *et al.* 1987). In the angler flatfish *Asterorhombus fijiensis* (Amaoka *et al.* 1994) and intermediate flounder *A. intermedius* (Manabe & Shinomiya 1998), a membranous lure resembling a small fish or crustacean on the end of the elongated first dorsal fin ray may attract prey to within striking distance. In some other species, the pectoral fin may be used in the same way (Randall, quoted in Amaoka *et al.* 1994). Wide-eyed flounder (Abel 1962; Weber 1965; personal observation) and *Arnoglossus* (Weber 1965) exploit the prey disturbed by the foraging of other benthic animals.

10.5.2.2 Encounter/detection

The clues used to detect and identify suitable prey also depend on the relative importance of the sense organs. For visual feeders, motion is important in the initial detection of prey at a distance (MacDonald 1983; Holmes & Gibson 1986; Shaheen *et al.* 2001) but motion may also stimulate the lateral line system of both visual and non-visual feeders. The characteristics of potential prey items that stimulate an initial inspection and attack have not been studied in detail but appendage movements and an elongated shape are important for turbot (Holmes & Gibson 1986). Appropriate size is also important and a combination of visual and chemical clues increases the attractiveness of model prey items for both visual and olfactory feeders (de Groot 1969b, 1971). Olfaction has an important role in attracting fish to bait from a distance in Pacific halibut *Hippoglossus stenolepis* (Kaimmer 1998) and directing most species to prey in the vicinity. The relative importance of the senses in prey detection depends on the species and corresponds with the three basic feeding types described above. Visual feeders rely principally on sight, mixed feeders use both vision and olfaction, whereas non-visual feeders mainly employ olfaction for detecting prey (de Groot 1969b). In the common sole (Mackie & Mitchell 1982) and winter flounder (Fredette *et al.* 2000), betaine acts as a feeding stimulant. Mechanoreception is used to differing degrees by most species (de Groot 1969b, 1971; Appelbaum & Schemmel 1983; Livingston 1987; Batty & Hoyt 1995).

10.5.2.3 Capture

Capture of active benthic prey by visual species is usually preceded by some form of orientation and approach (Stickney *et al.* 1973; Holmes & Gibson 1983). Approach and attack may be one rapid continuous movement or consist of a more measured stalking behaviour (Olla *et al.* 1972; Holmes & Gibson 1983; Minello *et al.* 1987). Conspicuousness is further reduced by the camouflage provided by the colour pattern of the upper surface of the body. Eventual attack consists of a rapid lunge forward while simultaneously opening the mouth and protruding the jaws. The attack and capture behaviour of fishes by piscivorous species does not seem to have been described but several also feed on crustaceans in midwater (e.g. summer flounder, Olla *et al.* 1972; Japanese flounder, Miyazaki *et al.* 2000; *Scophthalmus* spp., Holmes & Gibson 1983). In such species approach may also be relatively slow.

Capture is by suction (Bels & Davenport 1996) and in strictly benthic feeders the mouth is brought down directly on to sedentary prey at varying angles (Steven 1930; Macdonald 1983), whereas prey swimming off the bottom are engulfed by the protrusion of the jaws and forward motion of the predator. These differing modes of attack are paralleled by differences in jaw structure. Predators that capture motile prey (Psettodidae, Bothidae and Paralichthyidae in particular) have relatively symmetrical jaws with teeth on both sides, whereas those feeding on buried prey have very asymmetrical jaws and teeth only on the blind side (Soleidae and others). Mixed feeders (e.g. Pleuronectidae) have jaw structures intermediate in structure between these two extremes. Gibb (1996, 1997) has considered the general question of whether the capture mechanism of flatfishes is essentially different from that of symmetrical fishes. She concluded that dextral and sinistral forms of the same species have identical capture mechanisms and that throughout the group there is no stereotypical pattern of prey capture. Consequently, flatfish taxa are not more similar to one another than to symmetrical percomorphs except in the presence of varying degrees of functional asymmetry during prey capture.

10.5.2.4 Ingestion

Following successful capture, the prey item may be accepted or rejected. Pharyngeal teeth may assist further processing of food in the mouth (Yazdani 1969). In plaice, acceptable food is transported to the oesophagus by actions of the jaws and opercula that are similar to those used in prey capture. After swallowing, further respiratory and opercular movements cleanse the orobranchial and opercular cavities and the gills (Bels & Davenport 1996).

10.5.3 Factors modifying feeding behaviour

10.5.3.1 External factors

Whereas temperature is probably the major external factor controlling feeding intensity, the actual behaviour employed is unlikely to be greatly affected. In contrast, light intensity affects both intensity and behaviour, particularly in visual feeders. Light intensity affects the extent of visual fields and hence encounter rates and diel variation in light intensity imposes a diurnal rhythm of feeding in such species. Species that do not rely on sight for the primary sense during feeding tend to be nocturnal in their feeding habits (de Groot 1971). Field and laboratory observations of temporal variation in feeding intensity bear out these general conclusions (e.g. de Groot 1963; Kruuk 1963; Olla *et al.* 1972; Chen *et al.* 1999).

In the absence of light, or below intensities at which the eyes can no longer function effectively, visual feeders either stop feeding or change their searching behaviour to employ other sensory modalities (Minello *et al.* 1987; Batty & Hoyt 1995). Other effects on visual fields have similar consequences. In flounder (*Platichthys flesus*), for example, increasing turbidity decreases the distance at which prey can be seen and increases the time required for their capture. The size of the smallest prey that can be detected at a given distance also increases with increasing turbidity (Moore & Moore 1976). In contrast, southern flounder predation on shrimp increases with increasing turbidity because the shrimp bury less and hence are more available (Minello *et al.* 1987). As in many other fish species, increasing complexity of the environment caused by the presence of vegetation may also have an effect on the predator's ability to detect and/or capture prey. Predation by summer flounder on winter flounder in the laboratory is lower when seagrass and algae are present (Manderson *et al.* 2000).

The abundance of prey is an important factor that determines feeding rate and, in juvenile flounder at least, the capture rate of prey increases with increasing prey density until an asymptote is reached (Kiørboe 1978; Mattila & Bonsdorff 1998). The level of this asymptote is dependent on prey size and temperature (Kiørboe 1978) and, to some extent, prey type (Mattila & Bonsdorff 1998). Newly settled plaice spend more time on sediments with food than those without (Wennhage & Gibson 1998) and spatial differences in prey density may be responsible for many of the foraging patterns seen in the wild. Differences in prey abundance above and below low water mark in tidal areas, for example, may result in tidally synchronised feeding migrations as in flounders (Summers 1980), juvenile plaice (Gibson 1973a; Kuipers 1973; Burrows *et al.* 1994), summer flounder (Rountree & Able 1992) and winter flounder (Wells *et al.* 1973), or in movements into nursery areas along a prey density gradient (southern flounder; Burke 1995).

The activity of individual prey may also stimulate feeding activity. Prey activity directs the attention of the predator towards the prey so that prey activity level affects predation rate. This effect has been observed in several species, notably flounder (Moore & Moore 1976), summer flounder (Manderson *et al.* 2000) and winter flounder (Shaheen *et al.* 2001). Conversely, and for the same reason, the feeding activities of the flatfishes themselves expose them to a greater risk of predation and it has been observed that the presence of a predator reduces feeding activity in plaice (Burrows *et al.* 1994; Burrows & Gibson 1995) and that the predation rate of summer flounder on winter flounder is related to the activity level of the latter (Manderson *et al.* 2000).

Most flatfishes appear to show little territoriality or aggressive behaviour in relation to feeding. Captive Atlantic halibut (*Hippoglossus hippoglossus*) will compete for food in tanks (Rabben & Furevik 1993) and video recordings of Pacific halibut demonstrated rare attacks by one individual on another near baited hooks deployed in the sea (Kaimmer 1998). No interactions between individuals were seen in many hours of direct observation of juvenile plaice in the wild (Gibson 1980) but fin biting has been recorded at high densities in rearing tanks (Riley 1966). Rare instances of agonistic behaviour and fin damage have been recorded in captive greenback flounder (*Rhombosolea tapirina*), but the growth differences in this species are more likely to be the result of exploitative rather than interference competition (Carter *et al.* 1996; Shelverton & Carter 1998). In captivity, nipping of conspecifics (Sakakura & Tsukamoto 2002) and cannibalism is common in Japanese flounder (Furuta 1996).

10.5.3.2 Internal factors

The principal internal factor that controls feeding behaviour and intensity is hunger. Hungry fishes are more active than satiated ones (Macquart-Moulin *et al.* 1991; Champalbert & Le Direach-Boursier 1998; Furuta 1998; Neuman & Able 1998; Chen & Purser 2001) and this greater foraging activity increases the chance of encountering prey. Hungry fish are also likely to be less selective in their choice of prey items. Reproductive condition may also have a major effect on feeding intensity because several species do not feed during the spawning season (Chapter 4).

10.6 Predation and reactions to predators

Flatfishes, particularly in their youngest stages, are prey for a wide variety of invertebrates and fishes (Chapter 9) but their reactions to predators have not been well studied. Behaviour in relation to predators and predation risk is the converse of feeding and can also be related to the predation cycle (Fig. 10.4). Prey must avoid predators, evade and if possible escape once attacked and, if captured, resist ingestion in order to survive.

10.6.1 Avoidance of predators

Strategies for avoiding predation usually involve reducing encounter and detection probabilities in the short term and employing patterns of movement that minimise predation risk in the longer term.

10.6.1.1 Reducing encounter and detection probability

When on the bottom the flattened shape of flatfishes combined with their cryptic colouration and ability to match their background decreases their conspicuousness to predators. Their low profile is further enhanced by keeping the head flat on the bottom when inactive, in contrast to the 'head up' posture when alert and feeding. Movement is a major stimulus attracting the attention of predators, and the relative immobility of flatfishes and their camouflage probably contribute much to their survival because the presence of predators tends to reduce activity levels (Burrows *et al.* 1994; Burrows & Gibson 1995) and active fishes are more prone to predation than inactive ones (Furuta 1996). One other behaviour pattern that reduces detection by visual predators is to be nocturnal and many flatfishes, particularly those that do not rely on vision for feeding, employ this behaviour. Furthermore, species that spend a considerable amount of time swimming off the bottom when migrating often do so at night.

Flatfishes characteristically bury in the sediment and in the past it has been implicitly assumed that burial reduces predation. While such an assumption seems intuitively reasonable it has only recently been tested experimentally. The results have provided both direct and indirect evidence that burial does reduce predation risk. Experiments with reared common sole showed that burying behaviour is innate and improves with experience. Furthermore, reactive distance is less when fish are buried, indicating that burial is a cryptic behaviour (Ellis *et al.* 1997). Young winter flounder cover less of their body at night but cover more of it when tethered (Curran & Able 1998). More summer flounder bury in the presence of a fish predator (Keefe & Able 1994) and predator presence also greatly reduces both activity level and the frequency of burying behaviour in young plaice (Burrows & Gibson 1995). Lower sediment selectivity in the dark (Neuman & Able 1998; Gibson & Robb 2000) is also relevant to this argument. Predation risk from visual predators is likely to be lower in the dark and fish increase their activity and hence decrease their selectivity under these conditions. Conversely, predation by some predators may be higher in the dark (Sekai *et al.* 1993; Yamashita *et al.* 1994). Buried common sole have a lower heart and opercular rate (Peyraud & Labat 1962) when buried and both this species and Japanese flounder consume less oxygen (Howell & Canario 1987; Honda 1988) than those exposed. Presumably, therefore, their respiratory currents are reduced and buried fishes may thus provide less of a chemical signal to searching predators. Studies on juvenile plaice (Ansell & Gibson 1993), however, demonstrated that the presence of sediment in which the fish could bury did not reduce predation by Atlantic cod (*Gadus morhua*), shrimps and crabs, although it did have this effect when the plaice were exposed to the pelagic pollack (*Pollachius pollachius*). It was concluded that burial did not provide protection against predators with non-visual methods of detecting prey. Similarly, experiments with young winter flounder (Manderson *et al.* 1999) exposed to striped searobins (*Prionotus evolans*) showed that burying was relatively ineffective against detection by this predator. In this case, the predator searched for the prey using its free chemosensory pectoral fin rays and actually dug out buried fish from the sediment. In comparable experiments, sediment had a limited effect in reducing cannibalism in Japanese flounder (Dou *et al.* 2000) and did not affect the mortality of winter flounder in the presence of larger summer flounder (Manderson *et al.* 2000) nor of hatchery-reared summer flounder exposed to crabs (Kellison *et al.* 2000). However, Manderson *et al.* (2000) also warn of the dangers of extrapolating the results of long duration experiments in confined spaces to the field where predator density is likely to be lower and predation threat more transitory. Nevertheless, in the field it is likely that even partial burial combined with cryptic colouration acts as a first line of defence (Stoner & Ottmar 2003).

10.6.1.2 Burial and substratum selection

The ability to bury is usually considered as one of the primary causes of the frequently observed correlation between flatfish distribution and the presence of fine sediments (e.g. Scott 1982; Norcross *et al.* 1995; Abookire & Norcross 1998; McConnaughey & Smith 2000). The second primary cause is the presence of suitable food. The ability actively to select particular substrata has been tested in numerous laboratory experiments (Gibson & Robb 2000 and references therein; Phelan *et al.* 2001; Stoner & Abookire 2002; Stoner & Ottmar 2003). In most cases, fish selected sandy sediments over coarser ones and tended to avoid mud or areas of the experimental tank without sediment. The preferred sediments were those in which fish could bury. Detailed analysis of the behaviour involved in sediment selection by young plaice showed that burial reduces activity and thereby increases the time that fish spend on sediments in which they can bury. Activity level, which is dependent on light intensity and endogenous factors, therefore determines the degree of selectivity shown (Gibson & Robb 2000). If the ability to bury determines selection of particular range of sediments then selectivity should decrease with increasing body size. Conversely, if preference simply depended on matching body size with grain size, then all species in a given size range should be found on the same sediments. In some cases, experiments have confirmed this prediction (starry flounder, Moles & Norcross 1995; winter flounder, Phelan *et al.* 2001; Pacific halibut, Stoner & Abookire 2002; northern rock sole, *Lepidopsetta polyxystra*, Stoner & Ottmar 2003), whereas in others (plaice, Gibson & Robb 2000; rock sole, Moles & Norcross 1995) no change in selectivity with size was observed. In one species at least (yellowfin sole *Limanda aspera*, Moles & Norcross 1995), sediment grain size preference decreases with increasing body size, so that no generalisation seems possible in this respect. However, subsequent laboratory experiments and field observations of northern rock sole and Pacific halibut (Stoner & Titgen 2003) indicated their preference for a structured habitat rather than a smooth substratum. This strategy may increase survival rates by decreasing predator efficiency (Ryer *et al.* 2004).

10.6.2 Escape from predators following attack

The escape behaviour of flatfishes has not been well studied and most information relates to just a few species. Winter flounders respond by a fast C-start escape reaction aided by opercular jetting to reduce adhesion to the bottom (Brainerd *et al.* 1997). Similar escape reactions are shown by other flatfishes (Webb 1981; Gibson & Johnson 1995; Stoner & Ottmar 2003). Common sole react to an artificial predator by swimming away from it at a smaller distance when on sand than when sand is absent (Ellis *et al.* 1997). The reactive distance of winter flounder may also be shorter in the presence of sand (Manderson *et al.* 1999). Virtually all common sole responded to the artificial predator in the absence of sand but less than 30% did so in the presence of sand (Ellis *et al.* 1997). If wholly or partially buried, starry flounder (Orcutt 1950), plaice (personal observations), winter flounder (McCracken 1963), summer flounder (Olla *et al.* 1972) and northern rock sole (Stoner & Ottmar 2003) may also not attempt to escape and remain immobile.

10.6.3 Effect of size on vulnerability and avoidance of ingestion

The importance of small changes in the mortality rate of the early stages on the survival and recruitment of fish populations has prompted many investigations into predation on these stages. The role of growth and development in decreasing vulnerability has received particular attention and, from a behavioural point of view, the relative size of predator and prey is important in determining vulnerability (Sogard 1997). Growth reduces the number and range of predators to which

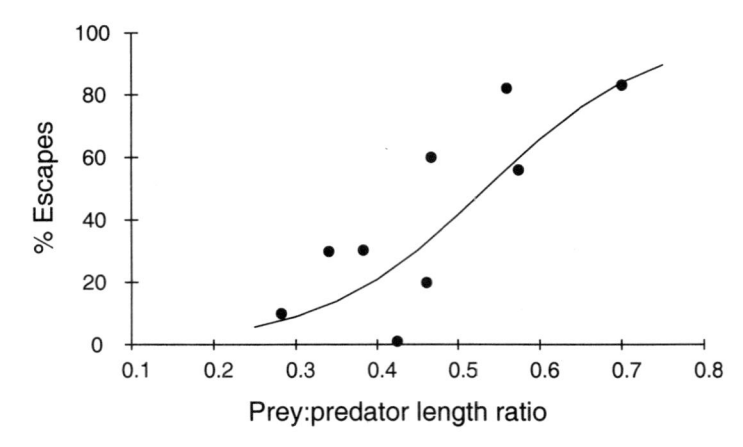

Fig. 10.5 The relationship between percentage escapes of small plaice from capture by shrimps and the prey:predator length ratio. Modified and reproduced with permission from Gibson *et al.* (1995).

an individual is vulnerable but may also bring it into the acceptable range of larger predators including birds (Raffaelli *et al.* 1990; Leopold *et al.* 1998) and, of course, man. Experiments have shown that mortality is strongly dependent on the predator:prey size ratio in plaice (Van der Veer & Bergman 1987; Gibson *et al.* 1995; Ellis & Gibson 1997; Arnott & Pihl 2000; Wennhage 2000) (Fig. 10.5), Japanese flounder (Sekai *et al.* 1993; Yamashita *et al.* 1994; Hossain *et al.* 2002), and summer and winter flounder (Witting & Able 1993; Fairchild & Howell 2001). Finally, although most flatfishes are palatable to their predators some species of the soleid genus *Pardachirus* possess poison glands at the base of their median fin rays. The secretions of these glands are toxic to a variety of teleosts and presumably prevent their ingestion (Clarke & George 1979).

10.7 Movements, migration and rhythms

Searching for food, avoiding predators and adverse environmental conditions, or seeking appropriate sites for spawning, frequently involves movement over a range of spatial and temporal scales. Throughout an individual fish's lifetime such movements may be complex (Harden Jones 1968; Neill 1984; Gibson 1997) (Fig. 10.6). In flatfishes, the movements of larvae (Chapter 5) and juveniles and adults (Chapter 8) are well known in many species but the behaviour underlying them is not, although several putative mechanisms have been proposed. These mechanisms are mostly described in terms of reactions to changes in environmental variables with which the movements are correlated (e.g. McCracken 1963; Rogers 1993; Armstrong 1997; Peterson-Curtis 1997) and which provide possible cues and clues for their timing and direction (Gibson 1997). The complexity of the reactions is illustrated by studies of common sole (Champalbert & Koutsikopoulos 1995) and Japanese flounder (Burke *et al.* 1995). In both species reactions to light and salinity, and in the common sole rheotaxis and responses to pressure, are involved in vertical movement in the water column and subsequent transport by currents to onshore nursery areas. Furthermore, these responses are not constant and change with development. The young stages of several other species use water currents for transport (Forward & Tankersley 2001) and presumably the mechanisms

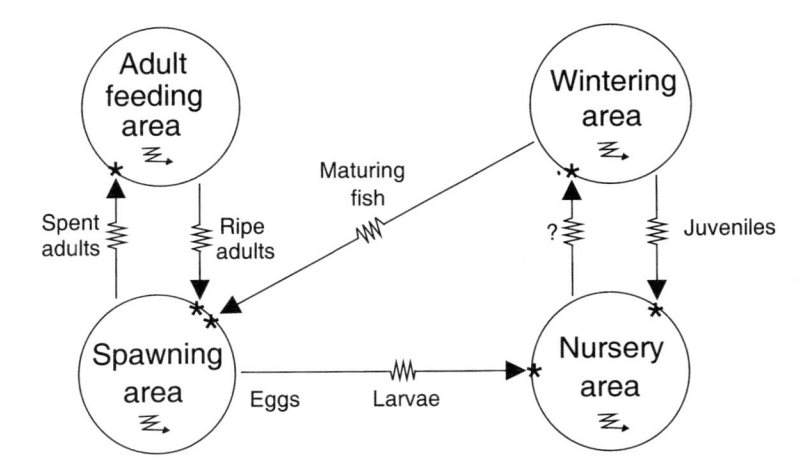

Fig. 10.6 Diagram illustrating movements of plaice throughout its life cycle. From the spawning area (bottom left) eggs and larvae drift towards the nursery area and the late larvae may use selective tidal stream transport (STST). Complex movement patterns on the nursery grounds are succeeded by an offshore movement in winter. Juveniles may return to the nursery ground in spring, also in some cases using STST. As the fish grow and mature there is a gradual offshore movement. Maturing fish eventually spawn and some may follow the adults in a 'dummy run'. Movement between feeding and spawning grounds is accomplished using STST. Movements on and off the bottom also occur in both areas. 'Decision points' occur when fish must recognise their destination. Reproduced with permission from Gibson (1997).

are similar. The adults of plaice and common sole use a comparable system of 'selective tidal stream transport' in the North Sea to migrate between feeding and spawning grounds and may make energy savings by doing so (Metcalfe & Arnold 1990). During this behaviour fish swim well off the bottom both day and night and time their ascents and descents with respect to the tidal and diel cycles (Fig. 10.7). The cues for timing movement into the water column include

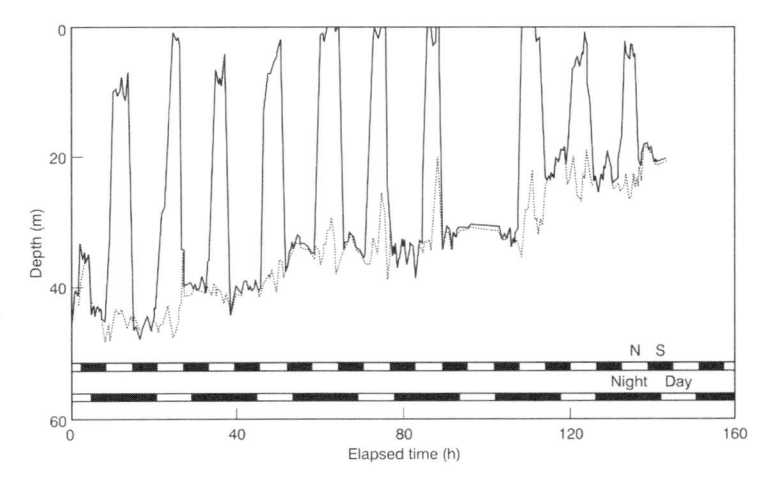

Fig. 10.7 The vertical movements (solid line) of a plaice tracked in the North Sea in relation to the depth of the seabed (dotted line) and the direction of the tidal stream. The bars along the abscissa indicate periods of north-going (N) and south-going (S) tidal streams and periods of day and night respectively. The fish ascended into the water column on north-going tides but remained on the bottom when the tidal stream was south-going. Reproduced with permission from Metcalfe *et al.* (1992).

hunger, pressure, currents, and turbulence and associated noise (Arnold & Cook 1984). Common sole can also use noise in directing their movements (Lagardère *et al.* 1994). Once in the water column plaice may navigate using magnetic fields or electric currents generated by water movement (Metcalfe *et al.* 1993). Recent developments in electronic tag technology have contributed much to the knowledge of the behaviour involved in these movements, often with unexpected results. Some plaice, for example, may visit more than one spawning ground in a season and the ability to track individuals provides much more information than can be gleaned from mark and recapture studies (Metcalfe & Arnold 1997). On a smaller scale, the diel movements of juvenile plaice in shallow water (Gibson *et al.* 1998) seem to be at least partly regulated by a response to changing light intensity (Burrows 2001).

Although the movements of many, if not all, species are partly timed and directed by responses to external biotic and abiotic stimuli, several are also known to possess an internal timing mechanism, a 'biological clock' (Ali 1992). Such 'clocks' are present in at least seven flatfish species (Gibson 1997) and are manifested in the laboratory as tidal and/or diel activity rhythms that persist in constant conditions. The adaptive function of endogenous rhythms is usually considered to be the anticipation of regular changes in the environment. The endogenous circatidal rhythms recorded in juvenile plaice, flounder and turbot (Gibson 1973b, 1976) in which maximum activity is phased with the ebb tide, may prevent the stranding of these species in unfavourable environments by the ebbing tide. The discovery that circatidal and circadian rhythmicity is also present in adult plaice (Page 1996) may provide a partial solution to the problem of the timing of ascents and descents of these fish during selective tidal stream transport. Endogenous rhythms are also likely to be implicated in the behavioural mechanisms involved in the recruitment of the young stages to estuarine environments (Boehlert & Mundy 1987; Champalbert & Koutsikopoulos 1995).

10.8 Behaviour in relation to fishing, aquaculture and stock enhancement

10.8.1 Fishing

Successful fishing requires the location and capture of fish. In both of these operations an understanding of the behaviour of the target species is vital. Until relatively recently, improvements in catch were the result of accumulated experience based on trial and error. Increasingly, however, field and laboratory studies have yielded greater insights into the interactions between fish, fishing gear and the environment. Essentially, all reactions to gear are related to the natural instincts of feeding and escape from predators. Behavioural studies can result in knowing where and when to fish, improved gear efficiency and improved estimates of abundance and stock structure. Much has been written on this topic (e.g. Wardle & Hollingworth 1993; Fernö & Olsen 1994) but flatfishes have received relatively little attention.

Initial qualitative observations of reactions to gear mostly described the tendency of flatfishes to swim ahead of advancing trawl gear before entering the net (Main & Sangster 1981; Wardle 1983; Walsh & Hickey 1993). Later quantitative analysis of herding and net entry involving mostly rock sole indicated that these species probably react initially to the sound of the approaching trawl. Certainly, some species are sensitive to sound in the region of ~50–500

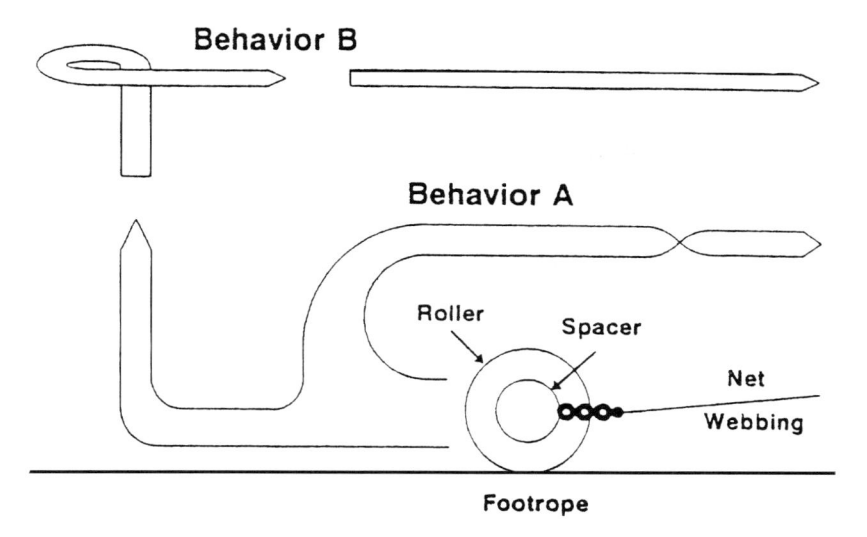

Fig. 10.8 Behaviour patterns shown by flatfishes during entry to trawls. In the commonest response (Behaviour A) fishes flip over the footrope in an inverted or sideways orientation. In the rarer response (Behaviour B) fishes rise slowly into the water column and either swim in the direction of tow or turn and swim into the net. Reproduced with permission from Bublitz (1996).

Hz (Chapman & Sand 1974; Fujieda *et al.* 1996) and even down to 0.1 Hz in plaice (Karlsen 1992). The fishes are subsequently herded ahead of the footrope and enter the trawl in ways that correspond to avoidance and escape behaviour (Bublitz 1996) (Fig. 10.8). This kind of information, when combined with a knowledge of factors affecting swimming speeds and endurance (e.g. Winger *et al.* 1999), can be used to modify net design to increase capture efficiency and/or selectivity (Main & Sangster 1982; Thomsen 1993; Bublitz 1996).

The frequently observed diel variation in flatfish catch rates (Casey & Myers 1998) can partly be explained by greater visual gear avoidance during the day but other factors may also be in operation. While there is no empirical evidence to suggest that trawls pass over buried fishes (Walsh 1991), it cannot be assumed that their reactions to gear are constant over the diel cycle. As described above, many flatfishes show tidal and diel rhythms in activity that may be expressed by swimming at different levels in the water column (Olla *et al.* 1972; Gibson *et al.* 1978; Metcalfe *et al.* 1992). Such behaviour will clearly affect their vulnerability to bottom trawls. Furthermore, reactivity may vary with time of day (Olla *et al.* 1969; Walsh & Hickey 1993).

Observations of flatfishes in relation to fixed gear seem to be limited to the study of Pacific halibut, whose response to baited hooks in the field is clearly a form of feeding behaviour (Kaimmer 1998), and winter flounder (He 2003). Laboratory experiments with Pacific halibut (Stoner 2003) serve to illustrate the complexity of the interactions between the factors that result in capture by baited fishing gear. Direct responses to many environmental factors (particularly light intensity) are involved that may vary with the age and the internal state of the fish. Understanding these complex interactions is essential for explaining variations in catch rate and improving fishing techniques for this and other species.

10.8.2 Aquaculture

The aim of aquaculture is to produce good quality fish for breeding, sale or stocking, and maximising growth and survival – particularly prior to metamorphosis – is central to this process. Rearing environments are predator-free and so survival is mainly dependent on effective feeding and disease control during the larval stages. Fishes are present in rearing conditions at much higher densities than in the wild and so the scope for interaction is greater. Aggression, even at high densities, appears to be rare (see Feeding behaviour above) but cannibalism does exist in hatcheries and can be reduced with a knowledge of the behavioural factors contributing to it (Dou *et al.* 2000). Observations of behaviour in cages (Martinez Cordero *et al.* 1994; Anraku *et al.* 1998) or ponds (e.g. Lagardère & Sureau 1989; Lagardère *et al.* 1995) can also contribute to an understanding of how cultured fishes adapt to these surroundings.

10.8.3 Stock enhancement

Survival of stocked fish is the main prerequisite for the success of stock enhancement programmes. To maximise survival, released fish must adapt to conditions in their new environment, feed successfully and avoid predation (Howell 1994). Most stocked fish are small in size and so avoidance of predation and starvation are key elements in this process. Several laboratory studies indicate that the relative size of predator and prey determines mortality rates (see Predation section above) and have allowed recommendations of suitable sizes for release (Yamashita *et al.* 1994; Fairchild & Howell 2001). Other behavioural observations have shown that the reactive distance of common sole to predators is the same in wild and reared fish and that the ability to bury is innate, the effectiveness of the latter increasing with experience. Furthermore, the ability to match the colour of a new substratum requires some time so that conditioning to a substratum matching that of the intended release area is likely to enhance survival (Ellis *et al.* 1997). The limited evidence available indicates that death by starvation is unlikely because released fishes rapidly adjust their diet to natural food (Iglesias & Rodriguez-Ojea 1994). However, starvation increases activity in common sole (Macquart-Moulin *et al.* 1991) and Japanese flounder (Miyazaki *et al.* 2000) and may make newly released fishes more susceptible to predation. All of these factors suggest that adaptation of fishes to as near natural conditions as possible and some form of behavioural training (Kellison *et al.* 2000; Brown & Laland 2001; Ellis *et al.* 2002; Hossain *et al.* 2002) are desirable before fishes are released.

Acknowledgements

I am most grateful to A.W. Stoner, J.M. Healey, R.S. Batty and L. Robb for their help in the preparation of this chapter.

References

Abel, E.F. (1962) Freiwasserbeobachtungen an Fischen im Golf von Neapel als Beitrag zur Kenntnis ihrer Ökologie and ihres Verhaltens. *Internationale Revue der Gesamten Hydrobiologie,* **47**, 219–290.

Abookire, A.A. & Norcross, B.L. (1998) Depth and substrate as determinants of distribution of juvenile flathead sole (*Hippoglossoides elassodon*) and rock sole (*Pleuronectes bilineatus*) in Kachemak Bay, Alaska. *Journal of Sea Research,* **39**, 113–123.

Ali, M.A. (1992) *Rhythms in Fishes*. Plenum Press, New York.

Amaoka, K., Senou, H. & Ono, A. (1994) Record of the bothid flounder *Asterorhombus fijiensis* from the western Pacific, with observations on the use of the first dorsal-fin ray as a lure. *Japanese Journal of Ichthyology,* **41**, 23–28.

Anraku, K., Matsuda, M., Shigesato, N., Nakahara, M. & Kawamura, G. (1998) Flounder show conditioned response to 200–800 Hz tone-bursts despite their conditioning to 300 Hz tone-burst. *Nippon Suisan Gakkaishi,* **64**, 755–758.

Ansell, A.D. & Gibson, R.N. (1993) The effect of sand and light on predation of juvenile plaice (*Pleuronectes platessa*) by fishes and crustaceans. *Journal of Fish Biology,* **43**, 837–845.

Appelbaum, S. & Schemmel, C. (1983) Dermal sense organs and their significance in the feeding behaviour of the common sole *Solea vulgaris*. *Marine Ecology Progress Series,* **13**, 29–36.

Armstrong, M.P. (1997) Seasonal and ontogenetic changes in distribution and abundance of smooth flounder, *Pleuronectes putnami*, and winter flounder, *Pleuronectes americanus*, along estuarine and salinity gradients. *Fishery Bulletin,* **95**, 414–430.

Arnold, G.P. (1969) The reactions of the plaice (*Pleuronectes platessa* L.) to water currents. *Journal of Experimental Biology,* **51**, 681–697.

Arnold, G.P. & Cook, P.H. (1984) Fish migration by selective tidal stream transport: first results with a computer simulation model for the European continental shelf. In: *Mechanisms of Migration in Fishes* (eds J.D. McCleave, G.P. Arnold, J.J. Dodson & W.H. Neill). pp. 227–261. Plenum Press, New York.

Arnold, G.P. & Weihs, D. (1978) The hydrodynamics of rheotaxis in the plaice (*Pleuronectes platessa* L.). *Journal of Experimental Biology,* **75**, 147–169.

Arnott, S.A. & Pihl, L. (2000) Selection of prey size and prey species by 1-group cod *Gadus morhua*: effects of satiation level and prey handling times. *Marine Ecology Progress Series,* **198**, 225–238.

Batty, R.S. & Hoyt, R.D. (1995) The role of sense organs in the feeding behaviour of juvenile sole and plaice. *Journal of Fish Biology,* **47**, 931–939.

Baynes, S.M., Howell, B.R., Beard, T.W. & Hallam, J.D. (1994) A description of spawning behaviour of captive Dover sole, *Solea solea* L. *Netherlands Journal of Sea Research,* **32**, 271–275.

Beamish, F.W.H. (1966) Swimming endurance of some northwest Atlantic fishes. *Journal of the Fisheries Research Board of Canada,* **23**, 341–347.

Bels, V.L. & Davenport, J. (1996) A comparison of food capture and ingestion in juveniles of two flatfish species, *Pleuronectes platessa* and *Limanda limanda* (Teleostei: Pleuronectiformes). *Journal of Fish Biology,* **49**, 390–401.

Blaxter, J.H.S. & Dickson, W. (1959) Observations on the swimming speeds of fish. *Journal du Conseil International pour l'Exploration de la Mer,* **24**, 472–479.

Boehlert, G.W. & Mundy, B.C. (1987) Recruitment dynamics of metamorphosing English sole, *Parophrys vetulus*, to Yaquina Bay, Oregon. *Estuarine, Coastal and Shelf Science,* **25**, 261–281.

Brainerd, E.L., Page, B.N. & Fish, F.E. (1997) Opercular jetting during fast-starts by flatfishes. *Journal of Experimental Biology,* **200**, 1179–1188.

Brown, C. & Laland, K. (2001) Social learning and life skills training for hatchery reared fish. *Journal of Fish Biology,* **59**, 471–493.

Bublitz, C.G. (1996) Quantitative evaluation of flatfish behavior during capture by trawl gear. *Fisheries Research,* **25**, 293–304.

Burke, J.S. (1995) Role of feeding and prey distribution of summer and southern flounder in selection of estuarine nursery habitats. *Journal of Fish Biology,* **47**, 355–366.

Burke, J.S., Tanaka, M. & Seikai, T. (1995) Influence of light and salinity on behaviour of larval Japanese flounder (*Paralichthys olivaceus*) and implications for onshore migration. *Netherlands Journal of Sea Research,* **34**, 59–69.

Burrows, M.T. (2001) Depth selection behaviour during activity cycles of juvenile plaice on a simulated beach slope. *Journal of Fish Biology,* **59**, 116–125.

Burrows, M.T. & Gibson, R.N. (1995) The effects of food, predation risk and endogenous rhythmicity on the behaviour of juvenile plaice, *Pleuronectes platessa* L. *Animal Behaviour,* **50**, 41–52.

Burrows, M.T., Gibson, R.N. & Maclean, A. (1994) Effect of endogenous rhythms and light conditions on foraging and predator-avoidance in juvenile plaice. *Journal of Fish Biology,* **45** (Suppl. A), 171–180.

Butler, G.W. (1897) Report on the spawning of the common sole in the aquarium of the Marine Biological Association Laboratory during April and May, 1895. *Journal of the Marine Biological Association of the United Kingdom,* **4**, 3–9.

Carter, C.G., Purser, G.J., Houlihan, D.F. & Thomas, P. (1996) The effects of decreased ration on feeding hierarchies in groups of greenback flounder (*Rhombosolea tapirina*: Teleostei). *Journal of the Marine Biological Association of the United Kingdom,* **76**, 505–516.

Carvalho, N., Alfonso, P. & Santos, R.S. (2003) The haremic mating system and mate choice in the wide-eyed flounder *Bothus podas*. *Environmental Biology of Fishes,* **66**, 249–258.

Casey, J.M. & Myers, R.A. (1998) Diel variation in trawl catchability: is it as clear as day and night? *Canadian Journal of Fisheries and Aquatic Sciences,* **55**, 2329–2340.

Champalbert, G. & Koutsikopoulos, C. (1995) Behaviour, transport and recruitment of Bay of Biscay sole (*Solea solea*) – laboratory and field studies. *Journal of the Marine Biological Association of the United Kingdom,* **75**, 93–108.

Champalbert, G. & Le Direach-Boursier, L. (1998) Influence of light and feeding conditions on swimming activity rhythms of larval and juvenile turbot *Scophthalmus maximus* L.: an experimental study. *Journal of Sea Research,* **40**, 333–345.

Chapman, C.J. & Sand, O. (1974) Field studies of hearing in two species of flatfish *Pleuronectes platessa* (L.) and *Limanda limanda* (L.) (Family Pleuronectidae). *Comparative Biochemistry and Physiology,* **47A**, 371–385.

Chen, W.M. & Purser, G.J. (2001) The effect of feeding regime on growth, locomotor activity pattern and the development of food anticipatory activity in greenback flounder. *Journal of Fish Biology,* **58**, 177–187.

Chen, W.M., Purser, J. & Blyth, P. (1999) Diel feeding rhythms of greenback flounder *Rhombosolea tapirina* (Gunther 1862): the role of light-dark cycles and food deprivation. *Aquaculture Research,* **30**, 529–537.

Clark, E. & George, A. (1979) Toxic soles, Pardachirus marmoratus from the Red Sea and P. pavoninus from Japan, with notes on other species. Environmental Biology of Fishes, **4**, 103–123.

Cole, F.J. & Johnston, J. (1901) *Pleuronectes. Liverpool Marine Biology Committee Memoirs*, VIII. Williams & Norgate, London.

Curran, M.C. & Able, K.W. (1998) The value of tethering fishes (winter flounder and tautog) as a tool for assessing predation rates. *Marine Ecology Progress Series,* **163**, 45–51.

de Groot, S.J. (1963) Diurnal activity and feeding habits of plaice. *Rapports et Procès-Verbaux des Réunions Conseil International pour l'Exploration de la Mer,* **155**, 48–51.

de Groot, S.J. (1969a) A review paper on the behaviour of flatfishes. In: Proceedings of the FAO conference on fish behaviour in relation to fishing techniques and tactics. *FAO Fisheries Report,* **62**, 139–167.

de Groot, S.J. (1969b) Digestive system and sensorial factors in relation to the feeding behaviour of flatfish (Pleuronectiformes). *Journal du Conseil International pour l'Exploration de la Mer,* **32**, 385–394.

de Groot, S.J. (1970) Some notes on the ambivalent behaviour of the Greenland halibut *Reinhardtius hippoglossoides* (Walb.) Pisces: Pleuronectiformes. *Journal of Fish Biology,* **2**, 275–279.

de Groot, S.J. (1971) On the interrelations between morphology of the alimentary tract, food and feeding behaviour in flatfishes (Pisces: Pleuronectiformes). *Netherlands Journal of Sea Research,* **5**, 121–196.

de Groot, S.J., Norde, R. & Verheijen, F.J. (1969) Retinal stimulation and pattern formation in the common sole *Solea solea* (L.) (Pisces: Soleidae). *Netherlands Journal of Sea Research,* **4**, 339–349.

Dou, S., Seikai, T. & Tsukamoto, K. (2000) Cannibalism in Japanese flounder juvenile, *Paralichthys olivaceus*, reared under controlled conditions. *Aquaculture,* **182**, 149–159.

Ellis, T. & Gibson, R.N. (1997) Predation of 0-group flatfishes by 0-group cod: handling times and size-selection. *Marine Ecology Progress Series,* **149**, 83–90.

Ellis, T., Howell, B.R. & Hughes, R.N. (1997) The cryptic responses of hatchery-reared sole to a natural sand substratum. *Journal of Fish Biology,* **51**, 389–401.

Ellis, T., Hughes, R.N. & Howell, B.R. (2002) Artificial dietary regime may impair subsequent foraging behaviour of hatchery-reared turbot released into the natural environment. *Journal of Fish Biology,* **61**, 252–264.

Evans, H.M. (1937) A comparative study of the brains in Pleuronectidae. *Proceedings of the Royal Society, Series B,* **122**, 308–342.

Fairchild, E.A. & Howell, W.H. (2001) Predator-prey size relationship between *Pseudopleuronectes americanus* and *Carcinus maenas*. *Journal of Sea Research,* **44**, 81–90.

Fernö, A. & Olsen, S. (1994) *Marine Fish Behaviour in Capture and Abundance Estimation.* Fishing News Books, Oxford.

Forster, G.R. (1953) The spawning behaviour of the plaice. *Journal of the Marine Biological Association of the United Kingdom,* **32**, 319.

Forward, R.B., Jr & Tankersley, R.A. (2001) Selective tidal-stream transport of marine animals. *Oceanography and Marine Biology: An Annual Review,* **39**, 305–353.

Fredette, M., Batt, J. & Castell, J. (2000) Feeding stimulant for juvenile winter flounders. *North American Journal of Aquaculture,* **62**, 157–160.

Fujieda, S., Matsuno, Y. & Yamanaka, Y. (1996) The auditory threshold of the bastard halibut *Paralichthys olivaceus*. *Nippon Suisan Gakkaishi,* **62**, 201–204.

Fujimoto, M., Satoh, Y. & Morishita, F. (1992) Measurement of background-visual field in flatfishes by CRT display method. *Marine Behaviour and Physiology,* **19**, 285–290.

Furuta, S. (1996) Predation on juvenile Japanese flounder (*Paralichthys olivaceus*) by diurnal piscivorous fish: field observations and laboratory experiments. In: *Survival Strategies in Early Life Stages of Marine Resources* (eds Y. Watanabe, Y. Yamashita & Y. Ooozeki). pp. 285–294. A.A. Balkema, Rotterdam.

Furuta, S. (1998) Basic studies on release techniques of hatchery-reared Japanese flounder. V – Effects of starvation on feeding behavior and predation vulnerability of wild Japanese flounder juvenile. *Nippon Suisan Gakkaishi,* **64**, 658–664.

Gibb, A.C. (1996) The kinematics of prey capture in *Xystreurys liolepis*: do all flatfish feed asymmetrically? *Journal of Experimental Biology,* **199**, 2269–2283.

Gibb, A.C. (1997) Do flatfishes feed like other fishes? A comparative study of percomorph prey-capture kinematics. *Journal of Experimental Biology,* **200**, 2841–2859.

Gibson, R.N. (1973a) The intertidal movements and distribution of young fish on a sandy beach with special reference to the plaice (*Pleuronectes platessa* L.). *Journal of Experimental Marine Biology and Ecology,* **12**, 79–102.

Gibson, R.N. (1973b) Tidal and circadian activity rhythms in juvenile plaice, *Pleuronectes platessa*. *Marine Biology,* **22**, 379–386.

Gibson, R.N. (1976) Comparative studies on the rhythms of juvenile flatfish. In: *Biological Rhythms in the Marine Environment*. Belle W. Baruch Library in Marine Science No. 4 (ed. P.J. DeCoursey). pp. 199–213. University of South Carolina Press, Columbia, SC.

Gibson, R.N. (1980) A quantitative description of the behaviour of wild juvenile plaice (*Pleuronectes platessa* L.). *Animal Behaviour,* **28**, 1202–1216.

Gibson, R.N. (1997) Behaviour and the distribution of flatfishes. *Journal of Sea Research,* **37**, 241–256.

Gibson, R.N. & Robb, L. (2000) Sediment selection in juvenile plaice and its behavioural basis. *Journal of Fish Biology,* **56**, 1258–1275.

Gibson, R.N., Blaxter, J.H.S. & de Groot, S.J. (1978) Developmental changes in the activity rhythms of the plaice (*Pleuronectes platessa* L.). In: *Rhythmic Activity of Fishes* (ed. J.E. Thorpe). pp. 169–186. Academic Press, London.

Gibson, R.N., Yin, M.C. & Robb, L. (1995) The behavioural basis of the predator-prey size relationship between shrimp (*Crangon crangon*) and juvenile plaice (*Pleuronectes platessa*). *Journal of the Marine Biological Association of the United Kingdom,* **75**, 337–349.

Gibson, R.N., Pihl, L., Burrows, M.T., Modin. J., Wennhage, H. & Nickell, L.A. (1998) Diel movements of juvenile plaice (*Pleuronectes platessa*) in relation to predators, competitors, food availability and abiotic factors on a microtidal nursery ground. *Marine Ecology Progress Series,* **165**, 145–159.

Gibson, S. & Johnston, I.A. (1995) Scaling relationships, individual variation and the influence of temperature on maximum swimming speed in early settled stages of turbot *Scophthalmus maximus*. *Marine Biology,* **121**, 401–408.

Harden Jones, F.R. (1968) *Fish Migration*. Edward Arnold, London.

He, P. (2003) Swimming behaviour of winter flounder (*Pleuronectes americanus*) on natural fishing grounds as observed by an underwater video camera. *Fisheries Research,* **60**, 507–514.

Healey, E.G. (1999) The skin pattern of young plaice and its rapid modification in response to graded changes in background tint and pattern. *Journal of Fish Biology,* **55**, 937–971.

Hiatt, R.W. & Strasburg, D.W. (1960) Ecological relationships of the fish fauna on coral reefs of the Marshall Islands. *Ecological Monographs,* **30**, 65–127.

Hill, S., Burrows, M.T. & Hughes, R.N. (2000) Increased turning per unit distance as an area-restricted search mechanism in a pause-travel predator, juvenile plaice, foraging for buried bivalves. *Journal of Fish Biology,* **56**, 1497–1508.

Holmes, R.A. & Gibson, R.N. (1983) A comparison of predatory behaviour in flatfish. *Animal Behaviour,* **31**, 1244–1255.

Holmes, R.A. & Gibson, R.N. (1986) Visual cues determining prey selection by the turbot, *Scophthalmus maximus* L. *Journal of Fish Biology,* **29** (Suppl. A), 49–58.

Honda, H. (1988) Displacement behaviour of Japanese flounder *Paralichthys olivaceus* estimated by the difference of oxygen consumption rate. *Nippon Suisan Gakkaishi,* **54**, 1259.

Hossain, M.A.R., Tanaka, M. & Masuda, R. (2002) Predator-prey interaction between hatchery-reared Japanese flounder juvenile, *Paralichthys olivaceus*, and sandy shore crab, *Matuta lunaris*: daily rhythms, anti-predator conditioning and starvation. *Journal of Experimental Marine Biology and Ecology*, **267**, 1–14.

Howell, B.R. (1994) Fitness of hatchery fish for survival in the sea. *Aquaculture and Fisheries Management*, **25** (Suppl. 1), 3–17.

Howell, B.R. & Canario, A.V.M. (1987) The influence of sand on the estimation of resting metabolic rate of juvenile sole, *Solea solea* L. *Journal of Fish Biology* **31**, 277–280.

Iglesias, J. & Rodriguez-Ojea, G. (1994) Fitness of hatchery-reared turbot, *Scophthalmus maximus* L., for survival in the sea: first year results on feeding growth and distribution. *Aquaculture and Fisheries Management*, **25** (Suppl. 1), 179–188.

Kaimmer, S.M. (1998) Direct observations on the hooking behavior of the Pacific halibut, *Hippoglossus stenolepis*. *Fishery Bulletin*, **97**, 873–883.

Karlsen, H.E. (1992) Infrasound sensitivity in the plaice (*Pleuronectes platessa*). *Journal of Experimental Biology*, **171**, 173–187.

Kawabe, R., Nashimoto, K., Hiraishi, T., Naito, Y. & Sato, K. (2003) A new device for monitoring the activity of freely swimming flatfish, Japanese flounder *Paralichthys olivaceus*. *Fisheries Science*, **69**, 3–10.

Keefe, M. & Able, K.W. (1994) Contributions of abiotic and biotic factors to settlement in summer flounder, *Paralichthys olivaceus*. *Copeia*, **1994**, 458–465.

Kellison, G.T., Eggleston, D.B. & Burke, J.S. (2000) Comparative behaviour and survival of hatchery-reared versus wild summer flounder (*Paralichthys dentatus*). *Canadian Journal of Fisheries and Aquatic Sciences*, **57**, 1870–1877.

Kiorbøe, T. (1978) Feeding rate in juvenile flounder in relation to prey density. *Kieler Meeresforschungen, Sonderheft No. 4*, 275–281.

Konstantinou, H.J. & Shen, D.C. (1995) The social and reproductive behaviour of the eyed flounder, *Bothus ocellatus*, with notes on the spawning of *Bothus lunatus* and *Bothus ellipticus*. *Environmental Biology of Fishes*, **44**, 311–324.

Kruuk, H. (1963) Diurnal periodicity in the activity of the common sole, *Solea vulgaris* Quensel. *Netherlands Journal of Sea Research*, **2**, 1–28.

Kuipers, B. (1973) On the tidal migration of young plaice, *Pleuronectes platessa*, in the Wadden Sea. *Netherlands Journal of Sea Research*, **6**, 376–388.

Lagardère, J.P. & Sureau, D. (1989) Changes in the swimming activity of the sole (*Solea vulgaris* Quensel, 1806) in relation to winter temperatures in a saltmarsh – observations using ultrasonic telemetry. *Fisheries Research*, **7**, 233–239.

Lagardère, J.P., Begout, M.L., Lafaye, J.Y. & Villotte, J.P. (1994) Influence of wind-produced noise on orientation in the sole (*Solea solea*). *Canadian Journal of Fisheries and Aquatic Sciences*, **51**, 1258–1264.

Lagardère, J.P., Anras, M.L.B., Breton, H. & Claret, J.B.C. (1995) The effects of illumination, temperature and oxygen concentration on swimming activity of turbot *Psetta maxima* (Linné 1758). *Fisheries Research*, **24**, 165–171.

Leopold, M.F., Van Damme, C.J.G. & Van der Veer, H.W. (1998) Diet of cormorants and the impact of cormorant predation on juvenile flatfish in the Dutch Wadden Sea. *Journal of Sea Research*, **40**, 93–107.

Lindsey, C.C. (1978) Form, function and locomotory habits in fish. In: *Fish Physiology*, Vol. 7 (eds W.S. Hoar & D.J. Randall). pp. 1–100. Academic Press, New York.

Livingston, M.E. (1987) Morphological and sensory specializations of five New Zealand flatfish species, in relation to feeding behaviour. *Journal of Fish Biology*, **31**, 775–795.

McConnaughey, R.A. & Smith, K.R. (2000) Associations between flatfish abundance and surficial sediments in the eastern Bering Sea. *Canadian Journal of Fisheries and Aquatic Sciences,* **57**, 2410–2419.

McCracken, F.D. (1963) Seasonal movements of the winter flounder, *Pseudopleuronectes americanus* (Walbaum), on the Atlantic coast. *Journal of the Fisheries Research Board of Canada,* **20**, 551–586.

Macdonald, J.S. (1983) Laboratory observations of feeding behaviour of the ocean pout (*Macrozoarces americanus*) and winter flounder (*Pseudopleuronectes americanus*) with reference to niche overlap of natural populations. *Canadian Journal of Zoology,* **61**, 539–546.

Mackie, A.M. & Mitchell, A.I. (1982) Further studies on the chemical control of feeding behaviour in the Dover sole, *Solea solea*. *Comparative Biochemistry and Physiology,* **73A**, 89–93.

Macquart-Moulin, C., Champalbert, G., Howell, B.R., Patriti, G. & Ranaivoson, C. (1991) La relation alimentation-fixation benthique chez les jeunes soles *Solea solea* L metamorphosées. Evidence experimentales. *Journal of Experimental Marine Biology and Ecology,* **153**, 195–205.

Main, J. & Sangster, G.I. (1981) A study of the fish capture process in a bottom trawl by direct observations from an underwater vehicle. *Scottish Fisheries Research Report,* **23**, 1–23.

Main, J. & Sangster, G.I. (1982) A study of a multi-level bottom trawl for species separation using direct observation techniques. *Scottish Fisheries Research Report,* **26**, 1–17.

Manabe, H. & Shinomiya, A. (1998) Use of the first dorsal fin ray as a lure by the bothid flounder, *Asterorhombus intermedius*. *Japanese Journal of Ecology (Tokyo),* **48**, 117–121.

Manabe, H., Ide, M. & Shinomiya, A. (2000) Mating system of the left eye flounder, *Engyprosopon grandisquama*. *Ichthyological Research,* **47**, 69–74.

Manderson, J.P., Phelan, B., Bejda, A.J., Stehlik, L.L. & Stoner, A.W. (1999) Predation by striped searobin (*Prionotus evolans*, Triglidae) on young-of-the-year winter flounder (*Pseudopleuronectes americanus* Walbaum): examining prey size selection and prey choice using field observations and laboratory experiments. *Journal of Experimental Marine Biology and Ecology,* **242**, 211–231.

Manderson, J.P., Phelan, B., Stoner, A.W. & Hilbert, J. (2000) Predator-prey relations between age-1+ summer flounder (*Paralichthys dentatus*, Linnaeus) and age-0 winter flounder (*Pseudopleuronectes americanus*, Walbaum): predator diets, prey selection, and effects of sediments and macrophytes. *Journal of Experimental Marine Biology and Ecology,* **251**, 17–39.

Marsh, E. (1977) Structural modifications of the pectoral fin rays in the order Pleuronectiformes. *Copeia,* **1977**, 575–578.

Martinez Cordero, F.J., Beveridge, M.C.M. & Muir, J.F. (1994) A note on the behaviour of adult Atlantic halibut, *Hippoglossus hippoglossus* (L.), in cages. *Aquaculture and Fisheries Management,* **25**, 475–481.

Mast, S.O. (1914) Changes in shade, color, and pattern of fishes, and their bearing on the problems of adaptation and behaviour, with especial reference to the flounders *Paralichthys* and *Ancylopsetta*. *Bulletin of the United States Bureau of Fisheries,* **34**, 177–238.

Mattila, J. & Bonsdorff, E. (1998) Predation by juvenile flounder (*Platichthys flesus* L.): a test of prey vulnerability, predator preference, switching behaviour and functional response. *Journal of Experimental Marine Biology and Ecology,* **227**, 221–236.

Metcalfe, J.D. & Arnold, G.P. (1990) The energetics of migration by selective tidal stream transport: an analysis for plaice tracked in the southern North Sea. *Journal of the Marine Biological Association of the United Kingdom,* **70**, 149–162.

Metcalfe, J.D. & Arnold, G.P. (1997) Tracking fish with electronic tags. *Nature,* **387**, 665–666.

Metcalfe, J.D., Fulcher, M. & Storeton West, T.J. (1992). Progress and developments in telemetry for monitoring the migratory behaviour of plaice in the North Sea. In: *Wildlife Telemetry. Remote Monitoring and Tracking of Animals* (eds I.G. Priede & S.M. Swift). pp. 358–366. Ellis Horwood, New York.

Metcalfe, J.D., Holford, B.H. & Arnold, G.P. (1993) Orientation of plaice (*Pleuronectes platessa*) in the open sea: evidence for the use of external directional clues. *Marine Biology,* **117**, 559–566.

Minello, T.J., Zimmerman, R.J. & Martinez, E.X. (1987) Fish predation on juvenile brown shrimp, *Penaeus aztecus* Ives: effects of turbidity on predation rates. *Fishery Bulletin,* **85**, 59–70.

Miyazaki, T., Masuda, R., Furuta, S. & Tsukamoto, K. (2000) Feeding behaviour of hatchery-reared juveniles of the Japanese flounder following a period of starvation. *Aquaculture,* **190**, 129–138.

Moles, A. & Norcross, B.L. (1995) Sediment preference in juvenile Pacific flatfish. *Netherlands Journal of Sea Research,* **34**, 177–182.

Moore, J.W. & Moore, A. (1976) The basis of food selection in flounders, *Platichthys flesus* (L.), in the Severn estuary. *Journal of Fish Biology,* **9**, 139–156.

Moyer, J.T., Yogo, Y., Zaiser, M.J. & Tsukahara, H. (1985) Spawning behaviour and social organisation of the flounder *Crossorhombus kobensis* (Bothidae) at Miyake-Jima, Japan. *Japanese Journal of Ichthyology,* **32**, 363–367.

Neill, W.H. (1984) Behavioural enviroregulation's role in fish migration. In: *Mechanisms of Migration in Fishes* (eds J.D. McCleave, G.P. Arnold, J.J. Dodson & W.H. Neill). pp. 61–66. Plenum Press, New York.

Nelson, J.S. (1994) *Fishes of the World*, 3rd edn. John Wiley & Sons, New York.

Neuman, M.J. & Able, K.W. (1998) Experimental evidence of sediment preference by early life history stages of windowpane (*Scophthalmus aquosus*). *Journal of Sea Research,* **40**, 33–41.

Nevitt, G.A. (1991) Do fish sniff? A new mechanism of olfactory sampling in pleuronectid flounders. *Journal of Experimental Biology,* **157**, 1–18.

Nichols, J.H. (1989) The diurnal rhythm in spawning of plaice (*Pleuronectes platessa* L.) in the southern North Sea. *Journal du Conseil Permanent International pour l'Exploration de la Mer,* **45**, 277–283.

Norcross, B.L., Müter, F-J. & Holladay, B.A. (1995) Habitat models for juvenile pleuronectids around Kodiak Island, Alaska. In: Proceedings of the International Symposium on North Pacific Flatfish, pp. 151–175. *Alaska Sea Grant College Program Report No. 95–04*. University of Alaska, Fairbanks, AK.

Norman, J.R. (1934) *A Systematic Monograph of the Flatfishes* (*Heterosomata*), Vol. 1. Psettodidae, Bothidae, Pleuronectidae. British Museum Natural History, London.

Olla, B.L., Wicklund, R. & Wilk, S. (1969) Behavior of winter flounder in a natural habitat. *Transactions of the American Fisheries Society,* **98**, 717–720.

Olla, B.L., Samet, C.E. & Studholme, A.L. (1972) Activity and feeding behaviour of the summer flounder (*Paralichthys dentatus*) under controlled laboratory conditions. *Fishery Bulletin,* **70**, 1127–1136.

Orcutt, H.G. (1950) The life history of the starry flounder, *Platichthys stellatus* (Pallas). *California Department of Natural Resources Division of Fish and Game Fish Bulletin*, No. 78, 1–64.

Page, N.S. (1996) *Rhythmic behaviour in two species of marine flatfish: solenette* (Buglossideum luteum Risso) *and plaice* (Pleuronectes platessa *L.*). PhD thesis, University of Birmingham, England.

Peterson-Curtis, T.L. (1997) Effects of salinity on survival, growth, metabolism, and behaviour in juvenile hogchokers, *Trinectes maculatus fasciatus* (Achiridae). *Environmental Biology of Fishes,* **49**, 323–331.

Peyraud, C. & Labat, R. (1962) Réactions cardio-respiratoire observé chez la sole au cours de l'ensablement. *Hydrobiologia,* **19**, 351–356.

Phelan, B., Manderson, J.P., Stoner, A.W. & Bejda, A.J. (2001) Size related shifts in the habitat associations of young-of-the-year winter flounder (*Pseudopleuronectes americanus*): field observations and laboratory experiments with sediments and prey. *Journal of Experimental Marine Biology and Ecology,* **257**, 297–315.

Rabben, H. & Furevik, D.M. (1993) Application of heart rate transmitters in behaviour studies on Atlantic halibut (*Hippoglossus hippoglossus*). *Aquacultural Engineering,* **12**, 129–140.

Raffaelli, D., Richner, H., Summers, R. & Northcott, S. (1990) Tidal migrations in the flounder (*Platichthys flesus*). *Marine Behaviour and Physiology,* **16**, 249–260.

Ramachandran, V.S., Tyler, C.W., Gregory, R.L., *et al.* (1996) Rapid adaptive camouflage in tropical flounders. *Nature,* **379**, 815–818.

Riley, J.D. (1966) Marine fish culture in Britain VII. Plaice (*Pleuronectes platessa* L.) post-larval feeding on *Artemia salina* L. nauplii and the effects of varying feeding levels. *Journal du Conseil International pour l'Exploration de la Mer,* **30**, 204–221.

Rogers, S.I. (1993) The dispersion of sole, *Solea solea*, and plaice, *Pleuronectes platessa*, within and away from a nursery ground in the Irish Sea. *Journal of Fish Biology,* **43** (Suppl.), 275–288.

Rountree, R.A. & Able, K.W. (1992) Foraging habits, growth and temporal patterns of salt-marsh creek habitat use by young-of-year summer flounder in New Jersey. *Transactions of the American Fisheries Society,* **121**, 765–776.

Ryer, C.H., Stoner, A.W. & Titgen, R.H. (2004) Behavioral mechanisms underlying the refuge value of benthic habitat structure for two flatfishes with differing anti-predator strategies. *Marine Ecology Progress Series,* **268**, 231–243.

Sakakura, Y. & Tsukamoto, K. (2002) Onset and development of aggressive behaviour in the early life stage of Japanese flounder. *Fisheries Science,* **68**, 854–861.

Scheuring, L. (1921) Beobachtungen und Betrachtungen uber die Beziehungen der Augen zum Nahrungserweb bei Fishen. *Zoologische Jahrbücher Abteilung für Allgemeine Zoologie und Physiologie der Tiere,* **38**, 113–136.

Scott, J.S. (1982) Selection of bottom type by groundfish of the Scotian shelf. *Canadian Journal of Fisheries and Aquatic Sciences,* **30**, 943–947.

Sekai, T., Kinoshita, I. & Tanaka, M. (1993) Predation by crangonid shrimp on juvenile Japanese flounder under laboratory conditions. *Nippon Suisan Gakkaishi,* **59**, 321–326.

Shaheen, P.A., Stehlik, L.L., Meise, C.J., Stoner, A.W., Manderson, J.P. & Adams, D.L. (2001) Feeding behaviour of newly settled winter flounder (*Pseudopleuronectes americanus*) on calanoid copepods. *Journal of Experimental Marine Biology and Ecology,* **257**, 37–51.

Shelverton, P.A. & Carter, C.G. (1998) The effect of ration on behaviour, food consumption and growth in juvenile greenback flounder (*Rhombosolea tapirina*: Teleostei). *Journal of the Marine Biological Association of the United Kingdom,* **78**, 1307–1320.

Sogard, S.M. (1997) Size-selective mortality in the juvenile stages of teleost fishes: a review. *Bulletin of Marine Science,* **60**, 1129–1157.

Steven, G.A. (1930) Bottom fauna and the food of fishes. *Journal of the Marine Biological Association of the United Kingdom,* **26**, 677–700.

Stickney, R.R., White, D.B. & Miller, D. (1973) Observations of fin use in relation to feeding and resting behaviour in flatfishes (Pleuronectiformes). *Copeia,* **1973**, 154–156.

Stoner, A.W. (2003) Hunger and light levels alter response to bait by Pacific halibut: laboratory analysis of detection, location and attack. *Journal of Fish Biology,* **62**, 1176–1193.

Stoner, A.W. & Abookire, A.A. (2002) Sediment preferences and size-specific distribution of young-of-the-year Pacific halibut in an Alaskan nursery. *Journal of Fish Biology,* **61**, 540–559.

Stoner, A.W. & Ottmar, M.L. (2003) Relationships between size-specific sediment preferences and burial capabilities in juveniles of two Alaska flatfishes. *Journal of Experimental Marine Biology and Ecology,* **282**, 85–101.

Stoner, A.W. & Titgen, R.H. (2003) Biological structures and bottom type influence habitat choices made by Alaska flatfishes. *Journal of Experimental Marine Biology and Ecology,* **292**, 43–59.

Stoner, A.W., Bejda, A.J., Phelan, B.A., Manderson, J.P., Stehlik, L.S. & Pessutti, J.P. (1999) Behavior of winter flounder (*Pseudopleuronectes americanus*) during the reproductive season: laboratory and field observations on spawning, feeding and locomotion. *Fishery Bulletin,* **97**, 999–1016.

Summers, R.W. (1980) The diet and feeding behaviour of the flounder, *Platichthys flesus* (L.) in the Ythan estuary, Aberdeenshire, Scotland. *Estuarine, Coastal and Marine Science,* **11**, 217–232.

Sumner, F.B. (1911) The adjustment of flatfishes to various backgrounds: a study of adaptive colour change. *Journal of Experimental Zoology,* **10**, 409–505.

Sumner, F.B. (1940) Quantitative changes in pigmentation resulting from visual stimuli in fishes and amphibia. *Biological Reviews,* **15**, 351–375.

Thomsen, B. (1993) Selective flatfish trawling. *ICES Marine Science Symposia,* **196**, 161–164.

Van der Veer, H.W. & Bergman, M.J.N. (1987) Predation by crustaceans on a newly settled 0-group *Pleuronectes platessa* population in the western Wadden Sea. *Marine Ecology Progress Series,* **35**, 203–215.

Videler, J. (1993) *Fish Swimming*. Chapman & Hall, London.

Walsh, S.J. (1991) Diel variation in availability and vulnerability of fish to a survey trawl. *Journal of Applied Ichthyology,* **7**, 147–159.

Walsh, S.J. & Hickey, W.M. (1993) Behavioural reactions of demersal fish to bottom trawls at various light conditions. *ICES Marine Science Symposia,* **196**, 68–76.

Wardle, C.S. (1983) Fish reactions to towed fishing gears. In: *Experimental Biology at Sea* (eds A. MacDonald & I.G. Priede). pp. 167–195. Academic Press, London.

Wardle, C.S. & Hollingworth, C.E. (1993) Fish behaviour in relation to fishing operations. *ICES Marine Science Symposia,* **196**, 1–215.

Webb, P.W. (1981) The effect of the bottom on the fast start of flatfish *Citharichthys stigmaeus*. *Fishery Bulletin,* **79**, 271–276.

Webb, P.W. (1989) Station-holding by three species of benthic fishes. *Journal of Experimental Biology,* **145**, 303–320.

Webb, P.W. & Gerstner, C.L. (2000) Fish swimming behaviour: predictions from physical principles. In: *Biomechanics in Animal Behaviour* (eds P. Domenici & R.W. Blake). pp. 59–77. Bios Scientific Publishers, Oxford.

Weber, E. (1965) Eine fakultative Fressgemeinschaft von Fischen und Stachelhäutern. *Zeitschrift für Tierpsychologie,* **22**, 567–569.

Weihs, D. (1973) Mechanically efficient swimming techniques for fish with negative buoyancy. *Journal of Marine Research,* **31**, 194–209.

Wells, B., Steele, D.H. & Tyler, A.V. (1973) Intertidal feeding of winter flounder (*Pseudopleuronectes americanus*) in the Bay of Fundy. *Journal of the Fisheries Research Board of Canada,* **30**, 1374–1378.

Wennhage, H. (2000) Vulnerability of settling plaice *Pleuronectes platessa* to predation: effects of developmental stage and alternative prey. *Marine Ecology Progress Series,* **203**, 289–299.

Wennhage, H. & Gibson, R.N. (1998) Influence of food supply and a potential predator (*Crangon crangon*) on settling behaviour of plaice (*Pleuronectes platessa*). *Journal of Sea Research,* **39**, 103–112.

Winger, P.D., He, P. & Walsh, S.J. (1999) Swimming endurance of American plaice (*Hippoglossoides platessoides*) and its role in fish capture. *ICES Journal of Marine Science,* **56**, 252–265.

Witting, D.A. & Able, K.W. (1993) Effects of body size on probability of predation for juvenile summer and winter flounder based on laboratory experiments. *Fishery Bulletin,* **91**, 577–581.

Yamashita, Y., Nagahora, S., Yamada, H. & Kitagawa, D. (1994) Effects of release size on survival and growth of the Japanese flounder *Paralichthys olivaceus* in coastal waters off Iwate Prefecture, northeastern Japan. *Marine Ecology Progress Series,* **105**, 269–276.

Yazdani, G.M. (1969) Adaptations in the jaws of flatfish (Pleuronectiformes). *Journal of Zoology (London),* **159**, 181–222.

Chapter 11
Atlantic flatfish fisheries

Richard Millner, Stephen J. Walsh and Juan M. Díaz de Astarloa

11.1 Introduction

Flatfish fisheries occur throughout the Atlantic from the deep Arctic to the coasts of southern Africa and South America. In 1998, flatfish landings from the Atlantic amounted to 0.4 million tonnes or nearly half of the total world flatfish catch, with the largest and most heavily exploited fisheries occurring in northern waters. However, even this production is small in comparison to Atlantic fisheries for other demersal species such as cod and hake from which over 4.5 million tonnes were landed in 1998 (FAO 2000).

In this chapter, the Atlantic fisheries have been divided into three regions covering the northeast, northwest and southern Atlantic (Fig. 11.1). The northwest covers the western Atlantic (FAO area 21) from Greenland across to Labrador and down to Cape Hatteras at 35°N and also includes the west central Atlantic American States (FAO area 31). The northeast Atlantic (FAO area 27) extends from east Greenland and the Barents Sea in the north down to Gibraltar at latitude 36°N. The third area considered covers the whole south Atlantic, including the coasts of Brazil, Uruguay and Argentina in the west and Angola, Namibia and South Africa in the east (FAO areas 41 and 47). Although there are productive flatfish fisheries in areas of the central Atlantic such as the shallow coastal region around the North African coast, the absence of detailed statistics has meant that they have not been considered further in this review. [Note that throughout the descriptions of the flatfish fisheries in the three areas of the Atlantic, 1998 was chosen as a representative year to compare economic importance across all regions.]

11.2 Main species and nature of the fisheries

11.2.1 Northwest Atlantic

In the northwest Atlantic, there are 51 species of flatfishes divided into 4 families: Bothidae (28), Pleuronectidae (7), Soleidae (5) and Cynoglossidae (11) (Miller *et al.* 1991). Of these, only 8 species (7 pleuronectids and 1 bothid) divided into 28 stocks (Table 11.1) and two flatfish complexes (mixed species) are under fisheries management control. These stocks range from sub-area 0 to 6 in the Northwest Atlantic Fisheries Organization (NAFO) convention area (Fig. 11.2). On the Greenland side of Baffin Bay and Davis Strait, the only flatfish fishery under management is the Greenland halibut. In Canadian and American waters, eight species are regulated: American plaice, yellowtail flounder, witch, winter flounder, summer

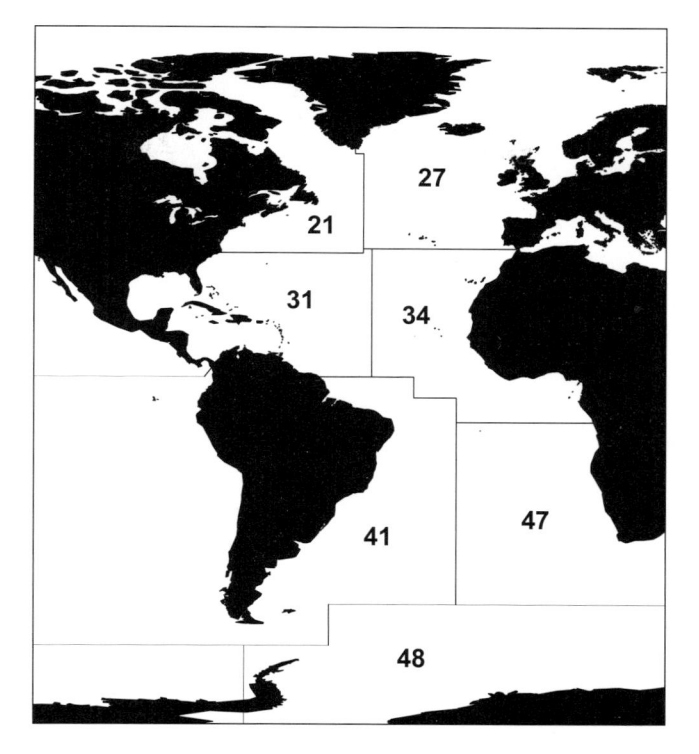

Fig. 11.1 The North Atlantic showing the FAO fishing areas, northwest (FAO 21 and 31), northeast (FAO 27) and southern Atlantic (FAO 41 and 47).

flounder (US only), windowpane (US only), Greenland halibut (Canada only) and Atlantic halibut (Table 11.1). The fisheries have supported average landings about one-third of the combined landings of gadoids and redfishes (Murawski *et al.*1997).

The flatfish fisheries occur on inshore and offshore banks over their entire range. The eight commercial species can be broadly categorised into three habitats: shallow, on top-of-continental shelf (bank) species; upper (mid-shelf) slope dwellers along the continental shelf and lower (deep) slope dwellers along the shelf edge. Most of the following descriptions are taken from Bigelow & Schroeder (1953) and Scott & Scott (1988). The choice of fishing gears used in these fisheries is often governed by nearness to shore and bottom depth of the resource.

Table 11.1 Landings of the main commercial flatfish species in the Northwest Atlantic Fisheries Organization (NAFO) Convention Area extending from West Greenland to Cape Hatteras, USA (landings are for the 1998 fisheries)

Common name	Species	Main area fished (see Fig. 11.2)	Landings (t)
American plaice	*Hippoglossoides platessoides*	Sub-area 2 to sub-area 5	7 210
Yellowtail flounder	*Limanda ferruginea*	Sub-area 3 to sub-area 6	8 809
Witch	*Glyptocephalus cynoglossus*	Sub-area 2 to sub-area 5	3 950
Winter flounder	*Pseudopleuronectes americanus*	Sub-area 4T to sub-area 5	6 698
Summer flounder	*Paralichthys dentatus*	Sub-area 5 and 6 and the South Atlantic	6 758
Windowpane	*Scophthalmus aquosus*	Sub-area 5	521
Greenland halibut	*Reinhardtius hippoglossoides*	Sub-area 0 to sub-area 4	12 222
Atlantic halibut	*Hippoglossus hippoglossus*	Sub-area 3 to sub-area 5	1 309
Other flatfishes			1 551

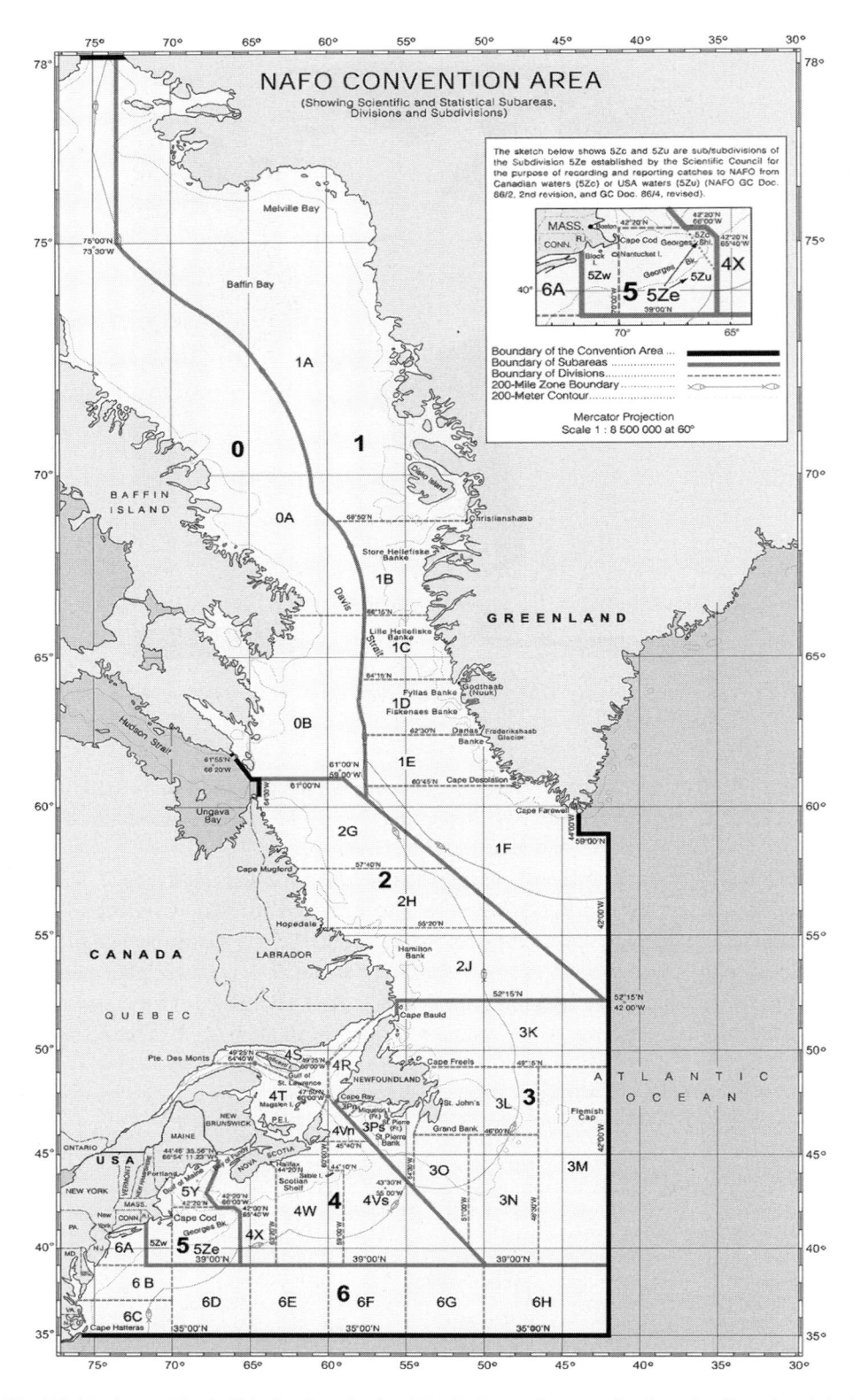

Fig. 11.2 Northwest Atlantic Fisheries Organization (NAFO) Convention Area showing scientific and statistical sub-areas (numbers), divisions (capital letters) and subdivisions (lowercase letters).

11.2.1.1 Shallow dwellers

The yellowtail flounder ranges along the continental shelf from Labrador to Chesapeake Bay with commercial concentrations found generally at depths of 40–70 m. The greatest commercial concentrations are found on the Grand Bank, Georges Bank and Southern New England Bank. The winter flounder is found mainly in coastal areas from Labrador to Georgia, in the Gulf of Mexico, with concentrations highest at depths of 20–40 m from the Gulf of St Lawrence to Chesapeake Bay. The windowpane is distributed both inshore and offshore along the continental shelf from the Gulf of St Lawrence to Florida. Commercial concentrations are found around 56 m on Georges Bank and Southern New England. The summer flounder ranges both inshore and offshore from the Gulf of Maine to South Carolina with concentrations highest in the Mid-Atlantic Bight (Cape Cod to Cape Hatteras) area. Commercial concentrations are found at depths of 15–50 m.

11.2.1.2 Mid-shelf dwellers

The American plaice is distributed along the continental shelf from West Greenland to Rhode Island in relatively deep waters with greatest commercial concentrations between 90 and 200 m on the Grand Banks, the Gulf of St Lawrence, the Gulf of Maine and Georges Bank. In recent years, small commercial concentrations have been fished off the eastern Grand Bank down to 1000 m. The witch is found along the deep-water shelf slopes from Hamilton Bank off southern Labrador to Cape Hatteras, North Carolina. Commercial concentrations are found between 110–350 m, especially on the Georges Bank and the Northeast Newfoundland Shelf.

11.2.1.3 Deep-slope dwellers

Atlantic halibut occurs from Disko Bay, West Greenland south to the coast of Virginia. Commercial concentrations are found mainly in depths of 200–500 m along the edge of the southern Grand Bank, St Pierre Bank, the Gulf of St Lawrence, Gulf of Maine and Georges Bank. Greenland halibut is distributed from Georges Bank in the south to Baffin Bay in the north. Commercial concentrations are mainly found at depths of 500–1000 m in Baffin Bay, along the edge of the Newfoundland Northeast Shelf and the eastern Grand Bank-Flemish Pass area.

The flatfish fisheries in the northwest Atlantic extend south from Baffin Bay to the Gulf of Mexico. Off west Greenland, a recent flatfish fishery began in the mid-1980s on the western side of Baffin Bay-Davis Strait. Off Atlantic Canada, the primary commercial flatfish fisheries extend from the eastern side of Baffin Bay, in the north, southward to the northeast corner of Georges Bank and include the Gulf of St Lawrence (Fig. 11.2). Off the Atlantic coast of the United States, flatfish fisheries extend south from the Gulf of Maine-Georges Bank area to the Gulf of Mexico. The main commercial fisheries, however, take place north of Cape Hatteras (Fig. 11.2). The fisheries occur mainly in depths ranging from less than 20 m to 1500 m.

The flatfish fisheries off Canada and the United States are generally harvested from mixed species aggregations which results in many technical interactions between gear types and target species. Harvesting is primarily by otter trawls, nevertheless there is an abundance of other gears used such as Danish and Scottish seines, bottom tending (sunken) gill nets, longlines, handlines and traps. Choice of fishing gear is generally dictated by the size of vessel,

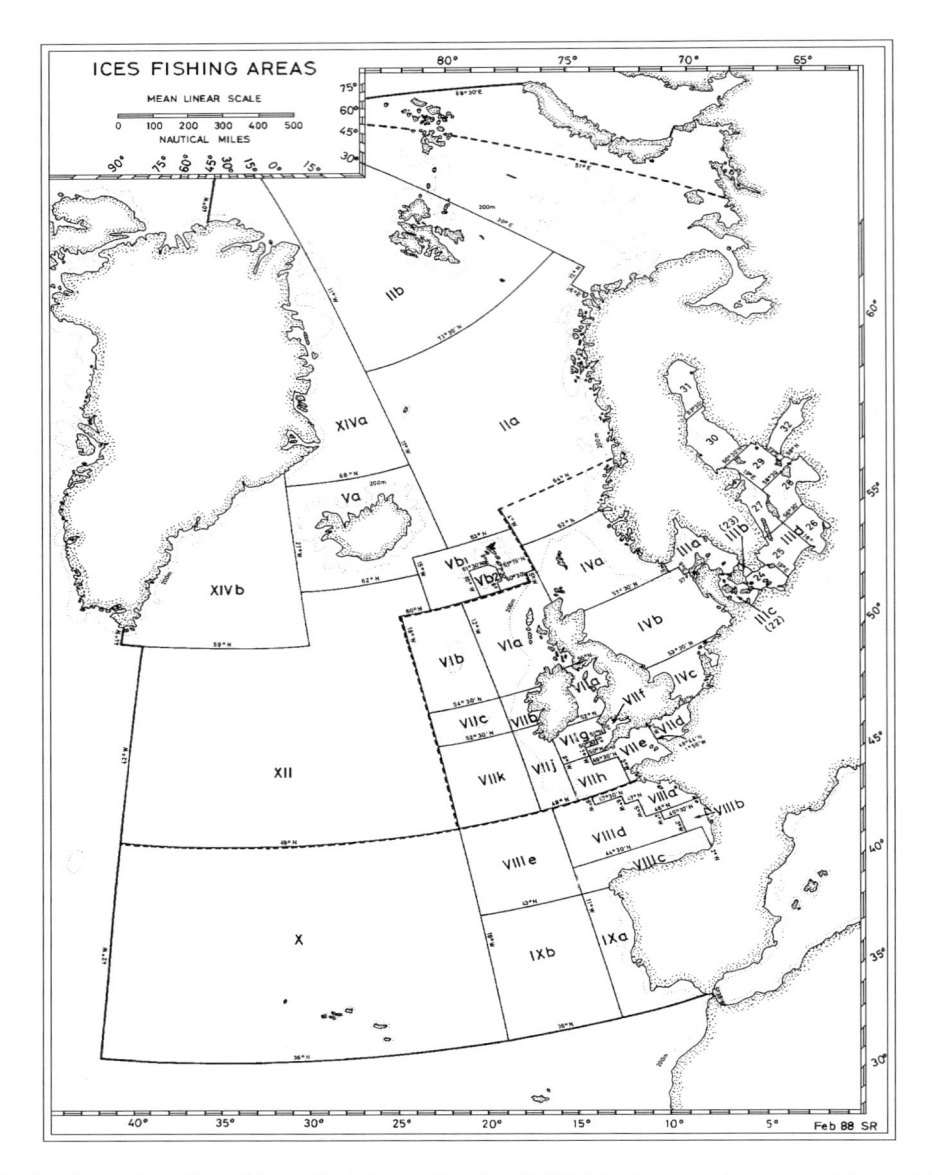

Fig. 11.3 International Council for the Exploitation of the Seas (ICES) fisheries area in the northeast Atlantic. Sub-areas (Roman numerals) and divisions (lower case letters).

the closeness of the resource to the coast, the mix of the demersal aggregation and the bottom habitat. Atlantic halibut fisheries in Canada and the United States are restricted to hook and line. The overall domestic effort in both Canada and the United States increased rapidly in all gear types after the Extended Economic Zones (EEZs) were declared in 1977.

11.2.2 Northeast Atlantic

In the northeast Atlantic (Fig. 11.3) abundant flatfish populations occur in areas where there are large shallow sandy or muddy sand sediments. The widest areas of shallow continental

Table 11.2 Landings of the main commercial flatfishes from the northeast Atlantic in 1998

Common name	Species	Main area fished (see Fig. 11.3)	Landings (t)
Plaice	*Pleuronectes platessa*	Sub-area I, IV, V, VII	104 671
Greenland halibut	*Reinhardtius hippoglossoides*	Sub-area I, II, V, XIV	34 867
Common sole	*Solea solea*	Sub-area III, IV, VII, VIII	31 194
Dab	*Limanda limanda*	Sub-area III, IV, V	22 004
Megrim	*Lepidorhombus* spp.	Sub-area IV, VII, VIII	20 365
Flounder	*Platichthys flesus*	Sub-area III, IV	14 478
Lemon sole	*Microstomus kitt*	Sub-area IV, V, VII, VIII	13 402
Witch	*Glyptocephalus cynoglossus*	Sub-area IV, V, VII	12 630
American plaice	*Hippoglossoides platessoides*	Sub-area I, II	10 333
Turbot	*Scophthalmus maximus*	Sub-area IV, VII	5 431
Brill	*Scophthalmus rhombus*	Sub-area IV, VII	2 888
Other flatfishes			24 541

shelf are found in the North Sea (570 000 km^2), the Barents Sea (550 000 km^2), the Baltic including the Kattegat and Skagerrak (420 000 km^2) and around the British Isles (380 000 km^2). Most flatfish fisheries occur in water depths shallower than 200 m. The main exceptions are those for Greenland halibut which can be caught at depths down to 2000 m and witch are often fished in depths between 200 and 500 m. Few commercially exploited flatfishes occur on the continental slope (>1000 m) and none on the continental rise below 3000 m (Merritt & Haedrich 1997).

The flatfish fisheries in the northeast Atlantic are dominated by species from three families (Table 11.2), the Pleuronectidae (plaice, Greenland halibut, flounder), the Soleidae (common sole) and Bothidae (turbot, brill and megrim). In 1998, landings of flatfish amounted to 296 800 tonnes, with three species, plaice, common sole and Greenland halibut contributing more than 50% to the total weight landed. An additional six species – dab, megrim, flounder, lemon sole, witch and American plaice – were also important on the basis of the quantity landed. Economically valuable species such as turbot and brill contribute significantly to the profitability of local fisheries even though caught in smaller quantities.

During the early part of the twentieth century, more than half the total landings of flatfishes in the northeast Atlantic consisted of plaice, which are widely distributed from the Barents Sea to the Mediterranean, although their abundance is much reduced south of the Bay of Biscay, 45°N. The main fisheries for plaice are those in the North Sea (ICES sub-area IV) and the adjacent areas including the Skagerrak/Kattegat (IIIa) and eastern English Channel (VIId) (Fig. 11.4). There are also substantial landings of plaice from around Iceland (Va). In the North Sea and around the British Isles, plaice are mainly caught by beam trawlers from the Netherlands, Belgium and the UK. The fishery occurs throughout the year. There is also a directed fishery by seine and gill net vessels fishing in the eastern North Sea and an important gill net fishery for common sole, plaice and turbot along the Danish coast. Off Iceland, plaice are taken as part of a mixed demersal fishery by stern and otter trawlers. The main landings occur in the south and southwest of the island from April or May through to October (Arnason 1995).

The next most common pleuronectid species in commercial landings in recent years has been the Greenland halibut. This species has a more northerly distribution (Fig. 11.4) and is mostly caught in water depths below 200 m. In the northeast Atlantic, Greenland halibut

246 *Flatfishes*

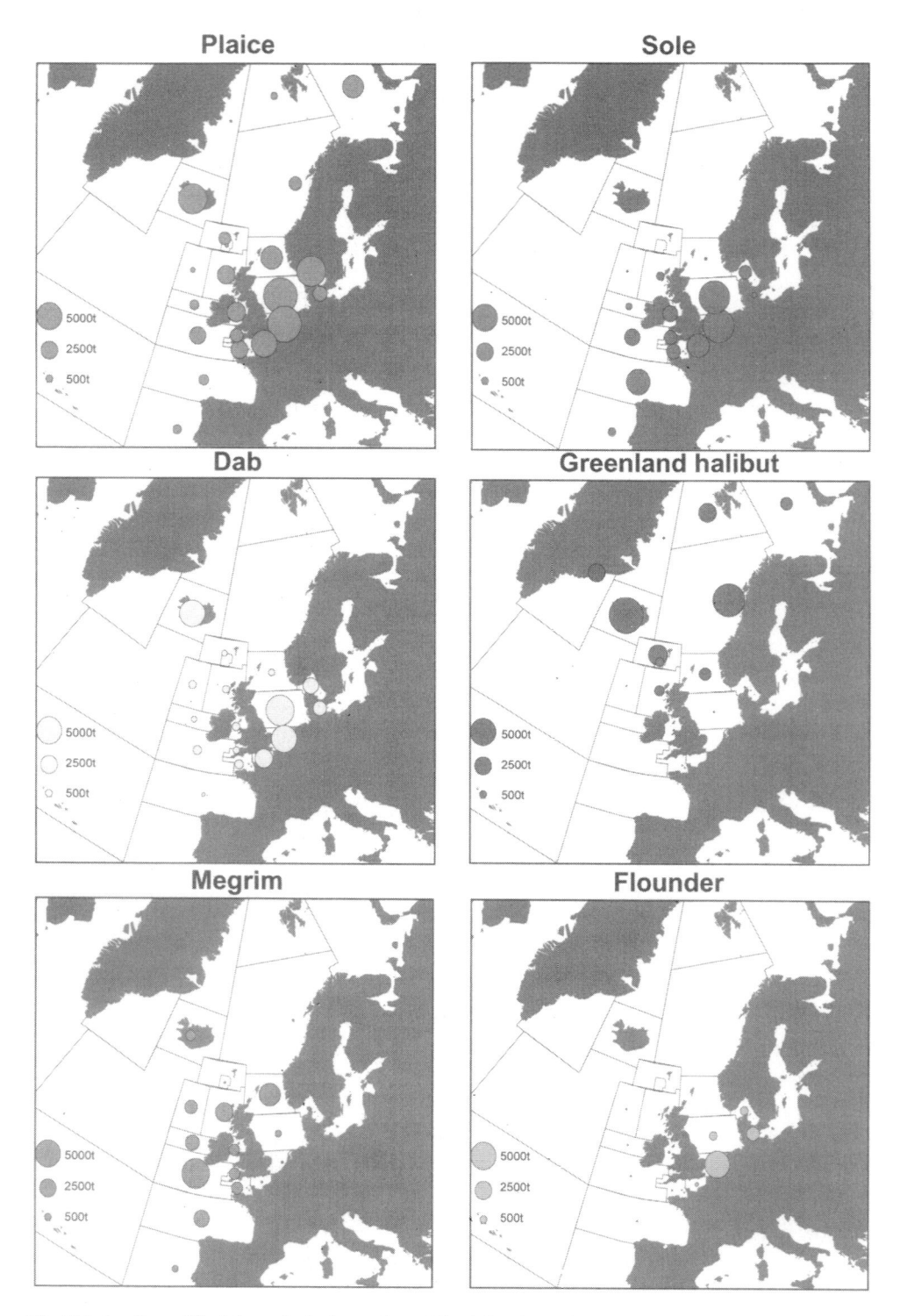

Fig. 11.4 Landings of flatfish species in the northeast Atlantic by ICES sub-area or division in 1998.

occurs in deep water from at least 60°N to the Arctic Ocean north of 70°N in water depths down to 2000 m (Bowering & Nedreaas 2000). It occurs in abundance on the continental slope off the coast of Norway and in the Barents Sea (Godø & Haug 1989; Albert et al. 1998; Bowering & Nedreaas 2000). Peak abundances in the northeast Atlantic are found between 400 m and 1000 m at water temperatures between 0°C and 6°C (Bowering & Nedreaas 2000). Large-scale fisheries occur in ICES division IIa along the Norwegian shelf and into the Barents Sea (Fig. 11.3). There are also extensive fisheries in deep water off the Icelandic shelf (Va), on the Faroes plateau (Vb1) and off east Greenland (XIVb). Fishing for Greenland halibut is by longline and increasingly trawling on the continental shelf. In the Norwegian Sea and Barents Sea the fishery is mainly by Norwegian vessels with minor landings from French and Russian Federation vessels. The Icelandic fishery is almost entirely prosecuted by vessels from Iceland, whereas off Greenland, vessels from Norway, Germany, Iceland and Greenland are all involved. Gill net fisheries for Greenland halibut occur off the coast of Greenland but are mainly concentrated in the northwest Atlantic.

In contrast to the distribution of plaice and Greenland halibut, the commercially valuable common sole has a more southerly distribution, reaching its northern limit around the coasts of England and Denmark and extending southwards to the Mediterranean and off North Africa. The largest population of common sole occurs near the northern limit of its range in the North Sea and in exceptionally cold winters, such as 1948 and 1963, large mortalities of juveniles and adult common sole can occur (Woodhead 1964). Targeted fisheries for common sole extend over much of its distribution area from east of Denmark in the Skagerrak and Kattegat to the coast of North Africa (Fig. 11.4). The main landings are from the southern and central North Sea (IVbc), which accounted for about 60% of the landings in 1998, eastern English Channel (VIId) and Biscay (VIII). There are also significant landings from the Irish Sea (VIIa). In the North Sea, the fishery for common sole overlaps those for plaice in most of the shallow areas south of 55–56°N. A large fleet of beam trawlers mainly from the Netherlands, Belgium and England targets common sole in the North Sea, the English Channel and during the spring spawning season in the area west of the British Isles. There is also a large Danish fishery for common sole in the eastern North Sea using fixed gill nets. In Biscay, common sole is fished by trawlers and by vessels using gill nets. Because common sole spawn inshore in many parts of its distribution it is an important part of the catch for inshore fishing vessels during the spring and early summer. Small vessels target common sole using trawls and fixed tangle nets, especially in the eastern English Channel (VIId).

A number of other species are landed in large quantities in the northeast Atlantic including megrim, witch, lemon sole, dab, flounder and American plaice. These species are usually taken as by-catch in flatfish or demersal fisheries. Megrim and witch occur down to depths of 500 m and are taken as part of a mixed demersal fishery with European hake (*Merluccius merluccius*) and angler fish (*Lophius* spp.) in the northern North Sea, and the areas off southwest Ireland and western Scotland and are often discarded because of their relatively low commercial value. Lemon sole, dab and flounder are caught throughout the northeast Atlantic north of Biscay, mainly as by-catch in trawl fisheries for plaice and common sole. Dab is common in shallow waters from Biscay to the Barents Sea and off Iceland, reaching highest abundance within the 50 m contour (Henderson 1998). In the North Sea, dab was estimated to be the third most abundant fish species after European sprat (*Sprattus sprattus*) and sandeel (Ammodytidae) (Yang 1981). Flounder occur in shallow waters from the Mediterranean to the Barents

Sea. They are a predominantly coastal species occurring abundantly in brackish water and are found penetrating deep into estuaries. American plaice range north from the Dogger Bank in the North Sea, along the Norwegian coast to Spitzbergen and the Barents Sea, westward to Iceland and East Greenland and eastward into Kiel Bay in the western Baltic (Walsh 1994). Populations are also found on the Faroe Bank, in the Clyde Sea, Celtic Sea and Irish Sea. It is taken as a by-catch and generally discarded at sea in most areas. Two further species, which are economically important in flatfish fisheries in the northeast Atlantic, are turbot and brill. Their distribution broadly overlaps that of plaice and common sole and they tend to be widely taken as by-catch in beam and otter trawl fisheries as well as in directed fisheries using tangle nets. Although the total quantities landed are relatively low compared with other commercial flatfish, their high market price makes them a valuable part of the landings.

11.2.3 Southern Atlantic

11.2.3.1 Southwest Atlantic

Approximately 45 flatfish species belonging to six families (Pleuronectidae, Achiridae, Bothidae, Paralichthyidae, Cynoglossidae and Achiropsettidae) occur in the southwest Atlantic southward to the Amazon river. Not all of them are of commercial importance due to their small size or low abundance. The paralichthyids are the most economically important and have a high price in the market (Table 11.3).

The Patagonian flounder, *Paralichthys patagonicus*, is widely fished on the continental shelf (Fig. 11.5) from Rio de Janeiro to at least as far south as northern Patagonia at 43°S (Díaz de Astarloa 1994). It is the most economically important paralichthyid flounder in the southwest Atlantic. Although flatfish species are not distinguished separately in commercial landings throughout the region, it was demonstrated that Patagonian flounder is the most abundant species caught. In Argentina, for example, it represents 70% of the flatfish landed (Cousseau & Fabré 1990; Fabré & Díaz de Astarloa 1996). Other species of Paralichthyidae taken commercially include *Paralichthys orbignyanus* which is a shallow-water flatfish occurring from Rio de Janeiro southward to Patagonia and *Xystreurys rasile* which is widely distributed from northeastern Brazil to at least as far south as the continental shelf off San Jorge Gulf (47°S). In Argentina, *X. rasile* and *Paralichthys isosceles* represent between 2.4% and 2.6% of the total fish species sold (Fabré 1992; Fabré & Díaz de Astarloa 1996) and both are fished offshore.

Table 11.3 Landings of the main commercial flatfishes from the southwest Atlantic in 1998 (FAO 2000)

Species	Main area fished (see Fig. 11.5)	Landings (t)
Paralichthys brasiliensis	North-northeastern Brazil (3–13°S)	643
Paralichthys spp.	Central-southeastern Brazil (13–25°S)	428
Paralichthys patagonicus, P. orbignyanus	Southern Brazil (25–33°S)	529
Paralichthys patagonicus, P. orbignyanus, P. isosceles	Uruguay	496
Paralichthys patagonicus, P. orbignyanus, P. isosceles, Xystreurys rasile	Northern Argentina (35–41°S)	7346
Paralichthys patagonicus, P. orbignyanus, P. isosceles, Xystreurys rasile	Northern and central Patagonia (41–47°S)	1411

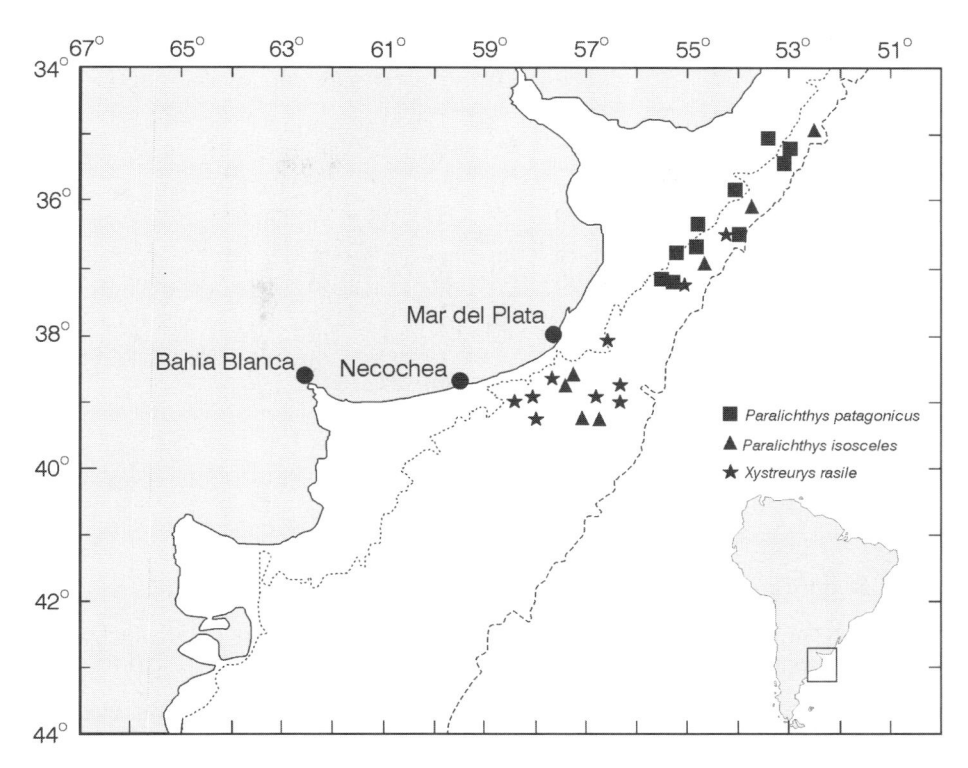

Fig. 11.5 Principal concentrations of paralichthyid flounders in northern Argentina and Uruguay. The dotted lines are depth contours.

The Brazilian flounder *Paralichthys brasiliensis* is a shallow-water species commonly inhabiting bays, beach zones and estuarine areas on muddy and sandy bottoms in northeastern Brazil, where it is regarded as one of the finest marketable flatfish species (Carvalho-Filho 1999).

In the southwest Atlantic, the main regions of flatfish captures in Brazil are the southeastern and southern regions. Twin-rig trawling, in which two nets are towed together by a single vessel, is extensively used in fishing for Patagonian flounder. Fishing is carried out at 20–80 m depth between the autumn and spring. In recent years, this method of fishing has resulted in considerable landings of Patagonian flounder and *P. orbignyanus* (Haimovici 1997). In Uruguay, twin-rig trawling has been shown to be highly effective, taking 54.5% of flounders compared with 0.1% by conventional otter trawl (Arena *et al.* 1992). Since the 1990s in the Rio Grande do Sul area of Brazil twin-rigs have contributed more than 65% of total flatfish captures, followed by pair trawl 11.2%, artisanal fisheries 9.6%, otter trawl 3.3% and entangling nets 1.6% (Haimovici & Mendonça 1996).

In Argentina, 94% of the flatfish landings come from harbours in the north of the country with Mar del Plata being the most important. Coastal vessels of 18–27 m in length and 60–100 HP take 60% of the flatfish landings. Trawling is the most important method used in fishing for flatfishes, although small landings are also made in central and northern Patagonia using entangling nets. The seasonality of landings is shown in Fig. 11.6. Maximum flatfish landings occur within the reproductive season of the flatfish species (between October and March) when the species aggregate to spawn.

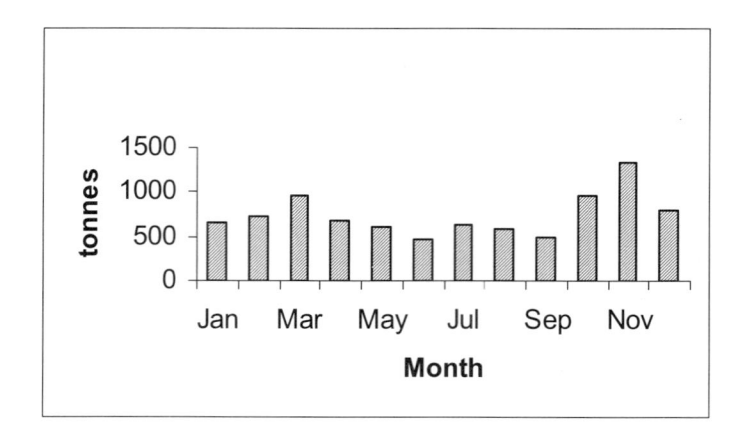

Fig. 11.6 Average monthly flatfish landings of *Paralichthys* spp. and *Xystreurys rasile* showing seasonality of catch over a 10-year period in Argentina.

11.2.3.2 Southeast Atlantic

Seven flatfish families (Psettodidae, Citharidae, Bothidae, Achiropsettidae, Paralichthyidae, Soleidae and Cynoglossidae) are present in the southeast Atlantic (FAO area 47), and they include approximately 35 species. Only the soleids, bothids and some species of cynoglossids constitute fishery resources of commercial importance in the area.

Of the soleids, species of *Austroglossus* are of notable commercial importance in the southeast Atlantic, especially in South Africa where the sole fishery is based on mud sole, *Austroglossus pectoralis* and the west coast sole, *A. microlepis*. Although soles account for a minor percentage of the annual South African trawlfish catch, they are the most valuable fish per unit weight landed and are therefore the principal target of the small trawlers, particularly along the south coast (Payne 1979). The mud sole attains a length of 58 cm and is caught along the south coast of South Africa, between longitudes 20° and 27°E. It is commercially the most important of the South African flatfishes (Heemstra & Gon 1995).

There are three geographically distinct stocks of west coast sole. Commercially exploitable quantities of an Angolan stock lie mainly off the mouth of the Zaire River. A northern stock occurs between 20° and 25°S, southwest and northwest of Walvis Bay, and a southern stock between 28° and 30°S, off the mouth of the Orange River. The northern stock is found in depths ranging from 75 to 300 m, whereas the southern stock has only been fished in depths of 50–100 m (Heemstra & Gon 1995). The depth range of the Angolan stock is unknown, but it is likely to be reasonably shallow (Payne & Badenhorst 1989).

The Bothidae include a number of species which are fished in southern Africa, although they are of relatively minor commercial importance. Species of *Arnoglossus* are moderate-sized flatfishes (15–25 cm TL) which are fished in Namibia. Other bothids, such as the wide-eyed flounder *Bothus podas* and the pelican flounder *Chascanopsetta lugubris,* attain larger sizes (40–45 cm TL). In the region, wide-eyed flounder is distributed as far south as Angola where it has commercial importance in fisheries as well as in aquaria. Although pelican flounder is also reported for Angola, Namibia and South Africa, only in Namibia is it considered of any commercial importance (Hensley 1995). The Zanzibar tonguefish *Cynoglossus zanzibarensis* is one of the few cynoglossids fished commercially. It grows to a length of 32 cm and is common around South Africa, particularly on the south coast. It is regarded by some as one of the best eating 'soles' in the area, and is caught as a small by-catch in the demersal trawl fishery of South Africa. However, because it is a small component of the demersal catch, no catch statistics separating it from true soles are available.

11.3 History of exploitation

11.3.1 Northwest Atlantic

The developments of flatfish fisheries in Atlantic Canada and United States have many similarities. In the United States, large-scale commercial bottom trawl fisheries began during the first half of the twentieth century. These fisheries evolved from sailing vessels using small trawls or hook and line to diesel- and gasoline-powered vessels towing otter trawls. Although hook and line fishing of Atlantic halibut has continued since the 1800s, the US fisheries for flatfishes began in the 1920s when commercial quantities of yellowtail and winter flounders were taken by otter trawls fishing for haddock (*Melanogrammus aeglefinus*) on Georges Bank (Murawski *et al.* 1997). Canada banned otter trawling on the Atlantic coast from 1928 until after World War II (1947) because inshore fishermen believed they would destroy the Atlantic cod (*Gadus morhua*) stocks. In the 1930s, longlining of plaice began in the Gulf of St Lawrence. Danish seining began in the 1940s on the south coast of Newfoundland and moved to the southern Gulf of St Lawrence by 1958, where it is still the primary fishing gear. After lifting the otter trawl ban in 1947, the Canadian offshore fisheries rapidly expanded during the 1950s and the 1960s. A similar expansion occurred in the US domestic fisheries.

Following War World II, there was an expansion of the 'distant-water' fishing fleets outside the 22-km national boundaries of both Canada and the United States until the late 1970s. These factory freezer trawler fleets originated mainly from the former USSR, German Democratic Republic, the Federal Republic of Germany, and Poland. Japan was also a player in the US fisheries, while France, Spain and Portugal were additional players in the Canadian fisheries. Landings were primarily of American plaice and to a lesser extent a mixture of other flatfishes from stocks on the Grand Bank, Scotian Shelf and Georges Bank. As a result of this huge expansion in effort by the 'distant-water' fleets off the Canadian and US coasts, landings – consisting primarily of Atlantic cod, haddock, flatfishes, redfishes (*Sebastes* sp.) and Atlantic herring (*Clupea harengus*) – greatly increased. Overall landings peaked during the late 1960s with an average of 4 million tonnes being taken during that decade (FAO 1997). In 1955, the total landings of flatfishes were 66 000 t and, by 1968, it had peaked at 315 000 t (41% American plaice). By 1974, a sharp decline in the abundance of many flatfish stocks had occurred followed by a subsequent decrease in landings to 223 000 t (Fig. 11.7).

Farther north, in the waters off Baffin Bay and the Davis Strait, between Canada and Greenland, an otter trawl fishery began in the 1950s primarily for Atlantic cod (FAO 1997). With declining catches of Atlantic cod, the fleets began exploiting the deeper water Greenland halibut in the Canadian zone (NAFO sub-area 0). From 1965 to 1984, landings of Greenland halibut were taken mainly by otter trawlers from the former USSR and the German Democratic Republic and, by the late 1980s, Faroe Islands and Norway (NAFO 2000). Canadian vessels, along with Russian and Japanese vessels under charter to Canadian fishing

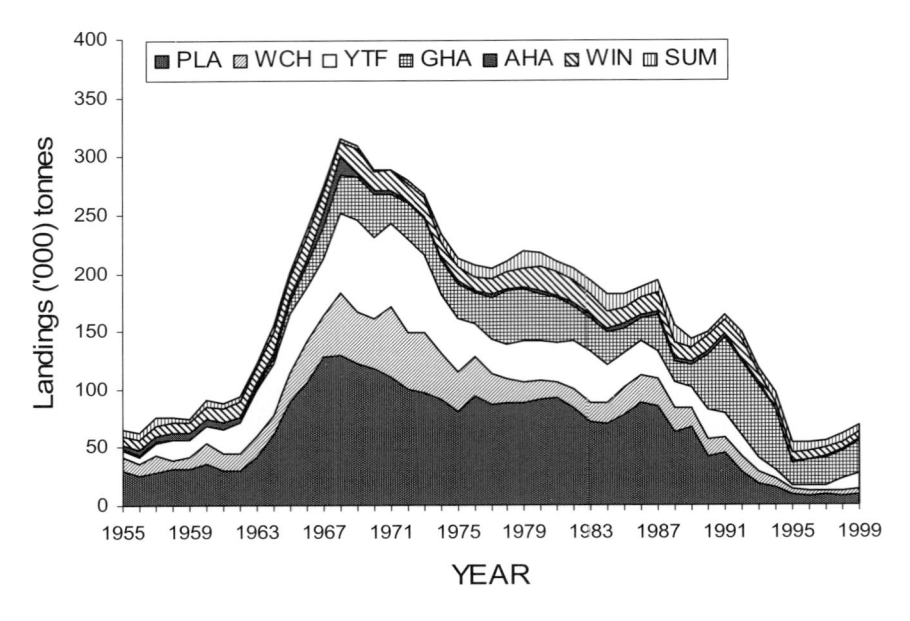

Fig. 11.7 Landings of flatfishes from Labrador to Cape Hatteras (NAFO sub-areas 2–6) from 1952 to 1999. Species are American plaice (PLA), witch (WCH), yellowtail flounder (YTF), Greenland halibut (GHA), Atlantic halibut (AHA), winter flounder (WIN) and summer flounder (SUM).

companies, began fishing in 1990. The offshore fishery on the west Greenland side of Davis Strait (NAFO divisions 1B–1F) began in 1987 with fleets from Greenland, Norway, Russia and Germany fishing primarily with otter trawls. During the 1990s catches remained fairly stable around 5000 t. Almost all of the fishery in this area occurs in deep water west of Nuuk (NAFO division 1D). Combined landings, primarily, in the offshore of Baffin Bay and Davis Strait rose from 17 000 t in 1965 to peak at 24 000 t in 1975 and declined to <1000 t in the 1980s (Fig. 11.8). Landings were relatively stable at around 10 000 t during the 1990s (NAFO 2000). In 1987, a domestic inshore Greenland halibut fishery began in the fjords of northwest Greenland with landings increasing from 7000 t in 1987 to 25 000 t by 1999 (Fig. 11.8).

By 1994, the landings of a majority of flatfish stocks in the northwest Atlantic had declined substantially to 96 000 t, a 70% decrease since the peak in 1968 (Fig. 11.7). Catches of American plaice, which peaked at 130 000 t in the mid-1960s, declined to 15 000 t by 1994. With the rapid decline in landings in the late 1980s/early 1990s in both Canada and the United States, Greenland halibut along the eastern edge of the Grand Bank now accounts for over half of the recent landings (Fig. 11.7) (NAFO 2000). Details of individual regions follow.

11.3.1.1 West Greenland

Fishing vessels from west Greenland began fishing the inshore and offshore areas for Greenland halibut on the Greenland side of Baffin Bay and Davis Strait with otter trawls and, in 1994, with longlines. Longlining catches constituted 28% of the offshore Greenlandic landings of 2600 t in 1999. Longlining is also used in the inshore areas along with gill nets and pound nets. The main commercial inshore fishing grounds for Greenland halibut are in the

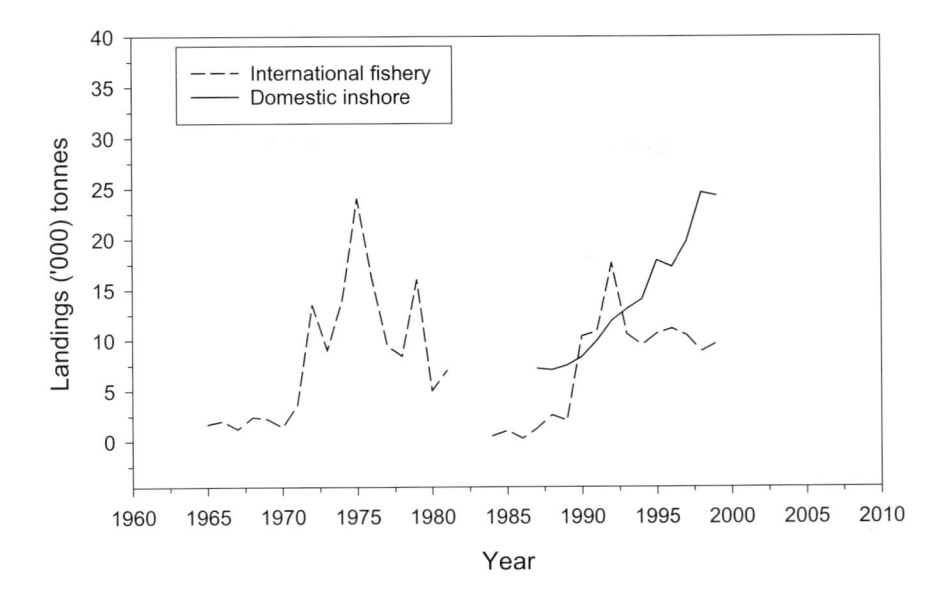

Fig. 11.8 Catches of Greenland halibut from the international fisheries in the Davis Strait and Baffin Bay area between Canada and Greenland (sub-area 0+1) and the domestic West Greenland inshore Greenland (NAFO division 1A).

northern area along the inner sections of the ice fjords of Disko Bay, Uummannaq and Upernavik (NAFO division 1A). Landings from this fishery rose steadily from 7000 t in 1987 to 25 000 t by 1999 (Fig. 11.8). Gill nets (minimum mesh size of 110 mm – half meshes) were introduced into the fishery in 1980 but, by 2000, were banned in favour of the use of longlines.

11.3.1.2 Canada

In the northern part of Atlantic Canada, there has been a small longlining fishery for Greenland halibut in the inshore area of Baffin Island since 1987 with catches ranging from 100 to 400 t (NAFO 2000). An otter trawl fishery for Greenland halibut began offshore (NAFO sub-area 0) in 1990 and, by 1995, a large mesh deep-water gill net fishery was also occurring. Most of the catch are taken by vessels from Newfoundland and Nova Scotia and are reported in the domestic landings in Fig. 11.8. By 1999, these catches consisted of 1900 t from otter trawls, 1900 t from gill nets and 400 t from longlines.

Farther south, the domestic catches of flatfishes have traditionally been taken primarily by otter trawls and, to some extent, by Danish and Scottish seines and gill nets in mixed fisheries on the Grand Bank, St Pierre Bank, in the Gulf of St Lawrence and on the Scotian Shelf. The increase in both capacity and effort of the offshore otter trawl fleets beginning in the 1950s resulted in increased landings throughout the 1960s. Landings peaked at 150 000 t in 1970 before declining to 103 000 t by 1975 (Fig. 11.9) (DFO 2001a). American plaice dominated the landings buoyed by catches from the Grand Bank. Canada's extended jurisdiction in 1977 and a major curtailment of the fishing activity by the 'distant-water' fleets led to great expectations among the national and local governments, and the fishing industry. The result

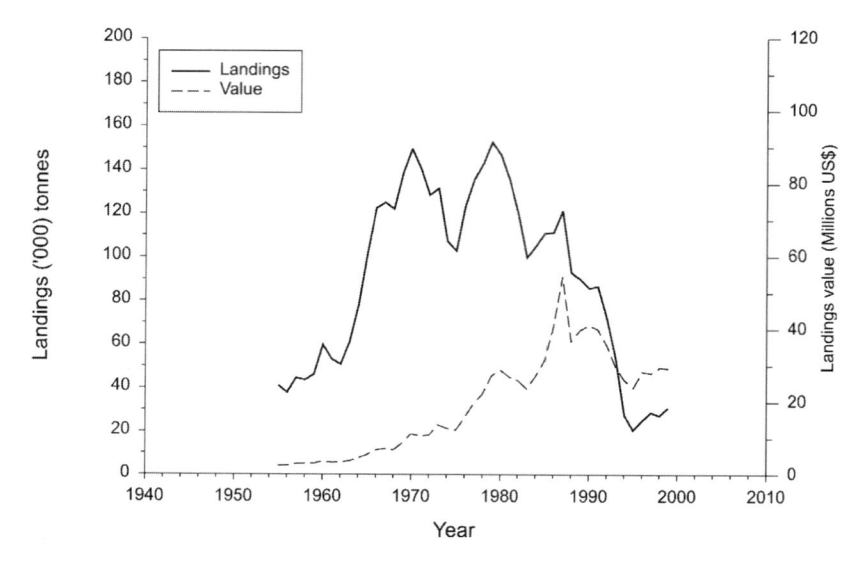

Fig. 11.9 Domestic flatfish landings and value ($US) to the Atlantic Canada.

was a rapid expansion in fishing capacity and effort in all domestic groundfish fisheries and all gear types. Some flatfish stocks had showed signs of recovery during the mid-1970s and landings rose to peak at 153 000 t by 1979. By 1983, landings had declined to 100 000 t and, despite a short-lived reversal in trend in the mid-1980s, underwent a further precipitous decline to 21 000 t by 1995. By the mid-1990s, three of the major flatfish fisheries were under moratorium while most other stocks had reduced total allowable catches (TACs) ranging from 1000 to 3000 t. By 1999, landings had risen to 31 000 t primarily due to improved recruitment during the mid-1990s of Greenland halibut, and the 1998 re-opening of the yellowtail flounder fishery on the Grand Bank.

11.3.1.3 USA

Flatfish landings in Atlantic United States come primarily from the Georges Bank, the Gulf of Maine and the southern New England mobile and static gears fisheries. From the 1940s to the 1960s, there was a build-up of fishing effort in the groundfish fishery. Landings increased sharply from a low of 37 000 t in 1955 to 61 000 t by 1965; however, by 1975 landings had declined to 47 000 t (Fig. 11.10). In 1977, there was rapid expansion in all fleets and gear sectors with the extension of national jurisdiction to 370 km. The otter trawl fishing effort doubled from 1976 to 1985 and remained high until 1993 (Murawski *et al.* 1997). Similar to flatfish stocks in Canadian waters, good recruitment led to a partial recovery in some stocks and a subsequent increase in landings from 1977 to 1983. After the 1983 peak landings of 84 800 t there was a precipitous decline to 21 000 t by 1999 (Fig. 11.10). By the mid-1990s, major stocks of plaice, yellowtail flounder and witch and their fisheries had collapsed.

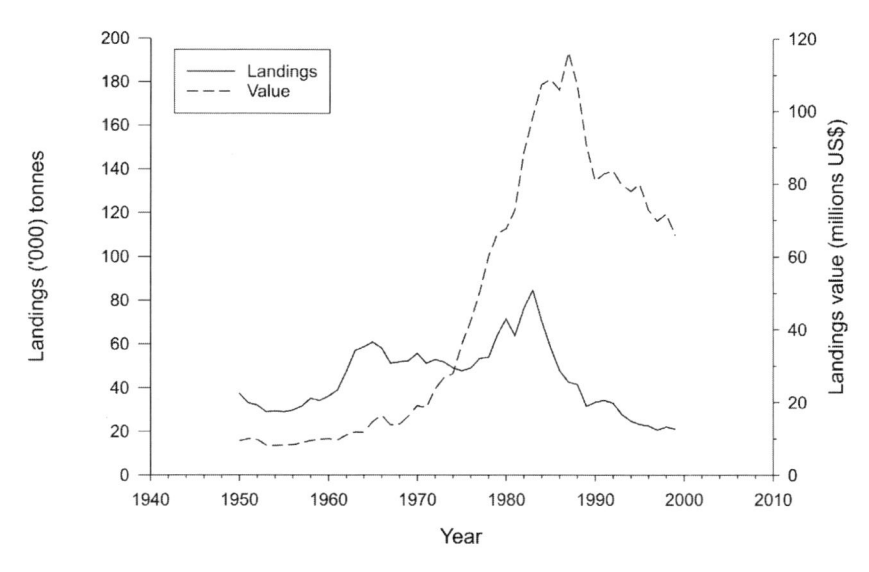

Fig. 11.10 Domestic flatfish landings and value ($US) in the Atlantic United States.

11.3.2 Northeast Atlantic

At the start of the nineteenth century, flatfish fishing was mainly confined to small vessels fishing the inshore waters using drift nets, longlines or handlines. The development of large-scale flatfish fisheries began with the exploitation of plaice and common sole in the North Sea in the 1830s. Trawling for flatfishes was carried out with beam trawls fished by sailing vessels, as the beam was the only method available to keep the mouth of the net open at slow speeds. Beam trawling was confined to shallow waters <100 m in depth because of the difficulty of retrieving the net by hand.

The high catch rates from largely unexploited stocks in the offshore waters of the southern North Sea led to a rapid expansion of fishing and by the late 1830s English sailing trawlers were fishing the whole of the southern Bight (Wood 1956). Over the next 40 years, fishing extended to the productive grounds of the Dogger Bank and then into the northern North Sea from Scotland to the Danish coast. Sailing trawlers in the North Sea were gradually replaced by steam-powered trawlers from the 1880s onwards. In the 1850s, Danish sailing vessels began using anchor seines to fish for plaice, first in the Kattegat, and by the 1900s this method was being widely used along the North Sea coast of Denmark (Holm 1996). In the early 1900s, Icelandic trawlers began exploiting the abundant Atlantic cod and plaice stocks off the coast of Iceland. These vessels were joined by more powerful foreign trawlers from England and other European countries which began fishing in Icelandic waters, the Faroes and along the Barents Sea coast. By 1928 the important fisheries of Bear Island and Spitzbergen in the Barents Sea were opened up. A more recent development has been the extension of fisheries into deeper waters for species such as Greenland halibut and megrim.

Landings of the five main species in the northeast Atlantic since 1950 are shown in Fig. 11.11. All five species show an increase in landings in the early 1970s and again in the late 1980s and all reflect a steep decline in recent years. Landings of plaice in the early 1900s

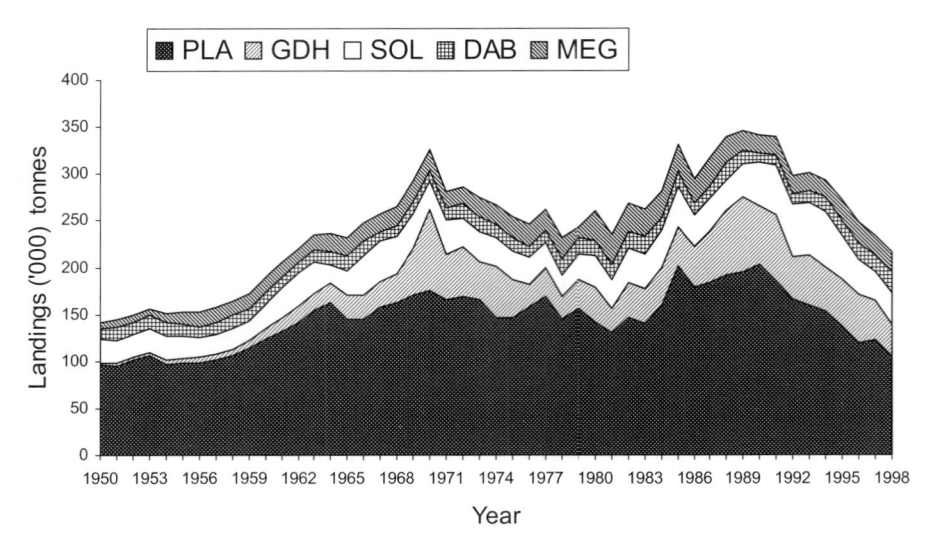

Fig. 11.11 Trends in landings of flatfish species from the northeast Atlantic (ICES sub-areas I–XIV) from 1950 to 1998. Plaice (PLA); Greenland halibut (GDH); common sole (SOL); dab (DAB); megrim (MEG) (FAO 2000).

was already as high as 50 000 t from the North Sea alone (Rijnsdorp & Millner 1996). They remained at around this level apart from the war years when access to the North Sea was severely restricted. Catches rose steeply after World War II, particularly in the North Sea, in line with the development of an intensive beam trawl fishery, and reached a peak of over 175 000 t by the early 1970s. There was a decline in landings during the 1970s before landings increased to a historically high level at 203 000 t in 1990. Since then landings have declined sharply, as stocks in the North Sea and adjacent areas have been reduced through over-exploitation.

The fishery for Greenland halibut is rather more recent in comparison with the long-established fisheries for plaice and common sole. Although a longline fishery had been in existence off the coast of Norway since the 1930s (Bowering & Nedreaas 2000), the fishery remained at a low level until trawlers capable of fishing at depths below 200 m began exploiting stocks of demersal species along the continental shelf edge in the 1960s. Russian vessels began fishing for Greenland halibut off Iceland in the 1960s (Troyanovsky & Lisovsky 1995). Landings peaked in 1970 at 86 500 t before declining steeply to below 30 000 t in the mid-1970s. Landings increased again up to 70 000 t in 1991. In northeast Arctic waters new management measures were introduced in 1992 to reduce the level of fishing on the stocks. Only a directed fishery by coastal longline and gill net vessels is now permitted. However, this has not prevented the continuing landings of Greenland halibut at levels which are in excess of the agreed TAC.

The development of an intensive targeted fishery for common sole began in the 1950s when vessels from the Netherlands began using beam trawls rigged with heavy chains to increase the common sole catch. In the following years there was a rapid increase in the size and power of vessels fishing for common sole, first in the North Sea and by the 1980s in all beam trawl fleets around the British Isles (Daan 1996; Rijnsdorp & Van Leuween 1996). Landings in the North Sea increased from <4000 t in 1905 to a peak of 33 000 t in 1968

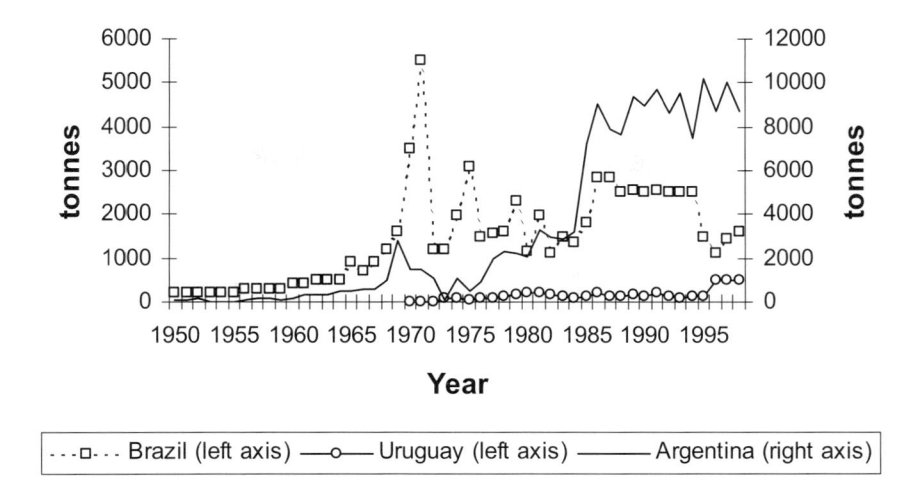

Fig. 11.12 Annual landings (1950–1998) of paralichthyid flounders in the southwestern Atlantic.

(Millner & Whiting 1996). Total landings from the northeast Atlantic (Fig. 11.11) reflect the changes in the North Sea. Landings increased to 43 000 t in 1967 before decreasing steadily to a minimum in 1978. Since then landings increased to historically high levels in the early 1990s. Over recent years, common sole landings have shown a steep decline as stocks in the North Sea and west of Britain have decreased.

11.3.3 Southern Atlantic

The development of flatfish fisheries in Brazil has been low in relation to the available area of the continental shelf. The fishery intensified in the late 1950s through pair trawlers landing in Santos (24°S) and increased in the late 1960s when the fishing industry started to be subsidised. In Brazil, paralichthyid flounders, mostly *P. patagonicus* and *P. orbignyanus,* are intensely exploited. Landings of these species reached 500 t annually in the late 1970s and early 1980s in southern Brazil (Fig. 11.12). They reached a maximum of 1892 t in 1989 but decreased considerably thereafter.

Landing of flatfish species in Uruguay were low up to the beginning of the 1990s, but have increased sharply to reach around 500 t annually since then. In Argentina, annual landings reached up to 3000 t in the early 1980s and considerably increased in the mid-1980s, exceeding 10 000 t annually between 1995 and 1997 (SAGPyA 1999).

Catches of soles in the southeast Atlantic since 1965 are shown in Fig. 11.13. Landings of west coast sole from Namibia have only been recorded separately since 1987 and have shown a sharp increase in the period since 1992. Flatfish landing statistics for Angola are given for all flatfish species, with no species discrimination. Since 1965 most annual landings have been <400 t but there have been periods of high flatfish landings (e.g. 700 t in 1970, 784 t in 1993, and a maximum of 928 t in 1998). Off South Africa, landings of west coast sole peaked at nearly 2000 t in 1972 and 1982 but have declined steeply since then. Landings of mud sole have remained relatively stable at around 800 t since 1984.

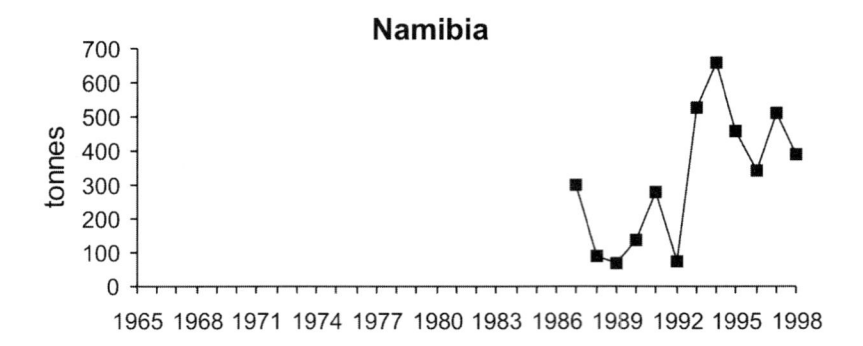

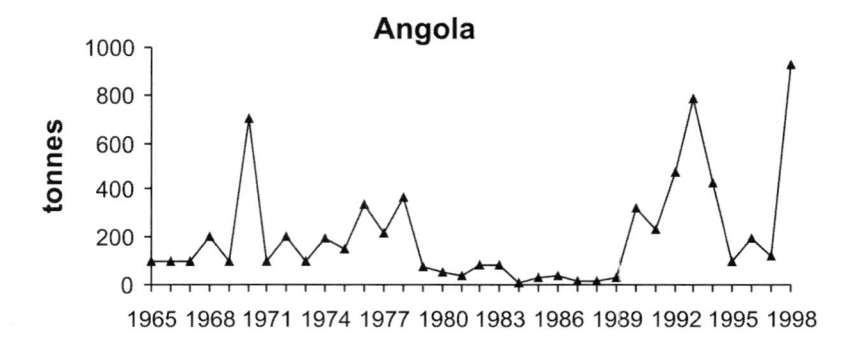

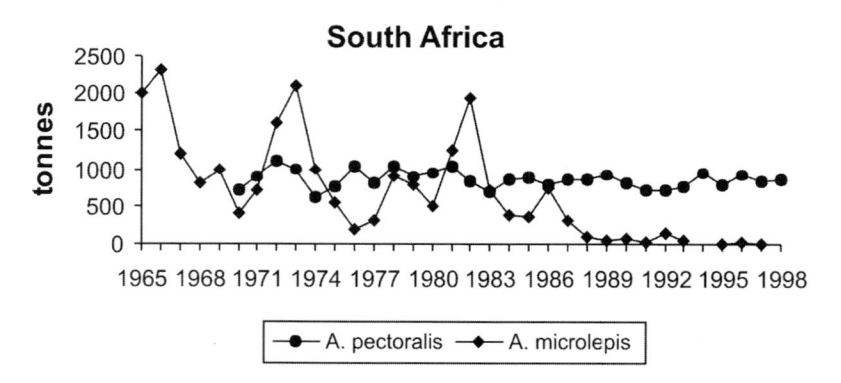

Fig. 11.13 Landings of flatfishes in the southeast Atlantic since 1965. Upper panel, west coast sole in Namibia; centre panel, mixed flatfish species in Angola; lower panel, mud (*A. pectoralis*) and west coast sole (*A. microlepis*) in South Africa.

11.4. Economic importance

11.4.1 Northwest Atlantic

The otter trawl fisheries for flatfishes and other groundfish species have produced the greatest percentage of total revenue earned in the harvest sector of Canada and the USA. The descriptions that follow use the year 1998 as a reference to present in detail the species composition by value ($US) for comparison across regions. Employment levels in the harvest sector directly related to flatfish landings are difficult to estimate due to the mixed nature of the otter trawl fishery.

11.4.1.1 West Greenland

Greenland harvesters in the offshore either process their catch onboard or ship it to another country for processing. No other countries fishing Greenland halibut land their catch in Greenland. Values are not available for Greenland inshore landings.

11.4.1.2 Canada

As the domestic fishery developed, particularly on the offshore banks, the value of flatfish landings rose steadily from $2.3 million in 1955 to a peak of $54.9 million in 1987 (Fig. 11.9) (DFO 2001a). From 1988 onward, the landed value declined sharply by 50% to $23.9 million by 1995.

In 1998, the flatfish fisheries represented 19% of the landed groundfish weight (146 000 t) and 22% of its value ($109 million). Greenland halibut constituted 45% of the flatfish catch and 51% of the landed value (Fig. 11.14). Although the landings of yellowtail flounder and American plaice together constituted 35% of the weight, the value of Atlantic halibut generated the second highest revenue ($5.5 million) in the 1998 fisheries (Fig. 11.14). In 1999, the landed value of the fishery had risen to $29.2 million. Most of Canada's production is processed as frozen block, fresh fillets, steaks and fish fingers both for home consumption and the commercial restaurant trade. The majority of this production is exported to the United States.

11.4.1.3 USA

The landed value of the domestic flatfish fisheries rose sharply from a low of $9.3 million in 1950 to a peak of $116 million in 1987, before declining to $65.8 million in 1999 (Fig. 11.10). In 1997, flatfish landings made up 18% of the total groundfish landed weight and 44% of the landed value ($134.7 million). Although most of the landings come from traditional mobile and static gear fisheries, there is also a large recreational angling fishery for some nearshore species. In 1990, commercial flatfish landings were valued at $80.7 million, while the recreational fishery (boat rental, bait, permits, lodgings, food, vessel rental, gas, etc.) for winter flounder and summer flounder generated $196 million for the economy (NOAA 1991).

Much of the domestic production is for the local market of which, in 1998, 31% of the flatfish landings and 37% of the value ($26.5 million) came from the summer flounder fishery (Fig. 11.14). The flatfish fisheries in the Gulf of Mexico area are relatively small and the catch is landed as flatfishes unspecified. In 1998, 293.5 t, mainly from otter trawls, were landed with a value of $1.1 million. In recent years Atlantic halibut landings have decreased and in 1998 the fishery yielded 8.4 t valued at $52.4 thousand. Similar to Canada, flatfishes are processed as frozen block, fresh fillets, steaks and fish fingers both for home consumption and the commercial restaurant trade.

260   *Flatfishes*

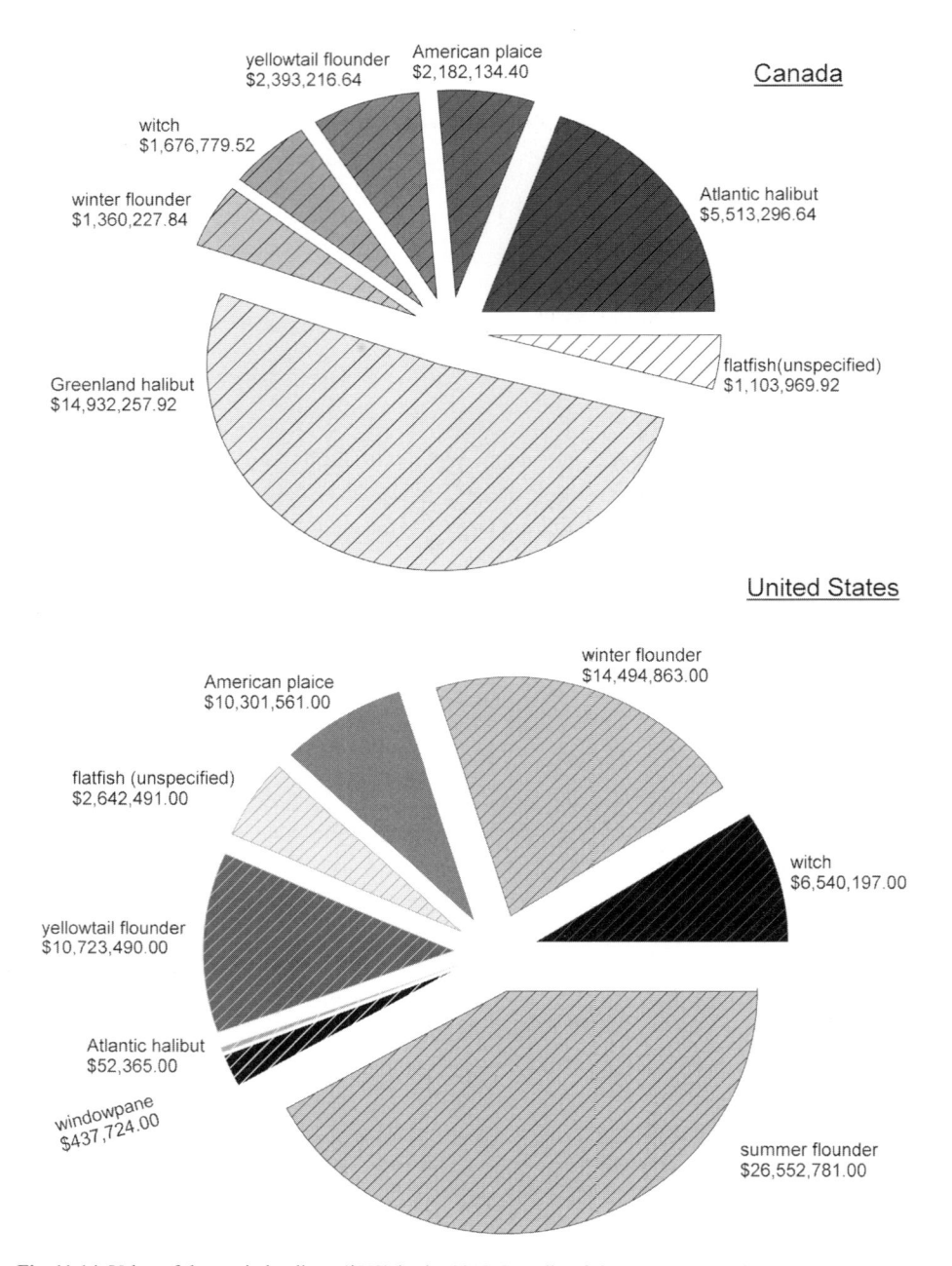

Fig. 11.14 Value of domestic landings ($US) in the 1998 Canadian fishery (upper panel) and in the 1998 United States fishery (lower panel).

11.4.1.4 Employment

Employment in the fish harvesting and processing industries of Atlantic Canada and the United States has been severely affected by the reduction in all finfish resources during the late 1980s and 1990s. Flatfish fisheries are generally a part of a mixed groundfish fishery in both

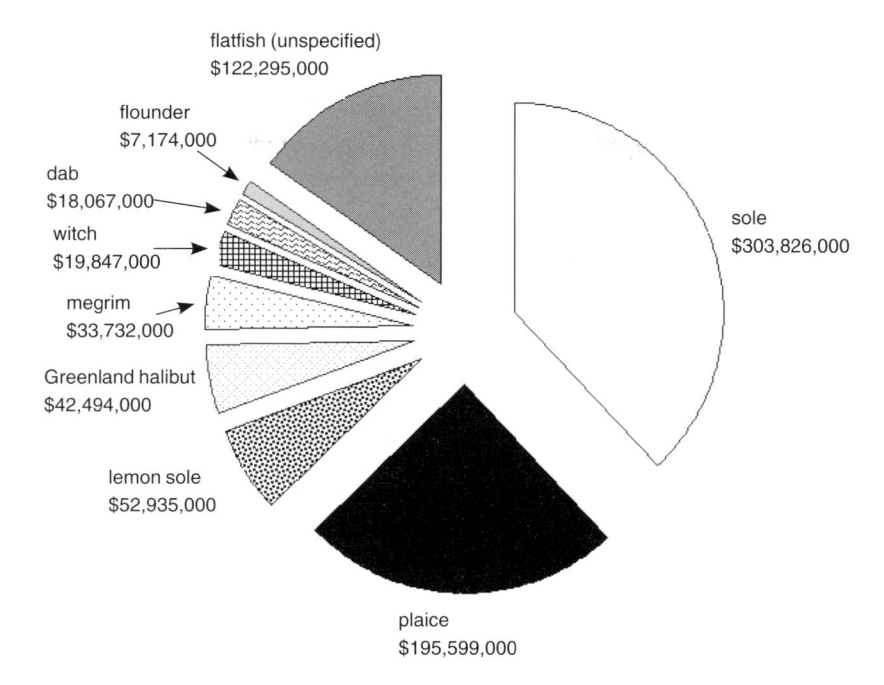

Fig. 11.15 Value of flatfish landings ($US) from the northeast Atlantic in 1998 (total = $796 million).

areas so it is difficult to obtain employment statistics directly related to their contribution. Such data are also not available for the Greenland fishery. However, some general economic trends in the groundfish fishery are considered here.

In Canada, the total finfish fishery landings declined in volume by 51% from 1.4 million tonnes in 1986 to 0.68 million tonnes by 1996. Several groundfish stocks, including three flatfish stocks, were placed under moratorium. As a result, employment in the industry dropped from 45 000 fishermen and fish plant workers in 1988 to about 38 000 by 1996 (DFO 2001b). This led to a reduction in the offshore fleet and the closure of several processing plants.

In the United States, the total finfish and shellfish fishery landings declined from 2.1 million tonnes in 1984 to 1.6 million tonnes by 1999. The flatfish fishery represented 4% of the overall landings in 1984; however, by 1999 it had dropped to 1%. As a consequence, the number of employees in the processing plants in the New England area and the area from Cape Cod to Cape Hatteras decreased by 40% from 17 485 persons in 1984 to 10 378 by 1993 (NOAA 1996). Since the 1980s, the employment in the fish harvesting sectors has also declined substantially (Murawski *et al.* 1997).

11.4.2 Northeast Atlantic

There are no reported figures for the value of flatfish landings covering the whole of the northeast Atlantic. However, an approximate figure can be derived from the total landings and average price per kilo on the UK market in 1998. The results are shown in Fig. 11.15 and amount to approximately $800 million. This compares with a total in excess of $6600 million

Table 11.4 Employment and vessel statistics for 1999 in EU waters (adapted from Anonymous 2000)

	Employment	Number of vessels	Fleet capacity (tonnage, '000s)
Belgium	700	125	22
Denmark*	4 500	1 468	90
France	15 500	5 906	167
Germany	2 800	2 305	68
Iceland*	6 200	1 928	188
Ireland	5 500	1 709	74
Netherlands	2 300	415	153
Norway*	21 300	13 199	293
Portugal	26 700	10 933	113
Spain	79 400	18 852	547
UK	14 700	7 500	219
All countries	179 600	64 340	1 934

*Data for 1998.

for landings of all fish species (Anonymous 2000). Common sole is the most valuable species, contributing 34% of the total value in the northeast Atlantic despite being only 10% of the landings by weight. Most of the flatfishes landed are for human consumption as fresh fish, fillets or added value fish products. The contribution of fishing to the Gross Domestic Product varies considerably within countries from 0.5% in Norway to 15% in Iceland.

11.4.2.1 Employment

Figures for employment within the flatfish sector are not available separately but information for all fisheries is shown in Table 11.4. The data do not cover all countries but provide some indication of the relative importance of the fisheries sectors in different countries. By far the largest fishing industry in terms of employment is Spain with nearly 80 000 people directly involved in the catching sector. Spain also has the largest fleet both in numbers of vessels and overall capacity.

11.4.3 Southern Atlantic

Flatfish represent a relatively small fraction of total landings in both the southwest and southeast Atlantic countries compared with species such as hake. For example, in the last 20 years, hake has contributed more than 60% of the total catch of the Argentine fleet (Bezzi *et al.* 1995), while flatfish landings represented between 1% and 2.1% of the catch in the same period. Nevertheless, since the 1980s the value of the flatfish catch has increased greatly from less than $20 million to more than $65 million in export income for 1995 (Fig. 11.16). This represents 7.2% of total fish export income (SAGPyA 1996). In the southeast Atlantic, hake accounts for about 70% of the weight landed, but soles remain the most important finfish species in terms of unit value (Payne 1985).

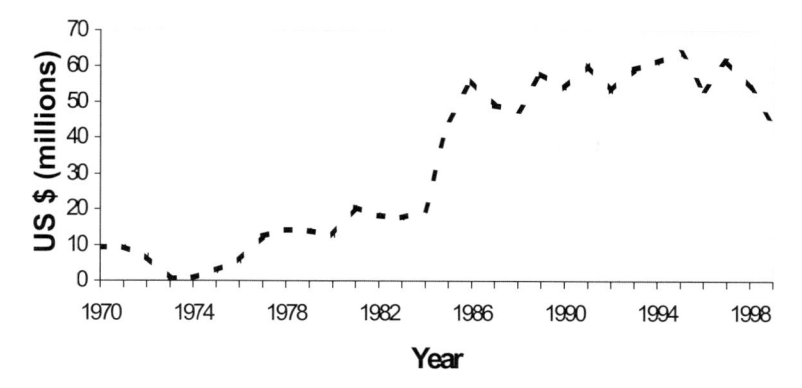

Fig. 11.16 Export income ($US) of flatfishes in Argentina.

11.5 Management

11.5.1 Northwest Atlantic

From 1952 to 1957, the management of fisheries beyond the 22-km national boundaries of Canada and the USA was carried out by the International Commission for the Northwest Atlantic Fisheries (ICNAF). During the early 1970s, ICNAF introduced increasingly restrictive management measures, through reduced quotas and increased mesh size regulations, to curtail the severe declines in stock size of many groundfish species. These actions together with the extension of the Canadian and American national jurisdictions to 370 km in 1977, which excluded the distant deep-water fishing fleets, led to a partial recovery of the flatfish resource (Murawski *et al.* 1997). However, the new EEZs also led to a build-up of domestic fishing effort in both countries and the recovery was short-lived. By 1982 landings were again declining (Figs 11.9 and 11.10). Off Canada, the EEZ did not take in all of the Grand Bank nor the Flemish Cap, a smaller bank east of the Grand Bank (Fig. 11.2). This resulted in transboundary groundfish fisheries which came under the management of the Northwest Atlantic Fisheries Organisation (NAFO), which replaced ICNAF in 1979. These fisheries include the Greenland halibut fisheries extending from the Davis Strait to the Southern Grand Bank-Flemish Cap area, the Grand Bank American plaice, yellowtail flounder and witch fisheries and the Flemish Cap American plaice fishery. During the mid-1980s to the early 1990s, there was an increase in fishing pressure on flatfishes, in particular juveniles, outside of Canada's 370-km limit on the southern Grand Bank (Walsh 1992; Walsh *et al.* 1995), contributing significantly to the collapse of these stocks.

A description of these management organisations and their use of technical measures to manage the resources is summarised below.

11.5.1.1 International Commission for the Northwest Atlantic Fisheries (ICNAF)

The ICNAF convention areas included most of the northwest Atlantic extending from west of the southern tip of Greenland, south to Cape Hatteras (Fig. 11.2). ICNAF used trawl mesh size regulations as the primary technical measure together with closed seasons and areas and fishing effort limits to control the pattern and level of exploitation of fish stocks (see Halliday & Pinhorn 1997). Prior to ICNAF, the otter trawl fisheries of the 1940s were mainly unregulated and mesh sizes in the codends were generally between 63 and 75 mm in Canada and the United States. In 1953, ICNAF increased codend mesh size to 114 mm in an effort to reduce the waste of juvenile haddock in the Georges Bank fisheries. By 1957, this regulation had expanded to all fisheries north to the Grand Banks and by 1968 it encompassed all fisheries from Labrador to West Greenland. Due to severe depletion of several groundfish stocks by the early 1970s, ICNAF increased the codend mesh to 130 mm in 1972. TACs were first introduced in 1970 for haddock and between 1971 and 1974 all stocks of flatfishes were under TAC control.

11.5.1.2 Canada

After extended jurisdiction of its national boundary out to 370 km in 1977, the management of fisheries within the boundary came under the Department of Fisheries and Oceans Act, later known as Canada's Fisheries Act. This act is administered by the federal authority of the Department of Fisheries and Oceans (DFO). DFO followed the pattern set by ICNAF and continued to use TACs and a variety of other technical measures to control the levels of exploitation.

By the early 1990s, most of the Canadian groundfish stocks, including the majority of the flatfishes, were severely depleted. DFO increased the regulated bottom trawl codend mesh size from 130 to 145 mm and higher in some fisheries. Between 1993 and 1994, Canada introduced Conservation Harvest Plans (CHPs). Each year, DFO and the fishing industry agree on a CHP where technical measures such as quotas, mesh sizes, and closures are set out for the different fisheries in different areas. Danish and Scottish seines, used predominately in the Gulf of St Lawrence witch and American plaice fisheries, mainly use 130–155 mm square mesh codends depending on area and species. Sunken gill nets, commonly used in the deep-water fisheries offshore and nearshore for Greenland halibut and other flatfishes, use a regulated mesh size ranging from 140 to 191 mm, depending on area and species being targeted. The Atlantic halibut fishery is restricted to longlining.

In 1988, Canada introduced the first minimum landed fish size for flatfishes by setting the limit at 81 cm for Atlantic halibut. In 1993, in an effort to further reduce the by-catch of all small fishes, DFO introduced real-time closures of specific fishing grounds when catches of small fishes exceed 15% by number along with a no-discarding regulation. By 1994, all commercial flatfish species had a regulated minimum landing size. The regulation stipulated that all Atlantic halibut <81 cm were to be returned to the sea alive.

11.5.1.3 Northwest Atlantic Fisheries Organization (NAFO)

After the Canadian and American extended jurisdictions in 1977, the US withdrew as a member of ICNAF. ICNAF operated in a modified form until 1979 when it was replaced by a new international commission, NAFO, which incorporated all of the statistical areas of ICNAF (Fig. 11.2). As a result of extended jurisdiction, seven trans-boundary stocks of flatfishes, as well as several stocks of Atlantic cod and redfishes, from Baffin Bay to the Grand Bank, including the Flemish Cap, were of shared interest among Canada, NAFO and Greenland.

NAFO continued the ICNAF precedent of using TACs as the primary measure to control fishing mortality. However, with the collapse of several flatfish stocks on the Grand Bank in the early 1990s, it invoked closed fisheries as a management tool to prevent further declines and to rebuild these stocks. Minimum mesh size remained at 130 mm for codends and gill nets. Minimum landing sizes for most flatfishes were introduced in the mid-1990s. Both the regulated minimum codend mesh size and the minimum landing sizes for flatfish fisheries in the NAFO Regulatory Area, i.e. outside the national boundary of Canada, are lower than those set by Canada for its fleets fishing the same stocks inside the national boundary. For example, the Canadian otter trawl fleet uses a minimum mesh size of 145 mm diamond in the codends while the minimum mesh size under NAFO is 130 mm diamond. Also the minimum landing size of plaice, yellowtail flounder and witch for the Canadian fleet is 30 cm while under NAFO regulations it is 25 cm and for Greenland halibut it is 43 cm and 30 cm, respectively.

11.5.1.4 USA

Following extended jurisdiction in 1977, the management of the US fisheries in coastal waters (out to 5.6 km) and within the 370-km limit came under the Magnuson–Stevens Fishery Conservation and Management Act in March of 1977. This act is administered by the Secretary of Commerce through the National Marine Fisheries Service (NMFS) which oversees eight regional fishery management councils. The New England and Mid-Atlantic Fishery Management councils are responsible for management of all Atlantic flatfish stocks from the Gulf of Maine-Georges Bank area to Cape Hatteras and includes those trans-boundary stocks it shares with Canada on Georges Bank. Because of the mixed nature of the fisheries, management uses mesh size regulations, area closures and minimum landing size to optimise yield. Initially, there were no direct controls on fishing mortality, such as quotas, until the rapid depletion of several resources during the 1980s and early 1990s. In 1994, a new multi-species management plan was introduced to target a 50% reduction in fishing effort over a 5–7 year range (Murawski et al. 1997). The 1994 plan also used rolling time and area fishing closures to rebuild stocks such as yellowtail flounder and American plaice. The management of summer flounder, however, uses quotas along with minimum fish size and mesh size to reduce mortality levels.

In the Gulf of Maine-Georges Bank and the southern New England areas, the otter trawl fleet uses a minimum mesh size of 152 mm diamond or 165 mm square throughout the entire net of the trawl. In the mid-Atlantic area (Cape Cod to Cape Hatteras) the regulated minimum mesh area is 140 mm diamond or 152 mm square mesh throughout the codend (NMFS 2001). Most of the inshore vessels fishing flatfishes use 152 mm square while the offshore uses 140 mm diamond mesh. Sunken gill nets are also commonly used in some inshore and offshore fisheries with a regulated minimum mesh size of 152 mm.

11.5.2 Northeast Atlantic

Prior to the extension of the EEZ to 370 km (200 miles), fisheries in western Europe within 19 km from the coast were managed exclusively by coastal states while outside this limit management was the responsibility of international commissions such as the Northeast Atlantic Fisheries Commission (NEAFC). After the establishment of a 370-km EEZ in January

1977, the European Union (EU) took over responsibility for managing joint fish stocks in cooperation with member states under the Common Fisheries Policy (CFP). The CFP became the largest single management regime in the northeast Atlantic and the third largest fishing group in the world after Japan and the former Soviet Union. The EU has also negotiated with other countries such as Norway and Iceland for access rights to fisheries and acted on behalf of member states in dealing with international commissions such as NEAFC.

The CFP was proposed in 1958 as part of the Treaty of Rome with the aims of increasing productivity, improving standards of living for fishing communities and stabilising markets, supplies and prices to consumers (OECD 1997). However, it was not until 1970 that the CFP was finally adopted. Initially, the policy was concerned with access to fishing grounds, markets and infrastructure. The main conservation regulations were established in 1983 within a framework for the conservation and management of resources. The basis of the regulation was the setting of annual TACs and the fixed allocation of these to member states based on historical fishing activity, a concept which became known as 'relative stability'. Further modifications in 1992 (Council Regulation No. 3760/92) were intended to:

- provide for the rational and responsible exploitation of the resources, while recognising the interests of the fisheries sector and taking into account the biological constraints with respect to the marine ecosystem;
- facilitate improvement in the selectivity of fishing methods in order to optimise utilisation of the biological potential and to limit discarding;
- establish measures in order to ensure the rational and responsible exploitation of the resources on a sustainable basis (OSPAR 2000).

The main technical conservation regulations were set out in 1998 (Council Regulation No. 850/98) and imposed minimum landing sizes and mesh sizes for use in a wide range of fisheries. From 1 January 2000, the EU established a further detailed series of technical conservation measures with the aim of minimising the capture and discarding of juvenile fishes. These regulations reviewed limits on minimum landing sizes and established mesh size ranges for both fixed and towed gear with agreed percentages for the target and by-catch species. The main mesh sizes for flatfishes are 70–99 mm, with not more than 30% of non-target species permitted in the landings. When flatfish are taken as part of mixed demersal fisheries mesh sizes of 100–120 mm are required, depending on the area fished. They also defined closed areas where fishing was prohibited with specified gears to protect juvenile fishes. The regulations included provisions to regulate the design of towed gear to minimise the capture of undersize fishes and in the small mesh fisheries for shrimp required that separator grids must be used. Since 2002, the EC has established additional technical measures to protect stocks considered to be depleted such as cod in the North Sea, Irish Sea and west of Scotland, and northern hake stocks in ICES areas V, VI, and parts of VII and VIII.

In formulating management proposals, both the EU and NEAFC rely on scientific advice from ICES. Since 1999, this advice has been given in terms of a precautionary approach to fisheries management. The basis of this advice is that in order for stocks and fisheries exploiting them to be within safe biological limits, that is fished sustainably, there should be a high probability that (1) the spawning stock biomass (SSB) is above the threshold where recruitment is impaired, and (2) the fishing mortality (F) is below that which will drive the spawning stock to the biomass threshold which must be avoided (ICES 1999).

The spawning stock biomass and fishing mortality thresholds identified for a range of flatfish stocks in 2000 are shown in Table 11.5. The status of the 18 assessed flatfish stocks in relation to the precautionary reference points is also presented. Three out of six plaice stocks and two of the common sole stocks are considered to be outside safe biological limits because the SSB is below the biomass precautionary reference point defined as B_{pa} and fishing mortality is above the precautionary reference point F_{pa}. [B_{pa} is the biomass below which a stock would be regarded as potentially depleted or overfished. F_{pa} is the upper bound on fishing mortality rate; fishing mortality rates above this point would be regarded as overfishing.] A further five stocks are above B_{pa} but fishing mortality remains too high and they are currently regarded as harvested outside safe biological limits. Of the remaining stocks, six are within safe biological limits while the status of the two Greenland halibut stocks is not known but both are regarded as being at low stock levels. In the case of 12 of the 18 stocks, recent management advice has been for a reduction in fishing mortality to allow the stocks to recover.

Outside the EU stocks are managed by individual countries, often with joint agreements for shared resources. For instance, in Norwegian waters, fisheries management is by TAC backed by a range of conservation measures including a ban on discarding, large minimum mesh sizes for demersal fish species and real-time closures to protect small fish. In Icelandic waters, management of demersal fisheries is by Individual Transferable Quotas (ITQs) which have been introduced progressively since 1976 (Arnason 1995). The quotas are based on a share of the TAC which is determined annually on the basis of scientific advice. In addition there are a number of regulations covering closed areas and mesh and minimum landing sizes.

11.5.3 Southern Atlantic

No specific assessment methods are performed in Argentina for flounder resources. The information available comes from research vessel surveys carried out for assessing Argentinian hake. Fabré (1992) estimated the MSY from the total mean biomass, using biological parameters (Sparre *et al.* 1989) and estimates of natural mortality. The results obtained were 3782 t yr^{-1} for *Xystreurys rasile* and 1931 t yr^{-1} for *Paralichthys isosceles*. Little information is given regarding the status of flatfish fisheries in southern Brazil. The fishing pressure on both *P. patagonicus* and *P. orbignyanus* is high, the stock abundance is unknown, and the stock is regarded as intensively exploited (Haimovici 1998). Despite high average landings of around 1360 t between 1990 and 1994, a decline in CPUE has implied a decrease in the abundance of the stock and a sharp decrease in future landings is expected (Haimovici 1998).

In the southeast Atlantic, there are few direct estimates of the level of exploitation of flatfish stocks. Biomass estimates are available for west coast sole off Namibia derived from demersal surveys for hake. The biomass estimates range from 400 t to 1100 t over the period 1983–1990 (Macpherson & Gordoa 1992). Estimates for west coast sole off the South African coast range from 37 t to 792 t over a 12-year period from 1985 to 1997 (Frances le Clus, personal communication).

Table 11.5 Status of major flatfish stocks in the northeast Atlantic assessed by ICES in 2000 (ICES 2001)

Species	Area	Status in 2000	SSB 2000 (000 t)	F 1999	B_{pa} (000 t)	F_{pa}
Plaice	IIIa	Within SBL	32	0.70	24	0.73
	IV	Outside SBL	271	0.32	300	0.30
	VIIa	Within SBL	4.9	0.32	3.1	0.45
	VIId	Harvested outside SBL	9.0	0.66	8.00	0.45
	VIIe	Outside SBL	1.9	0.61	2.5	0.45
	VIIf, g	Outside SBL	1.5	0.68	1.8	0.60
Common sole	IIIa	Within SBL	2.23	0.28	1.1	0.30
	IV	Harvested outside SBL	52.5	0.47	35	0.40
	VIIa	Harvested outside SBL	3.8	0.35	3.8	0.30
	VIId	Harvested outside SBL	11.3	0.43	8.0	0.40
	VIIe	Outside SBL	2.1	0.32	2.5	0.26
	VIIf, g	Outside SBL	1.6	0.58	2.2	0.37
	VIIIa, b	Within SBL	12.6	0.47	11.3	0.45
Greenland halibut	I & II	Not known	–	–	–	–
	V & XIV	Not known	–	–	–	–
Megrim	VI	?? Within SBL	–	–	–	–
	VII & VIIIa, b, d, e	Harvested outside SBL	62.0	0.32	55	0.30
	VIIIc & IXa*	Within SBL	5.39	0.23	6.5	0.30

The stock status in relation to safe biological limits (SBL) is defined in terms of the precautionary levels of fishing mortality (F_{pa}) and spawning stock biomass (B_{pa}); for descriptions of precautionary reference levels see text.
*Two species are assessed but data refer only to the fourspotted megrim (*L. boscii*).

Acknowledgements

Special thanks to Keith Brickley of DFO Statistics Branch in Ottawa, Steven Murawski of NMFS in Woods Hole, Ole Jørgensen at Greenland Institute of Natural Resources, Copenhagen and Tissa Armantunga at NAFO in Canada for access to data to make the northwest Atlantic review possible. Also to Stuart Rogers, Frans Van Beek and Rasmus Nielsen for their comments and advice on the northeast Atlantic. Thanks are also due to the following for their help with the southern Atlantic fisheries: Fernando Cervigón, Frances le Clus, Maria Berta Cousseau, Martine Desoutter, Inés Elías, Elizabeth Errazti, Daniel Figueroa, Manuel Haimovici, Raquel Perier and the late Guillermo Burgos.

References

Albert, O.T., Eliassen, J-E. & Hoines, A. (1998) Flatfishes of the Norwegian coasts and fjords. *Journal of Sea Research*, **40**, 153–171.

Anonymous (2000) *11th Report of the Scientific, Technical and Economic Committee for Fisheries. Commission of the European Union*, SEC (2001) 177, Brussels.

Arena, G., Barea, L., Barreiro, D., Beathyate, G. & Marín, Y. (1992) Utilización de redes de baja apertura en la pesca del lenguado (*Paralichthys* spp.). *Instituto Nacional de Pesca, Montevideo, Informe Técnico*, **37**, 1–22.

Arnason, R. (1995) *The Icelandic Fisheries: Evolution and Management of a Fishing Industry*. Fishing News Books, Oxford.

Bezzi, S.I., Verazay, G.A. & Dato, C.V. (1995) Biology and fisheries of Argentine hakes (*M. hubbsi* and *M. australis*). In: *Hake: Biology, Fisheries and Markets* (eds J. Alheit & T.J. Pitcher). pp. 239–267. Chapman & Hall, London.

Bigelow, H.B. & Schroeder, W.C. (1953) Fishes of the Gulf of Maine. *Fishery Bulletin*, **74**, 1–577.

Bowering, W.R. & Nedreaas, K.H. (2000) A comparison of Greenland halibut (*Reinhardtius hippoglossoides* (Walbaum)) fisheries and distribution in the Northwest and Northeast Atlantic. *Sarsia*, **85**, 61–76.

Carvalho-Filho, A. (1999) *Peixes: Costa Brasileira*, 3rd edn. Melro, São Paulo, Brazil.

Cousseau, M.B. & Fabré, N.N. (1990) Lenguados. In: *Muestreo Bioestadístico de Desembarque del Puerto de Mar del Plata Período 1980–1985* (ed. M.B. Cousseau). pp. 179–184. Contribución INIDEP, 585, Argentina.

Daan, N. (1996) TAC management in North Sea fisheries. *Journal of Sea Research*, **37**, 321–341.

DFO (2001a) Landings. Department of Fisheries and Oceans Statistical Services website url: http://intra01.ncr.dfo-mpo.gc.ca/policy/statistics/stat_e.

DFO (2001b) The fishing industry. In: *Canada's Ocean Industries: Contribution to the Economy 1988–1996*. Department of Fisheries and Oceans, Ottawa.

Díaz de Astarloa, J.M. (1994) *Las especies del género Paralichthys del Mar Argentino (Pisces, Paralichthyidae). Morfología y sistemática*. DPhil thesis, University of Mar del Plata, Argentina.

Fabré, N.N. (1992) *Análisis de la distribución y dinámica poblacional de lenguados de la Provincia de Buenos Aires (Pisces, Bothidae)*. DPhil thesis, Universidad Nacional de Mar del Plata, Argentina.

Fabré, N.N. & Díaz de Astarloa, J.M. (1996) Pleuronectiformes de importancia comercial del Atlántico sudoccidental, entre los 34° 30' y 55° S. Distribución y consideraciones sobre su pesca. *Revista de Investigación y Desarrollo Pesquero*, **10**, 45–55.

FAO (1997) Review of the state of world fishery resources: Marine Fisheries. B. Regional Reviews, 1. Northwest Atlantic, FAO Statistical Area 21. *FAO Fisheries Circular* N0. 920 FIRM C920.

FAO (2000) *FISHSTAT Plus: Universal software for fishery statistical time series*, version 2.3. FAO Fisheries Department, Fishery Information, Data and Statistics Unit, Rome.

Godø, A.R. & Haug, T. (1989) A review of the natural history, fisheries and management of Greenland halibut (*Reinhardtius hippoglossoides*) in the eastern Norwegian and Barents Seas. *Journal du Conseil International pour l'Exploration de la Mer*, **46**, 62–75.

Haimovici, M. (1997) Demersal and benthic teleosts. In: *Subtropical Convergence Environments: The Coast and Sea in the Southwestern Atlantic* (eds U. Seeliger, C. Odebrecht & J.P. Castello). pp. 129–136. Springer-Verlag, Berlin.

Haimovici, M. (1998) Present state and perspectives for the southern Brazil shelf demersal fisheries. *Fisheries Management and Ecology*, **5**, 277–289.

Haimovici, M. & Mendonça, J.T. (1996) Descartes da fauna acompanhante na pesca de arrasto de tangones dirigida a linguados e camarões na plataforma continental do sul do Brasil. *Atlântica*, **18**, 161–177.

Halliday, R.G. & Pinhorn, A.T. (1997) Policy frameworks. In: *Northwest Atlantic Groundfish: Perspectives on a Fishery Collapse* (eds J. Boreman, B.S. Nakashima, J.A. Wilson & R.L. Kendall). pp. 95–109. American Fisheries Society, Bethesda, MD.

Heemstra, P.C. & Gon, O. (1995) Family No. 262: Soleidae. In: *Smiths' Sea Fishes* (eds M.M. Smith & P.C. Heemstra). pp. 868–874. Southern Book Publishers, Johannesburg.

Henderson, P.A. (1998) On the variation in dab *Limanda limanda* recruitment: a zoogeographic study. *Journal of Sea Research*, **40**, 131–142.

Hensley, D.A. (1995) Family No. 259: Bothidae. In: *Smiths' Sea Fishes* (eds M.M. Smith & P.C. Heemstra). pp. 854–863. Southern Book Publishers, Johannesburg.

Holm, P. (1996) Catches and manpower in the Danish fisheries, c1200–1995. In: *The North Atlantic Fisheries, 1100–1976. National perspectives on a common resource* (eds P. Holm, D.J. Starkey & J. Th. Thor). pp. 177–206. Studia Atlantica, Esbjerg, Denmark.

ICES (1999) Report of the ICES Advisory Committee on Fisheries Management, 1998. *ICES Cooperative Research Report*, **229**, 446pp.

ICES (2001) Report of the ICES Advisory Committee on Fisheries Management, 2000. *ICES Cooperative Research Report*, **242**, 911pp.

Macpherson, E. & Gordoa, A. (1992) Trends in the demersal fish community off Namibia from 1983 to 1990. In: Benguela trophic functioning (eds A.I.L. Payne, K.H. Brink, K.H. Mann & R. Hilborn). *South African Journal of Marine Science*, **12**, 635–649.

Merritt, N.R. & Haedrich, R.L. (1997) *Deep-Sea Demersal Fish and Fisheries*. Chapman & Hall, London.

Miller, J.M., Burke, J.S. & Fitzhugh, G.R. (1991) Early life history patterns of Atlantic North American flatfish – likely (and unlikely) factors controlling recruitment. *Netherlands Journal of Sea Research*, **27**, 261–275.

Millner, R.S. & Whiting, C.L. (1996) Long-term changes in growth and population abundance of sole in the North Sea from 1940 to the present. *ICES Journal of Marine Science*, **53**, 1185–1195.

Murawski, S.A., Maguire, J.J., Mayo, R.K. & Serchuk, F.M. (1997) Groundfish stocks and the fishing industry. In: *Northwest Atlantic Groundfish: Perspectives on a Fishery Collapse* (eds J. Boreman, B.S. Nakashima, J.A. Wilson & R.L. Kendall). pp. 27–70. American Fisheries Society, Bethesda, MD.

NAFO (2000) *Northwest Atlantic Fisheries Organization Scientific Council Reports*. Dartmouth, Canada, ISSN-0250–6416: 303pp.

NMFS (2001) NE multispecies regulated mesh size area regulations. National Marine Fisheries Service, Northeast Region. *Information Sheet No. 1* (01/03/01).

NOAA (1991) National Oceanic and Atmospheric Administration. Our living oceans. *NOAA Technical Memorandum* NMFS-F/SPO.

NOAA (1996) National Oceanic and Atmospheric Administration. Our living oceans. *NOAA Technical Memorandum* NMFS-F/SPO.

OECD (1997) *Towards Sustainable Fisheries: Economic Aspects of the Management of Living Resources*. OECD, Paris.

OSPAR (2000) *Quality Status Report 2000. Region III – Celtic Sea*. OSPAR Commission, London.

Payne, A.I.L. (1979) A survey of the stock of the southern African west coast sole *Austroglossus microlepis* between 28° 30' S and 32° S. *Fishery Bulletin of the Division of Sea Fisheries of South Africa*, **12**, 26–34.

Payne, A.I.L. (1985) The sole fishery off the Orange River, Southern Africa. *International Symposium on the most important upwelling areas off Western Africa. Instituto de Investigaciones Pesqueras, Barcelona*, **2**, 1063–1079.

Payne, A.I.L. & Badenhorst, A. (1989) Other groundfish resources. In: *Oceans of Life off Southern Africa* (eds A.I.L. Payne & R.J.M. Crawford). pp. 148–156. Vlaeberg, Cape Town.

Rijnsdorp, A.D. & Millner, R.S. (1996) Trends in population dynamics and exploitation of North Sea plaice (*Pleuronectes platessa* L.) since the late 1800s. *ICES Journal of Marine Science*, **53**, 1170–1184.

Rijnsdorp, A.D. & Van Leuween, P.I. (1996) Changes in growth of North Sea plaice since 1950 in relation to density, eutrophication, beam-trawl effort, and temperature. *ICES Journal of Marine Science*, **53**, 1199–1213.

SAGPyA (1996) *Consumo de pescado en el mercado argentino*. Secretaría de Agricultura, Ganadería, Pesca y Alimentación, Subsecretaría de Pesca, Argentina.

SAGPyA (1999) *Estadísticas de capturas martimas pesqueras*. Secretaría de Agricultura, Ganadería, Pesca y Alimentación, Subsecretaría de Pesca, Argentina.

Scott, W.B. & Scott, M.G. (1988). Atlantic fishes of Canada. *Canadian Bulletin of Fisheries and Aquatic Sciences*, **219**, 1–741.

Sparre, P., Ursin, E. & Verema, S.C. (1989) Introduction to tropical fish stock assessment. Part 1. *FAO Fishery Technical Papers*, **306**, 1–337.

Troyanovsky, F.M. & Lisovsky, S.F. (1995) Russian (USSR) fisheries research in deep waters (below 500m) in the North Atlantic. In: *Deep Water Fisheries of the North Atlantic Ocean Slope* (ed. A.G. Hopper). pp. 357–365. Kluwer Academic Publishers, Netherlands.

Walsh, S.J. (1992) Commercial fishing practices on offshore juvenile flatfish nursery grounds on the Grand Banks of Newfoundland. *Netherlands Journal of Sea Research*, **27**, 423–432.

Walsh, S.J. (1994) Life history traits and spawning characteristics in populations of long rough dab (American plaice) *Hippoglossoides platessoides* (Fabricius) in the North Atlantic. *Netherlands Journal of Sea Research*, **32**, 241–254.

Walsh, S.J., Brodie, W.B., Bishop, C. & Murphy, E. (1995) Fishing on juvenile groundfish nurseries on the Grand Bank: a discussion of technical measures of conservation. In: *Proceedings of a Symposium on Marine Protected Areas and Sustainable Fisheries, Second International Conference on Science and the Management of Marine Protected Areas* (eds N.L. Shackell & J.H. Martin Willison). pp. 54–73. Dalhousie University, 16–20 May 1994. Acadia University, Nova Scotia.

Wood, H. (1956) Fisheries of the United Kingdom. In: *Sea Fisheries* (ed. M. Graham). pp. 10–79. Edward Arnold, London.

Woodhead, P.M.J. (1964) The death of North Sea fish during the winter of 1962/63, particularly with reference to sole, *Solea vulgaris*. *Helgoländer Wissenschaftliche Meeresuntersuchungen*, **10**, 283–300.

Yang, J. (1981) An estimate of fish biomass in the North Sea. *International Council for the Exploration of the Sea*, CM1981/G:15, 11pp.

Chapter 12
Pacific flatfish fisheries

Thomas Wilderbuer, Bruce Leaman, Chang Ik Zhang, Jeff Fargo and Larry Paul

12.1 Introduction

The largest and deepest of the world's five oceans, the Pacific Ocean covers 28% of the global surface (approx. 160 000 million sq km), and is larger than the total land area. This chapter covers the temperate (non-tropical) parts of the Pacific Ocean and the flatfish resources that are fished by the coastal countries of Australia and New Zealand in the southern hemisphere and Japan, the Republic of Korea, Russia, Canada and the United States in the northern hemisphere. Although species of flatfishes are also present along the narrow continental shelf of Chile, the large fishery there primarily comprises small pelagic or demersal fishes and flatfishes are not pursued.

12.2 Main species and nature of fisheries

Approximately 300 species of flatfishes inhabit the Pacific Ocean (Minami & Tanaka 1992), ranging from nearshore estuarine areas to deep waters of the continental slope. Of these, nearly 50 species are commercially important as food fishes (Tables 12.1–12.3). In areas where sub-tropical and sub-arctic waters overlap, there is a large diversity in flatfish commu-

Table 12.1 List of commercially important species in the Northeast Pacific Ocean and Bering Sea

Pacific halibut	*Hippoglossus stenolepis*
Yellowfin sole	*Limanda aspera*
Northern rock sole	*Lepidopsetta polyxystra*
Rock sole	*Lepidopsetta bilineata*
Flathead sole	*Hippoglossoides elassodon*
Alaska plaice	*Pleuronectes quadrituberculatus*
Dover sole	*Microstomus pacificus*
Rex sole	*Glyptocephalus zachirus*
English sole	*Parophrys vetula*
Greenland halibut	*Reinhardtius hippoglossoides*
Petrale sole	*Eopsetta jordani*
Starry flounder	*Platichthys stellatus*
Arrowtooth flounder	*Reinhardtius stomias*
California flounder	*Paralichthys californicus*

Table 12.2 List of commercially important species in the Northwest Pacific Ocean

Littlemouth flounder	*Pseudopleuronectes herzensteini*
Japanese flounder	*Paralichthys olivaceus*
Yellowfin sole	*Limanda aspera*
Blackfin flounder	*Glyptocephalus stelleri*
Large scale flounder	*Citharoides macrolepidotus*
Cinnamon flounder	*Pseudorhombus cinnamoneus*
Pointhead flounder	*Cleisthenes pinetorum*
Roughscale sole	*Clidoderma asperrimum*
Rikuzen flounder	*Dexistes rikuzenius*
Shotted halibut	*Eopsetta grigorjewi*
Stone flounder	*Platichthys bicoloratus*
Sand flounder	*Limanda punctatissima*
Marbled flounder	*Pseudopleuronectes yokohamae*
Slime flounder	*Microstomus achne*
Ridged-eyed flounder	*Pleuronichthys cornutus*
Willowy flounder	*Glyptocephalus kitaharai*
Spotted halibut	*Verasper variegatus*
Red tonguesole	*Cynoglossus joyneri*
Robust tonguefish	*Cynoglossus robustus*
Black cow-tongue	*Paraplagusia japonica*
Many-banded sole	*Zebrias fasciatus*

Table 12.3 List of commercially important species in the Southern Pacific Ocean

Longsnout flounder	*Ammotretis rostratus*
Yellowbelly flounder	*Rhombosolea leporina*
New Zealand flounder	*Rhombosolea plebeia*
New Zealand sole	*Peltorhamphus novaezeelandiae*
Southern lemon sole	*Pelotretis flavilatus*
Greenback flounder	*Rhombosolea tapirina*
Tudor's flounder	*Ammotretis lituratus*
Black flounder	*Rhombosolea retiaria*
New Zealand brill	*Colistium guntheri*
New Zealand turbot	*Colistium nudipinnis*

nities where catches comprise many species and individual species do not dominate. People throughout the countries bordering the Pacific Ocean as well as Europe and the eastern USA consume flatfishes from the Pacific Ocean, sometimes as a delicacy, due to their desirable flesh qualities combined with high protein and low fat content.

Flatfishes have undoubtedly been captured for centuries in aboriginal fisheries by the indigenous peoples of the coastal regions of the Pacific Ocean (Stewart 1977; Kalland 1995). These fisheries were typified by primitive hook and line gear, spearing and small-scale netting operations in nearshore areas (Fig. 12.1). In recent years the majority of the flatfish catch has come from professional commercial fishers using bottom trawls and longline gear in offshore waters, although some species of flatfishes are also highly sought by recreational fishers in some parts of the Pacific Ocean.

Following the extended jurisdiction of fisheries management zones throughout the Pacific Ocean, fisheries generally became more closely managed. Through the use of some form of national fisheries management plan, effort became limited to domestic fishermen and fishing effort became more closely controlled. In some countries increased emphasis was placed on

Fig. 12.1 Indian catch of halibut at Neah Bay, Washington, USA (ca. 1910). Photographed by A.H. Barnes; Hillary Irving of the Makah Tribe identified the location.

observing and monitoring the flatfish catch and seasonal controls were instituted. The combined flatfish harvests (excluding Pacific halibut) by the major flatfish fishing nations of the Pacific Ocean averaged over the most recent 10-year period with available catch has shown little variation (Table 12.4; Fig. 12.2) and are as follows: New Zealand 4200 t, Korea 19 900 t, Japan 87 100 t, Taiwan 270 t, US (Bering Sea 225 600 t, Gulf of Alaska 30 100 t, west coast 20 225 t), Canada 10 300 t and Russia 25 000 t (1984–1993). The 10-year average catch for Pacific halibut from the eastern Pacific Ocean is 26 900 t.

The contribution of flatfishes to the total fisheries catch in the temperate Pacific Ocean varies with geographical area. Flatfishes make up 25% of the total catch weight in Canada (excluding hake) and have ranged from 7 to 22% and 9 to 17% of the groundfish catch weight from 1990–1999 in the Gulf of Alaska and the Bering Sea, respectively. By contrast, they represented only 2% of the groundfish catch biomass in Tasmania and 1.5% of the total in Japan in 1998 (MAFF 2000). The relative commercial importance of catches identified to species in 1998 is shown in Table 12.5.

Due to the non-selective nature of bottom trawl gear, the principal method used to catch commercial quantities of flatfishes, many species are captured incidental to pursuing flatfishes in fishing operations. These species are often high-value fish and shellfish which have their own directed fishery such as Pacific halibut and red king crab (*Paralithodes camschaticus*) and tanner crabs (*Chionocetes* spp.). These incidental catches may have negative ecosystem

Table 12.4 Flatfish catch (t) of the major fishing nations of the temperate Pacific Ocean from 1950 to 1998

	Korea[1]	Japan[1]	Russia[1]	Taiwan[1]	N. Korea[1]	USA[2]	Canada[3]	New Zealand[4]	Australia[5]
1950	9.6	94	32.8	0.1	0	0	7.94	1.94	0
1951	9.6	94.3	34.6	0.1	0	0	5.56	1.68	0
1952	10.4	119.7	40.3	0.1	0	0	7.66	1.73	0
1953	10.2	119.9	53.8	0.1	0	0	4.5	2.21	0
1954	10.5	120.5	70.5	0.2	0	0	7.4	1.89	0
1955	10.9	125	119.2	0.3	0	0	8.64	1.66	0
1956	11.3	141.2	147.6	0.3	0	0	9.46	1.38	0
1957	11.4	140.4	140.4	0.3	0	0	8.97	1.08	0
1958	13.9	168.5	194.8	0.3	0	0	7.85	1.73	0
1959	11.4	252	195.1	0.4	0	0	7.81	1.76	0
1960	14.5	500.3	104.9	0.3	0	0	9.58	2.07	0
1961	12.3	583.4	118.4	0.4	0	0	7.78	1.43	0
1962	14.3	494.6	91	0.5	0	0	7.59	1.35	0
1963	17.1	226.1	84	0.7	0	0	7.59	1.31	0
1964	14.3	254.8	72.8	0.8	0	12.93	6.95	1.86	0
1965	17.6	207.1	87.1	1	0	12.84	7.63	2.14	0
1966	17.2	264.8	86.9	1.5	0	13.41	8.89	2.41	0
1967	22.4	279.2	77.3	2.2	0	11.57	8.45	2.64	0
1968	25.7	239.2	85.6	2.1	0	12.89	9.06	1.96	0
1969	24.3	272.4	90	2	0	14.34	8.44	2.08	0
1970	27.3	187.4	40.4	2.3	0	14.97	8.06	2.32	0
1971	26.3	221.8	45.9	2.6	0	14.7	7.02	1.86	0
1972	30.2	205.2	73.4	2.3	0	18.89	5.88	1.3	0
1973	31.4	223.2	76.2	1.7	0	18.79	6.56	1.64	0
1974	31.35	233.06	30.7	1.37	0	18.39	6.77	2.06	0
1975	32.81	237.86	37.51	2.33	0	19.99	7.9	1.7	0
1976	36.68	229.56	32.73	1.95	0	21.41	6.88	2.43	0
1977	33.68	173.08	57.44	2.24	0	18.92	5.65	3.13	0
1978	27.5	193.66	98.52	2.82	0	22.17	5.59	3.91	0
1979	31.99	171.13	52.45	3.24	0	25.47	6.33	3.94	0
1980	37.2	170.11	54.1	3.33	0	32.98	6.32	3.19	0
1981	35.46	168.91	58.03	2.31	0	43.91	5.42	3.69	0

Table 12.4 (*Continued*)

	Korea[1]	Japan[1]	Russia[1]	Taiwan[1]	N. Korea[1]	USA[2]	Canada[3]	New Zealand[4]	Australia[5]
1982	33.3	166.38	57	2.35	0	54.47	3.34	3.32	0
1983	33.3	149.26	66.52	2.56	0	63.64	2.93	4.75	0
1984	30.87	139.68	73.49	1.3	0	78.66	3.66	4.84	0
1985	34.5	135.17	112.38	1.77	0	202.18	3.42	4.47	0
1986	30.84	95.6	117.74	1.62	0	241.18	3.49	3.22	0
1987	32.57	91.09	108.71	0.36	0	275.39	4.06	2.75	0
1988	23.03	84.77	27.9	0.68	0	357.12	5.54	4.07	0
1989	22.22	83.3	24.01	0.7	0	265.18	7.13	4.3	0
1990	20.61	79.61	26.76	0.25	0	202.59	9.87	3.48	0.04
1991	19.94	79.89	29	0.25	0	269.76	10.19	2.98	0.04
1992	23.07	89.26	25.45	0.27	0	317.65	11.77	3.2	0.03
1993	22.68	90.14	11.41	0.12	0	274.91	14.03	5.09	0.03
1994	23.15	84.5	0	0.12	2.32	313.56	12.56	4.61	0.03
1995	24.3	86.39	0	0.15	2.95	279.95	12.74	4.58	0.03
1996	31.14	96.42	0	0.35	1.97	296.34	9.31	4.57	0.03
1997	47.34	93.62	0	0.35	6.97	359.5	6.88	4.79	0.03
1998	45.15	88.03	0	0.14	4	235.53	8.57	4.66	0.02

[1]Source: FAO catch statistics.

[2]Source: Gulf of Alaska catches 1964–1980: Murai *et al.* (1981). West coast catches: Pacific Fishery Management Council, Status of the groundfish fishery through 2000 and recommended acceptable biological catches in 2001 (document prepared for the Council and its entities). Pacific Fishery Management Council, Suite 224, 2130 SW Fifth Ave, Portland OR 97201. Bering Sea catches 1981–1998: Stock assessment and Fishery evaluation report for the groundfish resources of the Bering Sea/Aleutian Islands, November 2000, North Pacific Fisheries Management Council, 605 West 4th Ave, Suite 306, Anchorage, Alaska, AK 99501. Gulf of Alaska catches 1981–1998: Stock Asessment and Fishery evaluation report for the groundfish resources of the Gulf of Alaska, November 2000, North Pacific Fisheries Management Council, 605 West 4th Ave, Suite 306, Anchorage, Alaska, AK 99501.

[3]Landings for all species categories from 1950 to 1953 were taken from Ketchen (1976). Landings from 1954 through 1995 were taken from the archived groundfish catch database GFCATCH. Landings from 1996 to 2000 were taken from the relational database PacHarvest which resides on the Windows NT server PacStud at the Pacific Biological Station, 3190 Hammond Bay Road, Nanaimo, BC, Canada V9T 6N7.

[4]Landings compiled during the preparation of Paul (2000). Values to 1974 from Annual Reports on Fisheries; 1975 to 1983 from MAF data reports; 1984 onwards from Annual Fishery Stock Assessment Reports, New Zealand Ministry of Fisheries.

[5]Tasmania Scalefish Fishery Assessment 1999. Compiled by A.R. Jordan and J.M. Lyle, Tasmanian Aquaculture and Fisheries Institute, TAFI Marine Research Laboratories, PO Box 252-49, Hobart, TAS 7001, Australia.

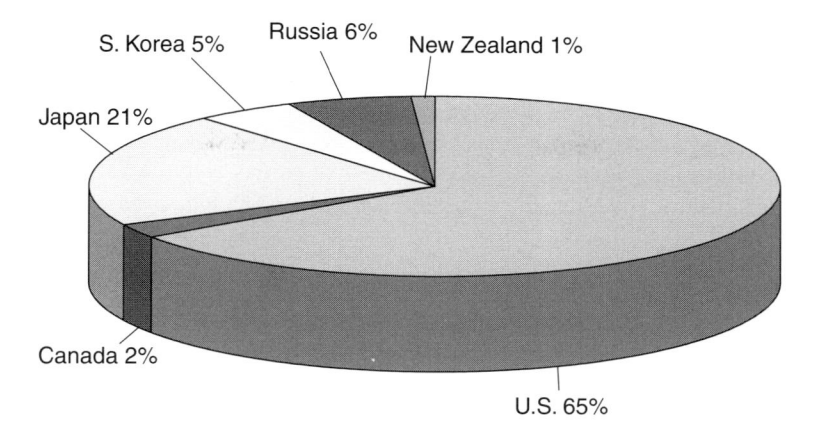

Fig. 12.2 10-year average national catch (most recent data where available) of combined flatfish species of the major Pacific Ocean fishing nations.

Table 12.5 Relative importance of flatfishes in Pacific catches (t) identified to species in 1998, where available (does not include Russia or catches reported as 'flounders')

Species	Catch
Pacific halibut	52351
Petrale sole	1817
Rex sole	2797
Flathead sole	26368
Rock sole (*Lepidopsetta* spp.)	36701
Dover sole	12911
English sole	1916
Yellowfin sole	101342
Greenland halibut	9196
Arrowtooth flounder	35001
Alaska plaice	14022
Butter sole (*Isopsetta isolepis*)	542
Pacific sand sole (*Psettichthys melanostictus*)	168
Japanese flounder	17222

impacts and have the potential to adversely affect the directed fishery of the by-catch species if they are not accounted for and are unregulated. In the US and Canadian waters of the North Pacific Ocean regulations are in place to limit the catch of certain high-value species in the directed flatfish fisheries. Furthermore, the total catch (numbers or weight) of the incidentally caught species are tallied and annual by-catch limits are in place which stop the directed fishery when they are attained. In the case of Pacific halibut, the condition of the fish at the time of release is also recorded by fishery observers to calculate the amount of discard mortality resulting from the by-catch process (Williams & Wilderbuer 1995). By-catch has been a major problem in North Pacific flatfish fisheries because there have been instances where they were restricted from reaching their annual quota of directed harvest because their by-catch quota was reached first. In recent years, by-catch avoidance programmes voluntarily instituted by the flatfish fleet have resulted in lower by-catch rates, leading to longer fishing seasons without reaching the by-catch limits.

12.3 History of exploitation

12.3.1 General account

The first record of the pursuit of flatfishes in offshore waters is in the San Francisco area where two sailboats towed paranzella nets in 1875 (Alverson *et al.* 1964). With the advent of steam-powered vessels in the 1880s, pair trawling for groundfish species (which include flatfishes) and hook and line fishing targeting Pacific halibut expanded along the California coast (Trumble *et al.* 1993; Trumble 1998). By the early 1900s otter trawling was reported off the west coast of North America and soon spread northward into Canadian waters. In 1918, 1600 t of flatfishes were landed in British Columbia ports but the development of these fisheries was constrained by limited markets for flatfishes (Alverson *et al.* 1964).

The North Pacific Ocean fishery for Pacific halibut began in 1888 when three sailing vessels from the Atlantic Ocean started fishing off the west coast of the USA (International Pacific Halibut Commission 1998) and has continued to the present. Halibut were harvested in both the western (Kodolov 1995) and eastern (Trumble *et al.* 1993) parts of the Pacific Ocean and made up the majority of flatfish landings prior to World War II. Fishing initially began using sailing schooners and sloops which launched small dories whose crews used handlines. By the 1920s these vessels were replaced with diesel-powered schooners which were designed to mechanically haul longline gear directly to the deck, negating the need for the hand operations from dories. Many of the schooners built in the 1920s are still used (with updated electronics and propulsion systems) in the current fishery, which operates with relatively small vessels employing crews ranging from one to eight people (Fig. 12.3). Pacific

Fig. 12.3 Traditional Pacific halibut schooner used from the 1920s to the present.

halibut have been managed separately from the other flatfish species in the northeast Pacific Ocean since the International Pacific Halibut Commission was formed in 1923.

In the 1930s, the harvest of flatfishes increased throughout the temperate Pacific Ocean from New Zealand to Peter the Great Bay (Ivankova 1995) and throughout the Bering Sea and the Gulf of Alaska (Alverson et al. 1964). Distant-water exploratory fishing was undertaken by fleets from Japan and Russia (former Soviet Union) into the Sea of Okhotsk and the Bering Sea to obtain information on the groundfish resources. These fleets generally consisted of catcher vessels associated with large mother ships that processed the catch. Following World War II, the build-up of fishing fleets by Japan, Russia, Korea and Taiwan in the North Pacific Ocean rapidly increased. These fleets used factory trawlers and mother ships with catcher vessels because there were no shoreside processing facilities in the current fishery. The Japanese fleet size increased from nine catcher-trawlers in 1954 to 173 in 1961. During this time flatfish species constituted about two-thirds of the total catch (Trumble 1998).

From 1960 to 1962, Japanese fishing fleets caught over 1.5 million t of flatfishes in the eastern Bering Sea, the majority of which were yellowfin sole which were frozen and returned to Japan as food fish. By 1964 the Alaska flatfish resource had declined from over-fishing, resulting in a shift in harvest to other species. Flatfish harvest thereafter continued at a limited scale until the stocks began to rebound in the 1970s. By the late 1970s most of the countries with Pacific Ocean shorelines extended their economic jurisdiction over the fisheries resources to 200 miles seaward of the coastline. This extension caused the decline in foreign nation distant-water fishing fleets and the start of domestic-only flatfish harvesting in each respective national area.

Details for the flatfish fisheries of the major harvesting nations are given below.

12.3.2 Republic of Korea

Nearshore flatfish harvesting in Korea has probably occurred for centuries and continues to the present. Distant-water catcher/processor stern trawlers began fishing operations in 1966 in the eastern Bering Sea and appeared in the Gulf of Alaska in 1972 (Murai et al. 1981). These large-scale commercial vessels returned frozen fish to Korea and continued through 1980 when joint-venture fisheries with US partners were established. The joint-venture fisheries lasted for 10 years before they ceased in 1989.

The present-day nearshore flatfish fishing is characterised by small boat operations delivering high-value live fish to local markets. Longlines and bottom gill nets are employed to capture flatfishes. The longline gears are baited with shrimp, herring, earthworms or shellfish and are fished at depths ranging from 7 m near shore to 40 m at 2–3 miles from the coast. Longline operations usually employ two to three people. The flatfish longline fishery typically catch all flatfishes and the gear is set in early morning and hauled in the afternoon of the same day. When fishing farther from shore, the catch is about 40% flatfishes and the gear is allowed to soak overnight. The bottom gill net or trammel net is set in the late afternoon and hauled in the next morning. The catch consists of about 80% flatfishes and is sold fresh to the local market.

12.3.3 Japan

As in most parts of the Pacific Ocean, nearshore flatfish harvesting has occurred in Japan for millennia. In the 1930s large-scale distant-water fishing began in the North Pacific-Bering Sea areas and continued until 1941 (Murai *et al.* 1981). Fishing in this area resumed in 1954 and expanded into the Gulf of Alaska in 1962. These large-scale commercial vessels caught large quantities of flatfishes which supplied the Japanese frozen fish market. This distant-water harvest continued through 1980 when joint-venture fisheries with US partners were established. The joint-venture fisheries lasted for 10 years before they ceased in 1990.

Fisheries information available from 1988 to 1998 indicates that flatfish made up only 1.5% (82 700 t) of the national catch in 1998 (MAFF 2000). The primary species, Japanese flounder, represented 7–10% of the annual flatfish catch over the 11-year period and the balance consisted of many species of flatfishes which are classified as 'other flatfish'. Trawling typically accounted for the majority of the flatfish catch and gill netting for about one-third, with trawling in the coastal areas by small boats and out to 200 miles offshore harvesting the most fish. Nearshore longline fisheries also operated, although they represented only 1% of the annual flatfish catch.

12.3.4 Russia (including the former Soviet Union)

Little is known about flatfish harvesting in the western North Pacific Ocean because available catch information is lacking. In Peter the Great Bay an intense harvest of 12 species of flatfishes was taken from 1930 to 1960 by the coastal trawl fleet (Ivankova 1995). Large fishing operations began in 1958 in the eastern Bering Sea and expanded to the Gulf of Alaska in 1972 (Murai *et al.* 1981). Fishing in Alaska ended in 1987 and fishing effort shifted to the western Bering Sea. Most of the Russian vessels were constructed for large-scale, high-seas fishing and their deployment in this area generated a rapid increase in fisheries exploitation (National Research Council 1996). Average annual catch for pleuronectid species increased from 12 000 t during 1976–1984 to 26 900 t during 1985–1993. Most of the fishing occurred along the Olutorsky shelf and the Kamchatka shelf with minor efforts along the continental slope. Kupriyanov (1996) reported that up to 90% of the shallow-water catch on the Kamchatka shelf consisted of yellowfin sole.

12.3.5 Canada

The first record of a commercial fishery for flatfishes off the coast of British Columbia is from about 1860. The fishery involved canoes and small vessels fishing with hook and line on grounds adjacent to the east coast of Vancouver Island near Victoria (Lord 1866). Pacific halibut were first recorded in Canadian commercial fishery landings in 1888 (Thompson & Freeman 1930). For the next 10 years this species was landed from fishing grounds on inshore banks. The first commercial trawling in British Columbia was in 1908. Various flatfish species were the targets of this fishery but it did not succeed because of the 'roughness of the bottom encountered' (Wallace 1945). However, in 1910 cold storage facilities were built in the Queen Charlotte Islands to process flatfishes and a number of other species (Forrester *et al.* 1978). Three steam trawlers, about 120 ft (36 m) in length, and operated by local fishermen provided the fishery. This operation was largely unsuccessful due to limited markets for the catch and the difficulty of transporting it. The first small-boat trawling took place in sheltered waters around Vancouver about 1911 (Hart 1953). These vessels were about 40 ft (12 m) in length, powered by 16-hp motors and fished in shallow water for flatfish species such as English sole. The small nets were initially hauled aboard with rope warp by hand (Hart 1953). By 1920, vessels using two-boat paranzella nets and otter trawls replaced this type of operation with mechanical systems. At this time both Canadian and US trawlers were fishing areas in the Strait of Georgia.

During the 1920s and 1930s the small-boat trawl fleet in British Columbia increased steadily in numbers (Forrester *et al.* 1978) and flatfish species were the dominant component in the landings from the commercial fishery (Ketchen 1952). By the mid-1940s flatfish landings in British Columbia amounted to about 3000 t with the fleet concentrating mainly on the harvest of petrale sole (Pruter 1966) because of the excellent quality of its fillets. By 1948 flatfish landings from the trawl fishery in British Columbia had increased to about 8000 t, of which petrale sole accounted for about 75%. Most of this catch was taken by US trawlers off all areas of the British Columbia coast (Ketchen & Forrester 1966). In the 1950s the Canadian trawl fleet underwent a rapid expansion and began fishing increasingly in all areas of the coast. The principal targets of the fishery between 1950 and the late 1960s were Pacific cod (*Gadus macrocephalus*) and various flatfish species, most notably petrale sole, rock sole and English sole. The trawl fishery during that period took place largely on inshore fishing grounds at depths between 10 and 100 fathoms (18–181 m). By the mid-1970s the intense fishery on petrale sole had depleted these stocks (Fargo 1997) and the fishery was expanding to greater depths. In the late 1970s Canada declared extended jurisdiction over its offshore resources and the Canadian fleet began to explore offshore grounds in deep water in earnest. A deep-water fishery for Dover sole was initiated in 1977 off the northwest coast of the Queen Charlotte Islands and off the west coast of Vancouver Island in the late 1980s (Fargo & Workman 1995). Landings from these fisheries reached a peak of about 3500 t by 1996. Since the late 1990s landings of flatfishes from the British Columbia trawl fishery have averaged about 8000 t (Fargo 2000) and all commercial stocks are fully exploited.

12.3.6 USA

From 1964 to 1980 most flatfish were caught off the west coast of the USA, with only small amounts from the Gulf of Alaska. By the mid-1980s, joint-venture operations with US catcher vessels delivering to foreign processor vessels resulted in large catches for US fishermen in the Bering Sea, whose contribution dominated the US Pacific Ocean flatfish catch.

Since 1990, when only domestic fishing and processing has been allowed in US waters, flatfishes (except Pacific halibut) have been caught primarily with bottom trawls. Flatfish trawling vessels are generally of two types, large offshore catcher-processor vessels (Fig. 12.4) and smaller (<38 m) inshore catcher vessels. Based on the 1997 fishing season, Bering Sea catcher-processors had an average catch size of 23 t haul^{-1}, in which yellowfin sole were >20% of the catch and catcher vessels averaged 7 t haul^{-1}. In the Gulf of Alaska, catcher vessels averaged 6 t total catch haul^{-1} when rock sole comprised >20% of the catch. Small coastal trawlers are also typical of the west coast flatfish fishers who primarily target Dover sole. In Alaskan waters, flatfishes are considered under-utilised due to the total groundfish catch

Fig. 12.4 Modern era flatfish fishing vessel in the Bering Sea.

limits and the by-catch constraints imposed on the fishery. However, flatfishes (except arrowtooth flounder) are considered fully utilised off the west coast of the USA (NMFS 1999).

A fishery targeting spawning yellowfin sole and rock sole has occurred in the eastern Bering Sea. The fishing for yellowfin sole took place in nearshore spawning areas to take advantage of the high catch rates associated with fishing on dense spawning concentrations. For rock sole, there is an annual high-value, short duration roe fishery that supplies a Tokyo market where rock sole flesh is consumed along with the roe (headed fish is grilled with roe skein still in the body cavity), the combination of which is considered a delicacy. This fishery is usually limited to about 40 000 t but makes up most of the annual Bering Sea catch in the US zone for rock sole.

12.3.7 New Zealand

Along with other inshore fishes, flatfishes have been taken by the indigenous Maori people for centuries, mostly by spear and nets in estuaries and shallow bays. They were traded or sold to European settlers in the 1800s, and because of their accessibility were reasonably important as commercial fishing developed in the early 1900s. Catches fluctuated between 1000 t and 2500 t between 1930 and 1975, then rose to almost 5000 t in the mid-1980s (Table 12.4). In 1986 a quota of 6050 t was introduced (6670 t from 1991), as a component of a broader management regime which limited catches of all the main marine fish species by means of individual transferable quotas (ITQs). Flatfish stocks were not considered to be over-fished (although other finfish were), and were assumed to be variable but resilient. The flatfish quota was set higher than historic catches in order not to restrict fishing for other fish species in years when flatfishes became unusually abundant and were taken as inevitable by-catch. The quota has not been reached and catches from 1987 to 2000 have varied between 3000 t and 5000 t.

Four species dominate commercial catches (Paul 2000). Yellowbelly flounder are targeted by setnet in northern harbours. Sand flounder are more widespread; some are taken by setnet in several regions, but most are trawl-caught in shallow water, particularly around the South Island, as a target species and as by-catch. New Zealand sole are also targeted and are a by-catch in slightly deeper central and southern coastal waters. Southern lemon sole are mainly a trawl by-catch, again being more commonly taken in the south. (Both these 'soles' are pleuronectid flounders.) New Zealand brill and New Zealand turbot form only a very small trawl by-catch in southernmost regions.

The four main commercial species are also highly sought by recreational fishers and they form about 10% of the commercial catch. They are caught with setnets, beach seines and by hand-spearing. Fishing usually occurs along sandy beaches, in small bays, harbours, at large river mouths and other shallow-water areas. Minimum sizes and bag limits apply.

12.3.8 Australia

The estuaries of south Australia have been fished for thousands of years by the aborigines (Olsen 1991). Presumably, flatfishes were a component of this catch as they have been in the Coorong estuary during the twentieth century (Evans 1991). The flatfish fishery in south Australia and Tasmania is small compared with other Pacific Ocean nations. The Tasmanian catch, primarily of three species (unspecified), has averaged 31 t annually from 1990 to 1999 (Jordan & Lyle 2000). Although a wide range of fishing gears are used, gill nets account for the majority of the flatfish catch.

The Tasmanian recreational fishery catch of flatfishes is estimated to exceed the commercial catch. Based on accounts from recreational fishers' diaries, the estimate is around 50 t annually (Lyle 2000). The flatfish harvest using spears is greater than that from gill nets (flatfish constitute 5% of the recreational gill net harvest).

12.4 Economic importance

Flatfishes are used in many product forms depending on the type of fishery and harvesting country. In Japan and Korea the fish are either kept alive or sold as fresh fish to the local market, thereby commanding a high price. The flatfish fishery in Japan was worth $US 471 million in 1998 (retail price, MAFF 2000) and the value of the Korean harvest to the Republic of Korea retail market was estimated at $US 78.7 million in 2000 (MOMAF 2001). In countries such as New Zealand, Canada, Russia and the USA, where trawling is the main harvest method, flatfishes are either processed onboard the vessel (and frozen) or kept on ice or in refrigerated seawater until delivered to port for processing. Products typically associated with trawling include headed and eviscerated fish, fillets, whole frozen fish, kirimi or fishmeal. With the exception of Pacific halibut, these are lower value flatfish products which are usually produced in large quantities to be profitable. Recently, a small-scale live flatfish trawl fishery has operated in the Straits of Georgia, Canada.

In contrast to Japan and the Republic of Korea, the large flatfish harvest in the US section of the Bering Sea (frozen at sea) averaged for 1999 and 2000 has the following estimated first wholesale value: rock sole $US 40–60 million, yellowfin sole $US 30–45 million, flathead sole $US 25–30 million. Rex sole and other flatfishes in the Gulf of Alaska are valued at about $US 15–20 million (Hiatt & Terry, 1999). The headed and eviscerated product form constituted about two-thirds of this value and whole fish accounted for 22%. The 1997 west coast US value was reported at $US 6.9 million for Dover sole and $US 6.1 million for petrale sole and others (NMFS 1999). The high-value Pacific halibut was valued at $US 116.9 million for Alaska catches in 1999 (SAFE 2000).

12.5 Management

General flatfish assessment and management are covered in Chapter 14, but are highly developed for most species of the eastern Pacific (Fargo 1997, 2000; Pacific Fishery Management Council 1998), particularly those of highest economic value. Statistical catch-at-age models form the basis for most stock assessments and target harvest limits for species are generally achieved. In the western Pacific, similar assessments are completed and harvest limits established for species where information is available.

Passive and active management measures are distinguished here as either those involving no annual or in-season changes or those involving annual adjustments of quotas, time-area closures, gear characteristics, or other actions, respectively. Contemporary management throughout most of the Pacific Ocean is largely an active process involving annual stock assessments and restricting harvest to specific or multi-species quotas. However, this has not always been the case.

12.5.1 Western North Pacific

In the western Pacific, most fisheries had already been reduced to a low level of abundance by fishing pressure before any passive measures were implemented. Since 1953, with the passage of the National Fisheries Act, Korea has relied on traditional control devices to manage its fisheries. Devices used include mesh size restriction, minimum size of fish limits, closed seasons/periods, boat licences and gear limitations. A number of problems were addressed using these traditional control devices. For example, mesh and minimum landing size regulations were adopted to avoid the dangers of harvesting fish before they reach full maturity.

In Japan, a regional management system has been in place since 1949 whereby fishing regulations range from simple to sophisticated depending on the management body in each prefecture and may include: minimum size limits, mesh size restrictions, closed areas and seasons, licence and effort limitation. In instances where resources are harvested by fishers from neighbouring prefectures, regional Fishery Coordination Committees are in place.

In the western Pacific, fishery management is more localised than its counterpart on the opposite side of the Pacific Ocean. The present coastal fishery management system in Japan resulted from a 1949 fishery law whereby 'community-based' fishery management plans were established for each prefecture through the formation of Fishery Coordination Committees. These Committees were established to promote the productivity and conservation of regional fishery resources through the regulation of fishing and aquaculture activities. Two separate fisheries management systems are used in Japan: a total allowable catch (TAC) system for offshore pelagic and migratory species and a community-based management system for small-scale coastal fisheries such as flatfish. This is similar to the governance structure of Korea where, after extending its fisheries jurisdiction in 1994, the Korean government adopted a TAC-based fisheries management system. This system was implemented in January 1999. However, the small-scale coastal fisheries such as flatfish are still managed by traditional control devices under the local governmental authority, although the TAC-based management system will be continuously extended eventually to include small-scale coastal fisheries.

12.5.2 Eastern North Pacific

In the eastern Pacific Ocean, active management of flatfish fisheries was instituted only relatively late in the history of most fisheries. While these fisheries have existed in the eastern Pacific for over 150 years and in the western Pacific for even longer, the first records of either passive or active management measures for flatfish fisheries, with the exception of Pacific halibut, date only from the 1940s (Harry 1956).

For most eastern Pacific flatfishes, measures to control landings were introduced initially not to achieve stock management goals but rather to achieve economic goals. For example, minimum mesh size restrictions were introduced primarily because fish plants did not want to process fish smaller than approximately 30 cm (12 in). Early restrictions also occurred because one of the primary markets for flatfishes was for feed used in farming minks for the fur trade and farm owners did not want juvenile fish in the feed (personal communication, R. Demory, Oregon Department of Fish and Wildlife). The fishing industry had previously accommodated such limitations through sorting and discarding of smaller-sized fish at sea, resulting in high levels of discard mortality for these smaller fish. Increasing the minimum mesh size in trawls resulted in altered selectivity parameters for the fishing gear and passive sorting of the catch prior to retrieval of the fishing gear. Indeed, the previously unaccounted mortality associated with discarding of fishes from the vessels was a limitation of the data collection programmes in place. Only landings of target species were monitored, rather than the catch, so that at-sea discards went undetected. With the exception of Pacific halibut, the first quota restrictions on eastern Pacific flatfishes were those imposed only after the extension of fisheries jurisdictions off the coasts of Canada and the USA in 1977 (French *et al.* 1982).

The structure of governance for eastern Pacific flatfish fisheries is generally one of limited-entry, domestic licensing, without individual harvester quota shares on a total quota. All US flatfish fisheries, with the exception of that for Pacific halibut, are managed in this manner. Fisheries are managed by regional Fishery Management Councils, which set quotas based on assessments conducted by State and Federal agencies. Enforcement of fishery management is accomplished through these agencies. Stakeholder and public views on assessment and management are obtained via the Council process. The Canadian and US fisheries for Pacific halibut, as well as all other Canadian flatfish fisheries, are managed on a restricted-entry quota-share basis. For Pacific halibut, overall quotas are determined by the International Pacific Halibut Commission and allocation among stakeholders is accomplished by the regional Councils in the USA and the Department of Fisheries and Oceans (DFO) in Canada. Stakeholder views on Pacific halibut management are provided within the Commission advisory framework. For other Canadian flatfish fisheries, quotas are set by the DFO based on scientific advice provided by DFO scientists and industry consultants. Stakeholder views are generally limited to extractive users and are obtained via an advisory committee framework.

The management of the Pacific halibut occupies a unique place in the history of eastern Pacific flatfishes (International Pacific Halibut Commission 1998). The stock and fishery have been managed by the International Pacific Halibut Commission, created by a treaty between the USA and Canada, since 1923. The Commission was created by the two governments in response to fishing industry concerns about low stock levels and over-exploitation of the resource. The Commission established a strong scientific research programme as well as an extensive data collection process. While the initial Commission regulatory actions concerned only season length, by 1930 it had moved into quota-based management of halibut by regulatory area, well in advance of actions by any other agency managing flatfishes. The long history of research and management by the Commission and commitment by the contracting parties have resulted in one of the most successful fishery management programmes in the world.

The history of Russian fisheries and their management is poorly documented (Pautzke 1997). While reporting of high-seas Soviet fisheries in the 1960s was extensive and translated into English, similar and contemporary documentation for the coastal fisheries targeting flatfishes is not available. In recent years, some reports of Russian catches and management measures are becoming available (Ivankova 1995; Kodolov 1995; Kodolov & Matveychuk 1995; Nikolenko 1995). All of these reports indicate major depletion of flatfish stocks in the Sea of Okhotsk and off the Kamchatka Peninsula in the late 1960s and 1970s. In most instances, complete closures of fisheries were required for 5–10-year periods before significant fisheries re-occurred. While flatfish yields are now in the tens of thousands of tons, it is not clear that fisheries are being managed in a sustainable manner.

The governance structure of Russian fisheries is a combination of regional and federal control, although it is somewhat difficult to determine the exact regulatory structure and its effectiveness. Nonetheless, annual assessments lead to quotas that undergo an elaborate allocation process involving domestic bidding for quotas among regional fishing districts and companies, as well as by foreign fishing companies. This structure may reflect a transition-type process from the previous centralised and command economy to the present market-driven economy.

12.5.3 Australia and New Zealand

The Australian Government passed fisheries legislation in 1991 to establish the Australian Fisheries Management Authority (AFMA) whereby a board was formed to provide policy advice and to focus on providing more secure access rights for fishers. Board membership included stakeholders from the government, natural resource management, fisheries science, marine ecology, business management and industry members with expertise in fishing and processing operations. Although the waters seaward of 3 nautical miles were managed by a

TAC with individual transferable quotas for stakeholders prior to the AFMA (Rayns 1997), the current management system has enabled stakeholders to have a greater voice in policy decisions and has contributed to the resolution of many contentious issues.

New Zealand's fishery management system represents one of the oldest examples of a tradable quota system. Initially, from 1938 to 1963, the domestic inshore fishery was managed by strict licensing and tight controls at a time when there was only minimal national interest in fishing. In 1963 the fishery was completely deregulated and offshore resources were fished by fleets from Korea, Japan and Russia. Fearing a collapse of the inshore fishery, New Zealand passed the Fisheries Act of 1983 whereby property rights to the allowable catch were introduced as transferable quotas (Clark *et al.* 1988). These transferable quotas, which also incorporated biological preservation and economic development into fisheries management, were expanded to the deep-water fisheries in 1986.

12.5.4 Data collection

For both the western and eastern Pacific, the lack of any management measures early in the history of flatfish fisheries resulted primarily from: (1) the absence of data collection on either total removals from the stock, fishery performance, or changes in the biological characteristics of the target species; (2) the monitoring of indices of performance or abundance that were insensitive to changes in the dynamics of the target species; or (3) the absence of a governance authority and structure for the target species.

Effective fisheries management is incumbent upon adequate data collection. In the eastern Pacific, although data collection on directed removals from the stocks of target species was instituted from the early stages of many fisheries (i.e. some of item (1) above), in most instances the shortcomings of both (2) and (3) apply to the history of these fisheries (Alverson *et al.* 1964). For fisheries where directed removals constituted the majority or the significant fishing mortality on the stock, data gathered for the directed fishery may have provided information for monitoring the gross dynamics of the stock, depending on the underlying age composition and age at recruitment. For species with early age at recruitment and few age groups in the exploited stock, such fishery statistics can reflect relative abundance if the entire range of the stock is exploited and monitored, and technological changes in the fishery which affect fishing power have been minimal. Conversely, fishery data from multi-year-class stocks with late age at recruitment, or where fishing technology and pattern of fishing have changed with exploitation history, often provide little insight into stock abundance. Unfortunately, technological change in fisheries generally precedes the detection of stock changes through fishery statistics, as harvesters act to maintain catches or catch rates in the face of declining abundance.

Data on landings of flatfishes from the eastern Pacific have been collected since the 1920s and in some instances even longer (Fielder 1928). Notable programmes include those coordinated by the Pacific Marine Fisheries Commission (later the Pacific States Marine Fisheries Commission), and drawing on programmes from within agencies of the Pacific coastal states of the USA, as well as that of the Department of Fisheries and Oceans Canada. Through the International Pacific Halibut Commission, the USA and Canada also collected extensive landing and fishery statistics for the Pacific halibut fisheries from the 1920s. In the multilateral arena, data collection programmes began much later. The International North

Pacific Fisheries Commission, created in 1953, was the organisation responsible for collation of fisheries statistics for fisheries in the northeastern Pacific. Although the mandate of the Commission concerned primarily the protection of salmon on the high seas, its data collection activities for flatfishes in the Bering Sea contributed much to the implementation of management programmes. However, data on landings alone do not capture all of the fishing mortality on the stock from either directed fishing or that arising from fishing effort on other target species. Indeed, the effectiveness of stock management measures in their early stages was severely limited by the lack of accounting for such incidental fishing mortality (Clark & Hare 1998; International Pacific Halibut Commission 1998; Sullivan *et al.* 1999). The data collected by observer programmes for distant-water fleets beginning in the 1960s, and for domestic fleets beginning in the 1980s, allowed calculation of these incidental mortality impacts. Contemporary stock assessments now incorporate many more sources of fishery-generated mortality than was possible in the early stages of these fisheries.

The major eastern Pacific fisheries for which data collection programmes existed from their early stages also benefited from the nearshore distribution of the target species of flatfishes. Almost all removals in the fisheries were therefore by domestic fleets, with the exception of trawl fisheries in the Bering Sea and Gulf of Alaska. Although statistics collection for some of these fisheries has been comprehensive, particularly by the Pacific States Marine Fisheries Commission and the Department of Fisheries and Oceans Canada, it has not often been accompanied by either active or passive management measures to ensure stock productivity. As with western Pacific fisheries, collection of such statistics alone has afforded scant protection from stock declines (Pacific Fishery Management Council 1998, 2000).

This nearshore distribution meant that fisheries were effectively under coastal state control, even prior to the advent of 200-mile extended fisheries jurisdiction. The major exceptions to this scenario concerned the major flatfish trawl fisheries of the Bering Sea and Gulf of Alaska, and the longline fishery for Pacific halibut throughout its range. In the case of the former fisheries, their distribution extended beyond 3- or 12-mile territorial waters and management required multilateral cooperative agreements (International North Pacific Fisheries Commission 1953). In the case of Pacific halibut, seasonal spawning migrations of adult fishes and movement of juvenile fishes across national boundaries of the USA and Canada necessitated bilateral management (International Pacific Halibut Commission 1998).

References

Alverson, D.L., Pruter, A.T. & Ronholt, L.L. (1964) *A study of demersal fishes and fisheries of the northeastern Pacific Ocean*. H.R. McMillan Lectures in Fisheries. Institute of Fisheries, University of British Columbia.

Clark, I., Major, P.J. & Mollett, N. (1988) Development and implementation of New Zealand's ITQ Management System. *Marine Resource Economics,* **5**, 325–349.

Clark, W.G. & Hare, S.R. (1998) Accounting for bycatch in management of the Pacific halibut fishery. *North American Journal of Fisheries Management,* **18**, 809–821.

Evans, D. (1991) The Coorong – a multi-species fishery. Part 2. Evolution of fishing methods, technology and personal experiences 1930–1966 and gear statistics 1972–1989. *Fisheries Research Paper, No. 22*, pp. 38–60. Department of Fisheries (South Australia), Adelaide.

Fargo, J. (1997) Flatfish stock assessments for the west coast of Canada for 1997 and recommended yield options for 1998. *Canadian Stock Assessment Secretariat Research Document 97/36.*

Fargo, J. (2000) Flatfish stock assessments for the west coast of Canada for 2000 and recommended catches for 2001. *Canadian Stock Assessment Secretariat Research Document 2000/199.*

Fargo, J. & Workman, G.D. (1995) Results of the Dover sole (*Microstomus pacificus*) biomass survey conducted off the west coast of Vancouver Island February 13–27, 1995. *Canadian Manuscript Report of Fisheries and Aquatic Sciences*, **2340**, 1–75.

Fielder, R.H. (1928) Fishery industries of the United States. *Bureau of Fisheries Document No. 1067.* US Department of Commerce.

Forrester, C.R., Beardsley, A.J. & Takahashi, Y. (1978) Groundfish, shrimp and herring fisheries in the Bering Sea and Northeast Pacific – historical catch statistics through 1970. *International North Pacific Fisheries Commission Bulletin*, **37**, 1–147.

French, R., Nelson, R. & Wall, J. (1982) Role of the United States observer program in management of foreign fisheries in the Northeast Pacific Ocean and Eastern Bering Sea. *North American Journal of Fisheries Management*, **2**, 122–131.

Harry, G.Y. (1956) *Analysis and history of the Oregon otter trawl fishery*. PhD thesis, University of Washington.

Hart, J.L. (1953) The trawl fishery on the Pacific coast of Canada. *Proceedings of 7th Pacific Science Congress (1949)*, **4**, 425–427.

Hiatt, T. & Terry, J. (2000) Economic status of the groundfish fisheries off Alaska, 1999. In: *Stock assessment and fishery evaluation report for the groundfish fisheries of the Gulf of Alaska and Bering Sea/Aleutian Islands area. Appendix C*. North Pacific Fishery Management Council, Anchorage, AK.

International North Pacific Fisheries Commission (INPFC) (1953) International Convention for the High Seas Fisheries of the North Pacific Ocean. Entered into effect, June 1953.

International Pacific Halibut Commission (1998) The Pacific halibut: biology, fishery, and management. *Technical Report No. 40*, 1–64.

Ivankova, Z.G. (1995) Dynamics of the flounder populations on the northwest Japan Sea. In: Proceedings of the International Symposium on North Pacific Flatfish. *Alaska Sea Grant College Program Report No. 95–04*, pp. 443–449. University of Alaska, Fairbanks, AK.

Jordan, A.R. & Lyle, J.M. (2000) *Tasmanian scalefish fishery assessment – 1999*. Tasmanian Aquaculture and Fisheries Institute, University of Tasmania, pp.1–56.

Kalland, A. (1995) *Fishing Villages in Tokugawa, Japan*. University of Hawaii Press, Honolulu.

Ketchen, K.S. (1952) The British Columbia trawl fishery. *Canadian Department of Fisheries Trade News*, **4**, 3–8.

Ketchen, K.S. (1976) Catch and effort statistics of the Canadian and United States trawl fisheries in waters adjacent to the British Columbia coast 1950–75. *Fisheries and Marine Service Data Record, No. 6*, 1–56.

Ketchen, K.S. & Forrester.C.R. (1966) Population dynamics of the petrale sole, *Eopsetta jordani*, in waters off western Canada. *Bulletin of the Fisheries Research Board of Canada*, **153**, 1–195.

Kodolov, L.S. (1995) Stock condition of Pacific halibut (*Hippoglossus stenolepis*) in the Northwestern Bering Sea. In: Proceedings of the International Symposium on North Pacific Flatfish. *Alaska Sea Grant College Program Report No. 95–04*, pp. 481–495. University of Alaska, Fairbanks, AK.

Kodolov, L.S. & Matveychuk, S.P. (1995) Stock condition of Greenland turbot (*Reinhardtius hippoglossoides matsuurae* Jordan et Snyder) in the Northwestern Bering Sea. In: Proceedings of the International Symposium on North Pacific Flatfish. *Alaska Sea Grant College Program Report No. 95–04*, pp. 451–465. University of Alaska, Fairbanks, AK.

Kupriyanov, S.V. (1996) Distribution and biological indices of yellowfin sole (*Pleuronectes asper*) in the southwestern Bering Sea. In: *Ecology of the Bering Sea: A Review of the Russian Literature, 1996* (eds O. Mathisen & K. Coyle). pp. 203–216. Alaska Sea Grant College Program, University of Alaska, Fairbanks, AK.

Lord, J.K. (1866) *The Naturalist in Vancouver Island and British Columbia. (1)*. Richard Bentley, London.

Lyle, J.M. (2000) Assessment of the licensed recreational fishery of Tasmania (Phase 2). *Final Report to Fisheries Research and Development* Corporation (96/161).

MAFF (2000) *Statistical Yearbook of Fisheries and Aquaculture Production*. Statistics and Information Division, Ministry of Agriculture, Forestry and Fisheries, Tokyo.

Minami, T. & Tanaka, M. (1992) Life history cycles in flatfish from the northwestern Pacific, with particular reference to their early life histories. *Netherlands Journal of Sea Research,* **29**, 35–48.

MOMAF (2001) *Statistical Yearbook of Korean Fisheries*. Ministry of Maritime Affairs and Fisheries, Seoul, Korea.

Murai, S., Gangmark, H.A. & French, R.R. (1981) All-nation removals of groundfish, herring, and shrimp from the eastern Bering Sea and northeast Pacific Ocean, 1964–80. *NOAA Technical Memorandum NMFS F/NWC-14*. US Department Commerce.

National Research Council, Committee on the Bering Sea Ecosystem (1996) *The Bering Sea Ecosystem*. National Academy Press, Washington, DC.

Nikolenko, L.P. (1995) Dynamics of abundance and biomass of Greenland turbot (*Reinhardtius hippoglossoides*) in western Kamchatka in 1976–1993. In: Proceedings of the International Symposium on North Pacific Flatfish. *Alaska Sea Grant College Program Report No. 95–04*, pp. 467–480. University of Alaska, Fairbanks, AK.

NMFS (1999) Our living oceans. US Department Commerce, National Oceanic and Atmospheric Administration, National Marine Fisheries Service. *NOAA Technical Memorandum, NMFS-F/SPO-41*.

Olsen, A.M. (1991) The Coorong – a multi-species fishery. Part 1. History and development. *Fisheries Research Paper Department of Fisheries (South Australia) Adelaide, No. 22*, 1–37.

Pacific Fishery Management Council (1998) *Status of the Pacific Coast Groundfish Fishery through 1998 and Recommended Acceptable Biological Catches for 1999. Appendix: Stock Assessment and Fishery Evaluation*. Pacific Fishery Management Council, Portland, OR, USA.

Pacific Fishery Management Council (2000) *Status of the Pacific Coast Groundfish Fishery through 2000 and recommended Acceptable Biological Catches for 2001. Appendix: Stock Assessment and Fishery Evaluation*. Pacific Fishery Management Council, Portland, OR.

Paul, L. (2000) *New Zealand Fishes: Identification, Natural History & Fisheries*. Reed Books, Auckland.

Pautzke, C.G. (1997) *Russian Far East Fisheries Management*. North Pacific Fishery Management Council, Anchorage, AK.

Pruter, A.T. (1966) Commercial fisheries of the Columbia River and adjacent ocean waters. *Fisheries Industrial Research,* **3**, 17–66.

Rayns, N. (1997) Partnerships in Australian commonwealth fisheries management. In: *Frontiers in Ecology: Building the Links* (eds N.I. Klomp & I.D. Lunt). Elsevier Science, Amsterdam.

SAFE (2000) *Stock assessment and fishery evaluation report for the groundfish resources of the Bering Sea/Aleutian Islands 2000*. Compiled by the Bering Sea/Aleutian Islands Plan Team, North Pacific Fishery Management Council, Anchorage, AK.

Stewart, H. (1977) *Indian Fishing*. University of Washington Press, Seattle, WA.

Sullivan, P.J., Trumble, R.J. & Parma, A.N. (1999) The Pacific halibut stock assessment of 1997. *International Pacific Halibut Commission Scientific Report No. 79*, 1–84.

Thompson, W.F. & Freeman, N.L. (1930) History of the Pacific halibut fishery. *International Fisheries Commission Report*, **5**, 1–61.

Trumble, R.J. (1998) Northeast Pacific flatfish management. *Journal of Sea Research,* **39**, 167–181.

Trumble, R.J., Neilson, J.D., Bowering, R.D. & McCaughran, D.A. (1993) Atlantic halibut (*Hippoglossus hippoglossus*) and Pacific halibut (*H. stenolepis*) and their North American fisheries. *Canadian Bulletin of Fisheries and Aquatic Sciences,* **227**, 1–84.

Wallace, F.W. (1945) Thirty years progress in Canada's fish industry, 1914–44. *Canadian Fisheries Manual*, 1945.

Williams, G.H. & Wilderbuer, T.K. (1995) Discard mortality rates of Pacific halibut bycatch: fishery differences and trends during 1990–1993. In: Proceedings of the International Symposium on North Pacific Flatfish. *Alaska Sea Grant College Program Report No. 95–04*, pp. 611–622. University of Alaska, Fairbanks, AK.

Chapter 13
Tropical flatfish fisheries

Thomas A. Munroe

13.1 Introduction

Flatfishes are common species in most tropical marine fish assemblages, especially those found on soft-bottom habitats in estuaries and on a variety of substrata on the inner continental shelf. Tropical seas represent one of the largest marine biomes on earth, and within these waters from nearshore to about 100 m depth on the continental shelf are found diverse communities of marine fish species, including some of the most diverse assemblages of flatfishes. Extensive demersal fisheries exploit resources of shallow tropical seas, and along tropical coastlines large human populations live, work and fish. So, it is within tropical nearshore seas and estuaries that harvesting of marine resources and other anthropogenic activities affect the greatest diversity of flatfish species found anywhere.

Although tropical flatfishes are frequently caught, are species-rich, and even sometimes numerically abundant, most are thin-bodied, small-sized species reaching only to 30–40 cm TL (Fig. 13.1). Seldom do flatfishes exceed 5% of fish biomass of tropical demersal fish communities and they contribute relatively little economic value to fisheries (Manickchand-Heileman 1994; Pauly 1994). High species diversity with low relative abundance and biomass for individual species is not unique to flatfish assemblages, rather, these features

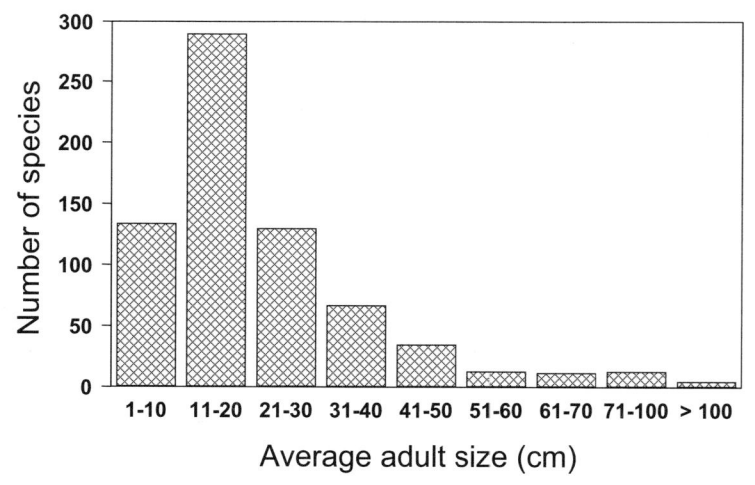

Fig. 13.1 Frequency distribution of relative adult size for 699 nominal species of flatfishes.

characterise demersal fish communities on tropical continental shelves in general (Longhurst & Pauly 1987; Manickchand-Heileman 1994). Even though flatfishes make only minor economic contributions to tropical fishery landings, subsistence and artisanal fishers, by their sheer numbers and intensity, harvest large numbers of flatfishes, and even larger numbers of tropical flatfishes are killed or damaged as by-products of industrial trawl fisheries operating in these waters.

Because faunas in most tropical shallow marine environments are increasingly affected or destroyed through a wide variety of anthropogenic activities including fishing, pollution and habitat degradation, conservation of these habitats and their faunas is of paramount importance. This chapter focuses on families and species of flatfishes harvested in tropical fisheries, relative landings of these fishes, and discusses impacts that tropical demersal fisheries have on their populations. It also addresses issues relative to difficulties incurred in attempting to regulate nearshore tropical demersal fisheries, management options proposed for conserving demersal fishes, and measures for protecting marine biodiversity in general. Detailed information on fisheries in tropical oceans appears in Longhurst & Pauly (1987) and Pauly & Murphy (1982) and references therein, while Blaber (1997) and referenced studies provide information on fish and fisheries of tropical estuaries. Landings data appearing in text and tables were extracted from FAO (1997) and FAO (2001).

13.2 Main species and nature of the fisheries

13.2.1 Habitats

Within tropical waters (Fig. 13.2), flatfishes are most abundant: (1) on soft, silty muds and sandy sediments typically found in shallow estuarine and mangrove habitats; (2) in rivers and nearby coastal waters; (3) on sand patches interspersed among submerged aquatic vegetation in coastal waters; (4) on a variety of sediments interspersed throughout coral and other reef-associated habitats at both continental and insular locations; and (5) on broad expanses of soft-bottom habitat on the continental shelf and upper continental slope to depths of about 2000 m. Relatively few flatfishes, including several tonguefishes (*Cynoglossus* spp.), achirid soles (*Achirus, Achiropsis*) and some soleids (*Brachirus* spp., *Dagetichthys lakdoensis*, *Achiroides* spp.), inhabit freshwaters. Few of these are harvested and marketed regionally and little is known regarding their general ecologies, magnitude of fishery catches, or relative importance to local fisheries.

The greatest diversity and largest populations (and majority of commercial landings) of flatfishes in tropical regions are those found on large, contiguous areas of muddy, sandy and calcareous sediments from the shoreline to about 50 m depth on the inner continental shelf. As nearly 30% of the world's total continental shelf area is found throughout tropical oceans (Longhurst & Pauly 1987), and many tropical continental shelves feature large areas dominated by soft substrata, considerable amounts of soft-bottom habitat are available to flatfishes in these regions. Widespread occurrence in the tropics of expansive areas of continental shelf dominated by various soft substrata, with large areas greatly affected by sediments discharged from large rivers, may be little realised and somewhat surprising given that most attention to tropical marine habitats focuses on coral reefs and mangroves (Longhurst & Pauly 1987).

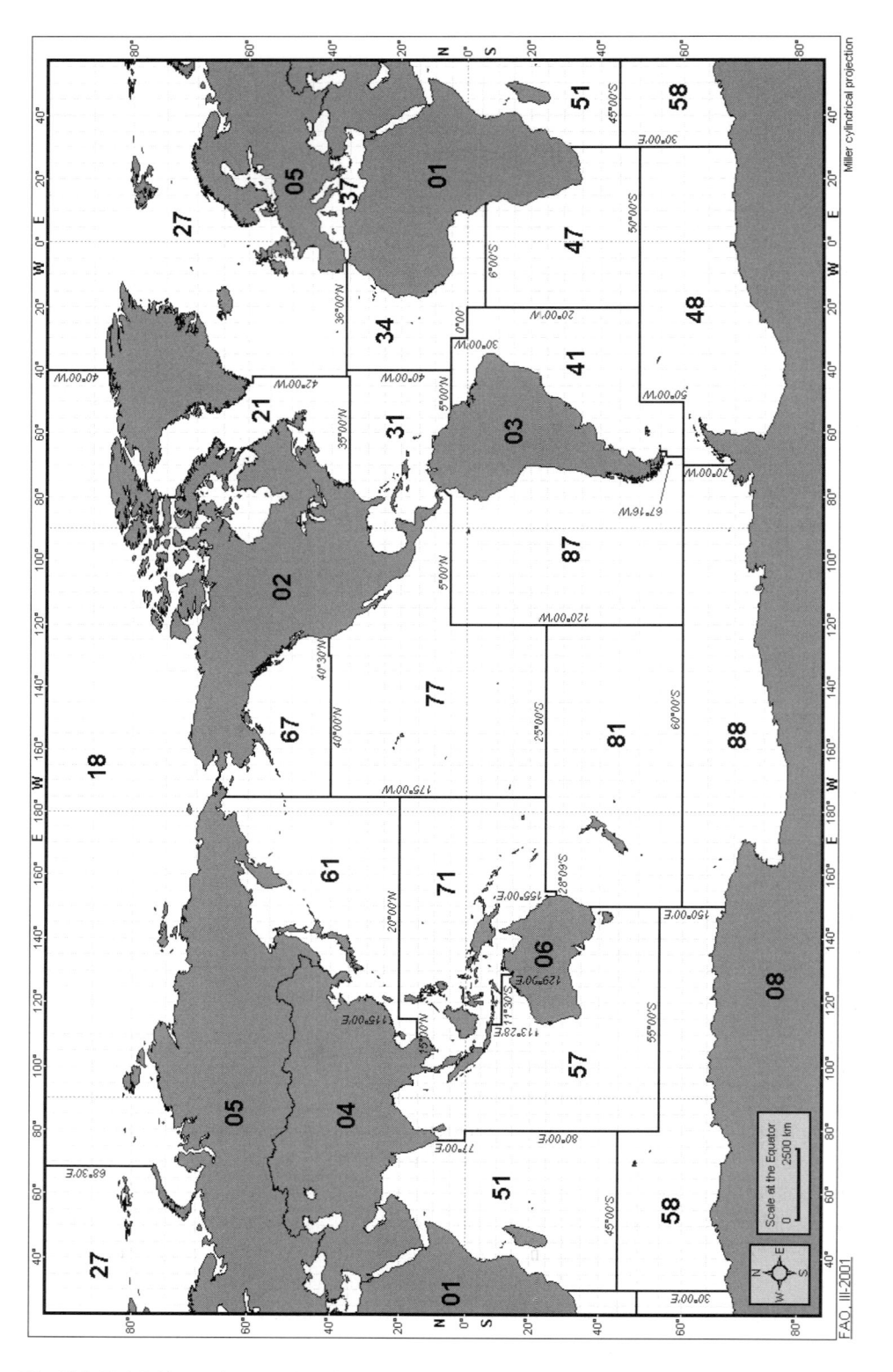

Fig. 13.2 FAO fishing regions mentioned in the text. The names of the numbered tropical areas are given in the footnote to Table 13.1.

Table 13.1 FAO fishery areas, annual flatfish landings for selected years, mean landings of flatfishes per area for 1990–1999, and associated statistics for latitudinal extent and relative continental shelf areas for tropical regions where flatfishes are captured

FAO region[a]	Latitude range[b]	Shelf area (km²10⁻³)[c]	Flatfish landings (mt)				Mean landings (mt yr⁻¹) (1990–1999)[d]
			1970	1980	1990	1999	
WCP	15°N–25°S	4610	6000	16601	15122	17856	15534
SWA	5°N–60°S	1950	5000	4452	12358	8821	11885
WIO	20°N–35°S	1620	13500	9323	25215	15888	24027
EIO	20°N–45°S	1380	1300	5944	12579	32686	22163
WCA	35°N–5°N	1370	2500	4917	1784	1673	2378
ECA	35°N–6°S	480	8900	31377	26567	40336	37426
ECP	40°N–5°N	450	6000	7393	8824	6470	8038

[a] WCP, Western Central Pacific (FAO area 71); SWA, Southwest Atlantic (FAO area 41); WIO, Western Indian Ocean (FAO area 51); EIO, Eastern Indian Ocean (FAO area 57); WCA, Western Central Atlantic (FAO area 31); ECA, Eastern Central Atlantic (FAO area 34); ECP, Eastern Central Pacific (FAO area 77).
[b] Approximate range of shelves (after Pauly 1994).
[c] Coastal waters, down to 200 m (Gulland 1971).
[d] Data from FAO (1997, 2001).

The largest tropical continental shelves (Table 13.1) are those in the western Central Pacific, western and eastern Indian Ocean, and western Central Atlantic (Longhurst & Pauly 1987). Much smaller shelf areas occur off east and west coasts of Africa, and in the tropical eastern Pacific and Red Sea. Continental shelves of the tropical western Pacific and Indian oceans feature long continental margins, small isolated continental blocks and abundant coral reefs. Large amounts of shelf habitat are in relatively shallow waters (<100 m), and here flatfish assemblages are especially diverse.

Throughout the tropics, flatfish assemblages inhabiting the outer continental shelf and upper continental slope are distinctly less diverse (ca. 171 species) than those occurring in shallower seas, and most are unavailable to the vast majority of fishers. Rarely, if ever, do deep-water species reach sufficient size or biomass, or occur in sufficient abundance, to contribute significantly to regional commercial landings. It is unlikely that significant, undiscovered and unexploited populations of large flatfishes are yet to be found in deep-water tropical habitats (Chapter 2).

A wide variety of tropical flatfishes occur in estuaries and adjoining nearshore habitats (Blaber 1997). Large tropical estuaries, including those of the Amazon, Orinoco, Zaire, Zambezi, Niger, Ganges and Mekong rivers, receive drainage from enormous catchments and owing to heavy rainfall they have tremendous outflows that strongly modify the morphology, sedimentary environment and coastal oceanography of adjoining shelf areas, which are usually dominated by soft substrata (Longhurst & Pauly 1987; Blaber 1997). Tropical estuaries also include contiguous shallow coastal waters that feature reduced salinities. This 'estuarisation' of tropical coastal oceans often creates conditions where no clearly defined boundary exists between estuary and sea (Longhurst & Pauly 1987). Shelf regions featuring low-salinity surface waters have high turbidity, muddy deposits and diverse flatfish assemblages. Shelf estuarisation is important in the Bay of Bengal, South China Sea, Gulf of Panama, Gulf of Guinea, and off northern South America (Lowe-McConnell 1987; Blaber 1997).

Reef-associated flatfishes, although not as well known as counterparts in other shallow-water habitats, are common and diverse in tropical waters, especially in the tropical Indo-West Pacific including Australia's Great Barrier Reef, the Indonesian Archipelago, the Philippines and New Guinea, and at many Central and West Pacific islands, in the Red Sea, Indian Ocean and Caribbean Sea. Local population densities of reef-associated flatfishes are generally low with individuals widely dispersed.

13.2.2 Commercially important species and/or taxa

Flatfishes landed in tropical fisheries are taxonomically different and significantly more diverse than those of temperate areas, a situation typical of tropical demersal fish communities in general (Longhurst & Pauly 1987). Although flatfishes are common elements of tropical marine fish assemblages found on soft-bottom habitats, most are only about 40 cm TL (Fig. 13.1) and are too small and occur in too low abundance to be directly targeted by industrial fisheries. Only a small proportion of the total diversity of flatfishes taken in regional tropical fisheries has commercial value as species marketed directly for human consumption. For example, of 77 flatfish species occurring off India, only the Malabar sole (*Cynoglossus macrostomus*) constitutes an important fishery species (Rajaguru 1992). Most smaller flatfishes captured in fisheries belong to diverse families such as the Soleidae, Cynoglossidae, Bothidae and Paralichthyidae. Many of the same species, and others in the Poecilopsettidae, Citharidae and Samaridae, are also common by-catch species in industrial fisheries where they are either discarded at sea after capture, or if landed, are processed into fishmeal or other products.

Larger-sized tropical flatfishes marketed for human consumption include the Indian and spottail spiny turbots (*Psettodes erumei* and *P. belcheri*, respectively), some paralichthyids (*Pseudorhombus* spp., *Paralichthys* spp., *Cyclopsetta* spp., *Syacium* spp.), bothids (especially *Bothus* spp.), a few soles (*Solea* spp., *Synaptura* spp., *Brachirus* spp., *Pardachirus* spp.), smaller numbers of achirids (*Trinectes* spp., *Achirus* spp.), tonguefishes (mainly *Cynoglossus* spp., especially Malabar sole and Senegalese tonguesole, *C. senegalensis*, Guinean tonguesole, *C. monodi*, and *Paraplagusia* spp.) and very few species of *Symphurus*. Even for larger flatfishes with commercial importance, such as Malabar sole and spiny turbots, most landings result from by-catch of other fisheries (Rajaguru 1992; Khan & Nandakumaran 1993; Jayaprakash & Inasu 1999; Jayaprakash 2000).

The taxonomic composition of flatfishes taken as by-catch in tropical fisheries and marketed commercially varies regionally and reflects both availability and relative abundance of the fauna found in that particular area, especially that occurring on shrimp-fishing grounds. Throughout the Indo-West Pacific, by-catch generally includes a large variety of flatfishes such as spiny turbot, numerous tonguefishes, soleids (*Brachirus* spp., *Synaptura* spp., *Pardachirus* spp., *Zebrias* spp.), bothids (especially *Bothus* spp.), paralichthyids (*Pseudorhombus* spp.) and several largescale flounders (citharids). The oriental sole, *Brachirus orientalis*, where it occurs in abundance, constitutes a substantial source of protein for coastal populations of Pakistan (Atiqullah 2001) and elsewhere. The Indian spiny turbot is of moderate commercial interest in fisheries of India (Pradham 1969; Hussain 1990; Mathew *et al.* 1992), Sri Lanka (De Bruin *et al.* 1995), and elsewhere. In the Persian Gulf (Carpenter *et al.* 1997), flatfish species usually marketed fresh are Indian spiny turbot, leopard flounder (*Bothus pantherinus*), largetooth flounder (*Pseudorhombus arsius*), deep flounder

(*P. elevatus*), Javan flounder (*P. javanicus*), Malayan flounder (*P. malayanus*), elongate sole (*Solea elongata*), convict zebra sole (*Zebrias captivus*) and Indian zebra sole (*Z. synapturoides*). Tonguefishes (largescale tonguesole, *Cynoglossus arel*; fourlined tonguesole, *C. bilineatus*; hooked tonguesole, *C. carpenteri*; shortheaded tonguesole, *C. kopsii*; and speckled tonguesole, *C. puncticeps*), oriental sole, finless sole (*Pardachirus marmoratus*) and Stanaland's sole (*Solea stanalandi*) are also marketed fresh and frozen, or are dried and salted. In Somalia, many of these same species, as well as Commerson's sole, *Synaptura commersonii*, tonguefishes (hooked tonguesole and doublelined tonguesole, *Paraplagusia bilineata*), and various less common species, appear in commercial markets. Tonguefishes are important species taken inshore on soft bottoms off Mozambique (Brinca *et al.* 1981) and off South Africa, where 4 of 13 species are caught in appreciable amounts (Booth & Walmsley-Hart 2000).

Off tropical West Africa (Fager & Longhurst 1968; Mgawe 1998), flatfish catches consist primarily of spottail spiny turbot in coastal and estuarine fisheries, large tonguefishes (Canary tonguesole, *Cynoglossus canariensis*, and Senegalese and Guinean tonguesoles) in shallow (20–40 m), soft-bottom areas, and Guinean and Portuguese soles (*Synaptura cadenati, S. lusitanica*, respectively) and other unidentified soles (*Solea* species), wide-eyed flounder (*Bothus podas*) and Atlantic spotted flounder (*Citharus linguatula*) at 30–40 m on the inner shelf.

Estuarine flatfishes contributing to subsistence and artisanal fisheries throughout the Indo-West Pacific include Indian spiny turbot, largetooth flounder and several other species of *Pseudorhombus*, several tonguefishes (*Cynoglossus* spp.) and ovate sole, *Solea ovata* (Hussain 1990; Blaber 1997). Largetooth flounder is the most abundant large flatfish in some estuaries of Northern Australia, followed by deep flounder (Blaber 1997). Black sole, *Achlyopa nigra,* is also taken in Northern Australian estuaries. Off Somalia, doublelined tonguesoles are commonly captured in fisheries operating in coastal and estuarine waters, and in Morrumbene Estuary, Central Mozambique, largetooth flounder and blackhand sole (*Solea bleekeri*) are commonly harvested from mangrove areas. Spottail spiny turbots are taken in larger estuaries of tropical West Africa (Fager & Longhurst 1968).

In the eastern Central Pacific and western Central Atlantic, most flatfishes composing fisheries landings belong to the same genera, but different species within each genus are landed in respective oceans. Landed species include larger paralichthyids (species of *Paralichthys, Syacium, Citharichthys, Cyclopsetta*) and bothids (primarily *Bothus* spp.) taken in deeper water on less muddy bottoms, while inshore a few, larger species of achirids (*Achirus* spp., *Trinectes* spp., *Gymnachirus* spp.) and some larger tonguefishes (*Symphurus* spp.) are numerous on soft, muddy substrata where they are frequently taken as by-catch in shrimp fisheries (Maharaj & Recksiek 1991; Manickchand-Heileman 1994; Tapia-García 1998).

Reef-associated flatfishes harvested in various tropical regions include larger bothids (especially *Bothus* spp.), paralichthyids (*Paralichthys* spp., *Pseudorhombus* spp.), soleids (*Brachirus* spp., *Solea* spp., *Pardachirus* spp.) and, to a lesser extent, reef-associated tonguefishes (mostly *Cynoglossus* spp.).

13.2.3 Nature of the fisheries

Tropical fisheries can be grouped into three categories: subsistence, artisanal and industrial fisheries. Subsistence fisheries are widespread throughout tropical regions, usually along coasts and estuaries of developing countries where they often constitute the main source of food for people living there (Blaber 1997). These fisheries use simple harvesting techniques and the entire catch is utilised by the local community, although in most developing countries subsistence fishing in marine waters is increasingly giving way to fishing that generates money (Sverdrup-Jensen 1999). Subsistence and artisanal fisheries conducted in shallow coastal and estuarine waters harvest a variety of flatfishes, including many smaller species, some of which have importance as food fishes in local markets. Subsistence fisheries have large numbers of participants, and although little quantitative information is available regarding extent or species composition of flatfishes taken in subsistence fisheries, landings in these fisheries are substantial (Blaber 1997).

Flatfishes in tropical waters are also landed as part of the multi-species catch of localised, artisanal fisheries, or as by-catch in large-scale industrial trawl fisheries for demersal finfishes and shrimps. Artisanal fisheries, small labour-intensive commercial operations, usually not focused to catch any one particular species, harvest a variety of flatfishes. Varying amounts of this catch are used locally, with more valuable species being sold in regional or international markets, and smaller flatfishes (<10 cm TL) being discarded at sea. Although artisanal fisheries do not use large boats or sophisticated gear, their impact on local resources is significant and their contribution to the marine fish food supply is considerable, as they account for the major part of fish landed for direct human consumption in some countries (Platteau 1989; Pauly 1997; Sverdrup-Jensen 1999). In African waters, artisanal canoes are by far the most numerous component of the fishing fleet (Mgawe 1998) and artisanal fishermen land more fish (80–90% of total landings off East Africa, 50–60% off West Africa) than do industrial fisheries operating in the same regions. Small-scale fisheries contribute significantly to income and employment, especially in areas without alternative employment outlets, with an estimated 100 million or more people wholly or partly economically dependent upon them (Sverdrup-Jensen 1999).

Industrial trawl fisheries are highly mechanised, sophisticated operations that use large fishing vessels and new technology which can target specific fish species or components of the fish community, as well as shrimps (Platteau 1989). Industrial trawl fisheries usually catch large amounts of unwanted by-catch, including smaller flatfish species and juveniles of larger species, most of which are discarded at sea (Andrew & Pepperell 1992; Alverson et al. 1994; Clucas & Teutscher 1998). Discarded by-catch is estimated to be about 25% of total marine catches, totalling >20 million tons per year (Sverdrup-Jensen 1999). In some areas, flatfish by-catch is periodically landed and utilised, whereas in other places the majority of flatfish landings are those taken as by-catch in shrimp fisheries. Extensive industrial fisheries have developed in coastal seas wherever shrimp and fish resources occur in abundance (Longhurst & Pauly 1987; Pauly 1994) including the Gulf of Thailand, other areas in southeast Asia, northern Australia, Indian Ocean, gulfs of Mexico, California and Tehuantepec, off northern South America, and off West Africa.

13.2.4 Types of gear employed

Tropical flatfishes are harvested using gear and techniques similar to those employed to catch flatfishes in other areas. Shrimp trawls, beam trawls and other benthic trawls are particularly successful and effective for capturing flatfishes. In shallow waters, they are captured with minitrawls, barrier or stake nets, seines, traps, lift nets, bottom setnets, gill nets, handlines, hook and line, and spear-fishing during snorkelling or SCUBA diving. In coastal fisheries, traditional gears used by indigenous fishers for catching tonguefishes and soles are steadily being replaced by more efficient trawls (Khan & Nandakumaran 1993). Live-bottom habitats are difficult, if not impossible, areas in which to trawl. More active and usually larger predatory flatfishes living in and among coral and other reef-associated environments are harvested using baited hooks, seines, setnets, spear-fishing and fish traps. Larger and smaller reef-associated species are harvested by spear-fishermen, are poisoned using ichthyocides, or are illegally harvested by the use of dynamite, which although banned in most places, is still utilised throughout tropical regions where little regulatory oversight takes place (Öhman *et al.* 1993).

13.2.5 Harvest on spawning concentrations, migrating stocks and impacts on recruitment

Quality life history information concerning movements, migrations, reproductive biology, population structure or recruitment remains unknown for the majority of tropical flatfishes, especially smaller species. Seasonal changes in catch rates of demersal fishes, including flatfishes, in inshore waters in the Arabian Gulf (Wright 1988), western Indian Ocean (Darracott 1977), Gulf of Carpentaria (Blaber *et al.* 1990), southern Caribbean (Lowe-McConnell 1987) and elsewhere indicate some degree of seasonal variation in availability of these animals. Inshore seasonal movements by tropical fishes during the rainy season are ascribed mainly to reproduction or recruitment to these areas (Blaber *et al.* 1990), although other factors may also influence these movements. Sudden increases in catches of Malabar sole along the entire Kerala coast suggest large-scale movements of these fishes into coastal waters following the monsoon season (Khan & Nandakumaran 1993; Jayaprakash 2000). These movements are partly correlated with cyclical monsoonal patterns of productivity and intensity of detritus, and settlement of food organisms (polychaetes, etc.) to the benthos in inshore waters. Seasonal movements into inshore waters, also directly related to estuarine processes, have also been noted for a variety of flatfishes including eastern Atlantic tonguefishes (Canary tonguesole, Senegalese tonguesole and Guinean tonguesole; Ajayi & Adetayo 1982), Pacific eyed flounder (*Bothus constellatus*; Tapia-García *et al.* 2000) and western Atlantic shoal flounder (*Syacium gunteri;* Sánchez-Gil *et al.* 1994).

Spawning and recruitment occur over a protracted season for some tropical flatfishes (Hussain 1990; García-Abad *et al.* 1992; Mathew *et al.* 1992; Rajaguru 1992; Khan & Nandakumaran 1993; Sánchez-Gil *et al.* 1994; Tapia-García *et al.* 2000; Reichert 2002), or is concentrated in one or two unequal periods per year for others (Longhurst & Pauly 1987; Manabe & Shinomiya 2001; Reichert 2002). Due to year-round harvesting in shallow waters by subsistence fisheries, and seasonal or constant activities of artisanal and industrial fisheries operating just offshore and on the inner shelf, tropical flatfishes are harvested or killed irrespective of their spawning condition or recruitment seasonality. The protracted nature of spawning seasons for some, and bimodal reproductive seasons for others, assure that flatfishes in spawning condition are captured during nearly any season when fishing occurs. Although many tropical flatfishes mature at small sizes (García-Abad *et al.* 1992; Rajaguru 1992; Pauly 1994; Sánchez-Gil *et al.* 1994; Munroe 1998; Reichert 1998), they are not protected from harvesting by small size, as large numbers of small fishes (mature or recruits) are taken in fine-meshed shrimp trawls and are either thrown back dead or utilised in regional markets. Harvesting practices in tropical fisheries afford little or no relief of fishing pressure on spawning fishes or new recruits. Fishing mortality on tropical flatfishes begins with new recruits and continues through adult stages. Both growth and recruitment over-fishing in tropical demersal fisheries (Pauly *et al.* 1989) are counterproductive measures leading ultimately to population reductions in these species.

13.2.6 Industrial versus artisanal characteristics of the fisheries

Historically, throughout the tropics, localised artisanal and subsistence fisheries were the primary harvesters of flatfishes. Even today, large areas of the tropics continue to support such fisheries. Beginning in the 1950s and 1960s, mechanised fisheries using trawls increased in both number and extent throughout tropical regions (Platteau 1989). Large trawling fleets developed wherever sizeable shrimp resources were found. With the onset of mechanised fisheries, both areal extent covered by fishers and actual numbers of participants also increased. Following rapid development of mechanised fisheries, catches of demersal finfishes initially increase, but landings then subsequently decrease as local resources become overfished. Areas that have undergone extensive trawling and heavy exploitation include the South China Sea, Java Sea, Gulf of Thailand, Eastern Indian Ocean, most Philippine waters, tropical eastern Pacific and tropical western Atlantic (Pauly & Murphy 1982; Longhurst & Pauly 1987; Pauly 1994). Off West Africa, heavy exploitation of demersal fish resources by fleets of small trawlers during the 1960s caused major changes in composition of demersal fish stocks (Fager & Longhurst 1968).

Impacts of rapid development of small inshore fleets using small-mesh trawls to catch shrimps are so great that these fisheries are considered to be principal agents of ecological change in tropical fish communities (Longhurst & Pauly 1987). Present-day views of tropical demersal fish communities bear little resemblance to former communities naturally occurring in these areas (Longhurst & Pauly 1987). Under high fishing pressure in temperate seas, flatfish populations, having less opportunity for spawning and replacement, have decreased substantially (Caddy & Rodhouse 1998). Populations of tropical flatfishes are also sensitive to fishing activity. As the Gulf of Thailand trawl fishery intensified over a 20-year period, sharp declines in catches of all flatfishes (spiny turbot, bothids and cynoglossids) were noted (Pauly 1994). Likewise, in the southeast Gulf of Carpentaria, Australia, where no commercial trawl fishery exists for fishes, but a shrimp fishery has developed, populations of flatfishes including soleids (*Zebrias* spp.), tonguefishes (*Cynoglossus* spp.) and largescale flounder (*Engyprosopon grandisquamum*) all decreased by 1985–1986, some 20 years after fishing commenced (Harris & Poiner 1991). Only populations of largetooth flounder had increased in abundance during this period.

Although attention has focused on dramatic impacts that mechanised fishing has on local marine resources, artisanal and subsistence fisheries also significantly impact local resources (Longhurst & Pauly 1987; Pauly 1997). Even if catches of individual fishers are small, the sheer numbers of subsistence/artisanal fishers plying their trade in some areas can be staggering (Platteau 1989; Pauly 1997). On the west coast of India and Pakistan about 65% of the fish landings come from artisanal fisheries, and along the west coast of India, an estimated 180 000 country crafts, 26 000 motorised traditional vessels, 34 000 mechanised boats, and a few large ships are involved in fishing (Dwivedi & Choubey 1998). Subsistence fisheries, even those using unsophisticated capture methods, can dramatically affect localised fish populations when harvesting non-migrating species, or when harvesting species with small populations.

13.3 History of exploitation

13.3.1 Commercial landings

Fisheries catch data and commercial landings reported from tropical regions reflect an unknown proportion of the actual total catch. Landings provide only gross estimates of actual amounts of fish taken, areal fishery production, or trends in catches for fisheries operating in tropical oceans (Longhurst & Pauly 1987). For flatfishes, in particular, landings data grossly underestimate total amounts captured or killed in trawl fisheries, especially shrimp fisheries, because they do not include discarded by-catch, which often far exceeds shrimp landings. Nor do landings data include catches of subsistence fisheries, which are largely or totally unreported for many areas. These fisheries catch substantial numbers of fishes, with important ecological consequences, especially for species with highly localised populations.

Fishery landings of developing countries located in the tropics increased substantially during the latter half of the past century (Platteau 1989; Garcia & Newton 1997; Sverdrup-Jensen 1999). Considerable diversification, extension and intensification of fisheries occurred there and elsewhere including vertical and horizontal extensions of traditional fishing grounds (Philip et al. 1984; Sivaprakasam 1986; Lowe-McConnell 1987; Sudarsan et al. 1987; Vijayakumaran & Philip 1990; Mathew 1994; Devaraj 1996), inception of new fisheries for previously unexploited resources (Moore 1999) and expansion of multi-species fisheries (Fischer 1989). During the 1970s, developing countries contributed slightly less than half to the world's total catches (Platteau 1989), whereas by 1993 landings reported from developing countries constituted about two-thirds of the world catch (Sverdrup-Jensen 1999). Although landings from tropical regions increased from the 1970s through the 1980s, recent data indicate that most demersal fish stocks are over-fished and landings for capture fisheries from many areas are either level or have decreased in recent years (Garcia & Newton 1997; Pauly et al. 1998). Most authorities now recognise that the finite nature of fisheries resources limits opportunities for further intensification of exploitation by marine capture fisheries (Garcia & Newton 1997; Pauly et al. 1998; Sverdrup-Jensen 1999).

Most landings data reported to FAO from tropical regions typically do not list statistics for individual flatfishes (except for Indian spiny turbot, see below). Therefore, determining contributions and relative importance of individual species in tropical fisheries is not possible. Flatfishes captured in tropical fisheries are often not identified even to genus or family level, rather, much of the catch is merely identified as 'Pleuronectiformes'. For FAO landings data reported during 1986–1999, 20–46% of annual flatfish catches consisted of Indian spiny turbot and unidentified species of Bothidae, Soleidae and Cynoglossidae (values represent minimum estimates because these flatfishes undoubtedly also contributed to annual landings reported as unidentified flatfishes). The remainder of catches for these years, representing 54–80% of total landings of tropical flatfishes, consisted of unidentified species. Recent

Table 13.2 Unidentified component of flatfish landings in metric tons (%) of the total flatfish landings reported annually to FAO from each of four tropical fishing regions

Year	Area			
	WIO	EIO	WCP	ECP
1994	27399	5310	3350	6104
	(87.0)	(30.3)	(22.6)	(99.9)
1995	18622	7862	3322	8466
	(85.7)	(37.9)	(20.1)	(99.9)
1996	17564	9583	3636	9220
	(84.7)	(31.4)	(24.0)	(99.9)
1997	21965	10238	7545	4596
	(86.8)	(31.0)	(30.1)	(51.4)
1998	17567	9445	3956	1830
	(85.0)	(28.6)	(24.7)	(37.7)
1999	12940	8292	4624	2893
	(81.4)	(25.4)	(25.9)	(44.7)

Area abbreviations are defined in Table 13.1.

(1994–1999) percentages of flatfishes taken in tropical areas and identified at least to family level has improved (Table 13.2). About 70–75% of flatfishes reported from the eastern Indian Ocean (EIO) and western Central Pacific (WCP) are now identified to family. In contrast, over 80% of annual catches from the western Indian Ocean (WIO) are not identified even to family. Likewise, in the eastern Central Pacific (ECP) most catches until 1997 were unidentified, whereas now over half of the catch is identified at least to family. Even with improved identifications, reported data are still woefully inadequate for assessment or management purposes. Due to low desirability, poor economic value and taxonomic complexity of the many closely related, difficult-to-identify species frequently taken together, no real incentives exist to warrant interest in identifying most tropical flatfishes taken in commercial fisheries. Data reported for individual species is not likely to be forthcoming. However, only when species harvested by fisheries are correctly identified will it be possible to critically evaluate ecological impacts on individual species or changes in biodiversity within demersal communities exploited by fisheries.

13.3.2 Geographic occurrence and historical landings

Annual landings reported for flatfishes taken in tropical fishing regions have increased over time (Table 13.3). An estimated 43 200 metric tonnes (mt) were landed in 1970, just over 80 000 mt in 1980, and landings increased to about 102 000 mt in 1990. Combined landings for tropical regions fluctuated between 111 428 and 128 723 mt yr^{-1} during 1991–1995, and increased during 1996–1998 to 123 675–143 989 mt yr^{-1}. Estimated 1999 landings for tropical flatfishes are projected at about 123 730 mt.

The relative proportion of tropical flatfishes in the world flatfish catch has increased over the past 30 years. Combined catches from tropical fishing regions represented only about 3.3% of total flatfish landings in 1970 (Table 13.3), increased to 7.4% in 1980, and remained approximately at this level (range 6.1–8.5%) during 1986–1990. During 1991–1995, the percentage of tropical flatfishes in the world flatfish catch increased to about 11.3% yr^{-1} (range

Table 13.3 Annual catches (mt) and relative percentage of catches (in parentheses) of flatfishes from FAO tropical fishing regions, including combined annual landings for tropical fishing regions and relative percentages of annual world flatfish catch contributed by tropical fisheries

Year	WCA	ECA	SWA	WIO	EIO	WCP	ECP	Combined landings	% of world catch
1970	2500 (5.6)	8900 (20.6)	5000 (11.6)	13500 (31.2)	1300 (3.0)	6000 (13.9)	6000 (13.9)	43 200	3.3
1980	4917 (6.1)	31377 (39.2)	4452 (5.6)	9323 (11.6)	5944 (7.4)	16601 (20.8)	7393 (9.2)	80 007	7.4
1986	3556 (3.9)	26693 (29.5)	12087 (13.4)	16636 (18.4)	10210 (11.3)	13924 (15.4)	7463 (8.2)	90 569	6.9
1987	3031 (3.3)	27292 (30.0)	11317 (12.5)	13465 (14.8)	10028 (11.0)	15055 (16.6)	10673 (11.8)	90 861	7.0
1988	3429 (4.2)	24618 (30.1)	10296 (12.6)	9827 (12.0)	9118 (11.2)	15773 (19.3)	8741 (10.7)	81 802	6.1
1989	2613 (2.9)	27630 (30.7)	12431 (13.8)	22426 (24.9)	13620 (15.2)	14542 (16.2)	8899 (9.9)	102 161	8.5
1990	1784 (1.7)	26567 (25.9)	12358 (12.1)	25215 (24.6)	12579 (12.3)	15122 (14.8)	8824 (8.6)	102 449	8.3
1991	2746 (2.4)	37235 (33.1)	14554 (12.9)	24872 (22.1)	12498 (11.1)	9914 (8.8)	10632 (9.4)	112 451	10.2
1992	2057 (1.7)	38921 (33.0)	12884 (10.9)	29460 (25.0)	13344 (11.3)	11820 (10.0)	9523 (8.1)	118 009	10.2
1993	2112 (1.9)	35210 (31.1)	13667 (12.3)	24898 (22.3)	15282 (13.7)	12937 (11.6)	7322 (6.6)	111 428	10.2
1994	4746 (3.7)	42503 (33.0)	11463 (8.9)	31509 (24.5)	17539 (13.6)	14852 (11.5)	6111 (4.8)	128 723	13.1
1995	2696 (2.3)	34717 (29.7)	11870 (10.2)	21718 (18.6)	20723 (17.8)	16545 (14.2)	8477 (7.3)	116 746	12.7
1996	2527 (2.0)	35088 (28.4)	10356 (8.4)	20730 (16.8)	30566 (24.7)	15179 (12.3)	9229 (7.5)	123 675	13.1
1997	1546 (1.1)	37801 (26.2)	11976 (8.3)	25299 (17.6)	33343 (23.2)	25087 (17.4)	8937 (6.2)	143 989	14.0
1998	1891 (1.4)	45882 (34.4)	10902 (8.2)	20679 (15.5)	33068 (24.8)	16028 (12.0)	853 (3.6)	133 303	14.3
1999	1673 (1.4)	40336 (32.6)	8821 (7.1)	15888 (12.8)	32686 (26.4)	17856 (14.4)	6470 (5.2)	123 730	12.9

Area abbreviations are defined in Table 13.1.

10.2–13.1%). Recently (1996–1999), tropical flatfish landings have again increased, and currently represent about 13.6% (range 12.9–14.3%) of the annual flatfish catch.

Several factors account for recent increases in flatfish landings reported from tropical regions. Tropical flatfish landings increased nearly threefold from 1970 to the 1990s, whereas world catches of flatfishes declined from over a million tons per year during this same period to about 931 000 tons in 1998 and 958 000 in 1999. Increased landings of tropical flatfishes reflect expansion of trawl fisheries and increased numbers of trawlers over time. Upward trends in landings may also signify changes in fishery practices reflecting increased willingness of fishermen to retain flatfishes. Earlier, when other species with greater commercial value or market appeal were more available, flatfishes in the by-catch were discarded at sea. With intensification of tropical fisheries over time, landings of demersal finfishes have trended downward and marketing of flatfishes (along with other small species) has become more necessary to fulfil local fish protein demands. Even as preference to land flatfishes continues to increase throughout heavily fished tropical regions, it is unlikely that flatfish landings will increase substantially because flatfish populations already suffer adverse effects of intensive fishing (see below).

Landings of flatfishes reported from each tropical region, except those from the WCA (Table 13.3) increased from 1970 to 1999. The largest increases were in the EIO, where annual catches during 1990–1999 increased by about 17 times over the 1970 value reported from this region. Recent landings in the eastern Central Atlantic (ECA) have increased about 4 times the 1970 value, while those in the WCP increased 2–3 times the 1970 value for that region. Recent flatfish landings in the southwest Atlantic (SWA) and other tropical areas (WIO and ECP) now average only about twice their 1970 values. In the WCA, flatfish landings either declined or increased only slightly over time.

Since 1980, the largest landings typically have been made in the ECA (Table 13.3). Landings there average about 37 000 mt yr^{-1} and contribute an estimated 30% to the annual total catch of tropical flatfishes. Senegal, among countries fishing in this region, records the highest landings (4257–11 857 mt yr^{-1}, average ca. 7000 mt yr^{-1}), with Morocco and Spain also reporting significant catches from this region.

About 19% of the annual catch of tropical flatfishes occurs in the WIO, where some 24 000 mt are landed (Table 13.3). India records the highest landings from this area (Table 13.4). The EIO is another important area for flatfishes with landings there averaging around 22 000 mt yr^{-1} and representing approximately 16% of the total annual catch of tropical flatfishes. India also records the highest annual landings of flatfishes from this region (Table 13.4), with Indonesia and Malaysia also capturing significant amounts of flatfishes.

The WCP region, where about 15 000 mt yr^{-1} are recorded (Table 13.3), contributes approximately 14% to annual catches of tropical flatfishes. Within this region, the Philippines reported their highest catches during the early 1990s (Table 13.4); however, since then their catches have steadily declined, and in 1997–1998, landings (about 650 mt yr^{-1}) were only one-third of their former values. Landings reported by Malaysia averaged about 1300–1400 mt yr^{-1} during most of the 1990s, increased to about 2180 mt in 1998, and are estimated at 2648 mt for 1999. Indonesia averaged only 600 mt yr^{-1} until 1995. In 1996, catches increased to 1112 mt, peaked in 1997 at 5243 mt, and have averaged about 1150 mt yr^{-1} for 1998–1999.

Table 13.4 Annual landings (mt) of flatfishes reported to FAO by India and the Philippines

	Area		Total landings	
Year	WIO	EIO	India	Philippines
1990	22 299	5675	27 974	1960
1991	19 446	4435	23 881	1853
1992	24 572	4932	29 504	1517
1993	20 724	4764	25 488	1186
1994	27 043	4084	31 127	1072
1995	18 533	6148	24 681	805
1996	17 456	6741	24 197	829
1997	21 696	7556	29 252	627
1998	17 463	6465	23 928	659
1999	12 840	5184	18 024	722

Area abbreviations defined in Table 13.1.

Table 13.5 Recent annual landings (mt) of flatfishes reported to FAO by Mexico and the USA from fishing area 77, the eastern Central Pacific

Year	Mexico	USA
1990	2814	5966
1991	3256	7366
1992	2202	7317
1993	1973	5343
1994	1192	4912
1995	1980	6486
1996	2342	6878
1997	2540	2056
1998	1165	665
1999	2071	822

New World catches of tropical flatfishes are relatively small compared with other tropical regions. Only about 8% of the annual tropical flatfish catch occurs each year in the ECP (including landings from subtropical and warm temperate areas). Of approximately 8000 mt landed annually, Mexico and the USA record the highest percentages (Table 13.5). Likewise, in the WCA, only about 2% of the annual catch of tropical flatfishes (about 2400 mt yr^{-1}) are landed (Table 13.3). The largest percentage of the landings are reported by the USA and Mexico. The small harvest of flatfishes in the WCA is somewhat surprising given that flatfishes are diverse and the continental shelf is relatively large in the region. Such small landings reported here indicate that much of the flatfish by-catch is discarded at sea.

Flatfish landings from the SWA (Table 13.3) constitute about 11% of annual catches (landings also include catches from subtropical and warm temperate areas). About 12 000 mt of flatfishes are recorded from this area annually, with Brazil reporting the largest landings at 1091–2556 mt yr^{-1} (average ca. 1600 mt).

Among tropical flatfishes, individual species data are reported only for Indian spiny turbot (Table 13.6), a large, easily identified species. With an extensive geographic range throughout the Indo-West Pacific from East Africa to the Philippines, the Indian spiny turbot is perhaps the most commercially important species of tropical flatfish. Annual landings (Table 13.6) usually represent 4–16% of flatfish catches reported from tropical regions. Prior to 1994, annual catches of this species ranged between 4000 and 8000 mt yr^{-1}. Since 1994, combined landings increased steadily, with 11 000–23 000 mt yr^{-1} reported during 1995–1999, with maximum landings of 23 309 mt in 1997. Significant landings are reported from the EIO, WIO and WCP (Table 13.6), with most landings typically from the EIO, especially during 1996–1999. Significant landings also take place in the WCP, although regional catches have been substantially less most recently. Reported catches from the WIO are lower than those of other regions, but this is misleading as India does not report statistics for this species, although spiny turbot is landed in quantity by trawl fisheries operating there, in the Bay of Bengal and off Burma. Significant landings of spiny turbots are reported from Tanzania, Yemen, Indonesia and Thailand. Off tropical West Africa (Fager & Longhurst 1968; Mgawe 1998), spottail spiny turbot landings are negligible in most years (1 and 2 mt in 1993 and 1997, respectively).

The Cynoglossidae is an important family of tropical flatfishes (Tables 13.7 and 13.8), although only relatively few of the larger of 150+ species in this family (primarily species

Table 13.6 Nominal fishery landings (mt) of Indian spiny turbot (*Psettodes erumei*) reported to FAO from three tropical fishing regions individually and for all three regions combined

Year	WIO	EIO	WCP	Total landings
1986	394	1662	5139	7195
1987	261	1589	5480	7330
1988	257	1892	4680	6829
1989	212	2138	4240	6590
1990	180	1952	4899	7031
1991	382	2162	1755	4299
1992	468	2319	3425	6212
1993	129	3654	4178	7961
1994	458	6372	3393	10223
1995	1021	6014	4053	11088
1996	880	10672	4018	15570
1997	811	12164	10334	23309
1998	945	12811	4625	18381
1999	850	13460	4860	19170

Area abbreviations are defined in Table 13.1.

Table 13.7 Recent combined annual catches (mt) for selected families of flatfishes from FAO tropical fishing regions

Year	Bothidae	Soleidae	Cynoglossidae
1986	158	2159	29164
1987	227	3487	27247
1988	184	2325	23306
1989	257	1624	25304
1990	209	1891	24905
1991	75	2534	28420
1992	137	4834	24842
1993	217	7079	22471
1994	122	8847	26530
1995	169	9042	27530
1996	62	12406	28467
1997	29	13122	29118
1998	39	15687	27821
1999	51	9132	31322

in *Cynoglossus* and *Paraplagusia*) are landed commercially. Tonguefishes are among the dominant families taken in inshore fisheries throughout most of the Indo-West Pacific region (Chong *et al.* 1990) and in coastal waters off ECA (Table 13.8), with several larger species of *Symphurus* from WCA and ECP regions also contributing smaller amounts to reported landings. Annually, tonguefish landings from the tropics are about 22 500–31 300 mt (Table 13.7), typically representing 20–33% of total tropical flatfish catches. These data underestimate amounts actually landed. Khan & Nandakumaran (1993), for example, reported that during 1986–1991 some 28 903 tons of tonguefishes were landed annually in India alone, with maximum landings of 34 991 tons reported for 1991. Based on these numbers, landings of tonguefishes in India during this time surpassed, or nearly surpassed, combined totals reported for this family from all other tropical regions.

Table 13.8 Recent landings (mt) of tonguefishes (Family Cynoglossidae) reported from FAO tropical fishing areas

Year	ECA	WIO	EIO	WCP	SWA	ECP
1986	9 136	1 513	4 733	6 447	–	–
1987	8 187	1 696	3 965	6 388	–	–
1988	7 308	1 774	2 365	6 626	–	–
1989	10 774	1 503	2 412	6 859	–	–
1990	7 606	2 186	4 058	6 356	488	–
1991	8 831	4 545	4 906	4 597	1 774	–
1992	5 162	4 064	5 302	4 974	1 348	–
1993	3 162	3 701	5 606	5 185	1 118	–
1994	3 871	3 652	5 808	7 994	1 124	–
1995	6 248	2 075	6 813	9 084	36	–
1996	5 579	2 286	10 285	7 487	23	–
1997	5 949	2 523	10 915	7 181	–	–
1998	5 629	2 158	10 786	7 417	–	510
1999	8 148	2 071	10 905	8 357	19	617

Area abbreviations defined in Table 13.1.

About 74–94% of the reported catch of tropical tonguefishes is taken in four (ECA, WCP, WIO, EIO) regions (Table 13.8). The largest catches (exclusively *Cynoglossus* spp.) typically were taken in the ECA during 1986–1991 when nearly 33% of the annual tropical catch of tonguefishes was landed (range 30.0–42.6% per year). From 1992 to 1999, tonguefish landings in the ECA decreased slightly, and in the late 1990s landings had declined to 14–26% of the tropical catch for the family.

Significant landings of tonguefishes were also reported from the WCP during the late 1980s and early 1990s. Landings there contributed on average about 24% towards the total catch (annual range 16.2–28.4%). Since 1991, annual landings of tonguefishes reported from this region were 20–33% (annual average 26.2%) of the total catch. Tonguefish landings in the EIO region during the late 1980s and early 1990s averaged only about 14% of the total annual catch of tonguefishes (range 9.5–17.3%). However, during 1992–1999, annual landings there increased and typically ranged between 21% and 38% (average 29.9%) for most of the 1990s. Since 1996, the largest landings of tonguefishes for any tropical region have come from the WCP, with Thailand and Malaysia recording much of these landings.

Landings from the WIO probably underestimate the amount of tonguefishes actually caught within this region. During 1986–1990, WIO landings represented only 5–9% of totals for the family. From 1991 to 1994, WIO landings nearly doubled those of earlier years, increasing to 13–16% of the total reported catch for cynoglossids. However, beginning in 1995 and continuing to 1999, reported tonguefish landings from this region declined to about 7–8% of the world total for this family. A large proportion of tonguefishes caught in the WIO may not be listed as such, rather, they are included in unidentified components of flatfish catches. Also, landings data reported by India do not list tonguefishes separately, although some of these are important fishery species in commercial catches off southern India and throughout the Bay of Bengal. An estimated 95% of the 25 000 mt of flatfishes landed at Kerala in 1999 consisted of Malabar sole (Jayaprakash 2000), considered to be a principal commercial tonguefish in the Indian Ocean (Seshappa & Bhimachar 1951; Khan & Nandakumaran 1993).

Soles (Soleidae), although taxonomically diverse in shallow, tropical marine waters, historically have constituted relatively minor components of fishery landings reported from these regions (Table 13.7). Soleid species inhabiting shallow marine, estuarine and mangrove habitats are probably very important in subsistence fisheries, although these landings are largely unreported. During the 1970s and 1980s, soles contributed only about 2–10% to tropical flatfish catches, but annual landings of soleids have steadily increased since the early 1990s (mid-1990s about 9000 mt yr^{-1}; by 1998 over 15 000 mt yr^{-1}).

Bothid and paralichthyid flatfishes (listed as Bothidae in FAO statistics and Table 13.7) represented only 0.01–0.2% (30–250 mt yr^{-1}) of flatfish landings reported from tropical regions. Less than 250 mt of bothids were reported in all but one of the most recent 14 years. Recent landings (1997–1999) indicate significant downward trends in bothid landings, with only 29, 39 and 51 mt reported for these years. Recent declines probably represent actual decreases in capture rates for these fishes, as bothid populations may be strongly susceptible to intensive exploitation. Pauly (1994), in summarising catch data (1961–1982) for flatfishes taken in trawl fisheries of the Gulf of Thailand, noted that 1961–1962 catch rates for bothids were 1.44 and 1.20 kg hr^{-1}. By 1966, catch rates had fallen to 0.63 kg hr^{-1} and continued to decline from 1967 to 1977. During 1978–1982, catch rates were essentially 0 kg hr^{-1}. Declines in fish landings in the Gulf of Thailand over time were attributed to dramatic increases (+60%) in fishing effort having occurred there during 1971–1973.

13.4 Importance

13.4.1 Economic importance

In fish markets throughout the tropics, large flatfishes, including spiny turbots, larger bothids, soles, tonguefishes, paralichthyids and achirid soles, are sold whole directly for human consumption. Regional differences in consumer preferences also determine whether or how flatfishes are utilised. Most small species discarded at sea as unsaleable by-catch by trawlers in Northern Australia are commercially valuable in similar fisheries throughout most of Southeast Asia and the Philippines (Blaber 1997). In Sri Lanka, smaller flatfishes caught as by-catch in shrimp fisheries are not especially desired by the local population, but are marketed as a source of fresh fish for consumption by tourists (De Bruin *et al.* 1995). And in some regions, as more desirable fish species have become scarcer due to over-fishing, more flatfishes find their way to local markets to meet increasing local demands for fish protein. In southwest India, for example, market values of cynoglossid flatfishes increased as stocks of other demersal fishes declined. Before intensive development of industrialised fisheries in this area, Malabar sole had some commercial importance but was consumed only by people of low income (Seshappa & Bhimachar 1955). In contrast, flatfishes, including tonguefishes (especially Malabar sole), are now important species along the Calicut coast (Khan & Nandakumaran 1993).

Flatfishes are landed and utilised in subsistence fisheries usually without regard to species or size. Smaller species and juveniles of larger species taken as by-catch in industrial fisheries are consumed fresh, are sun-dried (Philippines, *Cynoglossus* spp. and various soleids; Legaspi 1998), are processed for fishmeal for human consumption (Tapia-García & García-

Abad 1998), or become food for species in aquaculture (Legaspi 1998). In Senegal, soles caught in trawl fisheries are marketed as fresh or frozen fillets for European markets (Diakité 1998). Off Nigeria, when large and medium-sized tonguefishes and soles are caught they are handled carefully and iced generously, similar to the way shrimp are handled, to maximise their export potential after filleting to be marketed as 'sole fillets' to European Union countries and the USA (Akande 1998). In Tampico Port, Mexico, the Gulf of Mexico ocellated flounder (*Ancylopsetta ommata*), a species in the landed by-catch (Bojórquez 1998), is sold locally and also marketed in inland cities as fillet of sole. In the Philippines, small soles (5–10 cm) are dried so that the finished speciality product resembles a fan or round plate. The dried sole is pan-fried and served while crispy. Prepared in this way, dried soles cost around $US 5.00–7.50 kg^{-1}.

In Asian countries, a wide range of uses for fish by-catch has been encouraged due to decreasing catches of more desirable fishes and in response to increasing demands for fish products (Kungsuwan 1998). In the Philippines, changing food habits and eating patterns of Filipino people paved the way to almost maximum efficient utilisation of by-catch and trash fish produced by trawlers (Legaspi 1998). By-catch is sold to fish brokers, wholesalers and retailers for the fresh fish market. Eventual use, however, depends on quality of landed by-catch. Size and species are not primary criteria for fish to be destined for the fresh fish market, rather, their freshness determines marketability. Fishes not fit for human consumption are sold to processors for drying and fermenting as fish paste and sauce, or for aquaculture feeds and fishmeal manufacture. More than 2000 traditional fish processing plants operate in the Philippines, where they absorb all of the by-catch and trash fish that cannot be sold in fresh fish markets. In Nigeria, flatfishes in the by-catch are also utilised, but here, the dominant means of processing is by smoking (Akande 1998). Smoked soles have 2–3 times more value compared with other fish species that are smoked.

13.4.2 Human importance

Flatfishes are of only minor importance to fisheries conducted in tropical regions. In recent years, the quantity of flatfishes brought to markets has increased as greater amounts of by-catch are utilised and as changes in consumer preferences (both voluntary and necessitated) have led to greater consumption or utilisation of these fishes. Even with these changes in market demands it is unlikely that flatfishes will ever have more than limited importance as food throughout most of the tropics. Overall small size and generally low biomass prohibit flatfishes from contributing significant additional economic value to tropical fisheries landings. Additionally, declining populations of flatfishes as a result of over-fishing impose severe limitations for dramatically increasing future landings of quality fish. Until fishing pressure is relaxed and effective conservation measures are instituted, flatfish populations and their contribution to commercial landings are unlikely to increase in areas where over-fishing continues.

13.5 Management and conservation

13.5.1 Fishery conflicts, regulations and management

Tropical flatfishes are not usually directly targeted by commercial fishing interests, and there have been no fishery conflicts, regulations or management practices specifically concerning these fishes. However, as members of communities impacted by intense fisheries, the fate of tropical flatfish populations is intertwined with that of marine demersal communities in general. As human populations have continued to increase along tropical coastlines, so too have their protein demands. To meet these demands, fishing effort and fishing pressure on estuarine and nearshore resources have increased. Regulating or managing these largely unrestricted and open-access coastal fisheries, dispersed geographically and with numerous participants, poses many difficulties (Blaber 1997).

Most tropical estuarine fisheries are unmanaged and the situation in these fisheries has continued to decline because traditional resource allocation based on ownership or stewardship is decreasing in these fisheries as cultural values and systems change, and as human populations have increased (Blaber 1997). Development of mechanised trawl fisheries targeting shrimps in coastal regions also creates serious problems for traditional fisheries occurring in these areas. These new fisheries appropriate local sea space and exploit the same species as do local fisheries, but they adhere to different rules with respect to fishing (Cordell 1984). Trawl fisheries often result in reduction and over-fishing of demersal fish stocks because they capture or destroy large amounts of unwanted by-catch that includes juveniles of species important to other fishery sectors. Open-access fishing, whether by large-scale industrial trawlers or small-scale traditional fishermen, eventually leads to an over-capacity problem and over-fishing (Devaraj & Vivekanandan 1999).

Where industrialised fisheries have moved into regions traditionally fished by artisanal and subsistence fishers, numerous conflicts have arisen among user groups exploiting demersal fish resources (Longhurst & Pauly 1987; Bailey 1997; Bavinck 2001). Severe cases of competition between fishing groups has resulted in depletion or drastic reduction of quality resources in coastal waters such as in the Philippines (Cordell 1984), Malaysia (Anderson 1987) and India (Devaraj & Vivekanandan 1999). Sometimes conflicts between fishers result in violent injuries or even deaths (James 1992; Bavinck 2001).

In response to diminished stocks and to eliminate fishery conflicts, governments in developing countries have attempted to impose a variety of regulations to reduce competition between fishers and to try to reduce fishing pressure on inshore stocks (James 1992; Devaraj & Vivekanandan 1999). Regulations have largely been adopted to curb excesses of mechanised fisheries, although problems of over-fishing are not restricted to trawl fisheries (Pauly et al. 1989). Restrictions include registration and licensing of boats, mesh size regulation, prohibition of fishing methods and gears, delimitation of separate fishing zones for mechanised vessels and artisanal fishers, closed seasons, trawling bans, standardisation of fishing gears, restriction of fishing vessels permitted to fish in an area, catch restrictions or prohibitions, and regulation of allowable fishing time per day. Enforcement of most regulations has proven difficult due to lack of economic resources of governments in developing countries (Devaraj & Vivekanandan 1999). The most extreme regulation was that of the Indonesian government, which, frustrated by the inability to regulate trawler operations and concerned by widespread social unrest, imposed a total ban on all trawling (Bailey 1997). In most estuarine and nearshore fisheries in developing countries, strictly enforced legislation has met with little success for a variety of reasons, including socio-economic and political factors (Blaber 1997; Pauly 1997). The only success has been elimination of cyanide and dynamite fishing in many areas (Blaber 1997); otherwise, regulatory options proposed for these fisheries have only theoretical value, with very little else having been achieved (Pauly & Thia-Eng 1988).

Population-dynamic approaches used to regulate temperate fisheries may not be applicable for tropical fisheries because the multi-species stocks are composed mainly of short-lived species with complex biological–environmental interactions (see Pauly & Murphy 1982; Longhurst & Pauly 1987). Complex biological interactions in tropical multi-species stocks are not the major cause of changes in community structure of exploited demersal fishes (Pauly & Thia-Eng 1988). Rather, effects of fishing gear on both fish populations and habitats explain most observed changes in communities. Therefore, solutions to over-fishing and habitat destruction in tropical coastal regions will not be provided by traditional approaches to fisheries management attempting to regulate only resources (Pauly *et al.* 1989).

Underlying causes of over-exploitation in tropical inshore fisheries emanate from cultural, social or political environments, and management should focus on people, not fish, to find solutions for these fisheries' problems (Blaber 1997). Some fisheries-aid projects now integrate social and cultural components into their programmes, realising that unless fishermen themselves voluntarily cooperate, no amount of governmental regulation can be enforced that will protect local fishery resources. Options such as property rights systems, common property management and access control are becoming increasingly popular (Ahmed *et al.* 1999; Caddy 1999). Along with regulatory policies, aggressive education/outreach programmes concerning fishery conservation must be developed for local fishers. Important goals will include changing the culture of fishing to recognise that fishery resources are finite, that publicly owned resources need to be shared on a fair and equitable basis, and that regulations provide benefits towards conservation of resources (Vijayan *et al.* 2000; Baker 2001). By involving various fishing sectors in regulating and managing coastal resources, effectiveness of regulatory measures will likely increase, and in fact, this may be the only feasible way to establish regulations that actually are complied with.

Pauly *et al.* (1989) considered that poverty in rural communities is also a key issue preventing rational management of tropical inshore fisheries. Poverty along with misguided export-oriented development strategies brought about by globalisation of fishery products are root causes for destructive fishing techniques and environmental degradation occurring throughout many tropical regions. Because poverty is the basis for an array of fishery-related and other socio-economic problems, solutions to fishing problems will be forthcoming only when poverty itself has been resolved (Pauly & Thia-Eng 1988; Pauly *et al.* 1989). Fishes are particularly important to many poor people, often those at risk of food insecurity (Ahmed *et al.* 1999), and there is a striking link between poverty and food security. Changes in fisheries operations and management decisions influencing food availability also expose poor people in developing countries to insecurity with regard to fish-based food supplies (Ahmed *et al.* 1999). The rising export trade in fish (nearly one-third of total landings) shifts food resources away from the poor, who traditionally have diets highly dependent on fish and who find it difficult to compete for fish with other users. Lucrative markets for high-value species encourage fishing operations to move away from traditional lower-value fishing grounds, resulting also in reduced landings of lower-value fish in developing countries. In some areas, by-catch, including juveniles of species formerly essential as fish-for-food for urban and rural poor, is being diverted to feed high-value aquaculture species (Caddy 1999).

Over-fishing in coastal areas will continue to worsen in developing countries as human populations increase and fewer alternative employment opportunities are created to decrease the attraction of fisheries as the employment of last resort (Sverdrup-Jensen 1999). An entire range of solutions will have to be found to decrease poverty in coastal communities and to regulate numbers and activities of fishers participating in coastal fisheries. Such measures may include relocation of fishermen to other economic sectors, limiting entry into fisheries, establishing restricted areas or total bans on trawling in inshore waters where artisanal fishers are active, and requiring larger vessels to fish offshore or in deeper areas, thereby reducing pressure on over-fished and overcrowded coastal fisheries (Cordell 1984; Pauly *et al.* 1989; Bailey 1997; Ahmed *et al.* 1999; Devaraj & Vivekanandan 1999). Increasingly, developing countries will be obliged to make hard choices among long-term coastal employment, food security and foreign exchange earnings in international markets (Caddy 1999). Only if food security considerations prevail will management decisions favour local user rights and enhancement of small-scale fisheries. Given its importance, food policy research for poverty alleviation, ecological sustainability and food security will have to be incorporated into policy and management decisions for tropical coastal fisheries (Ahmed *et al.* 1999).

Besides over-fishing, factors beyond the fisheries sector, such as pollution, habitat destruction and environmental stress also negatively impact fishery resources (Ahmed *et al.* 1999). Sherman *et al.* (1998), and others, have noted that sustainability, health and biomass yields of marine resources can be enhanced by implementation of a more holistic and ecologically based strategy for assessing, monitoring and managing coastal ecosystems than has been generally practised during the past century. Ecosystem-based approaches to management (Sherman *et al.* 1998), utilising national and regional frameworks for conservation, management and sustainable development of coastal resources, may be important vehicles capable of addressing simultaneously the multiple environmental, political and socio-economic problems facing tropical inshore fisheries.

13.5.2 Conservation

Conserving biodiversity is a major issue confronting those involved with socio-economic planning of resource management systems interested in preserving sustainable resources (Agardy 1997; Costello 1998). Marine fisheries are among the most important activities impacting biological diversity in marine ecosystems (Caddy & Rodhouse 1998; Christensen 1998) and it is well established that fishing affects entire ecosystems (Corten 1995; Dayton *et al.* 1995; Jennings & Kaiser 1998; Hall 1999), with the impacts of fishing extending beyond targeted species (Pauly *et al.* 1998; Rogers *et al.* 1999; Kaiser & de Groot 2000). The fate of non-target species harvested, killed and discarded during fishery operations or impacted or killed in their habitat during fishing activities has attracted the attention of conservation organisations and the scientific community (Hall 1999).

Because tropical marine waters contain the greatest diversity of flatfishes found anywhere, and tropical flatfishes incur intense fishing pressure, with juvenile and adult life history stages suffering high mortalities, these fishes and the diversity they represent must receive protection. Measures that reduce fishing pressure and impacts on fishes and their habitats, such as limited entry, quotas, seasonal restrictions, trawling bans, area closures, and mandatory use of by-catch reduction devices, will help conserve demersal fish populations and mitigate habitat destruction (Pauly & Thia-Eng 1988; Kennelly 1995; Bailey 1997; Kungsuwan 1998; Robins et al. 1999; Broadhurst 2000). However, the best hope to conserve tropical flatfishes is in management practices that protect both critical habitat and the biodiversity contained within these habitats (Costello 1998; Caddy 1999). Creation of marine reserves should prevent regional extinction of exploited fish species and populations (Holland & Brazee 1996). Protective management holds much promise for low-cost management, especially of reef fisheries (Roberts & Polunin 1993), and management through reserves appears to be a reasonable alternative to effort management (Hastings & Botsford 1999). Habitat conservation may also be the best approach to manage species, such as flatfishes, that occupy patchy habitats and are relatively sedentary, or less wide-ranging, as adults. Also, saving habitats would result in conserving populations of many species simultaneously rather than trying to target individual species in regions where taxonomy is poorly known or difficult, and where insufficient ecological information prevents knowing how or where to protect most species from over-exploitation.

Acknowledgements

M. Nizinski and M. Vecchione kindly reviewed earlier drafts of the manuscript and provided critical comments. I thank M. Lamboeuf of the Species Identification and Data Programme (SIDP) of the Marine Resources Service, FAO of the UN, who kindly provided permission to use the world map figure.

References

Agardy, T.S. (1997) *Marine Protected Areas and Ocean Conservation*. Academic Press, San Diego.

Ahmed, M., Delgado, C., Sverdrup-Jensen, S. & Santos, R.A.V. (1999) Fisheries policy research in developing countries: issues, priorities and needs. *International Center for Living Aquatic Resources Management Conference Proceedings*, **60**, 1–112.

Ajayi, T.O. & Adetayo, J.A. (1982) On the fish bycatch and discard of the shrimp fishery of Nigeria. *Nigerian Institute of Oceanographic and Marine Research Technical Paper*, **5**, 1–28.

Akande, G. (1998) Bycatch utilization in Nigeria. In: *Report and Proceedings of FAO/DFID Expert Consultation on Bycatch Utilization in Tropical Fisheries, Beijing, China, 21–28 September 1998* (eds I. Clucas & F. Teutscher). pp. 241–52. FAO, Rome.

Alverson, D.L., Freeberg, M.H., Murawski, S.A. & Pope, J.G. (1994) A global assessment of fisheries bycatch and discards. *FAO Fisheries Technical Paper*, **339**, 1–233.

Anderson, E.N., Jr (1987) A Malaysian tragedy of the commons. In: *The Question of the Commons: The Culture and Ecology of Communal Resources* (eds B.J. McCay & J.M. Acheson). pp. 327–343. University of Arizona Press, Tucson, AZ.

Andrew, N.L. & Pepperell, J.G. (1992) The by-catch of shrimp trawl fisheries. *Oceanography and Marine Biology: An Annual Review*, **30**, 527–565.

Atiqullah, M. (2001) Study on length frequency distribution, relative condition factor and otolith weight of *Euryglossa orientalis* from Karachi coast, Pakistan. *Pakistan Journal of Marine Biology*, **7**, 73–79.

Bailey, C. (1997) Lessons from Indonesia's 1980 trawler ban. *Marine Policy*, **21**, 225–235.

Baker, J. (2001) Development of a draft inshore finfish fisheries management plan for the Queensland waters of the Gulf of Carpentaria, Australia. In: Aquaculture and fisheries resources management. Proceedings of the joint Taiwan-Australia aquaculture and fisheries resources management forum (eds I.C. Liao & J. Baker). pp. 11–13. *Taiwan Fisheries Research Institute, Conference Proceedings*, **4**.

Bavinck, M. (2001) *Marine Resource Management: Conflict and Regulation in the Fisheries of the Coromandel Coast*. Sage Publications India, New Delhi.

Blaber, S.J.M. (1997) *Fish and Fisheries of Tropical Estuaries. Fish and Fisheries Series 22*. Chapman & Hall, London.

Blaber, S.J.M., Brewer, D.T., Salini, J.P. & Kerr, J. (1990) Biomasses, catch rates and abundances of demersal fishes, particularly predators of prawns, in a tropical bay in the Gulf of Carpentaria, Australia. *Marine Biology*, **107**, 397–408.

Bojórquez, L.F.L. (1998) Shrimp bycatch utilization in Mexico. In: *Report and Proceedings of FAO/DFID Expert Consultation on Bycatch Utilization in Tropical Fisheries, Beijing, China, 21–28 September 1998* (eds I. Clucas & F. Teutscher). pp. 253–262. FAO, Rome.

Booth, A.J. & Walmsley-Hart, S.A. (2000) Biology of the redspotted tonguesole *Cynoglossus zanzibarensis* (Pleuronectiformes: Cynoglossidae) on the Agulhas Bank, South Africa. *South African Journal of Marine Science*, **22**, 185–197.

Brinca, L., Rey, F., Silva, C. & Saetre, R. (1981) A survey of marine fish resources of Mozambique. *Survey Report of the R/V Nansen*. Institute of Marine Research, Bergen, Norway, 1–58.

Broadhurst, M.K. (2000) Modifications to reduce bycatch in prawn trawls: a review and framework for development. *Reviews in Fish Biology and Fisheries*, **10**, 27–60.

Caddy, J.F. (1999) Fisheries management in the twenty-first century: will new paradigms apply? *Reviews in Fish Biology and Fisheries*, **9**, 1–43.

Caddy, J.F. & Rodhouse, P.G. (1998) Cephalopod and groundfish landings: evidence for ecological change in global fisheries? *Reviews in Fish Biology and Fisheries*, **8**, 431–444.

Carpenter, K.E., Krupp, F., Jones, D.A. & Zajonz, U. (1997) *FAO Species Identification Guide for Fishery Purposes. The Living Marine Resources of Kuwait, Western Saudi Arabia, Bahrain, Qatar, and the United Arab Emirates*. FAO, Rome.

Chong, V.C., Sasekumar, A., Leh, M.U.C. & D'Cruz, R. (1990) The fish and prawn communities of a Malaysian coastal mangrove system, with comparisons to adjacent mud flats and inshore waters. *Estuarine and Coastal Shelf Science*, **31**, 703–722.

Christensen, V. (1998) Fishery induced changes in a marine ecosystem: insight from models of the Gulf of Thailand. *Journal of Fish Biology*, **53** (Suppl. A), 128–142.

Clucas, I. & Teutscher, F. (1998) *Report and Proceedings of FAO/DFID Expert Consultation on Bycatch Utilization in Tropical Fisheries, Beijing, China, 21–28 September 1998*. FAO, Rome.

Cordell, J.C. (1984) Defending customary inshore sea rights. In: *Maritime Institutions in the Western Pacific* (eds K. Ruddle & T. Akimichi). pp. 301–326. Senri Ethnological Studies. 17. National Museum of Ethnology, Osaka.

Corten, A. (1995) Ecology and fisheries management: the way forward. *ICES Information*, **26**, 7–9.

Costello, M.J. (1998) To know, research, manage and conserve marine biodiversity. *Océanis*, **24**, 25–49.

Darracott, A. (1977) Availability, morphometrics, feeding and breeding activity in a multi-species, demersal fish stock of the western Indian Ocean. *Journal of Fish Biology*, **10**, 1–16.

Dayton, P.K., Thrush, S.F., Agardy, M.T. & Hofman, R.J. (1995) Environmental effects of marine fishing. *Aquatic Conservation: Marine and Freshwater Ecosystems,* **5**, 205–232.

De Bruin, G.H.P., Russell, B.C. & Bogusch, A. (1995) *FAO Species Identification Field Guide for Fishery Purposes. The Marine Fishery Resources of Sri Lanka.* FAO, Rome.

Devaraj, M. (1996) Deepsea fishing in Indian waters. In: *Proceedings of a Seminar on Fisheries – A Multibillion Dollar Industry* (eds B. Krishnamoorthi, K.N. Krishnamoorthi, P.T. Meenakshisundaram & K.N. Nayar). pp. 35–41. Aquaculture Foundation of India, Madras.

Devaraj, M. & Vivekanandan, E. (1999) Marine capture fisheries of India: challenges and opportunities. *Current Science,* **76**, 314–332.

Diakité, B. (1998) Bycatch in marine fisheries in Senegal. In: *Report and Proceedings of FAO/DFID Expert Consultation on Bycatch Utilization in Tropical Fisheries, Beijing, China, 21–28 September 1998* (eds I. Clucas & F. Teutscher). pp. 235–239. FAO, Rome.

Dwivedi, S.N. & Choubey, A.K. (1998) Indian Ocean large marine ecosystems: need for national and regional framework for conservation and sustainable development. In: *Large Marine Ecosystems of the Indian Ocean: Assessment, Sustainability, and Management* (eds K. Sherman, E.N. Okemwa & M.J. Ntiba). pp. 361–368. Blackwell Science, Malden, MA.

Fager, E.W. & Longhurst, A.R. (1968) Recurrent group analysis of species assemblages of demersal fish in the Gulf of Guinea. *Journal of the Fisheries Research Board of Canada,* **25**, 519–533.

FAO (1997) *Yearbook of Fishery Statistics, Catches and Landings for 1995,* Vol. 80. FAO, Rome.

FAO (2001) *Yearbook of Fishery Statistics for 1999,* Vol. 87. Part 1. *Capture Production.* FAO, Rome.

Fischer, W. (1989) The significance of FAO's Biosystematic Program in the enhancement of world fisheries. *Reviews in Aquatic Science,* **1**, 683–692.

Garcia, S.M. & Newton, C. (1997) Current situation, trends, and prospects in world capture fisheries. In: Global trends: fisheries management (eds E.L. Pikitch, D.D. Huppert & M.S. Sissenwine). pp. 3–27. *American Fisheries Society Symposium 20.* Bethesda, MD.

García-Abad, M.C., Yáñez-Arancibia, A., Sánchez-Gil, P. & Tapia-García, M. (1992) Distribución, reproducción y alimentación de *Syacium gunteri* Ginsburg (Pisces: Bothidae), en el Golfo de México. *Revista de Biología Tropical,* **39**, 27–34.

Gulland, J.A. (1971) *The Fish Resources of the Ocean.* Fishing News, West Byfleet, England.

Hall, S.J. (1999) *The Effects of Fishing on Marine Ecosystems and Communities.* Blackwell, Oxford.

Harris, A.N. & Poiner, I.R. (1991) Changes in species composition of demersal fish fauna of Southeast Gulf of Carpentaria, Australia, after 20 years of fishing. *Marine Biology,* **111**, 503–519.

Hastings, A. & Botsford, L.W. (1999) Equivalence in yield from marine reserves and traditional fisheries management. *Science,* **284**, 1537–1538.

Holland, D.S. & Brazee, R.J. (1996) Marine reserves for fishery management. *Marine Resource Economics,* **11**, 157–171.

Hussain, S.M. (1990) Biology of *Psettodes erumei* (Schneider, 1801) and *Pseudorhombus arsius* (Hamilton, 1822) from the northern Arabian Sea. *Indian Journal of Fisheries,* **37**, 63–66.

James, P.S.B.R. (1992) A review of the existing regulations in the maritime states of India in relation to exploitation of fishery resources and their conservation and management. *Journal of the Marine Biological Association of India,* **34**, 84–89.

Jayaprakash, A.A. (2000) Food and feeding habits of Malabar sole *Cynoglossus macrostomus* Norman. *Journal of the Marine Biological Association of India,* **42**, 124–134.

Jayaprakash, A.A. & Inasu, N.D. (1999) Age and growth of Malabar sole *Cynoglossus macrostomus* Norman off Kerala Coast. *Journal of the Marine Biological Association of India,* **40**, 125–132.

Jennings, S. & Kaiser, M.J. (1998) The effects of fishing on marine ecosystems. *Advances in Marine Biology*, **34**, 201–352.

Kaiser, M.J. & de Groot, S.J. (2000) *The Effects of Fishing on Non-target Species and Habitats. Biological, Conservation and Socio-economic Issues*. Blackwell, Oxford.

Kennelly, S.J. (1995) The issue of bycatch in Australia's demersal trawl fisheries. *Reviews in Fish Biology and Fisheries*, **5**, 213–234.

Khan, M.F. & Nandakumaran, K. (1993) Population dynamics of Malabar sole *Cynoglossus macrostomus* Norman along Calicut coast. *Indian Journal of Fisheries*, **40**, 225–230.

Kungsuwan, A. (1998) Bycatch utilization in Asia: an overview. In: *Report and Proceedings of FAO/DFID Expert Consultation on Bycatch Utilization in Tropical Fisheries, Beijing, China, 21–28 September 1998* (eds I. Clucas & F. Teutscher). pp. 19–39. FAO, Rome.

Legaspi, A.S. (1998) Bycatch utilization in the Phillippines. In: *Report and Proceedings of FAO/DFID Expert Consultation on Bycatch Utilization in Tropical Fisheries, Beijing, China, 21–28 September 1998* (eds I. Clucas & F. Teutscher). pp. 105–113. FAO, Rome.

Longhurst, A.R. & Pauly, D. (1987) *Ecology of Tropical Oceans*. Academic Press, San Diego, CA.

Lowe-McConnell, R.H. (1987) *Ecological Studies in Tropical Fish Communities*. Cambridge University Press, Cambridge.

Maharaj, V. & Recksiek, C. (1991) The by-catch from the artisanal shrimp trawl fishery, Gulf of Paria, Trinidad. *Marine Fisheries Review*, **53**, 9–15.

Manabe, H. & Shinomiya, A. (2001) Two spawning seasons and mating system of the bastard halibut, *Tarphops oligolepis*. *Ichthyological Research*, **48**, 421–424.

Manickchand-Heileman, S.C. (1994) Distribution and abundance of flatfish on the South American continental shelf from Suriname to Colombia. *Netherlands Journal of Sea Research*, **32**, 441–452.

Mathew, G. (1994) Exploratory trawl fishing off Tirunelveli coast. *Journal of the Marine Biological Association of India*, **36**, 152–160.

Mathew, G., Khan, M.F. & Nandakumaran, K. (1992) Present status of exploitation of fish and shellfish resources: flatfishes and flatheads. In: Monsoon fisheries of the west coast of India: prospects, problems and management (eds P.V. Rao, V.S. Murty & K. Rengarajan). pp. 197–204. *Central Marine Fisheries Institute Bulletin, No. 45*. Cochin, India.

Mgawe, Y.I. (1998) Utilization of bycatch from the fishing industry in Africa: an overview. In: *Report and Proceedings of FAO/DFID Expert Consultation on Bycatch Utilization in Tropical Fisheries, Beijing, China, 21–28 September 1998* (eds I. Clucas & F. Teutscher). pp. 41–52. FAO, Rome.

Moore, J.A. (1999) Deep-sea finfish fisheries: lessons from history. *Fisheries*, **24**, 16–21.

Munroe, T.A. (1998) Systematics and ecology of tonguefishes of the genus *Symphurus* (Cynoglossidae: Pleuronectiformes) from the western Atlantic Ocean. *Fishery Bulletin*, **96**, 1–182.

Öhman, M.C., Rajasuriya, A. & Lindén, O. (1993) Human disturbances on coral reefs in Sri Lanka: a case study. *Ambio*, **22**, 474–480.

Pauly, D. (1994) A framework for latitudinal comparisons of flatfish recruitment. *Netherlands Journal of Sea Research*, **32**, 107–118.

Pauly, D. (1997) Small-scale fisheries in the tropics: marginality, marginalization, and some implications for fisheries management. In: Global trends in fisheries management (eds E. Pikitch, D.D. Hubert & M.S. Sissenwine). pp. 40–49. *American Fisheries Society Symposium, 20*. Bethesda, MD.

Pauly, D. & Murphy, G.I. (1982) Theory and management of tropical fisheries. *International Center for Living Aquatic Resources Management Conference Proceedings, 9*. Manila.

Pauly, D. & Thia-Eng, C. (1988) The overfishing of marine resources: socio-economic background in Southeast Asia. *Ambio*, **17**, 200–206.

Pauly, D., Silvestre, G. & Smith, I.R. (1989) On development, fisheries and dynamite: a brief review of tropical fisheries management. *Natural Resource Modeling*, **3**, 307–329.

Pauly, D., Christensen, V., Dalsgaard, J., Froese, R. & Torres, F., Jr (1998) Fishing down marine food webs. *Science*, **279**, 860–863.

Philip, K.P., Premchand, B., Avhad, G.K. & Joseph, P.J. (1984) A note on the deep sea demersal resources of the Karnatka-North Kerala Coast. *Bulletin of the Fisheries Survey of India*, **13**, 23–30.

Platteau, J.-P. (1989) The dynamics of fisheries development in developing countries: a general overview. *Development and Change*, **20**, 565–597.

Pradham, M.M. (1969) Fishery biology of *Psettodes erumei* (Schn.) – an Indian Ocean flatfish. *Bulletin of the National Institute of Sciences, India*, **38**, 895–905.

Rajaguru, A. (1992) Biology of two co-occurring tonguefishes, *Cynoglossus arel* and *C. lida* (Pleuronectiformes: Cynoglossidae), from Indian waters. *Fishery Bulletin*, **90**, 328–367.

Reichert, M.J.M. (1998) *Etropus crossotus*, an annual flatfish species: age and growth of the fringed flounder in South Carolina. *Journal of Sea Research*, **40**, 323–332.

Reichert, M.J.M. (2002) *On the life history of the fringed flounder* (Etropus crossotus), *a small tropical flatfish in the South Atlantic Bight*. DPhil dissertation, Rijksuniversiteit Groningen, Holland.

Roberts, C.M. & Polunin, N.V.C. (1993) Marine reserves: simple solutions to managing complex fisheries? *Ambio*, **22**, 363–368.

Robins, J.B., Campbell, M.J. & McGilvray, J.G. (1999) Reducing prawn-trawl bycatch in Australia: an overview and an example from Queensland. *Marine Fisheries Review*, **61**, 46–55.

Rogers, S.I., Clarke, K.R. & Reynolds, J.D. (1999) The taxonomic distinctness of coastal bottom-dwelling fish communities of the North-east Atlantic. *Journal of Animal Ecology*, **68**, 769–782.

Sánchez-Gil, P., Arreguín-Sánchez, F. & García-Abad, M.C. (1994) Ecological strategies and recruitment of *Syacium gunteri* (Pisces: Bothidae) in the Southern Gulf of Mexico shelf. *Netherlands Journal of Sea Research*, **32**, 433–439.

Seshappa, G. & Bhimachar, F.S. (1951) Age determination studies in fishes by means of scales with special reference to the Malabar sole *Cynoglossus semifasciatus* Day. *Current Science*, **20**, 260–262.

Seshappa, G. & Bhimachar, F.S. (1955) Studies on the fishery and biology of the Malabar sole *Cynoglossus semifasciatus* Day. *Indian Journal of Fisheries*, **2**, 180–230.

Sherman, K., Okemwa, E.N. & Ntiba, M.J. (1998) *Large Marine Ecosystems of the Indian Ocean: Assessment, Sustainability, and Management.* Blackwell Science, Malden, MA.

Sivaprakasam, T.E. (1986) What is in store in the deep sea? Results of explorations into the demersal fishery resources of the Indian Exclusive Economic Zone. *Occasional Papers of the Fishery Survey of India*, **4**, 1–24.

Sudarsan, D., John, M.E. & Joseph, A. (1987) An assessment of demersal stocks in the southwest coast of India with particular reference to the exploitable resources in the outer continental shelf and slopes. National Symposium on Research and Development. *Central Marine Fisheries Research Institute Bulletin*, **44**, 266–272.

Sverdrup-Jensen, S. (1999) Policy issues deriving from the impact of fisheries on food security and the environment in developing countries. In: Fisheries policy research in developing countries: issues, priorities and needs (eds M. Ahmed, C. Delgado, S. Sverdrup-Jensen & R.A.V. Santos). pp. 73–91. *International Center for Living Aquatic Resources Management Conference Proceedings*, **60**.

Tapia-García, M. (1998) Evaluación ecológica de la ictiofauna demersal. In: *El Golfo de Tehuantepec: El Ecosistema y sus Recursos* (ed. M. Tapia-García). pp. 129–148. Universidad Autónoma Metropolitana-Iztapalapa, México.

Tapia-García, M. & García-Abad, M.C. (1998) Los peces acompañantes del camarón y su potencial como recurso en las costas de Oaxaca y Chiapas. In: *El Golfo de Tehuantepec: El Ecosistema y sus Recursos* (ed. M. Tapia-García). pp. 179–196. Universidad Autónoma Metropolitana-Iztapalapa, México.

Tapia-García, M., García, M.C. & Ladrón de Guevara, G.C. (2000) Reproduction, distribution and abundance of *Bothus constellatus* (Pisces: Bothidae), in the Gulf of Tehuantepec, Mexico. *Revista de Biología Tropical,* **48**, 205–213.

Vijayakumaran, K. & Philip, K.P. (1990) Demersal fishery resources off North Kerala, Karnataka and Konkan coasts. *Journal of the Marine Biological Association of India,* **32**, 177–186.

Vijayan, V., Edwin, L. & Ravindran, K. (2000) Conservation and management of marine fishery resources of Kerala State, India. *Naga (ICLARM Quarterly),* **23**, 6–9.

Wright, J.M. (1988) Seasonal and spatial differences in the fish assemblages of the non-estuarine Sulaibikhat Bay, Kuwait. *Marine Biology,* **100**, 13–20.

Chapter 14
Assessment and management of flatfish stocks

Jake Rice, Steven X. Cadrin and William G. Clark

14.1 Concepts and terms

The terms 'assessment' and 'management' have specific meanings in fisheries. Assessment refers to a scientific procedure by which present stock size, productivity (growth and recruitment) and mortality rate (at least mortality due to fishing) are estimated, and forecasts for at least a short term (1 or 2 years) are made. Historically assessments were based almost exclusively on data from monitoring fisheries and research surveys. They were conducted annually, and used some variant of age-structured population models (Hilborn & Walters 1992), which totalled up how large year-classes must have been, in order to have produced the quantities and age compositions observed in the catches and surveys. All these general features are evolving. More sources of information including environmental conditions and traditional ecological knowledge are contributing to evaluating stock status (Pitcher *et al.* 1998; Garcia *et al.* 1999). Also more diverse analytical methods, addressing uncertainty in data and model formulations, and risk in forecasts, are being applied (Schnute 1994; Patterson 1999).

The assessment estimates and forecasts form the basis of scientific advice on fishery management measures. The advice is framed primarily as recommended catch quotas but may address other aspects of the prosecution of fisheries, including gear usage, opening times and places, and many other features of fisheries that have biological consequences. Management comprises all regulations designed to achieve the biological and socio-economic objectives of the management authority, and how they are implemented. The desire to achieve short-term social and economic goals means that management may not always follow the scientific advice on conservation.

There is a natural complementarity between assessment and the biological aspects of management. Effective managers present clear and comprehensive requests for advice to scientists, specifying the management alternatives under consideration and aspects of status quo fishing which may be impeding achievement of objectives. Assessment scientists should provide clear answers to the requests from managers, fully addressing uncertainty in formats that communicate risk. They should also encourage managers to consider new alternatives and pay attention to new threats and additional types of information (Ulltang 1998; Rice 1999).

This complementarity between assessment and management is being structured more explicitly into the interactions among scientists and managers through increased use of the precautionary approach (see section 14.2.3) and associated biological reference points. For example, managers choose reference points such as maximum sustainable yield (MSY) as objectives, based partly on advice from scientists on the consequences of keeping stocks within the reference points. Then, to monitor the effectiveness of regulations for achieving the explicit objectives, assessments estimate the stock size and harvest rate that produce maximum sustainable yield (B_{MSY} and F_{MSY}), and the probability (or risk) that current estimates of stock size and harvest rate are consistent with the reference points.

For example, the fishery for Pacific halibut (*Hippoglossus stenolepis*) in the waters off Canada and the USA is managed by the International Pacific Halibut Commission. Each year the scientific staff of the Commission do a stock assessment (based on fishery and survey data) that produces an estimate of present stock size in each of several regulatory areas covering the coastal waters of the region. The Commission's harvest policy is to take 20% of the exploitable biomass in each area in each year's fishery, because scientific studies have shown that this rate of harvest will result in an average long-term yield near the maximum achievable by the stock, while at the same time maintaining spawning biomass within a range regarded as healthy (Sullivan *et al.* 1999). Consequently the Commission annually requests scientific advice for a set of assessment-based catch limits equal to 20% of the biomass estimates, but these figures are often accompanied by area-specific warnings about the reliability of the estimates and either higher or lower catch limits recommended by the staff. The Commission takes in the staff advice and industry comments, and then sets catch limits that may depart from the staff recommendations but seldom exceed them by much.

Flatfish stocks are relatively easy to assess because the fishes are widely distributed on the bottom where they can be surveyed effectively with standard fishing gear. (By contrast, species with highly contagious distributions or broad vertical distributions in the water column may be very difficult to survey with accuracy and precision.) The survey does not have to produce a direct estimate of absolute abundance, because a time series of some measure of relative abundance (such as catch per hectare swept by a survey trawl) can be used in conjunction with estimates of commercial catch at age to arrive at an estimate of absolute abundance. Many flatfish assessments also use commercial catch rates per standardised unit of effort (CPUE), although this index is not always a reliable indicator of stock size (Hutchings 1996). Like any other type of species, assessments of flatfishes will be affected by inaccuracies in data, including unreported discards or landings and catches misreported by species or location.

Flatfishes generally are easy to age and thereby are conducive to age-structured assessment methods. Alternative assessment approaches (e.g. length- or stage-based models) are commonly applied to other taxa that are difficult to age (such as invertebrates), but are less informative and are not common in flatfish assessments. On the other hand, the wide distribution, large populations and offshore nature of most flatfish stocks usually are not conducive to the information-rich individual-based models that are often applied to inland finfishes or marine mammal resources. As a result of the general assessment approach for flatfishes, management strategies usually reflect the information available from typical age-based assessments and rely on demographic features (e.g. recruitment, exploitable biomass, mature biomass, fishing mortality at age) of the current stock as well as forecasted conditions to evaluate management decisions.

The choice of a harvest policy is not so easy, because it depends critically on how rates of reproduction, natural mortality and growth change when stock size is reduced by fishing. These changes are extremely difficult to predict for several reasons. Functional relationships of each of these biological factors to stock size can be represented by several different model equations which may fit historic observations about equally well, but predict very different responses at much higher or lower stock sizes (Evans & Rice 1988). Moreover, the biological processes are strongly affected by environmental conditions, as well as stock size, so there is large uncertainty in estimating parameters to be used in modelling studies that explore consequences of alternative harvesting policies (Rosenburg & Restrepo 1994). Natural mortality, in fact, is very difficult to measure, and a change is nearly impossible to detect until long after it has occurred. As a result, although scientists often test the robustness of harvest policies with extensive simulation modelling (Hollowed & Megrey 1993; Punt *et al.* 1999) the choice of a harvest policy often rests on some strong assumptions.

Choice of a harvest policy presupposes, furthermore, that the goal of the policy is clear. Identifying and building consensus around operational biological, social and economic objectives, in fact, is becoming a major management task in itself. In a broad sense, objectives have to address biological conservation of the resource and economic sustainability of the harvesting community. These are very general objectives, yet even at this scale may not be inter-compatible on short timescales. Aggressive harvesting policies provide greater short-term economic benefits, but threaten conservation; whereas more cautious harvesting policies promote greater conservation, often at short-term failure to sustain existing harvesting levels (Charles 1997). The problem of setting management objectives is much more complex yet, because objectives must address many social considerations as well as strictly macro-economic ones (Garcia *et al.* 1999). Moreover, recent initiatives to apply a precautionary approach to promote risk-averse management increase needs for risk quantification in assessments, and explicit objectives in management (FAO 1996). In reality 'management' must address all these complexities in the concepts of sustainability (Mace 1997) and conservation, although for this chapter we consider only the subset of assessment and management issues with a biological basis.

In this chapter we first consider some aspects of flatfish population dynamics that are important to assessment and management. We focus particularly on the relationship between spawning stock size and reproductive success, because of its historic role as the cornerstone of evaluating stock dynamics and conservation targets and limits. We illustrate how traditional approaches are changing as a broader framework, acknowledging environmental forcing on recruitment and growth, is explored. We then summarise stock assessment results and harvest policies presently used in North America and Europe.

14.2 Population dynamics, assessment and management

Each year a stock of fish loses biomass to natural mortality (predation, disease, etc.) and gains biomass from the body growth of survivors and the addition of offspring spawned earlier. Offspring are usually counted as additions to a stock at an age when they are large enough to feed in the same way as older fishes and experience a similar rate of natural mortality. For most flatfish species, the transition from pelagic larvae to benthic habitats, associated with complete metamorphosis of eye location, marks recruitment to the benthic population. However, recruitment to the stock may be later due to greater predation on small fishes. At this age they are called recruits, and the number of fishes reaching this age in a given year is referred to as the recruitment to the stock in that year.

Historically, assessment and management both assumed equilibrium conditions; that is, in an unfished stock, gains from growth and recruitment just balance losses to natural mortality, so the stock varies around a stable average biomass. Fishing necessarily alters the balance, initially causing stock size to decrease. If the fishing mortality rate is not too high the stock size will stabilise at a new, lower equilibrium level where again gains balance losses even though total mortality is higher. The stock is able to establish a new equilibrium with a higher rate of loss because a number of biological processes may have density-dependent components that allow the rates of gain to increase in various ways:

(1) *Higher age-specific growth rates.* In some stocks, growth rates increase when abundance is reduced, presumably as a result of reduced competition for food. Such patterns of increased growth rates at reduced abundance have been observed for plaice (*Pleuronectes platessa*) (Bannister 1978; Rijnsdorp & Van Beek 1991; Modin & Phil 1994; Rijnsdorp 1994; Millner *et al.* 1996; Rijnsdorp & Van Leeuwen 1996), Atlantic halibut (*Hippoglossus hippoglossus*) (Haug & Tjemsland 1986), common sole (*Solea solea*) (Rijnsdorp & Van Beek 1991; Millner & Whiting 1996; Millner *et al.* 1996) and yellowtail flounder (*Limanda ferruginea*) (Ross & Nelson 1992).

(2) *Higher average growth rate due to stock juvenation.* Younger fishes generally have higher growth rates than older fishes, so even if age-specific rates do not increase, the higher mortality will increase the average growth rate of the stock by increasing the proportion of young fishes.

(3) *Lower natural mortality rate.* When abundance is reduced, natural mortality may decrease so total mortality does not increase as much as implied by the fishing mortality alone. Possible mechanisms include decreased cannibalism at lower abundance, a reduction in predation by other species if predators switch away from the species harvested by the fishery as it becomes more rare, or a decrease in mortality due to senescence as less of the population survives to older ages. In practice the rate of natural mortality is nearly impossible to estimate except where fishing is nil, so changes of this sort cannot be detected, and are rarely factored into assessments or management.

(4) *Higher average reproductive rate.* Few flatfish stocks mature before age 2 or 3 years, and many not until some years older; in each case with the proportion mature increasing over several years to 100%. Just as juvenation increases the average growth rate, it also reduces the average proportion mature. For the average reproductive rate of the stock to increase, therefore, the reproductive success (recruits per unit of spawning biomass) of mature fishes must increase substantially, or else age-specific maturation rate must respond to the reduced abundance. The former process will be addressed in the section on stock recruit relationships. The latter process has been observed in Atlantic halibut (Haug & Tjemsland 1986) and yellowtail flounder (Cadrin *et al.* 1998; Fig. 14.1), where the percentage of fish maturing at younger ages increased as total stock biomass declined.

The evidence from stocks that have sustained high rates of fishing mortality for long periods is that the higher reproductive rate is far and away the most important process enabling a

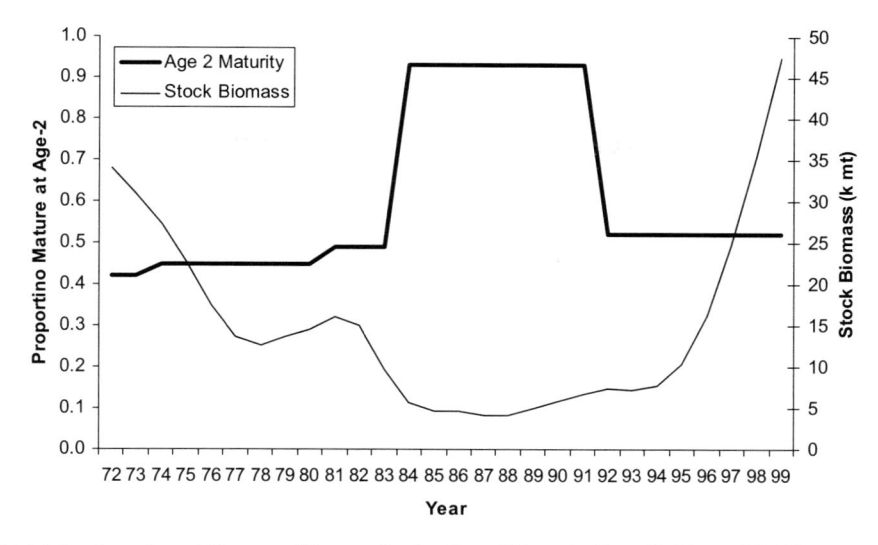

Fig. 14.1 Maturity and stock biomass of Georges Bank yellowtail flounder (from Cadrin *et al.* 2000).

stock to maintain itself while at the same time sustaining a fishery. Conversely, the stocks that have collapsed under moderate to heavy exploitation did so because the increase in mortality due to exploitation exceeded the ability of the stock to compensate with increased recruits per spawner.

In recent years results of research on marine ecosystems, from physical oceanography to top predators, are producing a new round of challenges to the historic equilibrium assumptions of assessment and management. Studies in both the North Pacific (Hollowed & Wooster 1995; Clark *et al.* 1999; Fargo & Wilderbuer 2000) and the North Atlantic (Daan *et al.* 1996; Serchuk *et al.* 1996; Alheit & Hagen 1997; Bowering *et al.* 1997; Drinkwater 1997) have found large-scale and multi-year environmental trends that can affect many of the biological processes discussed above, including growth rate, survivorship and recruitment. There are few instances where management has taken the changes in environmental conditions into account in harvesting strategies, and a few more where at least the scientific advice from assessments has called attention to these factors in interpreting assessment results. Work on flatfishes is at the forefront of accommodating these new environmental considerations and the precautionary approach in assessment and management, for example with ICES advice on North Sea plaice (ICES 2001), and NAFO's approach to Grand Banks yellowtail flounder (Cadrin & Walsh 1999; Walsh & Morgan 1999; Rivard & Walsh 2000). It is likely that these cases are just a beginning, and the future will see much more use of non-equilibrium approaches to growth, recruitment and mortality, and more acknowledgement of environmental conditions in management strategies.

14.2.1 Stock and recruitment

The relationship between stock size (S) and recruitment (R) has been described as 'the most important and generally most difficult problem in biological assessment of fisheries' (Hilborn & Walters 1992). The above section makes clear that the S-R relationship also is the key factor

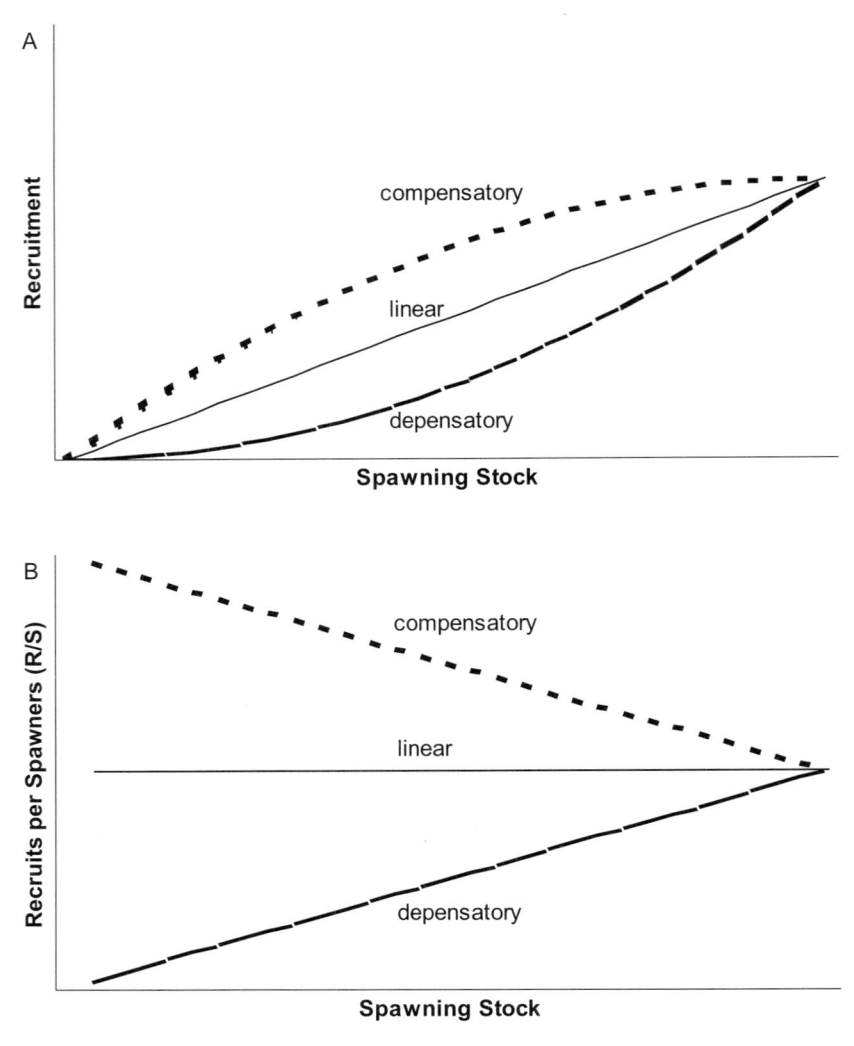

Fig. 14.2 Simple stock recruitment curves showing (a) linear, compensatory and depensatory relationships and (b) their associated trends in survival ratio (R/S).

in determining how a stock will respond to fishing – both how much yield it will produce at various fishing mortalities, and how alternative harvest policies will perform. The effect on yield arises from the curvature of the relationship, reflecting how productivity changes with stock size. Strong downward curvature (i.e. where R/S decreases) at high stock sizes suggests that the stock has increased productivity at intermediate stock sizes. This phenomenon, called compensation, provides a basis for sustainable yield theory and is typical for stocks supporting major fisheries. The effect of alternative harvesting policies on sustainability arises from the slope or steepness of the relationship, reflecting how resilient recruitment is to reduce mature biomass (Fig. 14.2). At low biomass a gradual linear slope or upward curvature (i.e. where R/S is initially low but increases) suggest that stocks are vulnerable to collapse when over-exploited and will recover slowly from depletion. The evidence for the latter pattern, termed depensation, is scarce, but its likelihood is vital to evaluating harvesting strategies.

Most flatfish produce large numbers of pelagic eggs (and subsequently larvae) that drift for months in the plankton before settling out and metamorphosing on more or less extensive nursery grounds. For many years it was a common view that even a relatively small spawning stock of flatfish can produce enough juveniles to seed the nursery grounds. Any excess will simply be lost to starvation or predation, so that the eventual recruitment will be the same over a wide range of spawning stock sizes. This pattern has in fact been observed for a number of heavily exploited flatfish stocks. In 1977 John Gulland, widely knowledgeable about the world's fish stocks, wrote, '... there are some stocks, including many flatfish, where any change in mean recruitment over the range of adult stocks observed in practice is so small that managers can safely accept the hypothesis that recruitment is independent of adult stock, ...' (Gulland 1977, p. 388).

Such views have been challenged in recent years on empirical grounds, although flatfishes are generally found to be more resilient than, for example, gadoid stocks (Myers & Barrowman 1996). Also, adoption of a precautionary approach (FAO 1996) to management focuses more attention on the risk of assuming that the size of the spawning biomass is unimportant to future recruitment and sustainability of harvesting strategies. Because of its critical importance in managing flatfish fisheries, we have compiled and analysed information on the relationship between spawning stock size and recruitment for a large number of reasonably well-studied stocks.

These results should be viewed with appropriate caution. Conducting stock-recruit analyses requires a long time series of demographic information over a wide range of stock size observations and decisions about how to model the relationship. The time series generally must be obtained from sound age-based assessments, requiring, in turn, long time series of age-disaggregated data from accurate catch reporting and research surveys. Several models have been developed to describe compensatory stock-recruit relationships. The most commonly used are the Beverton & Holt (1957) model, in which recruitment is an asymptotic function of stock size, and the Ricker (1954) model, in which recruitment can be a decreasing function of stock size at high stock sizes (Fig. 14.3). More generalised models and non-parametric descriptors have also been applied to quantify stock-recruitment relationships. S-R

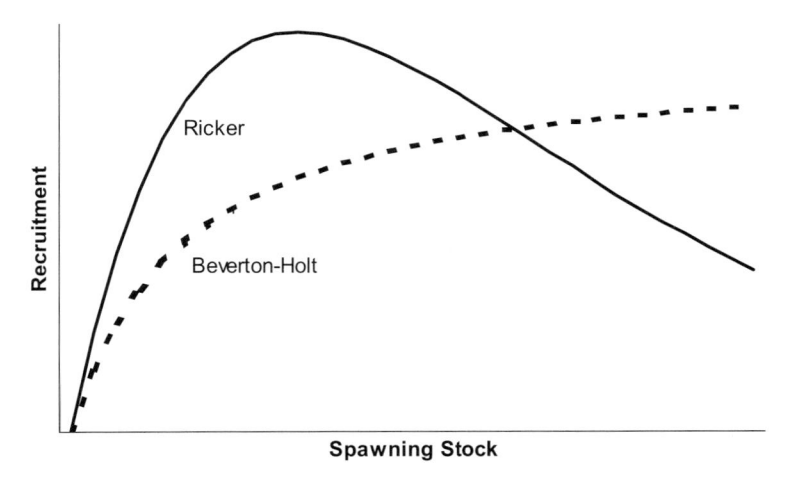

Fig. 14.3 Stock recruit curves derived with the Ricker and Beverton-Holt models.

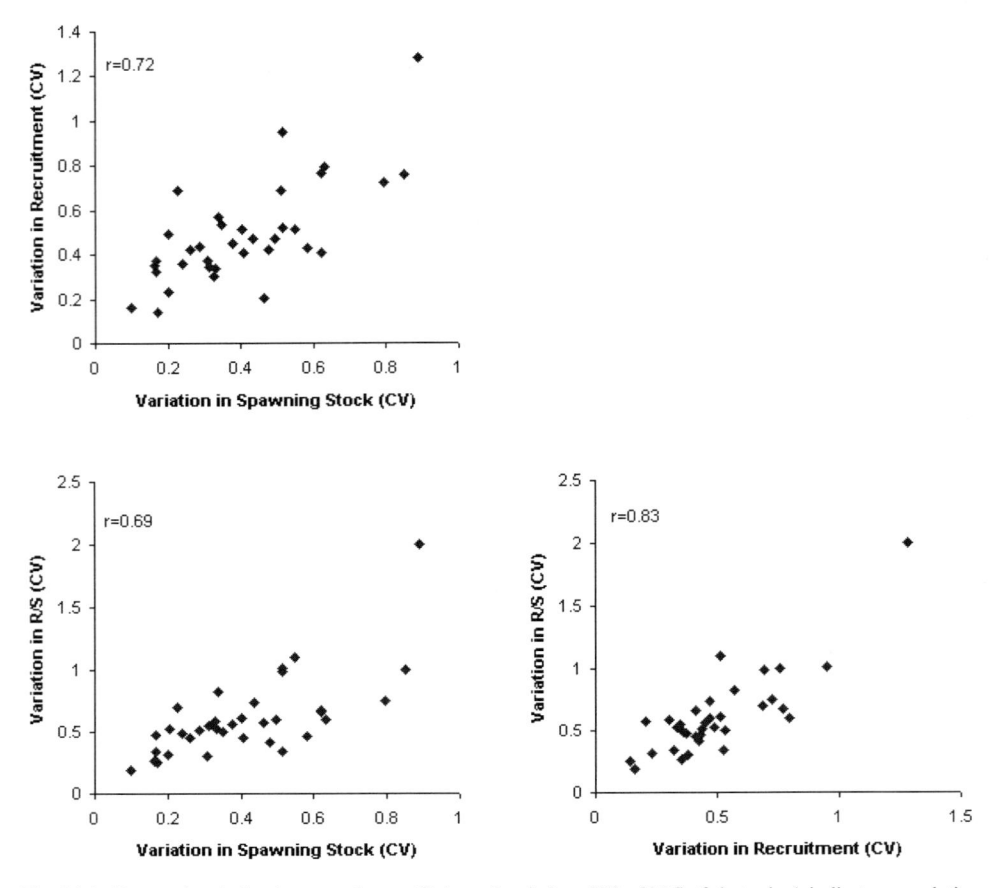

Fig. 14.4 Observed variation (measured as coefficient of variation: CV) of 35 flatfish stocks (r indicates correlation coefficient).

data often do not provide a good guide to choosing among the alternative formulations, yet the decision can have significant implications for conclusions about sustainable harvesting strategies. Model parameters can be estimated using non-linear regression, but the form of error assumed (e.g. normally distributed, lognormally distributed) is another analytical decision that may be difficult to justify empirically, but influential on conclusions. Autocorrelated observation of stock size is also a complication for conventional regression analysis.

This chapter summarises data published from assessments of 35 flatfish stocks in the north Atlantic and northeast Pacific. Several 'model-free' statistics were chosen to describe the data. The first three statistics describe how variable the observed time series are with respect to observed stock sizes, recruitment and survival ratio (R_t/S_t, where R_t is the cohort produced by S_t) using coefficient of variation (CV). The three CVs were strongly correlated among stocks (Fig. 14.4), that is stocks that had highly variable stock size also had highly variable recruitment and survival ratios. For example, yellowtail flounder in southern New England had CVs of 89% for spawning stock, 128% for recruitment and 200% for survival ratio (Table 14.1). Conversely, stocks that had low variability in stock size also had low variability in recruitment and survival ratios (e.g. common sole in the Bay of Biscay had CVs of 10% for

Table 14.1 Descriptive statistics of stock recruit data for 35 flatfish stocks, by geographic region, with eastern Atlantic above, western Atlantic middle, and eastern Pacific below

Species	Area	Number of years	CV(S)	CV(R)	CV(R/S)	r	r sig.	rankS (maxR)	rankS (minR)	R(Shi)/R(Slo)	Odds ratio	Odds ratio sig.	RankS (max(R/S))	RankS (min(R/S))	R/S(Shi)/R/S(S<Slo)
Flounder (*Platichthys flesus*)	Baltic Sea	27	0.20	0.23	0.31	0.02	ns	0.89	0.33	1.02	0.25	ns	0.19	0.96	*0.72*
Greenland halibut	Northeast Arctic	31	0.58	0.43	0.46	0.51	**	0.71	0.06	1.61	**8.25**	**	0.03	1.00	***0.61***
Megrim (*Lepidorhombus* spp.)	West Scotland-Rockall	16	0.17	0.14	0.24	−0.37		0.06	1.00	0.98	1.00	ns	0.06	1.00	***0.77***
Fourspotted megrim (*L. boscii*)	Iberian Region	14	0.17	0.32	0.34	0.15	ns	0.57	0.64	1.09	**6.25**	**	0.57	0.64	***0.86***
Megrim (*L. whiffiagonis*)	Iberian Region	14	0.26	0.42	0.45	0.32	ns	0.93	0.29	1.56	**6.25**	**	0.07	0.29	*1.02*
Plaice	Eastern English Channel	18	0.29	0.44	0.51	−0.12		0.56	0.06	1.13	**1.56**	ns	0.56	0.83	***0.83***
	Irish Sea	35	0.33	0.33	0.52	0.05	ns	0.03	0.74	1.06	**2.24**	ns	0.03	0.74	***0.60***
	North Sea	42	0.20	0.49	0.53	−0.01		0.60	0.67	0.98	0.38	ns	0.60	1.00	***0.71***
	Skagerrak-Kattegat	20	0.24	0.36	0.49	−0.51		0.45	0.90	0.71	0.06	ns	0.45	0.90	***0.50***
	Western English Channel	24	0.35	0.53	0.50	0.12	ns	0.79	0.50	1.47	**1.96**	ns	0.79	0.71	***0.84***
Common sole	Bay of Biscay	16	0.10	0.16	0.19	−0.13		0.25	0.38	0.97	0.33	ns	0.25	1.00	***0.85***
	Celtic Sea	29	0.33	0.30	0.58	−0.32		0.03	0.93	0.97	**3.60**	*	0.03	0.93	***0.55***
	Eastern English Channel	15	0.17	0.37	0.48	−0.26		0.20	0.73	0.79	0.75	ns	0.20	0.73	***0.59***
	Irish Sea	29	0.23	0.69	0.70	−0.06		0.45	0.17	0.80	1.14	ns	0.45	0.79	***0.59***
	North Sea	41	0.51	0.95	1.01	0.09	ns	0.78	0.24	**1.03**	0.33	ns	0.07	0.98	***0.44***
	Skagerrak-Kattegat	15	0.43	0.47	0.73	−0.41		0.53	0.80	0.82	0.24	ns	0.07	0.80	***0.38***
	Western English Channel	31	0.31	0.38	0.30	0.57	**	0.97	0.45	1.56	**8.13**	**	0.48	0.74	***0.98***
American plaice (*Hippoglossoides platessoides*)	Grand Bank	34	0.41	0.41	0.45	0.37	*	0.88	0.06	1.22	**2.0**	ns	0.15	0.68	***0.65***
	Gulf of Maine-Georges Bank	20	0.62	0.41	0.67	−0.35		0.20	0.50	0.75	0.2	ns	0.20	0.85	***0.40***
Greenland halibut	Greenland	10	0.16	0.35	0.26	0.69	*	0.70	0.40	1.49	**2.3**	ns	0.70	0.40	*1.15*
Summer flounder (*Paralichthys dentatus*)	US Atlantic coast	18	0.48	0.42	0.42	0.20	ns	0.78	0.28	1.41	**4.0**	**	0.06	1.00	***0.68***

Table 14.1 (Continued.)

Species	Area	Number of years	CV(S)	CV(R)	CV(R/S)	r	rankS r sig. (maxR)	rankS (maxR)	rankS (minR)	R(Shi) R(Slo)	Odds ratio	Odds ratio sig.	RankS (max(R/S))	RankS (min(R/S))	R/S(Shi)/ R/S(S<Slo)
Winter flounder	Georges Bank	15	0.49	0.47	0.60	−0.04		0.80	0.73	0.95	0.8	ns	0.07	0.73	*0.56*
	Southern New England–midAtlantic	17	0.51	0.52	0.34	0.78	**	0.88	0.35	2.10	**56.0**	**	0.06	0.35	0.86
Witch	Gulf of Maine–Georges Bank	15	0.51	0.69	0.98	−0.74		0.20	0.80	0.34	0.0	ns	0.20	0.80	*0.16*
Yellowtail flounder	Cape Cod	14	0.40	0.52	0.62	0.07	ns	0.36	0.14	0.85	6.3	**	0.36	0.79	*0.50*
	Georges Bank	26	0.63	0.80	0.60	0.55	**	0.85	0.88	2.44	**11.1**	**	0.19	0.88	*0.82*
	Southern New England	25	0.89	1.28	2.00	−0.01		0.72	0.20	1.65	**10.0**	**	0.28	0.84	*0.37*
Arrowtooth flounder (*Reinhardtius stomias*)	Bering Sea	28	0.85	0.76	1.00	−0.10		0.61	0.39	1.31	**3.2**	*	0.14	0.89	*0.42*
English sole (*Parophrys vetula*)	Gulf of Alaska	18	0.46	0.21	0.57	0.15	ns	0.44	0.22	1.04	**1.6**	ns	0.11	1.00	*0.42*
	British Columbia	50	0.31	0.35	0.55	0.10	ns	0.88	0.96	1.14	**2.3**	ns	0.02	0.96	*0.63*
Flathead sole	Bering Sea	21	0.55	0.51	1.11	−0.81		0.10	0.90	0.45	0.0	ns	0.10	0.90	*0.12*
Pacific halibut	IPHC 2+3A	61	0.38	0.45	0.56	0.25	*	0.87	**0.46**	0.96	0.5	ns	0.08	0.46	0.51
Rock sole (*Lepidopsetta bilineata*)	Bering Sea	19	0.80	0.72	0.75	−0.30		0.58	1.00	1.18	0.3	ns	0.37	1.00	*0.45*
	British Columbia	30	0.62	0.77	0.67	0.18	ns	0.57	0.80	1.88	**2.3**	ns	0.57	0.80	*0.85*
Yellowfin sole (*Limanda aspera*)	Bering Sea	19	0.34	0.57	0.82	−0.19		0.26	0.32	0.81	0.8	ns	0.05	0.74	*0.40*
Average			0.41	0.49	0.61	0.01		0.56	0.52	1.16	**4.2**		0.25	0.80	*0.62*

CV, coefficient of variation; S, spawning stock; R, recruitment; r, correlation coefficient; sig, significance; *, 5% significance; **, 1% significance; rankS(maxR), relative rank of S that produced maximum R; rankS(minR), relative rank of S that produced minimum R; R(Shi)/R(Slo), ratio of mean R produced when S>median to mean R produced when S<median; odds ratio, [odds R>median(S)]/[odds R>median(R) | S>median(S)]; rankS(max(R/S)), relative rank of S that produced maximum survival; rankS(min(R/S)), relative rank of S that produced minimum survival; R/S(Shi)/R/S(Slo), ratio of mean survival produced when S>median to mean survival produced when S<median; bold type indicates positive S-R relationship; italic bold type indicates compensation.

spawning stock, 16% for recruitment and 19% for survival ratio). As for these two example stocks, survival ratios were generally most variable among the three parameters, and spawning stock was generally least variable.

Five statistics were calculated to test for a positive relationship between spawning stock and recruitment: correlation coefficient between recruits and stock size over time, three statistics developed by Myers & Barrowman (1996) to indicate if (1) the highest recruitment occurred when the stock was abundant, (2) the lowest recruitment occurred when the stock was low, and (3) if recruitment is generally greater when stock size is high. Also, the odds of above median recruitment from above median stock size, relative to the odds of above median recruitment from below median stock size was estimated, as described by Brodziak et al. (2001). For example, an odds ratio of 2 indicates that there is twice the likelihood of strong recruitment when stock size is high.

Results generally indicate positive relationships between spawning stock and recruitment for most stocks. However, results vary by stock and the various statistics are equivocal within some stocks (Table 14.1). Eighteen stocks (51%) had a positive correlation between stock and recruitment, and seven of those were statistically significant. Twenty-two stocks (63%) had the greatest recruitment produced when spawning stock was high (i.e. greater than the median). Eighteen stocks (51%) had the minimum observed recruitment produced by a low stock size. Twenty stocks (57%) had stronger average recruitment when stock size was high than when stock size was low. The same number of stocks (57%) had increased likelihood of strong recruitment at high spawning stock (i.e. an odds ratio >1), and the increase was significant for 11 (31%) of those stocks.

Four stocks had positive indications from all statistics, suggesting that stock-recruit relationships are strong (Greenland halibut (*Reinhardtius hippoglossoides*) in the northeast Arctic, western English Channel common sole, American plaice (*Hippoglossoides platessoides*) on the Grand Bank, and winter flounder (*Pseudopleuronectes americanus*) off southern New England; Fig. 14.5a). On the other hand, five stocks did not have a positive indication from any statistic, suggesting a weak stock-recruitment relationship (megrim (*Lepidorhombus whiffiagonus*) and fourspotted megrim (*L. boscii*) off west Scotland and Rockall, Skagerrak-Kattegat plaice, eastern English Channel common sole, witch (*Glyptocephalus cynoglossus*) in the Gulf of Maine and Georges Bank, and flathead sole (*Hippoglossoides elassodon*) in the Bering Sea; Fig. 14.5b). However, it is important to note that indicating a positive relationship between stock and recruitment may require more observations than are available from many stock assessments. For example, three of the five stocks with weak relationships listed above have relatively short time series of observations (16 years or less), and the four stocks with strong relationships listed above have longer time series of observations.

Three statistics derived by Brodziak et al. (2001) were calculated to test for a negative relationship between spawning stock and survival ratio (i.e. density-dependent survival, or compensation). The three statistics indicate if (1) the highest survival occurred when the stock was abundant, (2) the lowest survival occurred when the stock was low, and (3) if survival is generally greater when stock size is high.

Results strongly suggest that survival ratios are density-dependent (i.e. there are generally fewer recruits per spawner produced when stock size is high). For 29 stocks (83%), the maximum survival ratio was observed when stock size was low (i.e. spawning stock was below median). Thirty-one stocks (89%) had minimum survival when stock size was high.

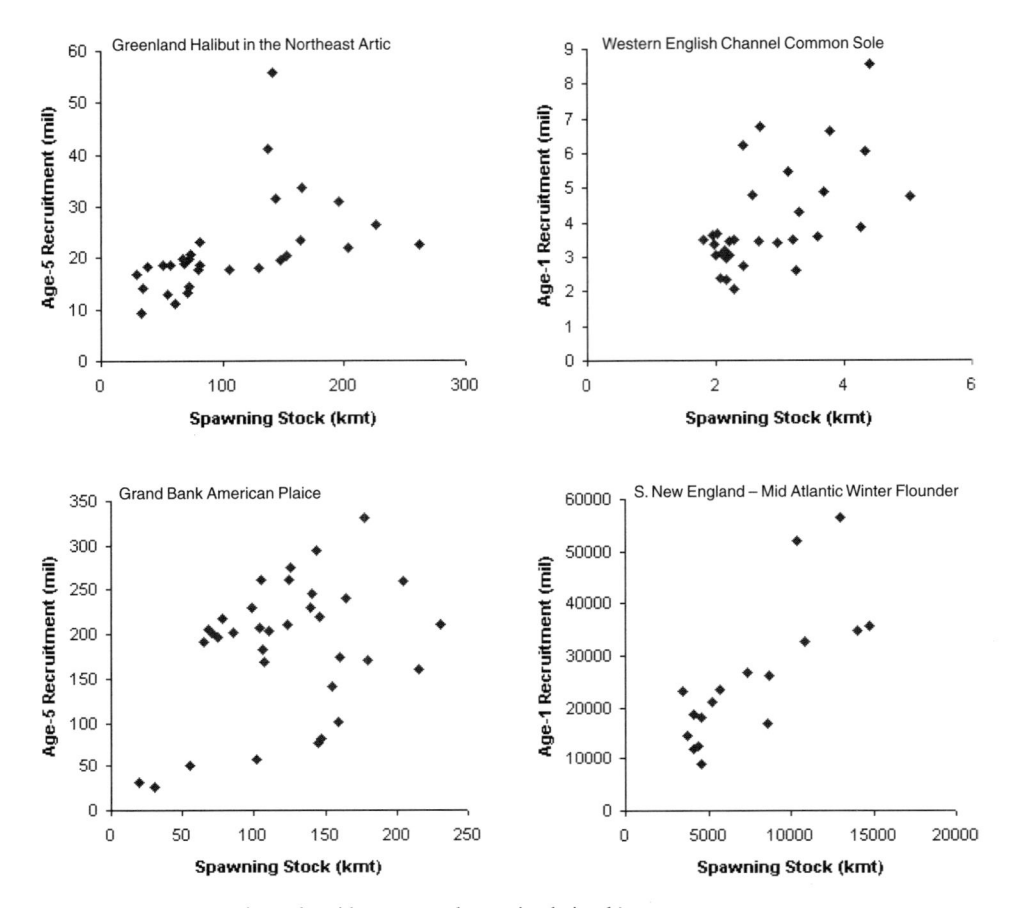

Fig. 14.5(a) Four example stocks with strong stock-recruit relationships.

Thirty-three stocks (94%) had greater survival on average when stock size was low than when stock size was high. Stock recruitment patterns appear to be stock-specific and vary widely among geographic areas within species. For example, western English Channel common sole appears to have a strong relationship, whereas eastern English Channel common sole has a weak relationship. Different stock recruitment patterns for adjacent resources of the same species may result from different environmental features affecting survival or retention of eggs or larvae and perhaps different habitats affecting density dependence.

In summary, results show that at least for most flatfish stocks, stock and recruitment are positively correlated, the weakest recruitment was produced by low levels of spawning stock biomass, or the strongest recruitment is produced by high levels of spawning stock biomass. Therefore, it is not generally true that for practical purposes recruitment of flatfishes is independent of spawning biomass. However, for the great majority of the stocks reviewed here, it appears that recruitment at lower levels of spawning biomass is not very different from recruitment at higher levels of spawning biomass. The pattern of R/S provides overwhelming evidence for compensatory survival, which dampens the stock-recruit relationship, but also offers resilience to exploitation.

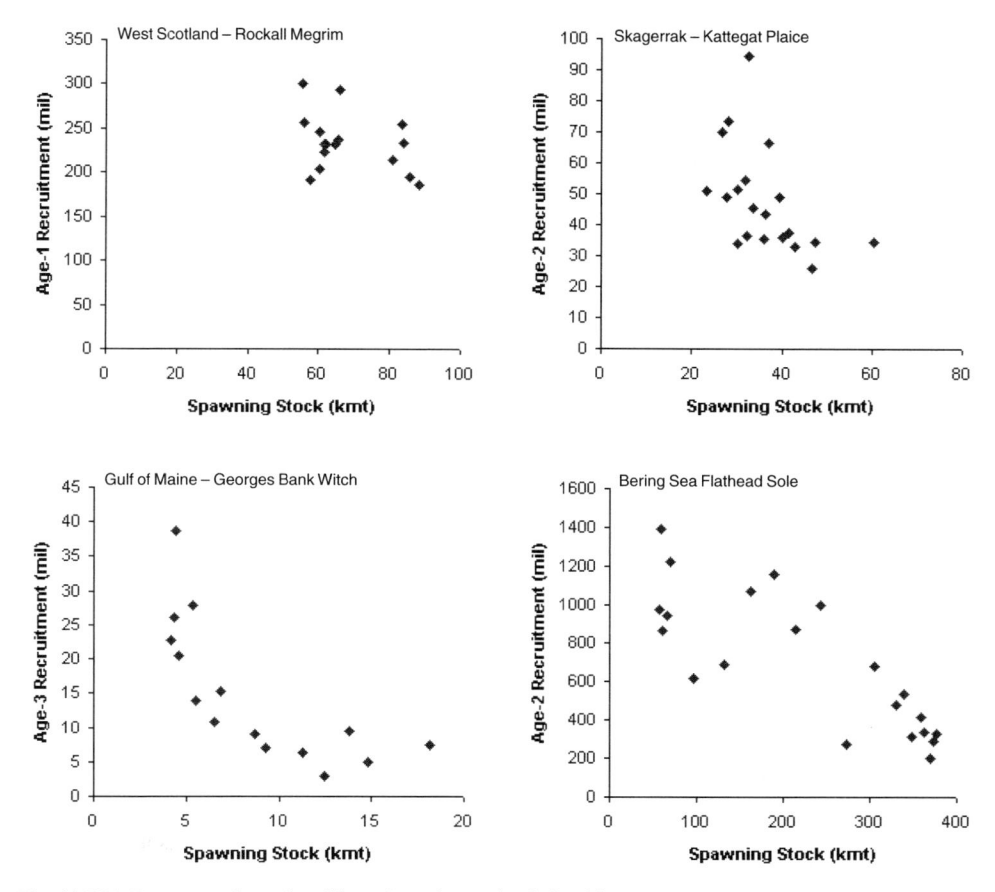

Fig. 14.5(b) Four example stocks with weak stock-recruit relationships.

Results for other fish groups provide a context for these updated and expanded analyses. Perhaps the most comparable group is the Gadidae (e.g. Atlantic cod, *Gadus morhua*; haddock, *Melanogrammus aeglefinus*; hakes, *Merluccius* spp.), which have similar ranges in life history attributes such as longevity, natural mortality, growth and maturity. Furthermore, many flatfishes are harvested in mixed species fisheries along with gadids. Myers & Barrowman (1996) analysed stock recruitment data for 68 gadid stocks. Forty-one stocks (60%, as compared with 63% for flatfishes) had maximum recruitment produced by high stock size, 43 stocks (63%, as compared with 51% for flatfishes) had minimum recruitment from low stock size, and 42 stocks (62%, as compared with 57% for flatfishes) had greater average recruitment when spawning stock was high. Therefore, the general strength of flatfish stock and recruitment relationships is comparable to another fish group with similar life histories.

Unfortunately, few other stocks have been analysed to detect compensation. Brodziak *et al.* (2001) used stock recruit data from four gadid stocks to study compensation. All four gadid stocks had maximum R/S at low stock size and minimum R/S at high stock size (83% and 89% respectively for flatfishes), and three had lower average R/S at low stock size (94% for flatfishes). These few analyses suggest that the frequency of compensation in flatfish stocks appears to be similar to that for gadid stocks.

The strength of the stock recruit relationship and the degree of compensation should be considered for several aspects of stock assessment and management. For example, some methods of deriving harvest targets and limits that assume equal recruitment at various levels of fishing mortality and stock size may not be appropriate for some flatfish stocks. Also, these results may be valuable for deciding how to project future recruitment for evaluating alternative management strategies.

14.2.2 Recruitment, environment, assessment and management

The strong influence of environmental conditions on stock productivity is well established for many fish stocks. Other chapters of this book document more fully the importance of these relationships for flatfishes, where some of the best documented examples occur. The effects are important to assessment and management; for example, the ICES advice on flatfish stocks in the northeast Atlantic specifically noted strong environmental effects on year-class strength in over half the flatfish stocks considered. One may question why, if the environmental influences on recruitment (and growth rates) are so strong, the assessment and management community focuses its attention on stock and recruitment, rather than on the environment. There are two types of responses to such questions.

First, it is the efficient way to do things. Although the environment is unquestionably important to stock productivity, beyond protecting habitat quality there is very limited management control over currents and water temperatures, for example, as contrasted with control over spawning biomass. Hence, management is pragmatic, and focuses on keeping harvesting activities sustainable to protect spawning biomass. This in fact, is farsighted, not myopic, in that failure to ensure adequate spawning biomass means that flatfish stocks are not positioned to benefit from favourable environmental conditions when they do occur, and ensuring a healthy spawning stock allows the stock to perform as well as allowed by the environmental conditions which do occur. Correspondingly, assessments in support of management focus on estimating stock sizes and mortality rates, as well as realised recruitment, to give management maximally useful information on which to base annual management plans.

Second, management and assessment do not ignore environmental conditions and their impacts on stock productivity. Recently, effects of environmental conditions on stock productivity have featured prominently in assessment, advice and management actions on many flatfish stocks. In the north Pacific, the importance of abrupt regime shifts is well established (Hollowed & Wooster 1995; McKinnell *et al.* 2001), and these regime shifts impact on productivity of Pacific halibut. In the late 1990s, the target exploitation rate for Pacific halibut was adjusted downward after the appearance of some large year-classes produced at a high level of spawning biomass lowered the estimated degree of density dependence in the stock-recruitment relationship (Clark *et al.* 1999).

In the North Sea, the large plaice stock has periodically produced outstanding year-classes. Although clearly not the result of suddenly increased SSB (Fig. 14.6), they have substantial impacts on the stock and fisheries in the succeeding years. The most recent of these, in 1996, led to a forecast of a strong increase in fishable biomass and, under the existing management approach, expectation of a large increase in yield and, subsequently, in SSB. However, starting in 1998, the scientific advice also included warnings about the potential for increased catches of undersized plaice, leading to increases in discarding and reduced potential yield

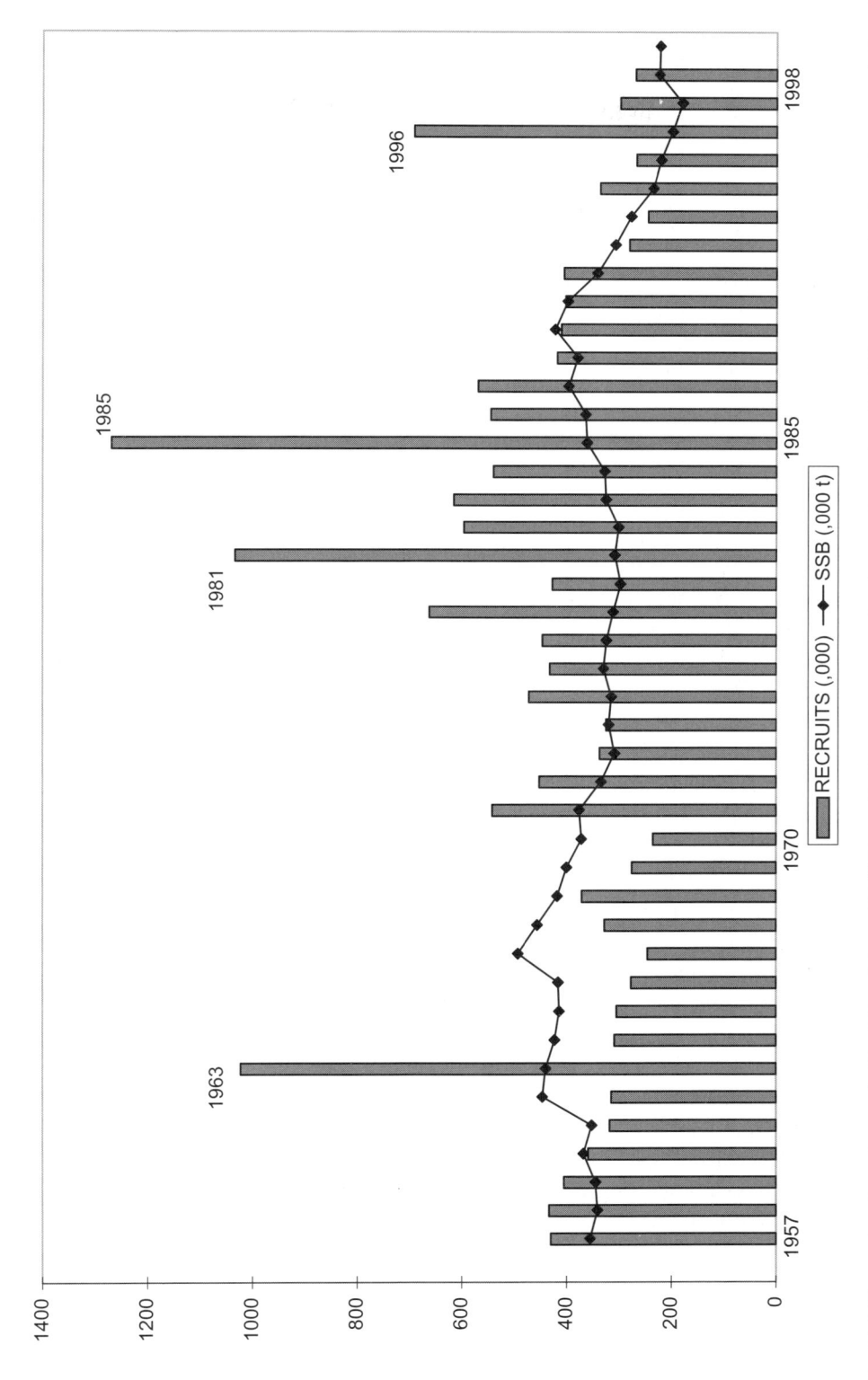

Fig. 14.6 Time series of recruitments for North Sea plaice, with SSB lagged by 1 year, to correspond to the year-class each produced. Years of exceptional recruitments are indicated in the figure.

and SSB from the year-class. Advice in subsequent years continued to emphasise that the cohort was growing slowly, maturing later than typical, suffering high discard mortality, and producing less yield with each successive assessment. The advice for 2001 highlighted both the lost opportunity from the 1996 year-class, and the impact of such occasional exceptional year-classes on ability to forecast expected trajectories of the stock under various management options (ICES 2001).

Hence assessments and advice sometimes do give importance to environmental influences on stock dynamics, but generally in responsive, rather than predictive modes. This is likely to remain the case, despite our growing understanding of stock–environment relationships, at least until ability to forecast environmental trends improves greatly.

14.2.3 *Assessment, management and uncertainty*

From the discussions of stock–recruit relationships and environmental influences on stock productivity, it is apparent that assessments include significant amounts of uncertainty in data, model formulations and future states of nature relevant to forecasts. Correspondingly, management must be risk-averse in the face of the uncertainty and the dangers of serious harm to stocks from over-fishing. Although the historic record of assessment and management is not outstanding in many parts of the world, through the 1990s the precautionary approach has provided a framework for addressing uncertainty and risk more effectively (Richards & Maguire 1998).

In fisheries, the precautionary approach has highlighted limit and target reference points, and pre-agreed harvest control rules (FAO 1996). Limit reference points define unacceptable conditions where conservation is not assured, such as spawning biomasses that are too low or fishing mortalities that are too high, and are set on the basis of biological knowledge. Target reference points designate desirable stock conditions of high productivity for achieving social and economic goals, and are set on the basis of both biological knowledge and the objectives of resource users. Harvest control rules are pre-agreed management measures invoked whenever the risk of approaching a limit reference point is unacceptably high. They are often supported by 'precautionary' or 'buffer' reference points that reflect the uncertainty in the assessments, and begin to trigger conservation actions by managers before the limit reference points are approached (FAO 1996).

The concepts in the precautionary framework are not novel. However, their systematic organisation of assessment activities around estimating likelihood of achieving targets and avoiding limits, and management activities around pre-agreed conservation measures, and their explicit treatment of uncertainty, have greatly altered practice in assessment and management in many regions of the world. The impact of the precautionary approach will be particularly apparent in the assessment and management in the northwest Atlantic and northeast Atlantic. NAFO and ICES, respectively, have incorporated the precautionary approach in their data collection and assessment activities, and the management agencies are applying a precautionary approach to a wide range of management and enforcement regulations.

14.3 Assessment and management summary

14.3.1 Northeast Pacific

Aside from Pacific halibut, which is managed by the bilateral (Canada–USA) International Pacific Halibut Commission, all flatfish stocks in the northeast Pacific are under national jurisdiction. All the analytical assessments use data from commercial catches and research trawl surveys. Analyses commonly use the stock synthesis method of Methot (1990), or similar many-parameter state-space models (Schnute 1994), often using the AD Model Builder software (Schnute *et al.* 1998); where AD stands for a flexible automatic-differentiation approach to model construction. Both approaches tend to estimate larger numbers of parameters than methods popular in circum-Atlantic jurisdictions (Tables 14.2–14.4), but also allow assessment models to be tailored very closely to the specific stocks being assessed. The specific reference points used among the three jurisdictions differ in mode of derivation, but are very comparable with regard to degree of conservation offered to stocks.

About half the stocks are fully exploited, and several of the Bering Sea stocks are only lightly harvested. The one stock which is depleted (petrale sole, *Eopsetta jordani*, off the West Coast of Vancouver Island) was over-fished between the 1940s and 1960s, and has never recovered. Fisheries oceanographic studies (Castille *et al.* 1994) indicate that the frequent and strong El Niños during the 1980s and 1990s were likely to produce poor recruitment for petrale sole in the area, so recovery is unlikely in the near future. However, for the other stocks, recent management regimes, combined with effective conservation reference points, have been able to sustain fisheries and generally keep flatfish stocks healthy.

14.3.2 Northwest Atlantic

Most coastal stocks of flatfishes (i.e. those in waters under national jurisdiction) are assessed and managed nationally by the fisheries agencies of Canada and the USA (Table 14.3). However, several stocks extend into international waters and are assessed and managed by the Northwest Atlantic Fisheries Organization (NAFO). The Georges Bank yellowtail flounder stock is distributed across national boundaries, is assessed cooperatively by the USA and Canada, and cooperative management is currently being planned. Management strategies vary with jurisdiction, with NAFO and Canadian stocks primarily managed through national catch quotas, size limits and gear regulations, and US stocks are managed by various methods including catch quotas, closed areas, effort limits and gear regulations. The Sustainable Fisheries Act of 1996 marks a distinct episode in assessment and management of US stocks. The Act mandates that fisheries be managed to maintain maximum sustainable yield, and national guidelines require that fishing mortality be limited to less than F_{MSY} and depleted stocks be rebuilt to B_{MSY}.

Assessment methods for northwest Atlantic flatfish resources vary according to the amount of information available for each stock. The most informative assessments apply age-based sequential population analysis, most commonly calibrated using ADAPT (Gavaris 1988). Less informative assessments monitor catch and survey indices of stock size. Biomass dynamics or surplus production models have recently had renewed application to northwest Atlantic stocks for their intermediate data requirements and ability to estimate MSY reference points for US management requirements (a common production model used is ASPIC; Prager 1994).

Flatfishes

Table 14.2 Assessment information for northeast Pacific flatfish stocks*

Species	Stock area	Statistical area(s)	Assessment reference	Jurisdiction	Assessment method(s)	Reference points	Present status
Arrowtooth flounder	Bering Sea		Wilderbuer & Sample 2000	USA (NPFMC)	Trawl survey, stock synthesis	F40%, B40%	Very lightly exploited
Dover sole	Gulf of Alaska		Turnock et al. 2000	USA (NPFMC)	Trawl survey, ADMB	F40%, B40%	Very lightly exploited
	Washington–Oregon		Brodziak et al. 1997	USA (PFMC)	Trawl survey, stock synthesis	F35%	Fully exploited
English sole	Hecate Strait	5CDE	Fargo 1999	Canada	Biomass dynamic (Fox)	$F_{0.1}$	Fully exploited, SSB ~ 40% B_0
	West Coast Vancouver Isl.	3CD	Fargo 1999	Canada	No analytical assessment		Fully exploited
	Hecate Strait	5CDE	Fargo 1999	Canada	Trawl survey, ADMB	$F_{0.1}$	$F < F_{0.1}$, Biomass uncertain
Flathead sole	Bering Sea		Spencer et al. 2000	USA (NPFMC)	Trawl survey, ADMB	F40%, B40%	Lightly exploited
Pacific halibut	NE Pacific		Clark & Hare 2001	IPHC	Setline survey, ADMB	Fmsy, SSBmin	High abundance
Petrale sole	West Coast Vancouver Isl.	3CD	Fargo 1999	Canada	No analytical assessment		Depleted; slow recovery
Rock sole	Bering Sea		Wilderbuer & Walters 2000	USA (NPFMC)	Trawl survey, ADMB	F40%, B40%	Lightly exploited
	Hecate Strait	5CDE	Fargo 1999	Canada	Trawl survey and commercial, ADMB	$F_{0.1}$	Biomass > long-term average. $F << F_{0.1}$
	Queen Charlotte Sound	5AB	Fargo 1999	Canada	No analytical assessment		Uncertain, likely declining
Yellowfin sole	Bering Sea		Wilderbuer & Nichol 2000	USA (NPFMC)	Trawl survey, ADMB	F40%, B40%	Lightly exploited

*Stock area is geographic range of stock; Statistical area from IPFC areas; Jurisdiction: country or regional fisheries management organization (NPFMC = North Pacific Fisheries Management Council, PFMC = Pacific Fisheries Management Council, IPHC = International Pacific Halibut Commission) with authority over the stock; Assessment methods used for the stock in most recent years; Reference point is management target adopted for the stock, see text for details on specific points; Present status as concluded in the assessment reference.

Table 14.3 Assessment information from the northwest Atlantic flatfish stocks

Species	Stock area	Statistical area(s)	Assessment reference	Jurisdiction	Assessment method(s)	Reference point(s)	Current status
American plaice	Labrador-northeast Newfoundland	NAFO 2,3K	Brodie & Morgan 2000	NAFO	Survey index	None	B99 = 6% historical average, no directed F
	Flemish Cap	NAFO 3M	Alpoim & Avila de Melo 2000	NAFO	Survey index	None	B99 record low
	Grand Bank	NAFO 3LNO	Morgan 2000	NAFO	ADAPT	None	B99 record low
	St Pierre Bank	NAFO 3Ps	Morgan et al. 1999	Canada	Survey index	None	B99 = 16% historical average, no directed F
	Gulf of St Lawrence	NAFO 4T	Morin et al. 1998	Canada	Survey index	None	B1997 record low
	Scotian Shelf	NAFO 4VWX	Stobo et al. 1997	Canada	Survey index	None	B97 = average
	Gulf of Maine-Georges Bank	NAFO 5,6	O'Brien & Esteves 2001	USA	ADAPT, YPR	$F_{0.1}$, Bmsy proxy	SSB99 = 58%Bmsy, F99 = 142%$F_{0.1}$
Atlantic halibut	Gulf of St Lawrence	NAFO 4RST	Archambault & Gregoire 1996	Canada	Landings	None	Unknown
	Grand Bank-Scotian Shelf	NAFO 3NOPs,4VWX	Zwanenburg et al. 1997	Canada	Survey index	None	B96 low
Greenland halibut	Davis Strait	NAFO 0, 1B-F	Jorgensen 2000	Canada-Denmark	Survey index	None	Unknown
	Greenland	NAFO 1A	Simonsen & Boje 2000	Denmark	SVPA	None	B99 average
	Labrador-Grand Bank	NAFO 2, 3KLMNO	Bowering & Brodie 2000	NAFO	XSA	None	B99 above average, F low
	Gulf of St Lawrence	NAFO 4RST	Morin & Bernier 2000	Canada	Survey index	None	B99 record high
Summer flounder	US Atlantic coast	NAFO 5,6	Terceiro 1999	USA	ADAPT, YPR	Fmax, Bmsy proxy	B99 = 39%Bmsy, F99 = 123%Fmax
Windowpane (*Scophthalmus aquosus*)	Gulf of Maine-Georges Bank	NAFO 5Y,5Ze	Hendrickson 1998	USA	Survey index	Fmsy, Bmsy	B99 = Bmsy, F99 = 39%Fmsy
	Southern New England-midAtlantic	NAFO 5Zw,6	Hendrickson 1998	USA	Survey index	Fmsy, Bmsy	B99 = 35%Bmsy, F99 = 38%Fmsy

Table 14.3 (Continued.)

Species	Stock area	Statistical area(s)	Assessment reference	Jurisdiction	Assessment method(s)	Reference point(s)	Current status
Winter flounder	Gulf of St Lawrence	NAFO 4T	Morin et al. 1999	Canada	Survey index	None	B98 record low
	Scotian Shelf	NAFO 4VWX	Stobo et al. 1997	Canada	Survey index	None	Unknown
	Gulf of Maine	NAFO 5Y	Cadrin et al. 1996	USA	Survey index	None	B94 below average, F94 = 200%Fmsy
	Georges Bank	NAFO 5Ze	Brown et al. 2000	USA	ADAPT, ASPIC	Fmsy, Bmsy	B99 = 76%Bmsy, F99 = 70%Fmsy
	Southern New England-midAtlantic	NAFO 5Zw, 6	Terceiro 2000	USA	ADAPT, ASPIC	Fmsy, Bmsy	B99 = 91%Bmsy, F99 = 44%Fmsy
Witch	Northeast Newfoundland	NAFO 2J, 3KL	Bowering 2000a	NAFO	Survey index	None	B99 record low
	Grand Bank	NAFO 3NO	Bowering 2000b	NAFO	Survey index	None	B99 below average
	St Pierre Bank	NAFO 3Ps	Bowering 1999	Canada	Survey index	None	B99 = 66% historical average
	Gulf of St Lawrence	NAFO 4RST	Swain et al. 1998	Canada	Survey index	None	SSB90s = 25% historical
	Scotian Shelf	NAFO 4VWX	McRuer et al. 1997	Canada	Survey index	None	Exploitable B97 = record low
	Gulf of Maine–Georges Bank	NAFO 5, 6	Wigley et al. 1999	USA	ADAPT, ASPIC	Fmsy, Bmsy	B99 = 102%Bmsy, F99 = 121%Fmsy
Yellowtail flounder	Grand Bank	NAFO 3LNO	Walsh et al. 2000	NAFO	ASPIC	Fmsy, Bmsy	B99 = 90%Bmsy, F99 = 50%Fmsy
	Gulf of St Lawrence	NAFO 4T	Poirier & Morin 1999	Canada	Survey index	None	B99 = average
	Scotian Shelf	NAFO 4VWX	Stobo et al. 1997	Canada	Survey index	None	B97 >average
	Cape Cod	NAFO 5Y	Cadrin et al. 1999	USA	ADAPT, ASPIC	Fmsy, Bmsy	B99 = 64%Bmsy, F99 = 78%Fmsy
	Georges Bank	NAFO 5Ze	Cadrin et al. 2000	USA–Canada	ADAPT, ASPIC	Fmsy, Bmsy	B99 = 92%Bmsy, F99 = 30%Fmsy
	Southern New England	NAFO 5Zw	Cadrin 2000	USA	ADAPT, ASPIC	Fmsy, Bmsy	B99 = 11%Bmsy, F99 = 58%Fmsy
	MidAtlantic	NAFO 6	Overholtz & Cadrin 1998	USA	Survey index	Fmsy, Bmsy	B99 = 2%Bmsy, F99 = 724%Fmsy

Columns as in Table 14.2; NAFO, Northwest Atlantic Fisheries Organization.

Table 14.4 Summary of assessment information about flatfish stocks in the northeast Atlantic

Stock	B_{pa} ('000t)	F_{pa}	Inside safe biological limits in January 2001?	Advised F to reach safe biological limits	Years with $B > B_{pa}$ and $F < F_{pa}$/total
Common sole in IIIa (Skagerrak & Kattegat)	1.06	0.30	Yes	Status quo	6/16
Plaice in IV (North Sea)	300	0.30	No	Reduce 20%	14/48
Common sole in IV (North Sea)	35	0.40	No	Reduce 14%	16/48
Common sole in VIId (Eastern Channel)	8.0	0.40	No	Reduce 6%	10/19
Plaice in VIId (Eastern Channel)	8.0	0.45	No	Reduce 30%	1/21
Plaice in VIIa (Irish Sea)	3.1	0.45	Yes	Could increase	10/37
Common sole in VIIa (Irish Sea)	3.8	0.30	No	Reduce 30%	0/31
Plaice in VIIfg (Celtic Sea)	1.8	0.60	No	Reduce 40%	6/24
Common sole in VIIfg (Celtic Sea)	2.2	0.37	No	Reduce 40%	9/30
Plaice in VIIe (Western Channel)	2.5	0.45	No	Reduce 22%	1/25
Common sole in VIIe (Western Channel)	2.5	0.26	No	Reduce 20%	7/32
Common sole in VIIIab (Bay of Biscay)	11.3	0.45	Yes	Status quo	9/16
Greenland halibut in I and II (northeast Arctic)	80.0	0.36	Uncertain	No increase	*
Megrim (*Lepidorhombus whiffiagonus* + *L. boscii*) in VI (West of Scotland)	*	*	Probably	Status quo	*
Megrims in VII and VIIIabde	55	0.30	No	Reduce 5%	7/15
Megrim in VIIIc and IXd (*L. boscii* portion)	4.7**	0.3**	Yes	Status quo	5/15
Megrim in VIIIc and IXd (*L. whiffiagonis*)	*	*	Probably	Status quo	*

Apart from Greenland halibut and megrim, all stocks are assessed with XSA using catch at age and CPUE information from one or more commercial fleets and one or more research surveys. Source for all assessments is ICES (2001).
*Insufficient information for analytical assessment and estimation of these values.
**Provisional precautionary reference points based on short time series of information.

Current status of northwest Atlantic flatfish stocks is variable, depending on species and geographic region (Table 14.3). For example, American plaice stocks from Labrador to the Gulf of St Lawrence are severely depleted, whereas the two southern stocks on the Scotian Shelf and in the Gulf of Maine are at more moderate stock sizes. Of the 34 flatfish stocks that are regularly monitored, the status of 3 (9%) is unknown, 8 (24%) are considered to be at moderate to high stock size, and 23 (68%) are depleted, some of which are at record low stock size. Despite the generally negative outlook for northwest Atlantic flatfish stocks, several resources that were depleted as recently as a decade ago have responded to harvest restrictions (e.g. Grand Banks yellowtail flounder, Georges Bank yellowtail flounder, southern New England winter flounder) or strong recruitment (e.g. Gulf of Maine–Georges Bank witch) and have grown to moderate stock size in a relatively short period of time.

14.3.3 Northeast Atlantic

The major flatfish stocks in the northeast Atlantic are all assessed by ICES. Management is by the European Union for all stocks except the widespread, northerly distributed Greenland halibut and North Sea stocks shared by formal agreements between the EU and Norway. For all stocks management is by overall quotas, based on advice from ICES, partitioned among countries with access rights to particular stocks, and minimum size limits supported by mesh size regulations. Enforcement of the size limits and quotas has often meant that discarding and misreporting are problems to varying degrees in almost all fisheries, with detrimental effects on both stocks and the accuracy of assessments. Recent programmes to place EU-funded observers on fishing vessels are providing information on magnitudes of discarding in some of the major fisheries, and may be reducing the practice to some degree.

All analytical ICES flatfish assessments are based on extended survivors analysis (XSA), an age-structured sequential population analysis (Shepherd 1992; Darby & Flatman 1994), using commercial catch and research survey data. In only some stocks are estimates of discards included. Starting in 1998, ICES advice was provided relative to explicitly identified reference points for spawning biomass and fishing mortality. Advised quotas are intended to provide a high probability of keeping the stock above the precautionary SSB reference point, and fishing mortality below the precautionary F reference point. (Table 14.4 does not contain some of the columns necessary in Tables 14.2 and 14.3. Where data are sufficient to support an analytical assessment of flatfish in the NE Atlantic, ICES uses the same methods consistently.)

In general the status of flatfish stocks in the northeast Atlantic is not good (Table 14.4). Only 4 of 14 stocks for which analytical assessments are possible were within safe biological limits entering 2001, where 'safe biological limits' is a technical phrase meaning biomass is estimated to be above B_{pa} and fishing mortality below F_{pa}. ICES advised reductions in fishing mortality of 20% or greater in more than half the stocks assessed. Moreover only one stock has been within safe biological limits for even half the years included in the respective assessment periods.

In recent years TACs have been set consistent with scientific advice, and conditions are stable or improving for the majority of stocks. However, some stocks have a long way to go. Also, high and variable discarding and inaccurate catch reporting continue to be a problem for some stocks. This contributes to continued poor stock condition in two ways; the first directly through excessive mortality. The second is indirectly, through providing inaccurate data for assessments, which consequently overestimate stock growth and underestimate the impact of the fishery, producing advice for overly optimistic catches in the subsequent year. The many advances in biological knowledge summarised in this book will contribute greatly to improved assessment and management. However, without effective enforcement and willing cooperation by harvesters, stock prognoses will continue to be pessimistic.

14.4 Conclusions

Reliable assessments require reasonable knowledge of the population dynamic processes and parameters of the flatfish stock, analytical models which capture those processes accurately, and reliable data from monitoring commercial fishery catches and research surveys. However, as Tables 14.2–14.4 document, even sophisticated assessments using several data sources do not ensure conservation of stocks and sustainability of harvesting, unless management is effective and respected by harvesters. It is unlikely to be a coincidence that stocks tend to be in better condition where industry and management agencies cooperate extensively. Not only are the data likely to be more extensive (e.g. industry-supported surveys to augment government-funded ones) and more reliable (accurate catch reporting), but harvesters seek ways to achieve the same goals as management, rather than ways to get around them.

Assessment of flatfish stocks has been at the forefront of advances in assessment methodology for many decades. Many of the advances in knowledge of flatfish biology presented in the previous chapters are already making their way into assessment methodology and scientific advice to management. There is much yet to be learned about flatfish biology, for example, how environmental variability impacts productivity. However, major strides in conservation of flatfish stocks will also require that constructive interactions between participants in management and harvesting similarly be at the forefront of thinking and innovation. Developments in several jurisdictions, such as the IPHC harvest strategy, application of the precautionary approach in management of NAFO and ICES stocks, and exploration of additional management tools such as closed areas on Georges Bank, are very encouraging, but in many areas there is much progress yet to be made.

Acknowledgements

The authors are pleased to thank the many assessment scientists working in North America and Europe, who provided all the results on which this chapter is based. We apologise to assessment scientists in many other parts of the world, whose work is also highly valuable, but simply more difficult for us to access, and covered much more superficially by this review. We also thank Isabelle Rondeau for help with manuscript preparation, and Robin Gibson for well-timed encouragement and insightful advice.

References

Alheit, J. & Hagen, E. (1997) Long-term climate forcing of European herring and sardine populations. *Fisheries Oceanography*, 6, 130–139.

Alpoim, R. & Avila de Melo, A. (2000) An assessment of American plaice (*Hippoglossoides platessoides*) in NAFO division 3M. *Northwest Atlantic Fisheries Organization Scientific Council Research Document 00/25.*

Archambault, D. & Grégoire, F. (1996) Revue des donnés historique de pêcheau flétan atlantique du golfe du Saint-Laurent (Divisions de l'OPANO 4RST (1893–1995). *Department of Fisheries and Oceans Atlantic Fisheries Research Document 96/56.*

Bannister, R.C.A. (1978) Changes in plaice stocks and plaice fisheries in the North Sea. *Rapports et Procès-Verbaux des Réunions Conseil International pour l'Exploration de la Mer*, 172, 86–101.

Beverton, R.J.H. & Holt, S.J. (1957) *On the Dynamics of Exploited Fish Populations. Fisheries Investigations Series 2*, Vol. 19. UK Ministry of Agriculture and Fisheries, London.

Bowering, W.R. (1999) Stock status of witch flounder in NAFO subdivision 3Ps. *Department of Fisheries and Oceans Canadian Stock Assessment Secretariat Research Document 99/144.*

Bowering, W.R. (2000a) Stock status update of witch flounder in Divisions 2J, 3K and 3L. *Northwest Atlantic Fisheries Organization Scientific Council Research Document 00/13.*

Bowering, W.R. (2000b) Resource status update of witch flounder in NAFO Divisions 3NO. *Northwest Atlantic Fisheries Organization Scientific Council Research Document 00/14.*

Bowering, W.R. & Brodie, W.B. (2000) An assessment of Greenland halibut in NAFO subarea 2 and divisions 3KLMNO. *Northwest Atlantic Fisheries Organization Scientific Council Research Document 00/43.*

Bowering, W.R., Morgan, M.J. & Brodie, W.B. (1997) Changes in the population of American plaice (*Hippoglossoides platessoides*) off Labrador and northeastern Newfoundland: a collapsing stock with low exploitation. *Fisheries Research*, 30, 199–216.

Brodie, W.B. & Morgan, M.J. (2000) An assessment of the American plaice in NAFO subarea 2 and division 3K. *Department of Fisheries and Oceans Atlantic Fisheries Research Document 2000/130.*

Brodziak, J., Jacobson, L., Lauth, R. & Wilkins, M. (1997) *Assessment of the Dover Sole Stock for 1997.* Pacific Fishery Management Council, Portland, OR.

Brodziak, J.K.T., Overholtz, W.J. & Rago, P.J. (2001) Does spawning stock affect recruitment of New England groundfish? *Canadian Journal of Fisheries and Aquatic Science*, 58, 306–318.

Brown, R.W., Burnett, J.M., Begg, G.A. & Cadrin, S.X. (2000) Assessment of the Georges Bank winter flounder stock, 1982–1997. *Northeast Fisheries Science Center Reference Document 00–16.*

Cadrin, S.X. (2000) Southern New England yellowtail flounder. In: Assessment of 11 Northeast Groundfish Stocks through 1999, pp. 65–82. *Northeast Fisheries Science Center Reference Document Reference Document 00–05.*

Cadrin, S.X. & Walsh, S.J. (1999) Surplus production analysis and potential precautionary reference points for Grand Banks yellowtail flounder (NAFO Div. 3LNO). *Northwest Atlantic Fisheries Organization Scientific Council Research Document 99/3.*

Cadrin, S., Howe, A., Correia, S., et al. (1996) An index based assessment of winter flounder *Pleuronectes americanus* populations in the Gulf of Maine. *Northeast Fisheries Science Center Reference Document 96–05a.*

Cadrin, S.X., Overholtz,W.J., Neilson, J.D., Gavaris, S. & Wigley, S.E. (1998) Stock assessment of Georges Bank yellowtail flounder for 1997. *Northeast Fisheries Science Center Reference Document 98–06.*

Cadrin, S.X., King, J. & Suslowicz, L.E. (1999) Status of the Cape Cod yellowtail flounder stock for 1998. *Northeast Fisheries Science Center Reference Document 99–04.*

Cadrin, S.X., Neilson, J.D., Gavaris, S. & Perley. P. (2000) Assessment of the Georges Bank yellowtail flounder stock for 2000. *Northeast Fisheries Science Center Reference Document 2000–10.*

Castille G.C., Li, H.W. & Golden, J.T. (1994) Environmentally induced recruitment variation in petrale sole *Eopsetta jordani. Fishery Bulletin,* **92**, 481–493.

Charles, A.T. (1997) The path to sustainable fisheries. In: *Peace in the Oceans: Ocean Governance and the Agenda for Peace* (ed. E.M. Borgese). pp. 201–213. International Oceanographic Congress Technical Services, Rome.

Clark, W.G. & Hare, S.R. (2001) Assessment of the Pacific halibut stock in 2000. *Report of Research and Assessment Activities 2000*, pp. 85–118. International Pacific Halibut Commission, Seattle.

Clark, W.G., Hare, S.R., Parma, A.M., Sullivan, P.J. & Trumble, R.J. (1999) Decadal changes in growth and recruitment of Pacific halibut (*Hippoglossus stenolepis*). *Canadian Journal of Fisheries and Aquatic Sciences,* **56**, 242–252.

Daan, N., Richardson, K. & Pope, J.G. (1996) Changes in the North Sea ecosystem and their causes: Århus 1975 revisited. *ICES Journal of Marine Science,* **53**, 879–883.

Darby, C.D. & Flatman, S. (1994) Virtual Population Analysis, version 3.1. Windows/DOS Users Guide. *Information Technical Series No. 1.* MAFF, Lowestoft, UK.

Drinkwater, K.F. (1997) Impacts of climate variability on Canadian fish and shellfish stocks. In: *Climate Change and Climate Variability in Atlantic Canada* (ed. R.W. Shaw). pp. 21–36. Occasional Report #1. Environment Canada, Ottawa.

Evans, G.T. & Rice, J.C. (1988) Predicting recruitment from stock without the mediation of a functional relationship. *Journal du Conseil International pour l'Exploration de la Mer,* **45**, 111–122.

FAO (1996) Precautionary approach to fisheries management, Parts 1 & 2. *Food and Agriculture Organization Fisheries Technical Paper No. 350 (1,2).* FAO, Rome.

Fargo, J.J. (1999) Flatfish stock assessments for the West Coast of Canada for 1998 and recommended yield options for 1999. *Department of Fisheries and Oceans Canadian Stock Assessment Secretariat Research Document 99/17.*

Fargo, J.J. & Wilderbuer, T.K. (2000) Population dynamics of rock sole (*Lepidopsetta bilineata*) in the North Pacific. *Journal of Sea Research,* **44**, 123–144.

Garcia, S.M., Cochrane, K.G., Van Santen, G. & Christy, F. (1999) Towards sustainable fisheries: a strategy for FAO and the World Bank. *Ocean and Coastal Management,* **42**, 369–398.

Gavaris, S. (1988) An adaptive framework for the estimation of population size. *Canadian Atlantic Fisheries Scientific Advisory Committee Research Document 88/29.*

Gulland, J.A. (1977) The problems of population dynamics and contemporary fishery management. In: *Fish Population Dynamics* (ed. J.A. Gulland). pp. 383–406. John Wiley, New York.

Haug, T. & Tjemsland, S. (1986) Changes in size- and age-distributions and age at sexual maturity in Atlantic halibut *Hippoglossus hippoglossus* caught in north Norwegian waters. *Fisheries Research,* **4**, 145–155.

Hendrickson, L. (1998) Windowpane. In: Status of fishery resources off the Northeastern United States for 1998 (ed. S.H. Clark). pp. 85–87. *National Oceanic and Atmospheric Administration Technical Memorandum NMFS-NE-115.*

Hilborn, R. & Walters, C.J. (1992) *Quantitative Fisheries Stock Assessment.* Chapman & Hall, New York.

Hollowed, A.B. & Megrey, B.A. (1993) Evaluation of risks associated with application of alternative harvest strategies for Gulf of Alaska walleye pollock. In: *International Symposium on Management Strategies for Exploited Fish Populations* (eds G. Kruse, D. Eggers, R.J. Marasco & T.J. Quinn). pp. 291–320. Alaska Sea Grant Publications, Fairbanks, AK.

Hollowed, A.B. & Wooster, W.S. (1995) Decadal-scale variations in the eastern subArctic Pacific. II: Responses of Northeast Pacific fish stocks. *Canadian Special Publication of Fisheries and Aquatic Sciences,* **121**, 373–385.

Hutchings, J.A. (1996) Spatial and temporal variation in the density of northern cod and a review of hypotheses for the stock's collapse. *Canadian Journal of Fisheries and Aquatic Sciences,* **53**, 943–963.

ICES (2001) Report of the ICES Advisory Committee on Fishery Management. *ICES Cooperative Research Report 242* (2 volumes). ICES, Copenhagen.

Jorgensen, O.A. (2000) Assessment of the Greenland halibut stock component in NAFO Subarea 0 + Div. 1A Offshore + Divisions 1B-1F. *Northwest Atlantic Fisheries Organization Scientific Council Research Document 00/38.*

Mace, P.M. (1997) Developing and sustaining world fisheries resources: the state of science and management. In: *Developing and Sustaining World Fisheries Resources* (ed. D.A. Hancock). pp. 1–20. Commonwealth Scientific and Industrial Research Organization, Victoria, Australia.

McKinnell, S.M., Brodeur, R.D., Hanawa, K., Hollowed, A.B., Polovina, J.J. & Zhang, C.-I. (2001) Pacific climate variability and marine ecosystem impacts from the tropics to the Arctic. *Progress in Oceanography,* **49**, 189–209.

McRuer, J., Halliday, R.G., Branton, R.M., Showell, M.A. & Morin, R. (1997) Status of witch flounder in div 4VWX in 1997. *Department of Fisheries and Oceans Canadian Stock Assessment Secretariat Research Document 97/106.*

Methot, R.D. (1990) Synthesis model: an adaptable framework for analysis of diverse stock assessment data. *International Pacific Fisheries Commission Bulletin,* **50**, 259–277.

Millner, R. & Whiting, C.L. (1996) Long-term changes in growth and population abundance of sole in the North Sea from 1940 to the present. *ICES Journal of Marine Science,* **53**, 1185–1195.

Millner, R., Flatman, S., Rijnsdorp, A.D., *et al.* (1996) Short communication: comparison of long-term trends in growth of sole and plaice populations. *ICES Journal of Marine Science,* **53**, 1196–1198.

Modin, J. & Phil, L. (1994) Differences in growth and maturity of juvenile plaice *Pleuronectes platessa* L. following normal and extremely high settlement. *Netherlands Journal of Sea Research,* **32**, 331–341.

Morgan, M.J. (2000) A stock status update of American plaice in NAFO Divisions 3LNO. *Northwest Atlantic Fisheries Organization Scientific Council Research Document 00/41.*

Morgan, M.J., Brodie, W.B. & Power, D. (1999) An assessment of the American plaice in subdivision 3Ps. *Department of Fisheries and Oceans Canadian Stock Assessment Secretariat Research Document 99/145.*

Morin, R. & Bernier, B. (2000) Assessment and biology of Greenland halibut (*Reinhardtius hippoglossoides*) in the Gulf of St. Lawrence (4RST) in 1999. *Department of Fisheries and Oceans Atlantic Fisheries Research Document 00/32.*

Morin, R., Chouinard, G.A., Forest-Gallant, I. & Poirier, G. (1998) Assessment of 4T American plaice in 1996 and 1997. *Department of Fisheries and Oceans Canadian Stock Assessment Secretariat Research Document 98/06.*

Morin, R., Forest, I. & Poirier, G. (1999) Assessment of NAFO division 4T winter flounder. *Department of Fisheries and Oceans Canadian Stock Assessment Secretariat Research Document 99/47.*

Myers, R.A. & Barrowman, N.J. (1996) Is fish recruitment related to spawner abundance? *Fishery Bulletin,* **94**, 707–724.

O'Brien, L. & Esteves, C. (2001) Update assessment of American plaice in the Gulf of Maine – Georges bank region. *Northeast Fisheries Science Center Reference Document 2001–02.*

Overholtz, W. & Cadrin, S.X. (1998) Yellowtail flounder. In: Status of the fishery resources off the Northeastern United States for 1998 (ed. S.H. Clark). pp 70–74. *Northwest Atlantic Fisheries Organization Technical Memorandum NMFS-NE-115.*

Patterson, K.R. (1999) Evaluating uncertainty in harvest control law catches using Baysian Markov chain Monte Carlo virtual population analysis with adaptive rejection sampling and including structural uncertainty. *Canadian Journal of Fisheries and Aquatic Sciences,* **56**, 208–221.

Pitcher, T.J., Hart, P.J.B. & Pauly, D. (1998) *Reinventing Fisheries Management. Fish and Fisheries Series No. 23.* Chapman & Hall, Kluwer Academic Publishers, Dordrecht, The Netherlands.

Poirier, G. & Morin, R. (1999) The status of yellowtail flounder in NAFO division 4T in 1998. *Department of Fisheries and Oceans Canadian Stock Assessment Secretariat Research Document 99/46.*

Prager, M.H. (1994) A suite of extensions to a nonequilibrium surplus-production model. *Fishery Bulletin,* **92**, 374–389.

Punt, A.F.E., Smith, A.D.M. & Musick, J.A. (1999) Management of long-lived marine resources: a comparison of feedback control management systems. *American Fisheries Society Symposium,* **23**, 243–265.

Rice, J.C. (1999) Stock assessment of target species. In: *Proceedings of the Conference on Integrated Fisheries Monitoring* (ed. P. Nolan). pp. 51–63. FAO, Rome.

Richards, L.J. & Maguire, J.-J. (1998) Recent international agreements and the precautionary approach: new directions for fisheries management science. *Canadian Journal of Fisheries and Aquatic Sciences,* **55**, 1545–1552.

Ricker, W.E. (1954) Stock and recruitment. *Journal of the Fisheries Research Board of Canada,* **11**, 559–623.

Rijnsdorp, A.D. (1994) Population-regulating processes during the adult phase in flatfish. *Netherlands Journal of Sea Research,* **32**, 207–223.

Rijnsdorp, A.D. & Van Beek, F.A. (1991) Changes in growth of North Sea plaice (*Pleuronectes platessa* L.) and sole (*Solea solea* L.) *Netherlands Journal of Sea Research,* **27**, 441–457.

Rijnsdorp, A.D. & Van Leeuwen, P.I. (1996) Changes in growth of North Sea plaice since 1950 in relation to density dependence, eutrophication, beam trawl effort, and temperature. *ICES Journal of Marine Science,* **53**, 1199–1213.

Rivard, D.R. & Walsh, S.J. (2000) Precautionary approach framework for yellowtail flounder in 3LNO in the context of risk analysis. *Northwest Atlantic Fisheries Organization Scientific Council Research Document 00/50.*

Rosenburg, A.A. & Restrepo, V.R. (1994) Uncertainty and risk evaluation in stock assessment advice for US marine fisheries. *Canadian Journal of Fisheries and Aquatic Sciences,* **51**, 2715–27230.

Ross, M.R. & Nelson, G.A. (1992) Influences of stock abundance and bottom temperature on growth dynamics of haddock and yellowtail flounder on Georges Bank. *Transactions of the American Fisheries Society,* **121**, 578–587.

Schnute, J.T. (1994) A general framework for developing sequential fishery models. *Canadian Journal of Fisheries and Aquatic Sciences,* **51**, 1676–1688.

Schnute, J.T., Richards, L.J., Olsen, N., *et al.* (1998) Fishery Stock Assessment Models. *Lowell Wakefield Fisheries Symposium Series No. 15,* pp. 171–184. American Fisheries Society, Washington, DC.

Serchuk, F.M., Kirkegaard, E. & Daan, N. (1996) Status and trends of the major groundfish, flatfish, and pelagic fish stocks in the North Sea: thirty-year overview. *ICES Journal of Marine Science,* **53**, 1130–1145.

Shepherd, J.G. (1992) Extended survivors analysis: and improved method for the analysis of catch-at-age data and catch-per-unit-effort data. *Working Paper No. 11.* ICES Multispecies Assessment Working Group. Copenhagen, Denmark.

Simonsen, C.S. &. Boje, J. (2000) An assessment of the Greenland halibut stock component in NAFO division 1A inshore. *Northwest Atlantic Fisheries Organization Scientific Council Research Document 00/47.*

Spencer, P.D., Walters, G.E. & Wilderbuer, T.K. (2000) Flathead sole. In: *Bering Sea Groundfish Plan Team. Stock Assessment and Fishery Evaluation Report for the Groundfish Resources of the Bering Sea/Aleutian Islands Regions*, pp. 327–346. North Pacific Fishery Management Council, Anchorage, AK.

Stobo, W. T., Fowler, G.M. & Smith, S.J. (1997) Status of 4X winter flounder, yellowtail flounder, and American plaice. *Canadian Stock Assessment Secretariat Research Document 97/105.*

Sullivan, P.J., Parma, A.M. & Clark, W.G. (1999) The Pacific halibut stock assessment of 1997. *International Pacific Halibut Commission Scientific Report 79.*

Swain, D.P., Poirier, G.A. & Morin, R. (1998) Status of witch flounder in NAFO divisions 4RST. *Department of Fisheries and Oceans Atlantic Fisheries Research Document 96/71.*

Terceiro, M. (1999) Stock assessment of summer flounder for 1999. *Northeast Fisheries Science Center Reference Document 99–19.*

Terceiro, M. (2000) Southern New England/mid-Atlantic winter flounder. In: Assessment of 11 Northeast Groundfish Stocks through 1999, pp. 153–162. *Northeast Fisheries Science Center Reference Document 00–05.*

Turnock, B.J., Wilderbuer, T.K. & Brown, E.S. (2000) Arrowtooth flounder. In: *Gulf of Alaska Groundfish Plan Team. Stock Assessment and Fishery Evaluation Report for the Groundfish Resources of the Gulf of Alaska*, pp. 137–168. North Pacific Fishery Management Council, Anchorage, AK.

Ulltang, O. (1998) Where is fisheries science heading? How can stock assessment be improved. *Journal of Northwest Atlantic Fisheries Science,* **23**, 133–141.

Walsh, S.J. & Morgan, M.J. (1999) Variations in maturation of yellowtail flounder (*Pleuronectes ferruginea*) on the Grand Bank. *Journal of Northwest Atlantic Fisheries Science,* **25**, 47–59.

Walsh, S.J., Morgan, M.J., Power, D., et al. (2000) The millennium assessment of Grand Bank yellowtail flounder stock in NAFO Divisions 3LNO. *Northwest Atlantic Fisheries Organization Scientific Council Research Document 00/45.*

Wigley, S.E., Brodziak, J. & Cadrin, S.X. (1999) Assessment of the witch flounder stocks in subareas 5 and 6 for 1999. *Northeast Fisheries Science Center Reference Document 99–16.*

Wilderbuer, T.K. & Nichol, D. (2000) Yellowfin sole. In: *Bering Sea Groundfish Plan Team. Stock Assessment and Fishery Evaluation Report for the Groundfish Resources of the Bering Sea/Aleutian Islands Regions*, pp. 213–254. North Pacific Fishery Management Council, Anchorage, AK.

Wilderbuer, T.K. & Sample, T.M. (2000) Arrowtooth flounder. In: *Bering Sea Groundfish Plan Team. Stock Assessment and Fishery Evaluation Report for the Groundfish Resources of the Bering Sea/Aleutian Islands Regions*, pp. 257–290. North Pacific Fishery Management Council, Anchorage, AK.

Wilderbuer, T.K. & Walters, G.E. (2000) Rock sole. In *Bering Sea Groundfish Plan Team. Stock Assessment and Fishery Evaluation Report for the Groundfish Resources of the Bering Sea/Aleutian Islands Regions*, pp. 291–326. North Pacific Fishery Management Council, Anchorage, AK.

Zwanenburg, K., Black, G., Fanning, P., Branton, R., Showell, M. & Wilson, S. (1997) Atlantic halibut (*Hippoglossus hippoglossus*) on the Scotian Shelf and southern Grand banks – evaluation of resource status. *Department of Fisheries and Oceans Canadian Stock Assessment Secretariat Research Document 97/50.*

Chapter 15
Aquaculture and stock enhancement

Bari R. Howell and Yoh Yamashita

15.1 Introduction

Flatfishes are among the most popular and most valuable fishes used for human consumption. They support valuable fisheries throughout the world and many species are among the most highly priced, features that make them attractive candidates for both stock enhancement and intensive cultivation. In Europe, the Atlantic halibut (*Hippoglossus hippoglossus*), turbot (*Scophthalmus maximus*) and common sole (*Solea solea*) are among the most valuable marine fishes, while elsewhere in the world many of the *Paralichthys* species are similarly highly regarded. This genus includes the Japanese flounder or hirame, *P. olivaceus*, a species that has been one of the main targets for aquaculture-related activities in Japan and elsewhere in Asia, and a number of other species that are proving attractive to aquaculturists particularly in North America.

Interest in developing culture techniques for flatfishes dates back to the end of the nineteenth century when procedures established for artificially fertilising salmonid eggs were successfully applied to marine fish. Shelbourne's review of marine fish culture (Shelbourne 1964) described how this advance stimulated the construction of numerous hatcheries for marine fish on both sides of the Atlantic with the aim of generating millions of eggs and hatched larvae for release into the sea. Although the primary aspiration of enhancing declining stocks was not achieved, the considerable amount of information generated on the early development of marine fishes and the experience of handling these delicate stages provided an important foundation for subsequent developments.

At this time, attempts to rear beyond the yolk-sac stages were largely unsuccessful. Although Fabre-Domerque & Biétrix (1905) claimed limited success in feeding common sole larvae, significant numbers were not reared through metamorphosis. The principal obstacle of food provision was eventually removed by the Norwegian fisheries biologist Gunner Rollefsen who discovered that plaice (*Pleuronectes platessa*) larvae would readily feed on the nauplii of the brine shrimp, *Artemia salina* (Rollefsen 1939). Further progress was to some extent delayed by World War II and it was not until the late 1950s and early 1960s that Shelbourne (1964) exploited Rollefsen's discovery in his pioneering work on the development of culture techniques for plaice and common sole. For the first time the feasibility of rearing marine fishes through their larval stages was demonstrated. This important advance led to renewed interest in the use of hatchery technology to support declining fisheries, because fishes could now be made available that were much less vulnerable to predation than those previously used. The ability to mass-produce juveniles also stimulated considerable interest in the intensive farming of marine fishes in shore-based tanks or cages in the coastal zone.

Since Shelbourne's early work there have been a number of major advances that have allowed the commercialisation of rearing methods for a wide range of species. The provision of adequate nourishment for the developing larvae and juveniles is probably the most significant, but other developments have also been of considerable importance. These include, for example, improvements in broodstock management that have permitted year-round availability of good quality eggs, a greater understanding of the importance of the microbial environment in improving the consistency of larval survival and of the effect of the rearing environment on important performance characteristics of the reared juveniles. Progress has also been made in some species towards improving growth rates by controlling sex.

15.2 Hatchery production of larvae and juveniles

15.2.1 Egg production

The availability of a regular supply of good quality eggs is clearly a prerequisite of commercial production of any species. Eggs artificially fertilised at sea supported much of the early research but it was not until it was shown that captive stocks could be sustained and will mature in captivity that large-scale production of juveniles became a realistic prospect. Shelbourne (1964) was able to maintain captive stocks of plaice and common sole, both species spawning spontaneously under the conditions provided. Subsequent experience showed that common sole would spawn naturally in a wide range of tank systems provided that certain basic environmental conditions, notably temperature and photoperiod, were met (Baynes *et al.* 1993). In Japan, a large number of flatfish species are produced to support stocking exercises and intensive farming (Table 15.1). Whereas all these can be induced to spawn naturally in captivity, the eggs of the spotted halibut, the willowy flounder and the stone flounder are generally poorly fertilised and are therefore artificially fertilised in mass-production operations. Other species, such as the turbot, generally do not spawn spontaneously in the tank systems commonly used for other species. This failure was considered to be due to the spatial limitation experienced by such a relatively large fish in normal holding tanks inhibiting the intricate pre-spawning behaviour displayed by flatfishes (Chapter 10). Some circumstantial support for this suggestion was obtained by Bromley *et al.* (1986) who found that turbot would spawn spontaneously when held at low density in relatively large tanks. For economic reasons, however, it is the preferred option of the industry to maintain the fish at higher density in smaller tanks and hand strip the gametes even though this approach increases the variability in the quality of the eggs (see below). The Atlantic halibut is an even larger flatfish than the turbot and natural spawning in captivity is rarely reported.

15.2.1.1 Manipulation of spawning time

The development of methods for obtaining year-round provision of eggs had considerable impact on commercial viability of farming. An understanding of the endocrine control of

Table 15.1 Summary of flatfish hatchery seedling production for stocking in Japan

Species	Spawning	Survival rate from hatching	Colour abnormality (%)	Morphological abnormality (%)	Size at release (mm, average of latest year)	Total no. released to 1999	Year***	Data source*
Pleuronichthys cornutus (ridge-eyed flounder)	Hormone use artificial, natural	14–17 (to 33)	72–90	High ratio	25	17000	1984–1986	Kitajima *et al.* 1987, 1988
Microstomus achne (slime flounder)	Natural	15–53 (to 32)	al 11–17	eye 23–30	40–45	29000	1989–1994	JASFA
Verasper variegattus (spotted halibut)	Artificial (natural)**	4–30 (to 30–38)	al 2–62 hy 100	eye 15–45	47–84	370000	1992–	FA, JASFA, Pref
Verasper moseri (barfin flounder)	Artificial (natural)**	14–58 (to 30–35)	al 3–52 hy 32–100	ver 21–67	71–138	660000	1987–	FA, JASFA, Pref
Eopsetta grigorjewi (shotted halibut)	Hormone use natural	2–18		eye 1–37	36–144	309000	1987–	JASFA
Glyptocephalus kitaharai (willowy flounder)	Artificial (natural)**	20–67 (to 20–25)	al 9–59	eye 26	26	10000	1999–	JASFA
Platichthys bicoloratus (stone flounder)	Artificial (natural)**	29–86 (to 20)	al 53–96	eye 41–53 ver 9	26–47	330000	1980–1987	Pref
Pseudopleuronectes schrenki (crested flounder)	Natural	37–90 (to 13–19)	al 75–100	eye 3–18	17–102	590000	1983–1987, 1999	JASFA
Pseudopleuronectes herzensteini (littlemouth flounder)	Natural	13–76 (to 18–24)	16–68		20–75	3036000	1984–	JASFA Aritaki *et al.* 1996
Pseudopleuronectes yokohamae (marbled flounder)	Natural	38.5–60.2 (to 20)	al 6–19 hy 2–97	eye 6–12 ver 100	20–170	37 million	1977–	Mutsutani 1988 Seikai 1985b, Pref
Paralichthys olivaceus (Japanese flounder)	Natural	10–80 (to 30)	al 0–15 hy 20–100	ver 10–80 eye 0	38–130	271 million	1977–	FA, JASFA, Pref

*All data on the number of stocked fish and size-at-release are from the 'Annual Report of Seed Production and Release Statistics' by Fisheries Agency & JASFA.
**Natural spawning is rare or not common.
***Experimental stocking started before 1977, but there are no available data.
JASFA, Japan Sea Farming Association Annual Report; Pref, Prefectural Hatchery Annual Report or unpublished data from prefectural governments; FA, Annual Reports on Stock Enhancement Projects published from Fisheries Agency; al, albinism; hy, hypermelanosis; eye, abnormal eye transformation; ver, vertebral deformity.

reproduction in fish and associated regulatory environmental cues has advanced considerably over the last two to three decades and has been reviewed by Pankhurst (1998). It became clear some years ago that significant shifts in spawning time could be achieved by the simple expedient of manipulating photoperiod and temperature, the two physical factors that control overall reproductive development and final gamete maturation respectively (Bye 1984; Lam & Munro 1987; Bromage *et al.* 1995). Control of spawning by this method is now widely practised by commercial hatcheries.

15.2.1.2 Egg quality

Variations in egg quality are of particular significance in species that do not spawn spontaneously in captivity. In turbot, for example, both fertilisation and hatching rates vary widely between egg batches. A large part of this variability, however, is attributable to the timing of fertilisation relative to ovulation rather than to the inherent quality of the eggs. In contrast to salmonids, the quality of marine fish eggs decreases rapidly after ovulation so that, for example, few larvae hatch from turbot eggs fertilised more than 24 hours after ovulation (McEvoy 1984). Acquiring the eggs from the fish as soon after ovulation as possible has to be the aim, but this is difficult to achieve in practice since in captive turbot ovulation may occur at any time of day and ovulatory periods vary between individuals (McEvoy 1984; Howell & Scott 1989). Similar problems of over-ripening have also been described in the Atlantic halibut (Bromage *et al.* 1994) and a wide range of other species. Such problems are exacerbated by suboptimal temperatures, a feature clearly demonstrated in the Atlantic halibut (Brown *et al.* 1995).

There is much less evidence of significant variations in egg quality among species maintained under an appropriate regime of environmental factors, particularly temperature, and conditions that permit spontaneous spawning. As Kjørsvik *et al.* (1990) observe in their review of egg quality in marine fishes, 'environmental factors (especially temperature) during gonad maturation and stripping of eggs at the correct time after ovulation seem to be important determinants of egg quality'. There is some evidence, however, that egg quality may improve with length of time in captivity (Bowers 1966; Devauchelle *et al.* 1987). The underlying causes of this improvement are not understood, but may in part be attributable to the traumatic effects of capture (Pankhurst 1998). A common feature of naturally spawned eggs is that an appreciable proportion may not be fertilised (Baynes *et al.* 1993). In groups of fish, however, it is not known whether the unfertilised eggs in a single egg collection originate from incompletely fertilised batches or from completely unfertilised batches mixed with completely fertilised batches. The former may be indicative of low inherent egg quality whereas the latter suggests that some aspect of the tank environment is inhibiting normal spawning behaviour of some individuals.

15.2.2 Food and feeding

15.2.2.1 Live food

The discovery that *Artemia* nauplii could successfully be used as food for marine fish larvae removed a considerable obstacle to the development of mass-production techniques. The larvae of many species, however, are too small to ingest *Artemia* nauplii as a first food and so the subsequent discovery that such larvae could ingest the readily cultured brackish water rotifer, *Brachionus plicatilis*, was an important step in increasing the range of species that could be reared (Lubzens *et al.* 1989). Rearing of many commercially important species, including the turbot and the Japanese flounder, depends on the use of rotifers as a live food for the early larvae. The particular advantage of these two organisms, neither of which forms part of the normal diet of marine fish larvae, is that they can be made readily available in large numbers. An important disadvantage, however, is that they commonly fail to fully meet the nutritional requirements of the fish to which they are fed. The most important deficiency is that of long-chain polyunsaturated fatty acids (PUFAs), particularly eicosapentaneoic acid (20:5 n-3) and docosahexaenoic acid (22:6 n-3) (DHA), that have an important role in maintaining the integrity of biological membranes and are particularly important in the normal development of neural tissue. Thus, for many species, including turbot and Japanese flounder, a dietary regime of rotifers followed by newly hatched *Artemia* nauplii was not sufficient to sustain survival. Some species, such as the plaice, common sole and lemon sole (*Microstomus kitt*) appear to have less stringent requirements for these nutrients as indicated by survival rates, but even for these species it seems that enhancement of dietary levels of PUFAs may be beneficial with regard to their physiological 'fitness' (Logue *et al.* 2000). Rectifying this deficiency was initially accomplished by feeding the live food organisms on algae rich in these fatty acids (Howell 1979). Subsequently, proprietary lipid emulsions and particulate diets became available for enriching live foods and are now widely, although not exclusively, used by commercial operators. However, problems of consistency remain, largely because *Artemia* does not retain DHA for very long, so that its nutritional quality decreases with increasing residence time in culture tanks (Evjemo *et al.* 1997).

15.2.2.2 Formulated feeds

Juvenile flatfishes can be successfully grown on natural or semi-natural foods such as the oligochaete worm, *Lumbricillus rivalis*, or mussels, *Mytilus edulis* (see for example Howell 1997). Commercial imperatives, however, demand the use of formulated feeds that are nutritious and readily available at the earliest opportunity in the production cycle. For turbot, the transfer to a formulated feed, or weaning as it has become known, can be readily achieved after metamorphosis using a fish-based moist or dry formulated feed offered by hand or automatically from feed dispensers. Initial results were highly variable, but the principal cause of this variability was attributable to the nutritional quality of the live food on which the fish had been reared (Bromley & Howell 1983) rather than to the attributes of the formulated feed. Such diets did not prove as acceptable to some species. The common sole, for example, is a nocturnal invertebrate feeder and is not attracted to fish-based diets. A feeding response can, however, be elicited by incorporating invertebrate tissue or chemical attractants into the diet (Howell 1997).

Complete replacement of live food with formulated feeds does not appear to be a realistic aspiration in the short term. Success appears to have been limited primarily by the poorly developed digestive system of marine fish larvae at first feeding and the relative unattractiveness of inanimate particles to these early stages. In the sea, food capture may be facilitated by the behaviour of the prey (Kolkovski *et al.* 1995) and digestion aided by the prey's own enzymes and high content of free amino acids (Fyhn 1989). It has also been suggested that certain chemical or physical factors that activate the digestive process may be absent in formulated feeds (Reitan et al. 1993).

15.2.3 Microbial environment

The importance of the microbial environment in larvae culture systems has become increasingly appreciated during the last decade or so. While the requirements of marine larvae with respect to the bacterial flora in their environment is not fully understood, it has become increasingly clear that the once common approach of water disinfection can lead to the development of conditions that are harmful to the larvae (Skjermo et al. 1997). Drastic reduction of the bacterial flora by whatever cause may permit the rapid development of potentially harmful opportunistic bacteria as the first step in ecological succession. Skjermo et al. (1997) postulated that water dominated by non-opportunistic bacteria would inhibit proliferation of opportunistic, and largely pathogenic, bacteria on the skin and gut surfaces of the larvae. This led to the development of a process known as microbial maturation (Vadstein et al. 1993) that was found to increase significantly the survival rate of yolk-sac Atlantic halibut larvae and induce a significantly higher growth rate of turbot larvae.

15.2.4 Juvenile quality

Reared fish are subject to a wide variety of physical deformities (Howell et al. 1998) but a phenomenon of particular importance in flatfishes is that of abnormal pigmentation. This condition was documented by Shelbourne (1964) for plaice and common sole and may range from total albinism to ambicoloration, conditions that may also be associated with skeletal deformities, particularly in the head region. The problem has been reported in all cultured flatfishes, including the Japanese flounder (see for example Kanazawa 1993) and Atlantic halibut (see Shields et al. 1999). Abnormal pigmentation is not confined to reared fishes because a similar range of abnormalities is evident in natural populations (de Veen 1969), although incidences are difficult to estimate because of the greater vulnerability of abnormally pigmented individuals to predation (Koshiisi et al. 1991).

The causes of these problems remain somewhat obscure although there is considerable evidence that nutritional factors are implicated. Dickey-Collas (1993), for example, described an association between PUFA deficiency and abnormal pigmentation in plaice. Kanazawa (1993) also linked abnormal pigmentation with lipid quality and postulated that melanin production in larval Japanese flounder was interrupted when rhodopsin formation in the eye was restricted by a dietary deficiency of vitamin A, docosahexaenoic acid and phospholipid. A positive effect of vitamin A supplementation in reducing the incidence of abnormal pigmentation was subsequently confirmed in this species by Takeuchi et al. (1995) but it was also noted that excessive amounts of this vitamin could be associated with skeletal abnormalities (Dedi et al. 1998). Viability and growth performance in captivity do not appear to be adversely affected by abnormal pigmentation, but market acceptability may be significantly decreased for species sold as whole fish. Abnormally pigmented fish are also considered unsuitable for release into natural water bodies because of their greater vulnerability to predation (Koshiisi et al. 1991).

The incidence of albinism on the ocular side of Japanese flounder was high (>50%) up to the early 1990s. Subsequently, the incidence has fallen to <10% through the increased use of rotifers and *Artemia* nutritionally enriched with fat-soluble vitamins, highly unsaturated fatty acids and diatoms (Seikai 1985a; Miki *et al.* 1990; Takahashi 1992; Kanazawa 1993). Hypermelanosis on the abocular side of Japanese flounder, however, remains a problem. The causes are poorly understood but are thought to be due to nutritional factors before the completion of metamorphosis (Takahashi 1992) and also to stress and light reflection after settlement (Suzuki 1994; T. Seikai, personal communication). The degree of hypermelanosis can be decreased by using a low rearing density, longer exposure to live food during the pelagic stage (Takahashi 1992) and by using porous ceramic beads as a substratum after settlement (T. Seikai, personal communication). However, complete prevention of hypermelanosis on the abocular side is still not possible and its incidence, including slight black spots on the body and/or caudal fin, remains extremely high (>95%) in almost all hatcheries. The permanency of black pigments on the abocular side (Uchino & Nakanishi 1983; Tominaga & Watanabe 1998) and the absence of such marks in wild fish means these can be used as a convenient visible marker to differentiate between hatchery and wild fish after release. The price on the markets of fish excessively pigmented on the abocular side may be reduced by 15–70%, so avoidance of this problem is of some economic importance.

It is now clear that behavioural deficits associated with the rearing process may significantly prejudice survivability in a natural environment. This effect was demonstrated by Olla *et al.* (1994) for salmonid fishes, reared fish being clearly less adept at avoiding predation and foraging than their wild counterparts. Recent work has demonstrated similar effects of the rearing environment on the behaviour of flatfishes. Ellis *et al.* (1997), for example, showed that naïve common sole were less adept at burying than individuals with experience of sand. Although their burying performance increased to that of experienced fish after about a week, significant losses through predation might be expected in that time. These effects may be critical to the success of stock enhancement exercises and particular consideration needs to be given to the way fish are reared and prepared for this purpose (see also Chapter 10).

15.2.5 Genetic improvement

Farmed marine fishes have to date been subjected to minimal genetic improvement and that which has been induced is mainly confined to *ad hoc* selection by the producers of the relatively few species currently being commercially produced. There is no doubt that considerable scope exists for improving those characteristics, particularly growth rate, that would enhance the economic viability of the industry. In many cultivated species, sex control, i.e. the production of single-sex broods, offers the most rapid way of achieving a significant increase in growth. In tilapias, in which the males are the fastest growing, all-male production has brought considerable benefits, whereas in salmonids the converse has been the case (MacLean 1998). In commercially produced flatfishes (turbot and Japanese flounder) the production of all-female broods would be an advantage because the males mature at a much smaller size than the females and long before a marketable size is reached. The method of producing all-female broods in salmonids depends on the sex determination mechanism being of a simple genetic type where the male is the heterogametic sex, i.e. the XX (female)-XY (male) type. The sex determining mechanism in flatfishes is not completely understood, but at least in some species it does appear to be complicated by the existence of a degree of environmental as well as genetic control. As shown for several fish taxa (e.g. Conover & Fleisher 1986; Strüssmann *et al.* 1996), temperature-dependent sexual determination has been reported in Japanese flounder, the marbled sole and barfin flounder. The former two species are known to be genetically XX-female and XY-male sex determined. In Japanese flounder genetic females reverse their sex to males if the water temperature is higher (>25°C) or lower (<15°C) than their normal temperature (around 20°C) (Yamamoto 1999). Similar effects have been demonstrated in the marbled sole (Goto *et al.* 2000) and barfin flounder (Goto *et al.* 1999). Genetic males in these three species are not affected by temperature, indicating that genetic males are sexually stable. The susceptible stage for sex reversal was reported to be between 20 and 32 mm TL in Japanese flounder, <25 mm TL in marbled sole and <30 mm TL barfin flounder.

While it is clear that in many species the production of all-female broods would be advantageous for intensive cultivation, the release of large numbers of temperature-induced phenotypic XX males could bring about alterations to the sex ratio of the target population. Rearing at a normal temperature during the susceptible period may therefore be important to inhibit spontaneous sex reversal to phenotypic males and avoid such effects.

15.3 Intensive farming

Many flatfish species have been, and are still considered to be, promising candidates for intensive farming, but the farming of only two species – the Japanese flounder and the turbot – has been fully commercialised. The most important of these is the Japanese flounder, which is now widely farmed in the Far East, notably in Japan, Korea and China. Production in Japan began in the early 1980s and having increased to around 7000 mt per annum by about 1990 has remained at that level subsequently (Fig. 15.1). Production in Korea began somewhat later but by 1999 was approximately three times that of Japan, giving a combined annual production from these two countries of 30 000 mt. This is the minimum estimate of production in the region, taking no account of the rapidly increasing production in China, for example, for which reliable information is not available.

Production of the turbot has been on a more modest scale. As with the Japanese flounder, commercial production began during the early 1980s and rapidly became centred on the northwest of Spain where water temperatures were most favourable (Howell 1998a). The rate of increase in production, however, has been relatively modest and has only recently exceeded 4000 mt per annum (Fig. 15.2). The turbot, however, is now recognised as an attractive species for farming elsewhere in the world and has been introduced into both Chile and China where commercial farming is becoming established.

Other species of flatfishes are on the verge of commercialisation or are considered to have considerable potential. In the cold waters of the north Atlantic, the Atlantic halibut is now being commercially farmed, albeit on a relatively small scale (Fig. 15.2), the challenging technical problems of rearing the eggs and larvae having largely been overcome (Shields *et al.* 1999). In Europe and elsewhere, there is currently considerable interest in farming soles following recent progress with the formulation of effective artificial feeds (Howell 1997). In Australia, the greenback flounder, *Rhombosolea tapirina*, is considered a potential species

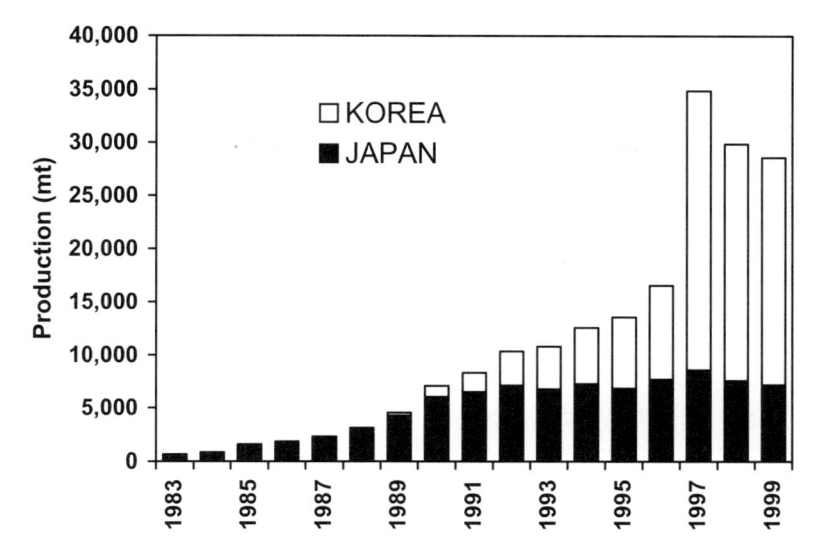

Fig. 15.1 Aquaculture production of the Japanese flounder in Japan and Korea from 1983 to 1999 (*Statistical Year Book of Maritime Affairs and Fisheries*, Ministry of Maritime Affairs and Fisheries Korea, Seoul, 1989–2000, and *Statistical Annual Report on Fisheries and Aquaculture Production*, Ministry of Agriculture, Forestry and Fisheries Japan, Tokyo, 1985–2001).

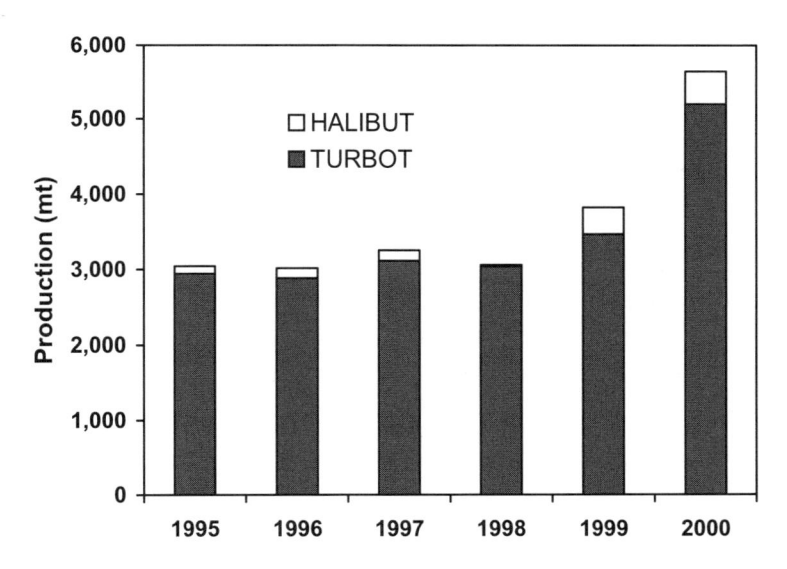

Fig. 15.2 Aquaculture production of turbot and Atlantic halibut in Europe from 1995 to 2000 (Federation of European Aquaculture Producers).

for aquaculture (Chen & Purser 2001). Elsewhere in the world, the many *Paralichthys* species probably offer the greatest promise for expansion. In the USA, for example, significant progress has already been made with the development of rearing techniques for the summer flounder, *P. dentatus* (Nardi 1998; Burke *et al.* 1999) and further south with the southern flounder, *P. lethostigma* (Jenkins & Smith 1999) and tropical, *P. tropicus* (Rosas *et al.*

1999). The further development of flatfish farming will not only result from adapting rearing methodologies to new species, but also on the introduction of new production technology. Recent developments in recirculation technology, for example, are already being exploited by commercial producers (Blancheton 2000). The benefits of this approach are considerable, including greater control of environmental conditions and hence increased freedom from climatic limitations, reduced dependency on prime coastal locations, and importantly reduced environmental impacts.

The production methods for the turbot, Japanese flounder, common sole and Atlantic halibut were outlined by Jones & Howell (1995). These and other flatfishes may be grown in tanks or cages but are most often grown in shore-based installations through which water is continuously pumped. The capital and running costs of tank systems are high, particularly those employing recirculation technology, and so productivity of even highly priced species also has to be high to ensure full cost recovery. Productivity is a function of both growth rate and stocking density and so these factors are two of the most critical determinants of the economics of this type of farming.

Growth rate varies considerably between flatfish species, as shown by Jones (1972) in his assessment of the most promising candidate species for intensive farming in the UK. Piscivorous species, such as turbot and Atlantic halibut, are relatively fast growing, contrasting with the relatively slow growth rates of benthic browsers such as the plaice, common sole and lemon sole.

The response of the fish to stocking density is equally critical and has also been shown to vary between species. The relationship between stocking density and growth rate is not a simple one. Jeon *et al.* (1993), for example, showed that in Japanese flounder increasing stocking density had a positive effect on growth rate at low stocking densities, but a negative effect once the total area of the fish was more than twice that of the tank bottom (i.e. 200% cover). Howell (1998b) described a similar effect in turbot, but found a much more marked negative effect of stocking density in the common sole. He suggested that this might be the result of the browsing habit of the sole, a mode of feeding that would perhaps increase competitive opportunities for food. A size hierarchy was evident in this trial because the reduction in growth with increasing stocking density was attributable to an effect on the smaller rather than the larger fish. This emphasises the importance of feeding strategy as a determinant of growth rate and size variation in cultured fishes.

15.4 Stock enhancement

Interest in stock enhancement dates back to the end of the nineteenth century when the belief that the release of the early development stages into the sea would arrest the decline in exploited marine fish stocks stimulated the development of hatchery techniques (for reviews, see Shelbourne 1964; Blaxter 2000). Interest in this approach was abandoned following the failure to demonstrate its effectiveness but was rekindled with the advent of techniques for the hatchery production of juvenile marine fishes in the 1960s. These developments allowed the release into the sea of development stages that were much less vulnerable to predation than eggs and yolk-sac larvae and would therefore have much greater survival potential.

15.4.1 Stock enhancement in Europe and the USA

The first reported trials were in the UK, where 15–20 mm long plaice generated by Shelbourne's mass-rearing exercises were released into both a sea loch and an artificial embayment on Scotland's west coast. All fish were lost either through predation or the effects of adverse environmental conditions (Blaxter 2000). Elsewhere in Europe and much more recently, Støttrup *et al.* (1998) released reared turbot off the Danish coast. The fish were much larger (10–15 cm) than the plaice referred to above but, although good growth rates during the first year were recorded and most fish remained within a few kilometres of the point of release, the recapture rate of 2.8% over 4 years was surprisingly low. This is more or less the limit to the practical exploration of this approach to the management of flatfish stocks in western Europe.

There have been some small-scale releases of summer and winter flounder in the USA but these studies are not yet fully analysed and only a few scientific papers are available (Burke *et al.* 2000; Fairchild & Howell 2000; Kellison *et al.* 2000). In contrast, the Japanese have been much more active than other countries, both with respect to the numbers of fish released and the range of species targeted. The following account reviews this work and identifies some of the important principles that have emerged from these studies.

15.4.2 Stock enhancement in Japan

15.4.2.1 Background

In Japan, there has been a significant decline in the catch of coastal fisheries, including those for flatfish, over the past 40 years (Fig. 15.3). This decline is mainly attributable to over-fishing and habitat degradation. Mitigation measures have included the regulation of commercial fisheries and conservation/improvement of the habitat, such as the construction of artificial

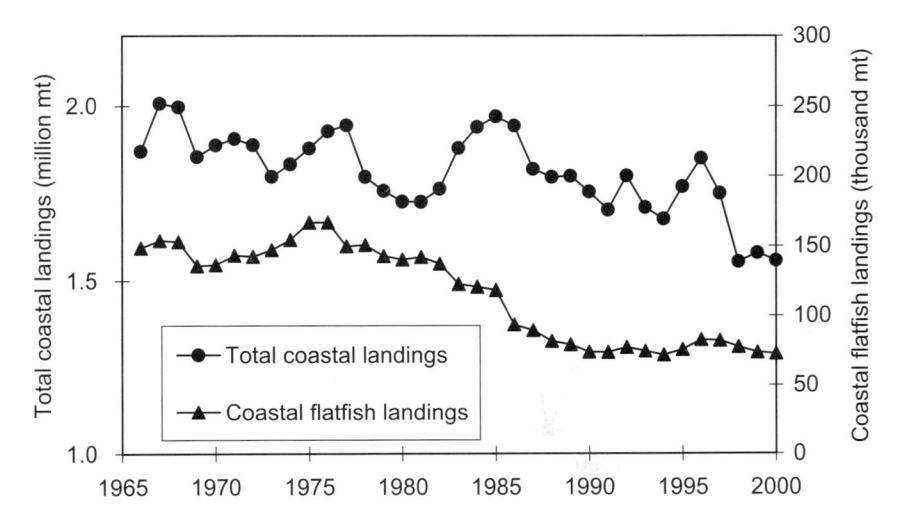

Fig. 15.3 Annual landings from coastal fisheries and those of coastal flatfishes in Japan (*Annual Report on Japanese Fisheries*, Fisheries Agency Japan, Tokyo, 2001, and *Statistical Annual Report on Fisheries and Aquaculture Production*, Ministry of Agriculture, Forestry and Fisheries Japan, Tokyo, 1968–2000).

nursery reefs and the creation of algal beds. However, there have been few clear indications that these countermeasures have been effective in fully restoring the living resources of these coastal areas. A third way to replenish depleted stocks is by stock enhancement.

Stocking of marine species has been carried out since 1963 under the Sea-Farming Programme of the Fisheries Agency (Ministry of Agriculture, Forestry and Fisheries) (Honma 1993; Imamura 1999). Stock enhancement projects are mainly undertaken by individual prefectural governments, each as a prefectural unit. There are a total of 61 prefectural hatcheries for marine stock enhancement, in addition to those for salmonids. Nine national fisheries research institutes and 16 national hatcheries are addressing the development of production technology and release strategies. In 1999, 15 301 million juvenile stage animals of 82 species were released into the coastal waters of Japan. Of these 82 species, 8 were flatfishes (Table 15.1).

Precise data on flatfish stock enhancement activities in Japan are not available for the period before 1976. In 1977, 1.29 million marbled flounder and 0.26 million Japanese flounder were stocked (Fig. 15.4). In 1999, a total of 33.9 million juveniles of eight flatfish species were reared in prefectural and national hatcheries and stocked in coastal waters. The number of released Japanese flounder juveniles each year has steadily increased and it has become the main target species for flatfish stocking. In 1999, approximately 30 million juveniles were stocked in 36 prefectures from Kagoshima (31°N) to Hokkaido (45°N), accounting for 90% of the total number of flatfishes released. Marbled sole was ranked second with 2.8 million being released. The production technology and release protocols of the other species are at an experimental stage.

15.4.2.2 Objectives and effectiveness of stock enhancement

The first objective of stock enhancement is as a commercial venture. The economic return rate (ERR), defined as the estimated financial return divided by the hatchery production and

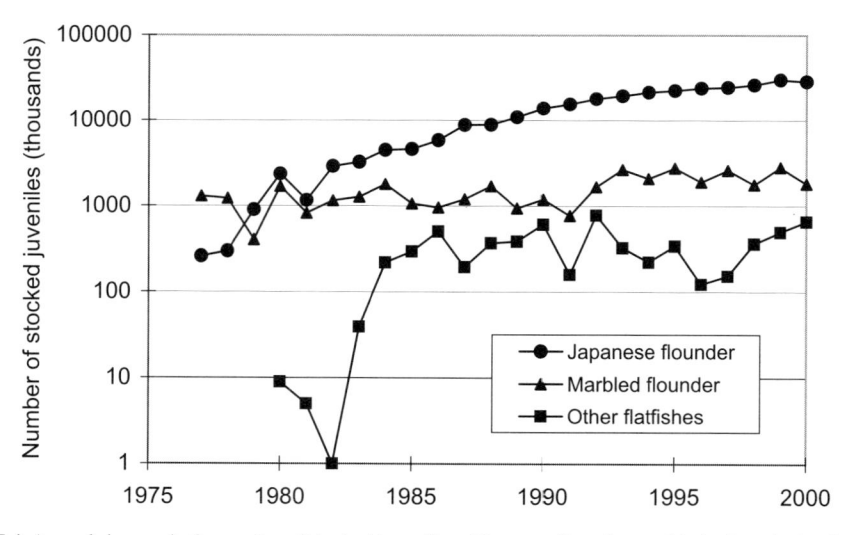

Fig. 15.4 Annual changes in the number of stocked juveniles of Japanese flounder, marbled sole and other flatfishes in the stock enhancement project in Japan (*Statistical Report on Seedling Production and Release for Sea Farming*, Japan Sea Farming Association, No 11, 1995, and *Annual Report on Seedling Production and Release for Sea Farming*, Japan Sea Farming Association, Tokyo, 1995–2001).

release costs, is usually used as an index of economic feasibility (Kitada 1999). These costs do not include harvesting costs by fishermen. Harvesting costs can be excluded in the case of flatfishes, because fishermen target several bottom fish species using trawl or gill nets and they operate the fishing gear regardless of the catch of the stocked fish. As systematic fish market surveys are required to estimate the ERR (Kitada et al. 1992) and prefectural governments are reluctant to report hatchery costs, the number of reports in which ERR data are available is somewhat limited.

Detailed data on the ERR for Japanese flounder have been reported from two prefectures. These range from 2.19 to 3.76 in Fukushima Prefecture (Fisheries Agency 1995) and from 0.46 to 1.96 for Miyako Bay area of Iwate Prefecture (Okouchi et al. 1999) (Table 15.2). Compared with the very high average ERR for a few other target species, for example 9.8 in chum salmon (Oncorhynchus keta) and 24.1 in scallop (Kitada 1999), economic profitability is not clear for Japanese flounder. As an economic enterprise, reductions in hatchery costs and improvements in post-release survival and hence return rates are required to raise the economic viability.

The second objective is to increase or stabilise living marine resources and their fisheries. A clear nationwide augmentation of stocks by the mass release of hatchery-reared fish has been accomplished for some species, such as chum salmon (Kaeriyama 1999) and scallop (Kitada 1999). Although such nationwide positive correlations have not been found for Japanese flounder, several cases of an increase of local stocks due to stocking have been reported. In Kagoshima Bay, landings of wild Japanese flounder increased after the commencement of stocking, indicating the effect of increased reproduction due to the addition of the stocked

Table 15.2 Examples of release, return and economic return rate (ERR) of Japanese flounder stock enhancement in Japan

Year of release	Number released	Mean release TL (mm)	Number of returns	Return rate	Total landed value (10 000 yen)	Release cost**	ERR
Fukushima Prefecture*							
1987	245 000	100	41 000	16.7	9 220	2 450	3.76
1988	336 000	83	97 000	29.0	9 934	2 789	3.56
1989	217 000	109	66 000	30.6	8 107	2 365	3.43
1990	389 000	81	76 000	19.5	8 683	3 151	2.76
1991	424 000	77	34 000	8.0	7 154	3 265	2.19
1992	428 000	84	44 000	10.2	10 246	3 595	2.85
Miyako Bay, Iwate Prefecture*							
1987	157 000	78	9 293	5.9	434	942	0.46
1988	145 000	82	10 071	6.9	720	870	0.83
1989	69 000	93	16 110	23.3	900	483	1.86
1990	80 000	90	12 750	15.7	772	560	1.38
1991	96 000	74	20 783	21.6	1 131	576	1.96
1992	64 000	78	8 734	13.6	516	384	1.34

*Data for Fukushima Prefecture are from Fisheries Agency (1995) and those for Miyako Bay are from Okouchi et al. (1999).
**Release cost was 100 yen for a 100-mm fish in the Fukushima Prefectural Hatchery (Fujita et al. 1993); so 70 yen for a 70-mm fish was assumed in this paper. Costs of 45 yen for a 60-mm fish and 80 yen for a 100-mm fish were used in the Miyako Hatchery of JASFA (Okouchi et al. 1999).

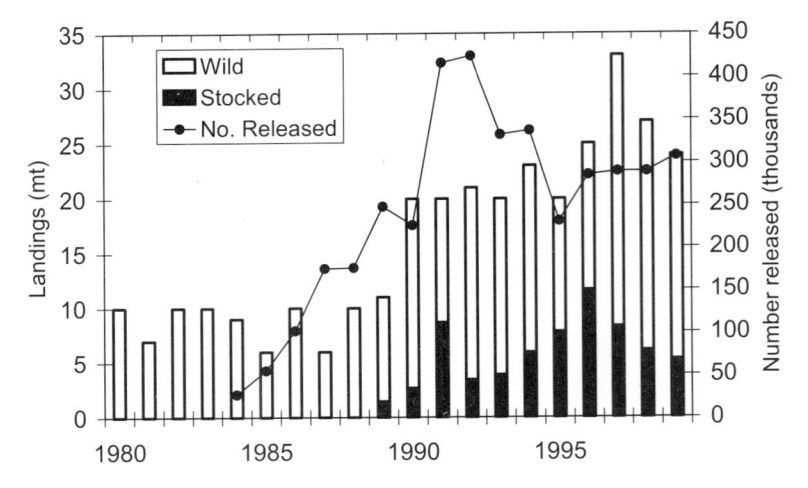

Fig. 15.5 Annual changes in Kagoshima Bay of the market landings of wild and stocked Japanese flounder and the number of stocked juveniles (Kagoshima Prefectural Government, unpublished data).

fish (Fig. 15.5). In addition, the total market landings of Japanese flounder, red seabream (*Pagrus major*) and black seabream (*Acanthopagrus schlegeli*) have been stable. These are the three most important target fish species for stock enhancement and more than 5 million juveniles are released every year. In contrast, those of non-targeted coastal pleuronectids and 10 other commercially important non-flatfish coastal fishes have declined continuously since the 1970s (Fig. 15.6). Although the landings of these three intensively stocked species did not increase after the commencement of stocking, the stable landings during the last 20 years may indicate that the mass release of juveniles has contributed to sustaining the recruitment to the commercial fisheries. The reproduction of stocked fish may play an important role in

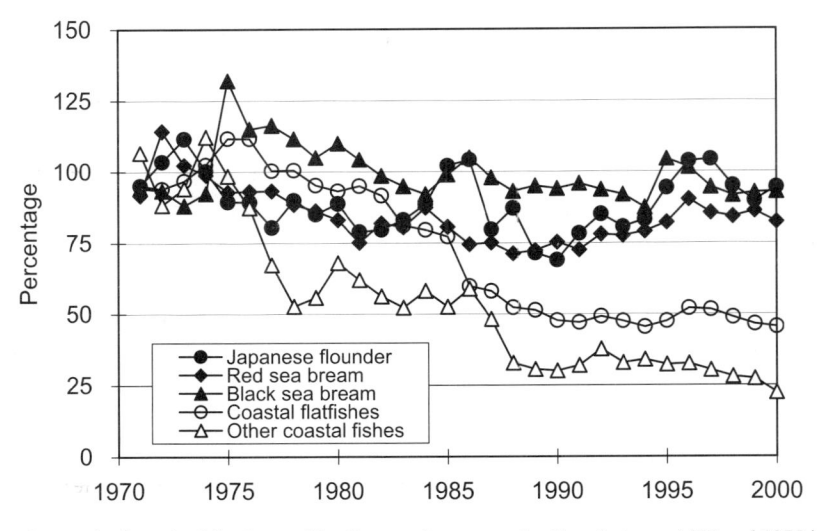

Fig. 15.6 Changes in the ratio (%) of annual landings to the average landings between 1971 and 1975 in Japan. Other coastal fishes include 10 dominant taxa (*Statistical Annual Report on Fisheries and Aquaculture Production*, Ministry of Agriculture, Forestry and Fisheries Japan, Tokyo, 1973–2000).

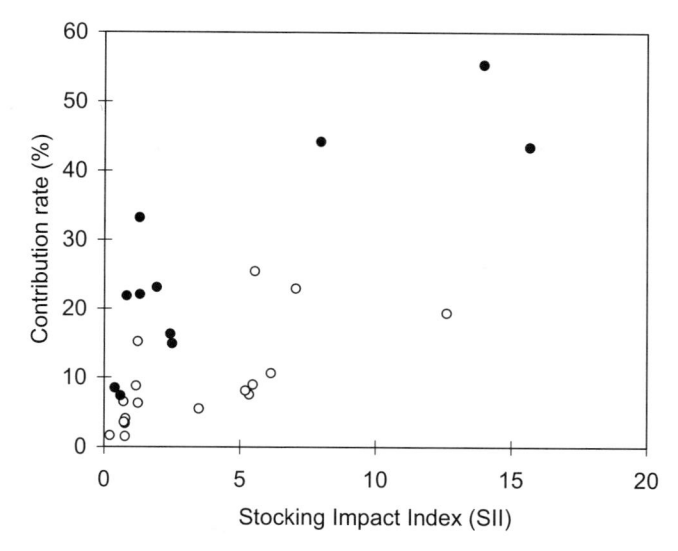

Fig. 15.7 Relationship between the stocking impact index and the contribution rate (returned released fish/total market landings, in number) in Japanese flounder. Data are primarily prefectural averages of 3 years between 1993 and 1995 for the SII and between 1994 and 1996 for the contribution rate, because flounder recruits to fishery stock at 1 year old. Closed circles indicate data from semi-enclosed bay or inland sea areas and open circles from open inshore areas (Fisheries Agency Japan, unpublished data).

sustaining the population level. However, the level of contribution of stocked fish to reproduction in the wild is almost unknown for flatfishes and several studies have just started using the release of genetically tagged hatchery-reared Japanese flounder juveniles.

The impact of stocking on wild populations is evaluated using the stocking impact index (SII), defined as the number of released juveniles divided by the total number of fish of the same species landed at markets in a unit area. This index shows the relative scale of the number of released fish against the total resource (released + wild) (Kitada 1999). Assuming an average fish weight at market of 500 g, the SII value of Japanese flounder in the whole of Japan in 1998 is 1.73. This value is much smaller than that for chum salmon (33.1), Kuruma prawn (*Penaeus japonicus*) (12.6), swimming crab (*Portunus trituberculatus*) (8.9) and abalone (*Haliotis* spp.) (4.7) (data from Kitada 1999). The contribution rate, defined as the proportion by number of reared fish in the landings, increased with SII showing the evident effect of stocking in Japanese flounder (Fig. 15.7). In addition, contribution rates in semi-enclosed bay areas and inland seas tended to be higher than those in open inshore areas. Almost all Japanese flounder landings are made up of 1–3-year-old fish (>95%). This age distribution indicates that the retention rate of released fish is higher in semi-enclosed areas compared with those in open inshore areas for fish up to 3 years old.

The annual number of marbled flounder juveniles released since 1977 has ranged from 0.4 to 2.8 million. However, reports on the effectiveness of stocking of this species are few. Because wild fish commonly have black pigment spots on the abocular side, pigments are of no use as a visible tag. Release size, usually about 30 mm TL, is too small to allow the application of external tags. In addition, very low SII values of 0.08 (average of four prefectures) indicate that the stocking impact is so small that the effectiveness cannot be evaluated. The estimated contribution rate (i.e. proportion of reared fish in the landings) in the Tokyo Bay area of Chiba Prefecture over an 8-year period (1992–1999) ranged from 0.3 to 1.6%.

The third objective is to replenish depleted stocks. Barfin flounder and spotted halibut are high-priced commercial fishes but, because the market landings of these species are too small to be recorded in fish market statistics, available landing data are restricted to a few prefectures. Landings of both species have drastically decreased in recent years. For example, landings of barfin flounder in Iwate Prefecture declined from 480 kg in 1987 to 16 kg in 1994 and that of the spotted halibut in Fukushima Prefecture from about 5 metric tonnes (mt) in the mid-1980s to 1 mt in 1993. Small-scale stocking started in Hokkaido in 1987 for barfin flounder and a total of 148 000 juveniles were released in two prefectures (Hokkaido and Iwate) in 1999. In Iwate Prefecture, where the stocking of barfin flounder has been conducted since 1993, landings have recovered from 16 kg in 1993 to 440 kg in 2000 (Fig. 15.8) and the current contribution rate of stocked fish in the landings was more than 95%. In Hokkaido, where market statistics for this species are not available before 1993, the landings increased from 1.2 mt in 1994 to 3.5 mt

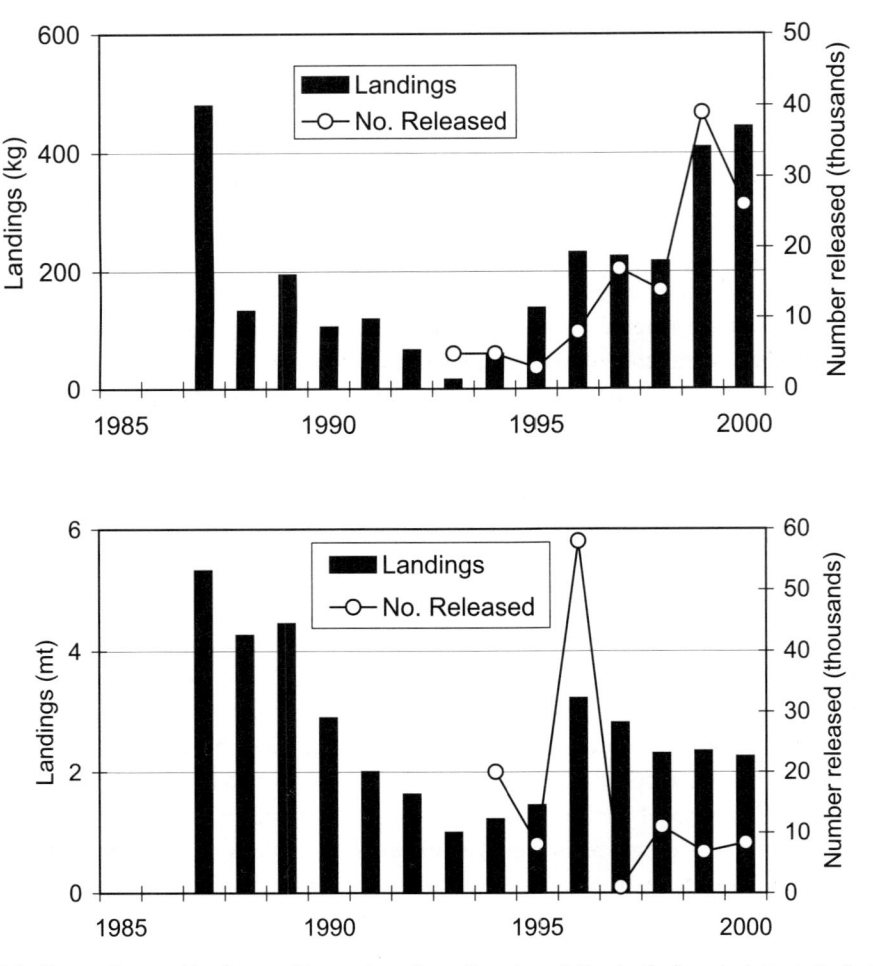

Fig. 15.8 Changes in annual landings and the number of juveniles released. Top: barfin flounder in Iwate Prefecture; bottom: spotted halibut in Fukushima Prefecture (Iwate Prefectural Government, Fukushima Prefectural Government, unpublished data).

in 1999 and the recent contribution rate is estimated to be >95%. Market return rate (MRR), defined as the number (or weight) of returned fish divided by the number of stocked fish, in Hokkaido ranged from 0.5 to 10.2%. The experimental release of the spotted halibut started in 1992 and 167 000 juveniles were released from three prefectures (Miyagi, Fukushima and Kanagawa) in 1999. The landing of the spotted halibut in Fukushima Prefecture has recovered from 0.8 mt in 1993 to 2.5 mt in 1996, mainly due to stock enhancement (Fig. 15.8). Contribution rate ranged from 23 to 62% during this period (Nemoto *et al.* 1999) and MRR ranged from 5.2% (1996 release) to 24.8% (1995 release). Nakamura & Yamada (1999) reported that the ERR of the spotted halibut with a size-at-release of 106 mm TL stocked in Tokyo Bay was 4.0. Because stocks of these two species were so depleted and the market value high, the effectiveness of stock enhancement is clear. In addition, the high growth rate, growing to 500 g body weight after 15–20 months from hatching in the wild for both species (Nakamura & Yamada, 1999; Watanabe *et al.* 2001), seems to enhance the effectiveness.

One of the positive consequences of the perceived contribution of fish releases to catches is that it stimulates a responsible attitude towards stock management initiatives among the fishing community. Having recognised the effectiveness of releases of Japanese flounder, the fishermen's unions decided to contribute to the hatchery costs by donating a percentage of the selling price of the fish. In addition, because they are financially supporting the flounder hatchery costs, the fishermen's unions have recently set a minimum landing size, usually 30–35 cm in TL, which they strictly observe by returning under-sized fish to the sea as quickly after capture as possible.

15.4.2.3 Technology for stock enhancement

The relationship between advancements in technology and stocking effectiveness has only been intensively studied for Japanese flounder (Yamashita & Yamada 1999). Thus, only an outline of the criteria important in enhancing efficiency is summarised here. These criteria are:

- quality of stocked fish;
- their genetic diversity and suitability;
- release habitat;
- release season;
- fish size at release;
- release method;
- carrying capacity of the habitat;
- management of the released stock, and
- impacts on the ecosystem.

Stocking effectiveness is most commonly evaluated using the market return rate (MRR). MRR used in this chapter is based on numbers rather than weight. MRR is also a convenient index for assessing the effectiveness of the technology. Currently, studies on release technology are carried out based on comparison of survival to market size of fish caught by efficient gears such as those used in commercial fisheries. This is equivalent to MRR. In release-recapture studies, almost all released flounder (>95%) can be clearly distinguished from wild flounder by means of pigmentation characteristics. If necessary, cultured fish are tagged by fin removal, anchor tags, chemical or fluorescent dyes to provide information such as size, time and place of release. The use of alizarin complexone to mark otoliths and scales (Tsukamoto *et al.* 1989, Yamashita *et al.* 1994) is valuable because in Japanese flounder the mark retention is very high and is detectable in scales under UV light (Nakamura & Kuwada 1994). Therefore, it is only necessary to remove a sample of scales rather than purchase the fish or the fish head to check for fluorescent marks on otoliths.

The main objective in developing release technology is to minimise the post-release mortality and therefore increase the return rate of market-sized fish. Recent studies have indicated that mortality of released Japanese flounder juveniles is mainly caused by predation (Yamashita *et al.* 1993, 1994; Furuta 1996; Furuta *et al.* 1998) with high predation mortality occurring within 1–2 weeks of release (Yamashita & Yamada 1999). The probable dominant predators on released flounder juveniles are piscivorous fishes during the daytime (Yamashita *et al.* 1993, 1994; Furuta 1996; Furuta *et al.* 1998) and carnivorous crabs and cuttlefish at night (Furuta 1994; Yamashita, unpublished observations). A temporary period of low nutritional intake and unsuccessful adaptation to the natural environment just after release may increase their susceptibility to predation.

Determination of the optimal season, size-at-release and release environment based on the ecology of the target species, food organisms and its predators has been shown to be significantly effective in increasing the economic return rate (Yamashita & Yamada 1999). Size-at-release is one of the most important factors in a stocking project, because it is closely associated with not only post-release survival but also the cost of resources for culturing juveniles in the hatchery (Tsukamoto *et al.* 1989; Svåsand & Kristiansen, 1990; Leber 1995; Munro & Bell 1997). The relationship between size-at-release and the MRR based on large-scale release studies on Japanese flounder is summarised in Fig. 15.9. Although there is a tendency for the MRR to increase with the size at release, the release of large-sized fish does

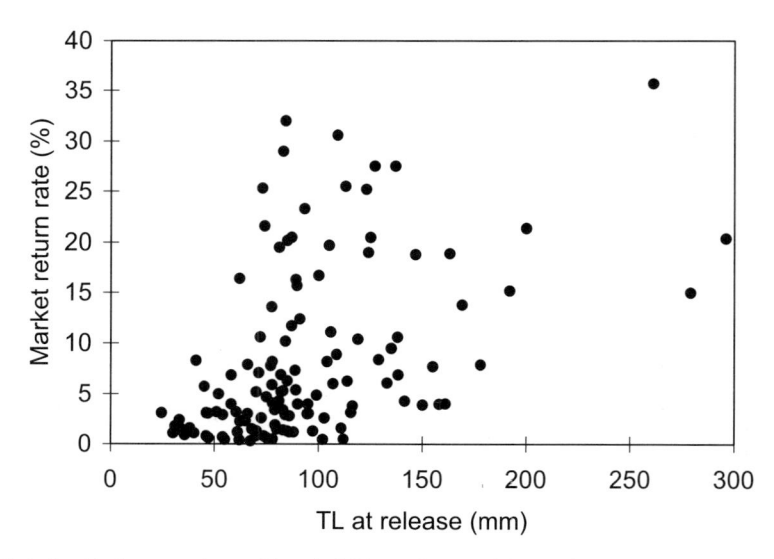

Fig. 15.9 Relationship between the total length (TL) at release and the market return rate of commercial sized Japanese flounder (redrawn from Yamashita & Yamada 1999).

not always ensure a high MRR. The relationship between MRR and size-at-release may be greatly affected by several other factors such as quality of the seeds, release season, release habitat and fishing intensity. The quality of stocked fish is directly related to feeding capability and predator avoidance. Release season and habitat are associated with food availability and predation pressure. The maximum MRR at a given size-at-release does not increase with the release size after reaching around 100 mm TL (Fig. 15.9). Predatory fishes in the wild can consume released flounder when the prey is smaller than one-third of the predator's TL (Yamashita et al. 1993). Large (>35 cm) piscivorous fishes are not common in the shallow nursery grounds of juvenile flounder, and there may be differences in post-release survival rate for flounder >100 mm TL at release. The optimal size-at-release in stock enhancement projects should be determined to obtain the highest economic effectiveness by considering both the market return and hatchery costs (Yamashita & Yamada 1999).

In summary, there are many biological and ecological problems that need to be addressed for successful and responsible stock enhancement. Of these, health management in the hatchery, conservation of the ecosystem, and the genetic conformity and diversity of the stocked fish are critical (see Blankenship & Leber 1995; Munro & Bell 1997).

15.4.3 Future perspectives

Hatcheries for stock enhancement are usually managed using a government budget or both a government budget and funds raised from the sale of the landed fish at markets. Considering stock enhancement as an economic enterprise, stocking projects should be financially self-supporting without the requirement for government subsidies. However, this will be difficult except for a few species such as scallop and chum salmon. Some prefectural governments have started or are preparing to collect fees from angling parties, which mainly target red seabream or Japanese flounder, to support the budget of hatcheries. However, this income from commercial and recreational fishing is not sufficient to fully cover hatchery costs. At present, the introduction of a licence system for recreational fishing as implemented in some states of the USA is not considered feasible in Japan.

In addition to commercial fishermen, a sizeable part of the Japanese population recreationally fish the coastal fish resources, and also industrialisation and development of cities for the last few decades have led to the deterioration of the coastal environment and destruction of fish nursery grounds. Taking these factors into account, using government budgets (taxes) for hatchery operation is probably the most realistic solution at present and collection of licence fees from commercial and recreational fishermen is a goal for the future.

15.5 Conclusions

Hatchery techniques for rearing flatfishes have advanced to the point that juveniles of most species can now be reared in large numbers. This capability has been exploited by the development of intensive cultivation systems in which fish are grown to market size in shore-based tank systems. Significant tonnages of Japanese flounder and turbot are now produced in Asia and Europe respectively, and other species – including Atlantic halibut, common soles and various *Paralichthys* species – are poised for commercial exploitation.

The availability of juveniles has also rekindled interest in the use of hatchery-reared fish for enhancing natural stocks. In Japan, extensive trials with several species have produced some encouraging results and induced a positive attitude to stock management among fishermen. However, economic indicators, such as the economic return rate (defined as the ratio of the financial return and the production and release costs), are much less favourable than for either homing or sedentary species such as chum salmon or scallops. Thus, full cost recovery without government subsidies remains a remote aspiration, although research addressing a wide range of factors that determine the survival of released fish will undoubtedly lead to improved cost-effectiveness.

Acknowledgements

We thank the staff of the prefectural government of Kagoshima, Oita, Kanagawa, Chiba, Fukushima, Iwate and Hokkaido, the Japan Sea-Farming Association and the Fisheries Agency for showing us their unpublished data.

References

Aritaki, M., Seikai, T. & Kobayashi, M. (1996) Reduction of morphological abnormalities in brown sole by larval rearing with higher temperature and early feeding of *Artemia* nauplii. *Nippon Suisan Gakkaishi*, **62**, 857–864.

Baynes, S.M., Howell, B.R. & Beard, T.W. (1993) A review of egg production in the sole, *Solea solea* (L.). *Aquaculture and Fisheries Management*, **24**, 171–180.

Blancheton, J.P. (2000) Developments in recirculation systems for Mediterranean fish species. *Aquaculture Engineering*, **22**, 17–31.

Blankenship, H.L. & Leber, K.M. (1995) A responsible approach to marine stock enhancement. *American Fisheries Society Symposium*, **15**, 167–175.

Blaxter, J.H.S. (2000) The enhancement of marine fish stocks. *Advances in Marine Biology*, **38**, 1–54.

Bowers, A.B. (1966) Marine fish cultivation in Britain. VI. The effect of the acclimatisation of adult plaice to pond conditions on the viability of eggs and larvae. *Journal du Conseil. Permanent International pour l'Exploration de la Mer*, **30**, 196–203.

Bromage, N., Bruce, M., Basavaraja, N., *et al.* (1994) Egg quality determinants in finfish: the role of overripening with special reference to the timing of stripping in the Atlantic halibut, *Hippoglossus hippoglossus*. *Journal of the World Aquaculture Society*, **25**, 13–21.

Bromage, N.R., Randall, C.F., Porter, M.J.R. & Davies, B. (1995) How do photoperiod, the pineal gland, melatonin and circannual rhythms interact to co-ordinate seasonal reproduction in salmonid fish? In: *Reproductive Physiology of Fish 1995* (eds F.W. Goetz & P. Thomas). pp. 164–166. Fish Symposium 1995, Austin, Texas.

Bromley, P.J. & Howell, B.R. (1983). Factors influencing the survival and growth of turbot larvae, *Scophthalmus maximus* L., during the change from live to compound feeds. *Aquaculture*, **31**, 31–40.

Bromley, P.J., Sykes, P.A. & Howell, B.R. (1986) Egg production of turbot (*Scophthalmus maximus* L.) spawning in tank conditions. *Aquaculture*, **53**, 287–293.

Brown, N.P., Bromage, N.R. & Shields, R.J. (1995) The effect of spawning temperature on egg viability in the Atlantic halibut (*Hippoglossus hippoglossus*). In: *Reproductive Physiology of Fish 1995* (eds F.W. Goetz & P. Thomas). p. 181. Fish Symposium 1995, Austin, Texas.

Burke, J.S., Seikai T., Tanaka, Y. & Tanaka, M. (1999) Experimental intensive culture of summer flounder, *Paralichthys dentatus*. *Aquaculture*, **176**, 135–144.

Burke, J.S., Monaghan, J.P. & Yokoyama, S. (2000) Efforts to understand stock structure of summer flounder (*Paralichthys dentatus*) in North Carolina, USA. *Journal of Sea Research*, **44**, 111–122.

Bye, V.J. (1984) The role of environmental factors in the timing of reproductive cycles. In: *Fish Reproduction: Strategies and Tactics* (eds G.W. Potts & R.J. Wootton). pp. 187–206. Academic Press, London.

Chen, W.-M. & Purser, G.J. (2001) The effect of feeding regime on growth, locomotor activity pattern and the development of food anticipatory activity in greenback flounder. *Journal of Fish Biology*, **58**, 177–187.

Conover, D.O. & Fleisher, M.H. (1986) Temperature-sensitive period of sex determination in the Atlantic silverside, *Menidia menidia*. *Canadian Journal of Fisheries and Aquatic Sciences*, **43**, 514–520.

Dedi, J., Takeuchi, T., Hosoya, K., Watenabe, T. & Seikai, T. (1998) Effect of vitamin A levels in *Artemia* nauplii on the caudal skeleton formation of Japanese flounder, *Paralichthys olivaceus*. *Fisheries Science*, **64**, 344–345.

Devauchelle N., Alexandre, J.C., Le Corre, N. & Letty Y. (1987) Spawning of sole (*Solea solea*) in captivity. *Aquaculture*, **66**, 125–147.

de Veen, J.F. (1969) Abnormal pigmentation as a possible tool in the study of populations of the plaice, *Pleuronectes platessa* L. *Journal du Conseil. International pour l'Exploration de la Mer*, **32**, 344–383.

Dickey-Collas, M. (1993) The occurrence of juvenile pigmentation abnormalities in plaice (*Pleuronectes platessa*) larvae fed on enriched and unenriched *Artemia salina* nauplii. *Journal of Fish Biology*, **42**, 787–795.

Ellis, T., Howell, B.R. & Hughes, R.N. (1997) The cryptic responses of hatchery-reared sole to a natural sand substratum. *Journal of Fish Biology*, **51**, 389–401.

Evjemo, J.O., Coutteau, P., Olsen, Y. & Sorgeloos, P. (1997) The stability of docosahexaenoic acid in two *Artemia* species following enrichment and subsequent starvation. *Aquaculture*, **155**, 135–148.

Fabre-Domerque, P. & Biétrix, E. (1905) Développment de la sole (*Solea vulgaris*). Travail du Laboratoire de Zoologie Maritime de Concarneau. Vuibert et Nony, Paris.

Fairchild, E.A. & Howell, W.H. (2000) Predator-prey size relationship between *Pseudopleuronectes americanus* and *Carcinus maenas*. *Journal of Sea Research*, **44**, 81–90.

Fisheries Agency (1995) *Summary Report of Stocking Technique Development Project for the Japanese Flounder 1990–1994, Pacific Ocean Block*. Fisheries Agency Japan, Tokyo.

Fujita, T., Mizuno, T. & Nemoto, Y. (1993) Stocking effectiveness of Japanese flounder *Paralichthys olivaceus* fingerlings released in the coast of Fukushima Prefecture. *Saibai Gijutsukenkyu*, **22**, 67–73 (in Japanese).

Furuta, S. (1994) Predation on released Japanese flounder at night. *Contributions to the Fisheries Researches in the Japan Sea Block*, **30**, 43–51 (in Japanese).

Furuta, S. (1996) Predation on juvenile Japanese flounder (*Paralichthys olivaceus*) by diurnal piscivorous fish: field observations and laboratory experiments. In: *Survival Strategies in Early Life Stages of Marine Resources* (eds Y. Watanabe, Y. Yamashita & Y. Oozeki). pp. 285–294. A.A. Balkema, Rotterdam.

Furuta, S., Watanabe, T. & Yamada, H. (1998) Predation by fishes on hatchery-reared Japanese flounder, *Paralichthys olivaceus*, juveniles released in the coastal area of Tottori Prefecture. *Nippon Suisan Gakkaishi*, **64**, 1–7 (in Japanese with English abstract).

Fyhn, H.J. (1989) First feeding of marine fish larvae: are free amino acids the source of energy? *Aquaculture*, **80**, 111–120.

Goto, R., Mori, T., Kawamata, K., *et al.* (1999) Effects of temperature on gonadal sex determination in barfin flounder *Verasper moseri*. *Fisheries Science*, **65**, 884–887.

Goto, R., Kayaba, T., Adachi, S. & Yamauchi, K. (2000) Effects of temperature on sex determination in marbled sole *Limanda yokohamae*. *Fisheries Science*, **66**, 400–402.

Honma, A. (1993) *Aquaculture in Japan*. Japan FAO Association, Tokyo.

Howell, B.R. (1979) Experiments on the rearing of larval turbot, *Scophthalmus maximus* L. *Aquaculture*, **18**, 215–225.

Howell, B.R. (1997) A re-appraisal of the potential of the sole, *Solea solea*, for commercial cultivation. *Aquaculture*, **155**, 355–365.

Howell, B.R. (1998a) Development of turbot farming in Europe. *Bulletin of the Aquaculture Association of Canada*, **98**, 4–10.

Howell, B.R. (1998b) The effect of stocking density on growth and size variation in cultured turbot, *Scophthalmus maximus*, and sole, *Solea solea*. *International Council for the Exploration of the Sea CM 1998/L:10*.

Howell, B.R. & Scott, A.P. (1989) Ovulation cycles and post-ovulatory deterioration of eggs of the turbot (*Scophthalmus maximus* L.). *Rapports et Procès-Verbaux des Réunions Conseil International pour l'Exploration de la Mer*, **191**, 21–26.

Howell, B.R., Day, O.J., Ellis, T. & Baynes, S.M. (1998) Early life stages of farmed fish. In: *Biology of Farmed Fish* (eds K.D. Black & A.D. Pickering). pp. 27–66. Sheffield Academic Press, Sheffield, UK.

Imamura, K. (1999) The organization and development of sea farming in Japan. In: *Stock Enhancement and Sea Ranching* (eds B.R. Howell, E. Moksness & T. Svåsand). pp. 91–102. Fishing News Books, Oxford.

Jenkins, W.E. & Smith, T.I.J. (1999) Pond nursery production of southern flounder (*Paralichthys lethostigma*) and weaning to commercial diets. *Aquaculture*, **176**, 173–80.

Jeon, I.G., Min, K.S., Lee, J.M., Kim, K.S. & Son, M.H. (1993) Optimal stocking density for olive flounder, *Paralichthys olivaceus*, rearing in tanks. *Bulletin of the National Fisheries Research Development Agency of Korea*, **48**, 57–70.

Jones, A. (1972) Marine fish farming: an examination of the factors to be considered in the choice of species. *Laboratory Leaflet, Fisheries Laboratory Lowestoft, New Series*, **24**, 1–16.

Jones, A. & Howell, B.R. (1995) The marine flatfishes. In: *World Animal Science, C8. Production of Aquatic Animals: Fishes* (eds C.E. Nash & A.J. Novotny). pp. 329–38. Elsevier, Amsterdam.

Kaeriyama, M. (1999) Hatchery programmes and stock management of salmonid populations in Japan. In: *Stock Enhancement and Sea Ranching* (eds B.R. Howell, E. Moksness & T. Svåsand). pp. 153–167. Fishing News Books, Oxford.

Kanazawa, A. (1993) Nutritional mechanisms involved in the occurrence of abnormal pigmentation in hatchery-reared flatfish. *Journal of the World Aquaculture Society*, **24**, 162–166.

Kellison, G.T, Eggleston, D.B. & Burke, J.S. (2000) Comparative behavior and survival of hatchery-reared versus wild summer flounder (*Paralichthys dentatus*). *Canadian Journal of Fisheries and Aquatic Sciences*, **57**, 1870–1877.

Kitada, S. (1999) Effectiveness of Japan's stock enhancement programmes: current perspectives. In: *Stock Enhancement and Sea Ranching* (eds B.R. Howell, E. Moksness & T. Svåsand). pp. 103–131. Fishing News Books, Oxford.

Kitada, S., Taga, Y. & Kishino, H. (1992) Effectiveness of a stock enhancement program evaluated by a two-stage sampling survey of commercial landings. *Canadian Journal of Fisheries and Aquatic Sciences*, **49**, 1573–1582.

Kitajima, C., Hayashida, G. & Yasumoto, S. (1987) Ambicoloration and albinism in hatchery-reared frog flounder juveniles, *Pleuronichthys cornutus*. *Bulletin of the Nagasaki Prefectural Institute of Fisheries*, **13**, 17–23 (in Japanese with English abstract).

Kitajima, C., Hayashida, G. & Yasumoto, S. (1988) Early development of the laboratory-reared flounder, *Pleuronichthys cornutus*. *Japanese Journal of Ichthyology* **35**, 69–77 (in Japanese with English abstract).

Kjørsvk, E., Mangor Jensen, A. & Holmefjord, I. (1990) Egg quality in marine fishes. *Advances in Marine Biology*, **26**, 71–113.

Kolkovski, S., Arieli, A. & Tandler, A. (1995) Visual and olfactory stimuli are determining factors in the stimulation of microdiet ingestion in gilthead seabream, *Sparus auratus*, larvae. In: *Larvi 95* (eds P. Lavens, E. Jaspers & I. Roelants). pp. 289–292. European Aquaculture Society, Special Publication 24, Gent.

Koshiisi, Y., Itano, H. & Hirota, Y. (1991) Artificial stock-size improvement of the flounder, *Paralichthys olivaceus*: present status of technological achievement. In: Marine Ranching: Proceedings of the Seventeenth US-Japan Meeting on Aquaculture, Ise, Mie Prefecture, Japan, 16–18 October, 1988 (ed. R.S. Svrjcek). pp. 33–43. *NOAA Technical Report NMFS 102*.

Lam, T.J. & Munro, A.D. (1987) Environmental control of reproduction in teleosts: an overview. In: *Reproductive Biology of Fish 1987* (eds D.R. Idler, L.W. Crim & J.M. Walsh). pp. 279–88. Memorial University of Newfoundland, St John's.

Leber, K.M. (1995) Significance of fish size-at-release on enhancement of striped mullet fisheries in Hawaii. *Journal of the World Aquaculture Society*, **26**, 143–153.

Logue, J.A., Howell, B.R., Bell, J.G. & Cossins, A.R. (2000) Dietary n-3 long-chain polyunsaturated fatty acid deprivation, tissue lipid composition, *ex vivo* prostaglandin production, and stress tolerance in juvenile Dover sole (*Solea solea* L.). *Lipids,* **35**, 745–755.

Lubzens, E., Tandler, A. & Minkoff, G. (1989) Rotifers as food in aquaculture. *Hydrobiologia*, **186/187**, 387–400.

McEvoy, L.-A. (1984) Ovulatory rhythms and over-ripening of eggs in cultivated turbot, *Scophthalmus maximus* L. *Journal of Fish Biology*, **24**, 437–448.

MacLean, N. (1998) Genetic manipulation of farmed fish. In: *Biology of Farmed Fish* (eds K.D. Black & A.D. Pickering). pp. 327–354. Sheffield Academic Press, Sheffield, UK.

Miki, N., Taniguchi, T., Hamakawa, H., Yamada, Y. & Sakurai, N. (1990) Reduction of albinism in hatchery-reared flounder 'hirame', *Paralichthys olivaceus* by feeding on rotifer enriched with vitamin-A. *Suisan Zoshoku*, **38**, 147–155 (in Japanese with English abstract).

Munro, J.L. & Bell, J.D. (1997) Enhancement of marine fisheries resources. *Reviews in Fisheries Science*, **5**, 185–222.

Mutsutani, K. (1988) Growth and metamorphosis of the marbled sole larvae *Limanda yokohamae* (Gunter) in culture. *Suisan Zoshoku*, **36**, 27–32 (in Japanese).

Nakamura, R. & Kuwada, H. (1994) Detection of alizarin complexone label in scales in the mass marking system of larval and juvenile fish. *Saibai Gijutsukenkyu*, **23**, 53–60 (in Japanese).

Nakamura, R. & Yamada, A. (1999) Seeding of hatchery produced juveniles of spotted halibut (*Verasper variegatus* (Temminck et Schlegel)) in Tokyo Bay. Growth and migration after release and estimating effects of seeding. *Bulletin of Kanagawa Prefectural Fisheries Research Institute*, **4**, 27–36 (in Japanese with English abstract).

Nardi, G.C. (1998) Commercial promise for *Paralichthys*: summer flounder culture at Great Bay Aquafarms. *Bulletin of the Aquaculture Association of Canada*, **98**, 18–20.

Nemoto, Y., Fujita, T. & Watanabe, M. (1999) Studies on spotted halibut (*Verasper variegatus* Temminck & Schlegel). *Bulletin of Fukushima Prefectural Fisheries Experimental Station*, **8**, 5–16 (in Japanese).

Okouchi, H., Kitada, S., Tsuzaki, T., Fukunaga, T. & Iwamoto, A. (1999) Numbers of returns and economic return rates of hatchery-released flounder *Paralichthys olivaceus* in Miyako Bay – evaluation by fish market census. In: *Stock Enhancement and Sea Ranching* (eds B.R. Howell, E. Moksness & T. Svåsand). pp. 573–582. Fishing News Books, Oxford.

Olla, B.L., Davis, M.W. & Ryer, C.H. (1994) Behavioural deficits in hatchery-reared fish: potential effects on survival following release. *Aquaculture and Fisheries Management*, **25** (Suppl. 1), 19–34.

Pankhurst, N.W. (1998) Reproduction. In: *Biology of Farmed Fish* (eds K.D. Black & A.D. Pickering). pp. 1–26. Sheffield Academic Press, Sheffield, UK.

Reitan, K.I., Rainuzzo, J.R., Øie, G. & Olsen, Y. (1993) Nutritional effects of algal addition in first-feeding of turbot (*Scophthalmus maximus* L.) larvae. *Aquaculture*, **118**, 257–275.

Rollefsen, G. (1939) Artificial rearing of fry of sea water fish. Preliminary communication. *Rapports et Procès-Verbaux des Réunions. Conseil International pour l'Exploration de la Mer*, **109**, 133.

Rosas, J., Arana, D., Cabrera, T., Millan, J. & Jory, D. (1999) The potential use of the Caribbean flounder *Paralichthys tropicus* as an aquaculture species. *Aquaculture*, **176**, 51–54.

Seikai, T. (1985a) Reduction in occurrence frequency of albinism in juvenile flounder *Paralichthys olivaceus* hatchery-reared on wild zooplankton. *Nippon Suisan Gakkaishi*, **51**, 1261–1267.

Seikai, T. (1985b) Effect of population density and tank color on the abnormal coloration of hatchery-reared mud dub, *Limanda yokohamae*. *Suisan Zoshoku*, **33**, 119–128 (in Japanese with English abstract).

Shelbourne, J.E. (1964) The artificial propagation of marine fish. *Advances in Marine Biology*, **2**, 1–83.

Shields, R.J., Gara, B. & Gillespie, M.J.S. (1999) A UK perspective on intensive hatchery rearing methods for Atlantic halibut (*Hippoglossus hippoglossus*). *Aquaculture*, **176**, 15–26.

Skjermo, J., Salvesen, I., Øie, G., Olsen, Y. & Vadstein, O. (1997) Microbially matured water: a technique for selection of a non-opportunistic bacterial flora in water that may improve performance of marine larvae. *Aquaculture International*, **5**, 13–28.

Støttrup, J.S., Lehmann, K. & Nicolajsen, H. (1998) Turbot, *Scophthalmus maximus*, stocking in Danish waters. In: *Stocking and Introduction of Fish* (ed. I.G. Cowx). pp. 301–318. Fishing News Books, Oxford.

Strüssmann, C.A., Moriyama, S., Hanke, E.F., Calsina Cota, J.C. & Takashima, F. (1996) Evidence of thermolabile sex determination in pejerrey. *Journal of Fish Biology*, **48**, 643–651.

Suzuki, N. (1994) Ultrastructure of the skin on reverse side of hatchery-reared Japanese flounder, *Paralichthys olivaceus* with reference to the pigmentation. *Bulletin of the Nansei National Fisheries Research Institute*, **27**, 113–128 (in Japanese with English abstract).

Svåsand, T. & Kristiansen, T.S. (1990) Enhancement studies of coastal cod in western Norway. Part II. Mortality of reared cod after release. *Journal du Conseil. International pour l'Exploration de la Mer*, **47**, 30–39.

Takahashi, Y. (1992) *Occurrence and prevention of color abnormality in hatchery-reared flounder*. Special Report 3, Japan Sea-Farming Association (in Japanese).

Takeuchi, T., Dedi, J., Ebisawa, C., *et al.* (1995) The effect of β-carotene and vitamin A enriched *Artemia* nauplii on the malformation and color abnormality of larval Japanese flounder. *Fisheries Science*, **61**, 141–148.

Tominaga, O. & Watanabe, Y. (1998) Geographical dispersal and optimum release size of hatchery-reared Japanese flounder *Paralichthys olivaceus* released in Ishikari Bay Hokkaido, Japan. *Journal of Sea Research*, **40**, 73–81.

Tsukamoto, K., Kuwada, H., Hirokawa, J., et al. (1989) Size-dependent mortality of red sea bream, *Pagrus major*, juveniles released with fluorescent otolith-tags in News Bay, Japan. *Journal of Fish Biology*, **35**, 59–69.

Uchino, K. & Nakanishi, M. (1983) Results of tagging experiments of hatchery-reared flounder, *Paralichthys olivaceus*, in the western part of Wakasa Bay (Tango-kai). *Bulletin of the Kyoto Institute of Oceanic and Fisheries Science*, **7**, 17–27 (in Japanese with English abstract).

Vadstein, O., Øie, G., Olsen, Y., Salvesen, I. & Skjermo, J. (1993) A strategy to obtain microbial control during larval development of marine fish. In: *Fish Farming Technology* (eds H. Reinertsen, L.A. Dahle, L. Jørgensen & K. Tvinnereim). pp. 69–75. Balkema, Rotterdam.

Watanabe, K., Suzuki, S. & Nishiki, A. (2001) Migration, growth and stock enhancement effect of hatchery-reared barfin flounder *Verasper moseri* juveniles released in Akkeshi Bay. *Saibaigiken*, **28**, 93–99 (in Japanese with English abstract).

Yamamoto, E. (1999) Studies on sex-manipulation and production of cloned populations in hirame, *Paralichthys olivaceus* (Temminck et Schlegel). *Aquaculture*, **173**, 235–246.

Yamashita, Y. & Yamada, H. (1999) Release strategy for Japanese flounder fry in stock enhancement programmes. In: *Stock Enhancement and Sea Ranching* (eds B.R. Howell, E. Moksness & T. Svåsand). pp. 191–204. Fishing News Books, Oxford.

Yamashita, Y., Yamamoto, K., Nagahora, S., et al. (1993) Predation by fishes on hatchery-raised Japanese flounder *Paralichthys olivaceus*, fry in the coastal waters of Iwate Prefecture, northeastern Japan. *Suisan Zoshoku*, **41**, 497–505 (in Japanese with English abstract).

Yamashita, Y., Nagahora, S., Yamada, H. & Kitagawa, D. (1994) Effects of release size on survival and growth of Japanese flounder *Paralichthys olivaceus* in coastal waters off Iwate Prefecture, northeastern Japan. *Marine Ecology Progress Series*, **105**, 269–276.

Appendix 1

List of scientific and common names of living flatfishes used in the book. Scientific and common names are those given by FishBase except where indicated otherwise. The families Paralichthodidae, Rhombosoleidae and Poecilopsettidae are listed as subfamilies in FishBase. *Current valid scientific name of Pleuronectidae following revision by Cooper & Chapleau (Cooper, J.A. & Chapleau, F. (1998) Monophyly and intrarelationships of the family Pleuronectidae (Pleuronectiformes) with a revised classification. *Fishery Bulletin,* **96**, 686–726) and Cooper (personal communication) (see Appendix 2 for common synonyms). **No common name listed by FishBase, name given here is that used by chapter author(s). Common names in parentheses are those frequently used in the literature but differ from the FishBase common name. ***Japanese flounder is preferred here over the FishBase 'bastard halibut' because the former is more accurately descriptive and almost universally used by authors.

Scientific name	Common name	Family
Achiropsetta tricholepis	Finless flounder	Achiropsettidae
Achirus lineatus	Lined sole	Achiridae
Achlyopa nigra	Black sole	Soleidae
Ammotretis lituratus	Tudor's flounder	Rhombosoleidae
Ammotretis rostratus	Longsnout flounder	Rhombosoleidae
Ancylopsetta ommata	Gulf of Mexico ocellated flounder	Paralichthyidae
Arnoglossus kessleri	Scaldback	Bothidae
Arnoglossus laterna	Scaldfish	Bothidae
Arnoglossus multirastris	No common name	Bothidae
Arnoglossus scapha	Mahue**, megrim**	Bothidae
Arnoglossus thori	Thor's scaldfish	Bothidae
Aseraggodes herrei	Herre's sole	Soleidae
Asterorhombus fijiensis	Angler flatfish	Bothidae
Asterorhombus intermedius	Intermediate flounder	Bothidae
Austroglossus microlepis	West coast sole	Soleidae
Austroglossus pectoralis	Mud sole (Agulhas sole)	Soleidae
Azygopus flemingi	Banded-fin flounder	Rhombosoleidae
Azygopus pinnifasciatus	Banded-fin flounder	Rhombosoleidae
Bathysolea profundicola	Deepwater sole	Soleidae
Bothus constellatus	Pacific eyed flounder	Bothidae
Bothus ellipticus	No common name	Bothidae
Bothus lunatus	Plate fish	Bothidae
Bothus mancus	Flowery flounder	Bothidae
Bothus myriaster	Indo-Pacific oval flounder	Bothidae
Bothus ocellatus	Eyed flounder	Bothidae
Bothus pantherinus	Leopard flounder	Bothidae
Bothus podas	Wide-eyed flounder	Bothidae
Brachirus orientalis	Oriental sole	Soleidae
Buglossidium luteum	Solenette	Soleidae
Chascanopsetta lugubris	Pelican flounder	Bothidae
Citharichthys arctifrons	Gulf Stream flounder	Paralichthyidae

Scientific name	Common name	Family
Citharichthys sordidus	Pacific sanddab	Paralichthyidae
Citharichthys spilopterus	Bay whiff	Paralichthyidae
Citharichthys stigmaeus	Speckled sanddab	Paralichthyidae
Citharoides macrolepidotus	Largescale flounder	Citharidae
Citharus linguatula	Atlantic spotted flounder	Citharidae
Cleisthenes herzensteini*	Sôhachi	Pleuronectidae
Cleisthenes pinetorum*	Pointhead flounder**	Pleuronectidae
Clidoderma asperrimum	Roughscale sole	Pleuronectidae
Colistium guntheri	New Zealand brill**	Rhombosoleidae
Colistium nudipinnis	New Zealand turbot**	Rhombosoleidae
Crossorhombus kobensis	Kobe flounder	Bothidae
Cynoglossus arel	Largescale tonguesole	Cynoglossidae
Cynoglossus bilineatus	Fourlined tonguesole	Cynoglossidae
Cynoglossus canariensis	Canary tonguesole	Cynoglossidae
Cynoglossus carpenteri	Hooked tonguesole	Cynoglossidae
Cynoglossus joyneri	Red tonguesole	Cynoglossidae
Cynoglossus kopsii	Shortheaded tonguesole	Cynoglossidae
Cynoglossus macrostomus	Malabar sole	Cynoglossidae
Cynoglossus monodi	Guinean tonguesole	Cynoglossidae
Cynoglossus puncticeps	Speckled tonguesole	Cynoglossidae
Cynoglossus robustus	Robust tonguefish**	Cynoglossidae
Cynoglossus senegalensis	Senegalese tonguesole	Cynoglossidae
Cynoglossus zanzibarensis	Zanzibar tonguefish	Cynoglossidae
Dagetichthys lakdoensis	No common name	Soleidae
Dexistes rikuzenius	Rikuzen flounder	Pleuronectidae
Dicologlossa cuneata	Wedge sole	Soleidae
Embassichthys bathybius	Deepsea sole	Soleidae
Engyprosopon grandisquama	Largescale flounder	Bothidae
Eopsetta grigorjewi	Shotted halibut	Pleuronectidae
Eopsetta jordani	Petrale sole	Pleuronectidae
Etropus crossotus	Fringed flounder	Paralichthyidae
Etropus microstomus	Smallmouth flounder	Paralichthyidae
Glyptocephalus cynoglossus	Witch (witch flounder)	Pleuronectidae
Glyptocephalus kitaharai*	Willowy flounder	Pleuronectidae
Glyptocephalus stelleri	Blackfin flounder	Pleuronectidae
Glyptocephalus zachirus*	Rex sole	Pleuronectidae
Hippoglossina macrops	Bigeye flounder	Paralichthyidae
Hippoglossoides dubius	Flathead flounder	Pleuronectidae
Hippoglossoides elassodon	Flathead sole	Pleuronectidae
Hippoglossoides platessoides	American plaice (long rough dab)	Pleuronectidae
Hippoglossus hippoglossus	Atlantic halibut	Pleuronectidae
Hippoglossus stenolepis	Pacific halibut	Pleuronectidae
Isopsetta isolepis*	Butter sole**	Pleuronectidae
Lepidopsetta bilineata*	Rock sole	Pleuronectidae
Lepidopsetta mochigarei*	Dusky sole	Pleuronectidae
Lepidopsetta polyxystra	Northern rock sole**	Pleuronectidae
Lepidorhombus boscii	Fourspotted megrim	Scophthalmidae
Lepidorhombus whiffiagonis	Megrim	Scophthalmidae
Limanda aspera*	Yellowfin sole	Pleuronectidae
Limanda ferruginea*	Yellowtail flounder	Pleuronectidae
Limanda limanda*	Dab	Pleuronectidae
Limanda punctatissima*	Sand flounder	Pleuronectidae
Lyopsetta exilis*	Slender sole	Pleuronectidae
Mancopsetta maculata	Antarctic armless flounder**	Achiropsettidae
Mancopsetta milfordi	Armless flounder	Achiropsettidae
Microchirus variegatus	Thickback sole	Soleidae
Microstomus achne	Slime flounder	Pleuronectidae

Scientific name	Common name	Family
Microstomus kitt	Lemon sole	Pleuronectidae
Microstomus pacificus	Dover sole	Pleuronectidae
Nematops macrochirus	Long-fin righteye flounder	Poecilopsettidae
Oncopterus darwini	Remo flounder	Rhombosoleidae
Paralichthodus algoensis	Peppered flounder	Paralichthodidae
Paralichthys brasiliensis	Brazilian flounder	Paralichthyidae
Paralichthys californicus	California flounder (California halibut)	Paralichthyidae
Paralichthys dentatus	Summer flounder	Paralichthyidae
Paralichthys isosceles	No common name	Paralichthyidae
Paralichthys lethostigma	Southern flounder	Paralichthyidae
Paralichthys microps	Small-eyed flounder**	Paralichthyidae
Paralichthys oblongus	American fourspot flounder	Paralichthyidae
Paralichthys olivaceus	Japanese flounder*** (hirame)	Paralichthyidae
Paralichthys orbignyanus	No common name	Paralichthyidae
Paralichthys patagonicus	Patagonian flounder	Paralichthyidae
Paralichthys tropicus	Tropical flounder	Paralichthyidae
Paraplagusia bilineata	Doublelined tonguesole	Cynoglossidae
Paraplagusia japonica	Black cow-tongue	Cynoglossidae
Pardachirus marmoratus	Finless sole	Soleidae
*Parophrys vetula**	English sole	Pleuronectidae
Pelotretis flavilatus	Southern lemon sole	Rhombosoleidae
Peltorhamphus novaezeelandiae	New Zealand sole	Rhombosoleidae
Peltorhamphus scapha	New Zealand sole**	Rhombosoleidae
Phrynorhombus regius	Eckström's topknot	Scophthalmidae
Plagiopsetta glossa	Tongue flatfish	Samaridae
*Platichthys bicoloratus**	Stone flounder	Pleuronectidae
*Platichthys flesus**	Flounder	Pleuronectidae
Platichthys stellatus	Starry flounder	Pleuronectidae
*Pleuronectes glacialis**	Arctic flounder	Pleuronectidae
*Pleuronectes pinnifasciatus**	Far Eastern smooth flounder	Pleuronectidae
*Pleuronectes platessa**	Plaice	Pleuronectidae
*Pleuronectes putnami**	American smooth flounder	Pleuronectidae
Pleuronectes quadrituberculatus	Alaska plaice	Pleuronectidae
Pleuronichthys cornutus	Ridged-eyed flounder	Pleuronectidae
Pleuronichthys decurrens	Curlfin sole	Pleuronectidae
*Pleuronichthys guttulatus**	Diamond turbot	Pleuronectidae
Pleuronichthys ritteri	Spotted turbot	Pleuronectidae
Psammodiscus ocellatus	Indonesian ocellated flounder	Rhombosoleidae
Psettichthys melanostictus	Pacific sand sole	Pleuronectidae
Psettodes belcheri	Spottail spiny turbot	Psettodidae
Psettodes erumei	Indian spiny turbot	Psettodidae
*Pseudopleuronectes americanus**	Winter flounder	Pleuronectidae
*Pseudopleuronectes herzensteini**	Littlemouth flounder (brown sole)	Pleuronectidae
*Pseudopleuronectes obscurus**	Kurogarei**	Pleuronectidae
*Pseudopleuronectes schrenki**	Cresthead flounder	Pleuronectidae
*Pseudopleuronectes yokohamae**	Marbled flounder (marbled sole)	Pleuronectidae
Pseudorhombus arsius	Largetooth flounder	Paralichthyidae
Pseudorhombus cinnamoneus	Cinnamon flounder	Paralichthyidae
Pseudorhombus elevatus	Deep flounder	Paralichthyidae
Pseudorhombus javanicus	Javan flounder	Paralichthyidae
Pseudorhombus malayanus	Malayan flounder	Paralichthyidae
Pseudorhombus oligolepis	No common name	Paralichthyidae
Pseudorhombus pentophthalmus	Fivespot flounder	Paralichthyidae
Pseudorhombus triocellatus	Three spotted flounder	Paralichthyidae
*Reinhardtius evermanni**	Kamchatka flounder	Pleuronectidae
Reinhardtius hippoglossoides	Greenland halibut	Pleuronectidae
*Reinhardtius stomias**	Arrowtooth flounder	Pleuronectidae

Scientific name	Common name	Family
Rhombosolea leporina	Yellowbelly flounder	Rhombosoleidae
Rhombosolea plebeia	New Zealand flounder	Rhombosoleidae
Rhombosolea retiaria	Black flounder	Rhombosoleidae
Rhombosolea tapirina	Greenback flounder	Rhombosoleidae
Scophthalmus aquosus	Windowpane	Scophthalmidae
Scophthalmus maximus	Turbot	Scophthalmidae
Scophthalmus rhombus	Brill	Scophthalmidae
Solea bleekeri	Blackhand sole	Soleidae
Solea elongata	Elongate sole	Soleidae
Solea impar	Adriatic sole	Soleidae
Solea lascaris	Sand sole	Soleidae
Solea ovata	Ovate sole	Soleidae
Solea senegalensis	Senegalese sole	Soleidae
Solea solea	Common sole (sole)	Soleidae
Solea stanalandi	Stanaland's sole	Soleidae
Syacium gunteri	Shoal flounder	Paralichthyidae
Symphurus civitatium	Offshore tonguefish	Cynoglossidae
Symphurus plagiusa	Blackcheek tonguefish	Cynoglossidae
Synaptura cadenati	Guinean sole	Cynoglossidae
Synaptura commersonnii	Commerson's sole	Cynoglossidae
Synaptura lusitanica	Portuguese sole	Cynoglossidae
Tarphops oligolepis	No common name	Paralichthyidae
Tephrinectes sinensis	Flower flounder**	Uncertain
Trinectes maculatus	Hogchoker	Achiridae
Verasper moseri	Barfin flounder	Pleuronectidae
Verasper variegatus	Spotted halibut	Pleuronectidae
Xystreurys rasile	No common name	Paralichthyidae
Zebrias captivus	Convict zebra sole	Soleidae
Zebrias fasciatus	Many-banded sole	Soleidae
Zebrias synapturoides	Indian zebra sole	Soleidae
Zeugopterus punctatus	Topknot	Scophthalmidae

Appendix 2

Common synonyms of those Pleuronectidae mentioned in the text.

Name used in text	Common synonyms used in recent literature
Cleisthenes herzensteini	*Hippoglossoides herzensteini*
Cleisthenes pinetorum	*Hippoglossoides pinetorum*
Clidoderma asperrimum	
Dexistes rikuzenius	
Eopsetta grigorjewi	
Eopsetta jordani	
Glyptocephalus cynoglossus	
Glyptocephalus kitaharai	*Tanakius kitaharai*
Glyptocephalus stelleri	
Glyptocephalus zachirus	*Errex zachirus*
Hippoglossoides dubius	
Hippoglossoides elassodon	
Hippoglossoides platessoides	
Hippoglossus hippoglossus	
Hippoglossus stenolepis	
Isopsetta isolepis	*Pleuronectes isolepis*
Lepidopsetta bilineata	*Pleuronectes bilineatus*
Lepidopsetta mochigarei	*Pleuronectes mochigarei*
Lepidopsetta polyxystra	
Limanda aspera	*Pleuronectes aspera*
Limanda ferruginea	*Pleuronectes ferruginea*
Limanda limanda	*Pleuronectes limanda*
Limanda punctatissima	*Pleuronectes punctatissima*
Lyopsetta exilis	*Eopsetta exilis*
Microstomus achne	
Microstomus kitt	
Microstomus pacificus	
Parophrys vetula	*Pleuronectes vetulus, Parophrys vetulus*
Platichthys bicoloratus	*Kareius bicoloratus*
Platichthys flesus	*Pleuronectes flesus*
Platichthys stellatus	
Pleuronectes glacialis	*Liopsetta glacialis*
Pleuronectes pinnifasciatus	*Liopsetta pinnifasciatus*
Pleuronectes platessa	*Pleuronectes platessus*
Pleuronectes putnami	*Liopsetta putnami*
Pleuronectes quadrituberculatus	
Pleuronichthys cornutus	
Pleuronichthys decurrens	
Pleuronichthys guttulatus	*Hypsopsetta guttulata*

Name used in text	Common synonyms used in recent literature
Pleuronichthys ritteri	
Psettichthys melanostictus	
Pseudopleuronectes americanus	*Pleuronectes americanus*
Pseudopleuronectes herzensteini	*Pleuronectes herzensteini*
Pseudopleuronectes obscurus	*Pleuronectes obscurus, Liopsetta obscura*
Pseudopleuronectes schrenki	*Pleuronectes schrenki, Limanda schrenki*
Pseudopleuronectes yokohamae	*Pleuronectes yokohamae, Limanda yokohamae*
Reinhardtius evermanni	*Atheresthes evermanni*
Reinhardtius hippoglossoides	
Reinhardtius stomias	*Atheresthes stomias*
Verasper moseri	
Verasper variegatus	

Index of scientific and common names

To find all entries to a species consult both scientific and common names. The scientific and corresponding common names for flatfishes are listed in Appendix 1.

abalone, 361
Acanthopagrus schlegeli, 360
Achiridae, 6, 12, 13, 16, 17, 19, 22, 23, 27, 28, 49, 55, 58, 59, 191, 248, 373, 376
Achiroides, 49, 293
Achiropsetta, 19
 tricholepis, 58, 373
Achiropsettidae, 12, 13, 16, 19, 22, 23, 26, 58, 62, 214, 248, 250, 373, 374
Achiropsis, 293
Achirus, 293, 296, 297
 lineatus, 143, 373
Achlyopa nigra, 297, 373
Adriatic sole, 72, 376
Agulhas sole, 373
Alaska plaice, 107, 109, 128, 186, 192, 272, 277, 375
 pollock, 103, 193, 194
American angler, 198
 fourspot flounder, 192, 193, 200, 202, 375
 plaice, 7, 57, 69, 71, 75, 80, 82, 84, 141, 144, 150, 151, 188, 190, 192, 195, 199, 200, 201, 202, 240, 241, 243, 145, 247, 148, 251, 252, 253, 259, 260, 263, 264, 265, 327, 329, 330, 337, 340, 374
 smooth flounder, 68, 84, 375
 soles, 6
Ammodytes, 193
Ammodytidae, 193, 247
Ammotretis lituratus, 273, 373
 rostratus, 139, 273, 373
Amphiprion melanopus, 97
Anarhichas lupus, 202
anchovy, 124
Ancylopsetta ommata, 309, 373

anglerfishes, 1, 247
angler flatfish, 220, 373
Antarctic armless flounder, 48, 374
Arctic flounder, 46, 57, 84, 375
Arenicola, 148
armless flounder, 62, 374
Arnoglossus, 14, 30, 45, 220, 250
 kessleri, 46, 373
 laterna, 125, 187, 191, 373
 multirastris, 46, 373
 scapha, 191, 373
 thori, 187, 373
arrowtooth flounder, 145, 174, 187, 192, 193, 194, 272, 277, 282, 328, 336, 375
Artemia, 7, 350, 351, 353
 salina, 347
Aseraggodes, 48, 58, 62
 herrei, 48, 373
Asterorhombus fijiensis, 220, 373
 intermedius, 220, 373
Atheresthes evermanni, 378
 stomias, 378
Atlantic cod, 103, 112, 124, 125, 127, 130, 194, 197, 198, 202, 224, 240, 251, 255, 264, 331
 halibut, 58, 78, 122, 141, 142, 143, 147, 148, 192, 193, 194, 200, 202, 203, 223, 241, 243, 244, 252, 259, 260, 264, 322, 337, 347, 348, 350, 352, 354, 355, 356, 365, 374
 herring, 103, 123, 193, 197, 202, 251
 mackerel, 202
 spotted flounder, 187, 297, 374
Aurelia aurita, 197
Austroglossus, 250
 microlepis, 250, 258, 373
 pectoralis, 250, 258, 373

Azygopus flemingi, 48, 373
　pinnifasciatus, 48, 373

banded-fin flounder, 48, 186, 373
barfin flounder, 349, 354, 362, 376
Bathysolea, 48
　profundicola, 58, 373
bay whiff, 166, 373
bigeye flounder, 187, 374
black cow-tongue, 293, 375
　flounder, 187, 376
　seabream, 360
　sole, 297, 373
blackcheek tonguefish, 150, 166, 376
blackfin flounder, 273, 374
blackhand sole, 169, 170, 191, 196, 297, 376
bluefish, 198
Bothidae, 6, 11, 13, 16, 17, 18, 19, 22, 23, 27, 28, 45, 58, 60, 61, 62, 69, 152, 165, 187, 191, 221, 240, 245, 248, 250, 296, 301, 306, 308, 373, 374
Bothinae, 18
Bothus, 46, 62, 218, 296, 297
　constellatus, 299, 373
　ellipticus, 218, 373
　lunatus, 218, 373
　mancus, 46, 220, 373
　myriaster, 46, 373
　ocellatus, 175, 218, 373
　pantherinus, 49, 296, 373
　podas, 141, 187, 218, 250, 297, 373
Brachionus plicatilis, 351
Brachirus, 48, 49, 293, 296, 297
　orientalis, 296, 373
Brachypleura, 19, 45
Brazilian flounder, 249, 375
brill, 169, 174, 192, 215, 245, 248, 376
Brosme brosme, 202
brown shrimp, 167
　sole, 375
Buglossidium, 48
　luteum, 48, 70, 71, 125, 150, 190, 373
butter sole, 277, 374

California flounder, 139, 145, 166, 173, 191, 195, 272, 375
　halibut, 375
Canary tonguesole, 297, 299, 374
capelin, 124, 193, 202
Carcinus maenas, 197
Chascanopsetta, 45, 58, 62
　lugubris, 250, 373
Chionectes, 274

Chromis atripectoralis, 97
chum salmon, 359, 365, 366
cinnamon flounder, 273, 375
Citharichthys, 46, 60, 297
　arctifrons, 168, 187, 373
　sordidus, 96, 187, 191, 374
　spilopterus, 166, 374
　stigmaeus, 214, 374
Citharidae, 11, 13, 16, 17, 22, 23, 26, 45, 187, 250, 296, 374
Citharoides, 45
　macrolepidotus, 273, 374
Citharus, 45
　linguatula, 187, 298, 374
Cleisthenes herzensteini, 187, 374, 377
　pinetorum, 273, 374, 377
Clidoderma asperrimum, 273, 374, 377
Clupea harengus, 103, 193, 251
　pallasii, 193
Colistium guntheri, 273, 374
　nudipinnis, 142, 273, 374
Commerson's sole, 297, 376
common sole, 2, 48, 69, 71, 73, 74, 75, 76, 77, 78, 80, 82, 98, 109, 111, 113, 125, 128, 139, 140, 145, 142, 143, 147, 151, 153, 169, 173, 174, 190, 216, 220, 221, 224, 225, 226, 227, 228, 230, 245, 246, 247, 248, 255, 256, 261, 262, 268, 322, 326, 327, 329, 330, 339, 347, 348, 351, 352, 353, 356, 365, 376
convict zebra sole, 297, 376
cormorants, 198
Cottidae, 198
Crangon, 187, 189, 190, 192, 198
　affinis, 167
　crangon, 167, 197
cresthead flounder, 68, 375
Crossorhombus kobensis, 218, 374
curlfin sole, 96, 375
Cyclopsetta, 8, 296, 297
Cynoglossidae, 6, 12, 13, 14, 15, 16, 17, 19, 20, 21, 22, 23, 27, 28, 35, 49, 55, 58, 60, 62, 69, 70, 152, 165, 187, 214, 240, 248, 250, 296, 301, 305, 306, 307, 374, 375, 376
Cynoglossus, 20, 29, 49, 55, 293, 296, 297, 306, 307, 308, 309
　arel, 297, 374
　bilineatus, 297, 374
　canariensis, 297, 374
　carpenteri, 297, 374
　joyneri, 187, 273, 374
　kopsii, 297, 374
　macrostomus, 296, 374

monodi, 296, 374
puncticeps, 297, 374
robustus, 273, 374
senegalensis, 296, 374
zanzibarensis, 250, 374

dab, 84, 110, 125, 127, 147, 151, 169, 171, 174, 186, 188, 190, 191, 196, 197, 198, 245, 246, 247, 256, 261, 374
Dagetichthys, 48
  lakdoensis, 293, 374
deep flounder, 297, 375
deepsea sole, 58, 374
deepwater sole, 58, 373
Dexistes rikuzenius, 273, 374, 377
diamond turbot, 191, 195, 375
Dicologlossa, 48
  cuneata, 190, 374
dolphins, 58
doublelined tonguesole, 297, 375
Dover sole, 58, 71, 84, 99, 103, 112, 140, 145, 146, 150, 168, 169, 171, 173, 186, 189, 190, 195, 272, 277, 281, 284, 336, 374
dusky sole, 374

Eckström's topknot, 220, 375
elongate sole, 297, 376
Embassichthys bathybius, 58, 374
English sole, 8, 71, 80, 82, 86, 98, 100, 147, 151, 186, 189, 272, 277, 281, 328, 336, 375
Engraulis encrasicolus, 124
  japonicus, 103
Engyprosopon, 30, 46
  grandisquama, 72, 218, 300, 374
Eobothus minimus, 14
Eobuglossus eocenicus, 14
Eopsetta exilis, 377
  grigorjewi, 187, 273, 349, 374, 377
  jordani, 71, 102, 192, 272, 335, 374, 377
Errex zachirus, 377
Etropus crossotus, 150, 166, 374
  microstomus, 173, 374
European hake, 247
  sprat, 247
eyed flounder, 218, 373

Far Eastern smooth flounder, 188, 375
finless flounder, 58, 62, 373
  sole, 216, 297, 375
fivespot flounder, 191, 375
flathead flounder, 188, 192, 374
  sole, 86, 106, 107, 143, 192, 193, 272, 277, 284, 328, 329, 331, 336, 374

flounder, 1, 75, 87, 94, 109, 111, 127, 138, 139, 147, 167, 172, 186, 189, 191, 196, 222, 228, 245, 246, 247, 261, 327, 375
flower flounder, 17, 45, 376
flowery flounder, 46, 50, 220, 373
fourlined tonguesole, 297, 374
fourspotted megrim, 58, 147, 268, 327, 329, 374
fringed flounder, 150, 166, 374

Gadidae, 131, 331
Gadiformes, 25
Gadus macrocephalus, 281
  morhua, 103, 124, 194, 224, 251, 331
Glyptocephalus, 110
  cynoglossus, 58, 80, 101, 110, 143, 167, 187, 241, 245, 329, 374, 377
  kitaharai, 273, 349, 374, 377
  stelleri, 273, 374, 377
  zachirus, 71, 97, 101, 110, 188, 272, 374, 377
greenback flounder, 73, 139, 143, 223, 273, 354, 376
Greenland halibut, 46, 57, 58, 72, 75, 96, 97, 189, 192, 296, 215, 240, 241, 243, 245, 246, 247, 251, 252, 253, 254, 256, 259, 260, 261, 263, 264, 265, 267, 268, 272, 277, 327, 329, 330, 337, 339, 340, 375
grenadiers, 202
grey seal, 198
Guinean sole, 297, 376
  tonguesole, 296, 374
Gulf of Mexico ocellated flounder, 309, 373
Gulf Stream flounder, 168, 169, 187, 193, 373
Gymnachirus, 297

haddock, 127, 130, 201, 202, 251, 264, 331
hakes, 193, 240, 262, 266, 267, 274, 331
halibuts, 6, 14, 46, 185
Halichoerus grypus, 198
Haliotis, 361
Hemitripterus americanus, 198
Herre's sole, 48, 373
Hippoglossina, 46
  macrops, 187, 374
Hippoglossoides dubius, 188, 192, 374, 377
  elassodon, 86, 106, 143, 192, 272, 329, 374, 377
  herzensteini, 377
  pinetorum, 377
  platessoides, 7, 57, 69, 70, 71, 102, 128, 141, 188, 190, 192, 241, 245, 329, 374, 377
Hippoglossus hippoglossus, 58, 78, 97, 122, 141, 192, 223, 241, 322, 347, 374, 377

stenolepis, 6, 33, 58, 70, 71, 102, 105, 168, 188, 192, 221, 272, 320, 374, 377
hirame, 375
hogchoker, 68, 166, 169, 170, 191, 195, 214, 376
hooked tonguesole, 297, 374
Hypsopsetta guttulata, 377

Indian spiny turbot, 45, 49, 296, 297, 301, 305, 306, 375
    zebra sole, 297, 376
Indonesian ocellated flounder, 48, 375
Indo-Pacific oval flounder, 46, 50, 373
intermediate flounder, 220, 373
Isopsetta isolepis, 277, 377

Japanese anchovy, 103
    flounder, 46, 60, 143, 145, 173, 187, 192, 193, 215, 221, 223, 224, 226, 230, 273, 277, 280, 347, 349, 351, 352, 353, 354, 355, 356, 358, 359, 360, 361, 363, 364, 365, 375
Javan flounder, 297, 375

Kamchatka flounder, 192, 193, 375
Kareius bicoloratus, 377
Kobe flounder, 218, 374
kurogarei, 68, 99, 375
Kuruma prawn, 361

largescale flounder(s), 17, 72, 218, 273, 296, 300, 374
    tonguesole, 297, 374
largetooth flounder(s), 191, 296, 297, 300, 375
lefteye flounders, 6
lemon sole, 71, 131, 186, 189, 214, 245, 247, 261, 351, 356, 374
leopard flounder, 46, 49, 296, 373
Lepidoblepharon, 45
Lepidopsetta, 277
    bilineata, 68, 99, 168, 188, 190, 192, 220, 272, 328, 374, 377
    mochigarei, 68, 99, 374, 377
    polyxystra, 29, 99, 225, 272, 374, 377
Lepidorhombus, 245, 327
    boscii, 58, 147, 268, 327, 329, 339, 374
    whiffiagonis, 153, 191, 327, 329, 339, 374
Leucoraja erinacea, 201
Limanda, 109
    aspera, 101, 109, 188, 192, 225, 272, 273, 328, 374, 377
    ferruginea, 6, 78, 101, 110, 141, 167, 188, 241, 322, 374, 377
    limanda, 84, 101, 110, 125, 147, 169, 188, 190, 191, 245, 374, 377

punctatissima, 273, 374, 377
schrenki, 378
yokohamae, 378
lined sole, 143, 373
Liopsetta glacialis, 377
    obscura, 378
    pinnifasciatus, 377
    putnami, 377
little skate, 201
littlemouth flounder, 100, 103, 104, 105, 175, 189, 273, 349, 375
long rough dab, 7, 374
long-fin righteye flounder, 47, 375
longsnout flounder, 139, 273, 373
Lophius, 247
    americanus, 198
Lumbricilius rivalis, 351
Lyopsetta exilis, 189, 374, 377

mackerel, 130
Macrouridae, 202
mahue, 191, 373
Malabar sole, 296, 299, 307, 308, 374
Malayan flounder, 297, 375
Mallotus villosus, 124, 193
Mancopsetta, 19
    maculata, 48, 58, 374
    milfordi, 62, 374
many-banded sole, 273, 376
marbled flounder, 99, 110, 175, 273, 349, 354, 358, 361, 375
    sole, 375
Marleyella, 47
megrim, 153, 191, 245, 246, 247, 256, 261, 168, 327, 329, 331, 339, 373, 374
Melanogrammus aeglefinus, 127, 201, 251, 331
Merlucciidae, 193
Merluccius, 331
    bilinearis, 202
    merluccius, 247
Microchirus, 48
    variegatus, 109, 139, 374
Microstomus achne, 273, 349, 374, 377
    kitt, 70, 71, 131, 189, 214, 245, 351, 375, 377
    pacificus, 58, 70, 71, 99, 102, 140, 168, 189, 190, 272, 336, 375, 377
Monochirus, 48
Monolene, 58, 62
mud sole, 250, 257, 258, 373
Mytilus edulis, 351

Nematops, 47
    macrochirus, 47, 375

Neoachiropsetta, 19
New Zealand brill, 273, 283, 374
    flounder, 191, 273, 376
    sole, 190, 273, 283, 375
    turbot, 142, 273, 283, 374
northern rock sole, 29, 33, 99, 225, 272, 374

ocean perch, 202
offshore tonguefish, 166, 376
Oncopterus darwini, 48, 375
Oncorhynchus keta, 359
    mykiss, 76
Ophidiiformes, 25
oriental sole, 296, 297, 373
ovate sole, 297, 376

Pacific cod, 281
    eyed flounder, 299, 373
    halibut, 6, 71, 106, 107, 112, 113, 168, 188, 192, 193, 194, 196, 221, 223, 225, 229, 272, 274, 277, 278, 280, 281, 283, 284, 285, 286, 288, 320, 328, 332, 335, 336, 374
    herring, 193
    sand sole, 277, 375
    sanddab, 96, 97, 187, 191, 373
Pagrus major, 360
Parabothus, 45
Paradicula, 48
Paralichthodidae, 12, 13, 16, 18, 22, 23, 47, 375
Paralichthodinae, 18
Paralichthodus algoensis, 19, 47, 375
Paralichthyidae, 6, 11, 13, 16, 17, 18, 22, 23, 26, 27, 46, 60, 70, 102, 152, 165, 167, 187, 191, 221, 248, 250, 296, 373, 374, 375, 376
Paralichthyinae, 17
Paralichthys, 18, 33, 46, 60, 103, 109, 248, 250, 296, 197, 347, 355, 365
    brasiliensis, 248, 249, 375
    californicus, 139, 166, 191, 272, 375
    dentatus, 70, 102, 109, 139, 166, 187, 191, 215, 241, 327, 355, 375
    isosceles, 248, 375
    lethostigma, 143, 167, 187, 191, 220, 355, 375
    microps, 187, 191, 375
    oblongus, 192, 375
    olivaceus, 46, 97, 102, 109, 143, 173, 187, 192, 215, 273, 347, 349, 375
    orbignyanus, 151, 248, 249, 257, 375
    patagonicus, 248, 257, 375
    tropicus, 355, 375
Paralithodes camtschaticus, 274

Paraplagusia, 20, 49, 296, 306
    bilineata, 297, 375
    japonica, 273, 375
Pardachirus, 48, 226, 296, 297
    marmoratus, 216, 297, 375
Parophrys vetula, 8, 70, 71, 80, 98, 101, 147, 189, 272, 328, 375, 377
    vetulus, 377
Patagonian flounder, 248, 249, 375
pelican flounder, 250, 373
Pelotretis flavilatus, 190, 273, 375
Peltorhamphus novaezeelandiae, 273, 375
    scapha, 190, 375
Penaeus japonicus, 361
peppered flounder, 19, 47, 375
Perciformes, 25
petrale sole, 71, 192, 272, 277, 281, 284, 335, 336, 374
Phrynorhombus regius, 220, 375
Phyllichthys, 48
Plagiopsetta glossa, 47, 375
plaice, 2, 3, 46, 70, 71, 72, 73, 74, 75, 76, 77, 78, 79, 80, 81, 82, 84, 85, 94, 95, 96, 98, 99, 100, 103, 104, 110, 111, 113, 121, 122, 123, 124, 125, 127, 128, 138, 139, 140, 141, 142, 143, 144, 145, 147, 148, 151, 153, 166, 167, 169, 171, 172, 173, 174, 189, 191, 195, 197, 198, 199, 215, 217, 218, 221, 222, 223, 224, 225, 226, 227, 228, 229, 245, 246, 247, 248, 251, 254, 255, 256, 261, 265, 268, 322, 323, 327, 329, 331, 332, 333, 339, 347, 348, 351, 352, 356, 357, 375
plate fish, 218, 373
Platichthys bicoloratus, 78, 100, 149, 167, 192, 273, 348, 349, 375, 377
    flesus, 1, 70, 94, 97, 101, 127, 138, 167, 189, 191, 222, 245, 327, 375, 377
    stellatus, 1, 70, 71, 97, 140, 189, 190, 192, 220, 272, 375, 377
Pleurobrachia pileus, 197
Pleuronectes americanus, 378
    aspera, 377
    bilineatus, 377
    ferruginea, 377
    flesus, 377
    glacialis, 46, 84, 375, 377, 377
    herzensteini, 378
    isolepis, 377
    limanda, 377
    mochigarei, 377
    obscurus, 378
    pinnifasciatus, 188, 375, 377

platessa, 2, 46, 70, 71, 94, 97, 101, 121, 138, 166, 189, 191, 215, 245, 322, 347, 375, 377
platessus, 377
punctatissima, 377
putnami, 68, 375, 377
quadrituberculatus, 101, 109, 189, 192, 272, 375, 377
schrenki, 378
vetulus, 377
yokohamae, 378
Pleuronectidae, 6, 8, 11, 12, 13, 16, 17, 18, 19, 20, 22, 23, 26, 58, 59, 69, 70, 101, 110, 142, 152, 165, 167, 187, 190, 191, 192, 221, 245, 248, 374, 375, 376, 377
Pleuronectiformes, 1, 4, 10, 14–16, 17, 18, 19, 21–5, 28–34, 35, 45, 46, 49–54, 57, 185, 301
Pleuronectinae, 17, 18, 19
Pleuronectoidei, 14, 17
Pleuronichthys, 96
  cornutus, 273, 349, 375, 377
  decurrens, 96, 375, 377
  guttulatus, 191, 375, 377
  ritteri, 96, 375, 378
Poecilopsetta, 30, 47, 62
Poecilopsettidae, 12, 13, 16, 19, 20, 23, 26, 27, 47, 58, 60, 62, 296, 375
Poecilopsettinae, 17, 18
pointhead flounder, 273, 374
Pollachius pollachius, 224
pollack, 224
Pomacentrus amboinensis, 97
Pomatomus saltatrix, 198
Portuguese sole, 297, 376
Portunus trituberculatus, 361
Potamotrygonidae, 59
Prionotus evolans, 224
Psammodiscus ocellatus, 48, 375
Psettichthys melanostictus, 277, 375, 378
Psettodes, 6, 17
  belcheri, 45, 296, 375
  erumei, 45, 296, 306, 375
Psettodidae, 11, 13, 16, 17, 22, 23, 26, 221, 250, 375
Psettoidei, 17
Pseudopleuronectes americanus, 46, 68, 99, 101, 139, 167, 189, 191, 214, 241, 329, 375, 378
  herzensteini, 100, 101, 175, 189, 273, 349, 375, 378
  obscurus, 68, 99, 375, 378
  schrenki, 68, 375, 378

  yokohamae, 68, 97, 99, 101, 110, 175, 273, 349, 375, 378
Pseudorhombus, 18, 46, 296, 297
  arsius, 191, 296, 375
  cinnamoneus, 273, 375
  elevatus, 297, 375
  javanicus, 297, 375
  malayanus, 297, 375
  pentophthalmus, 191, 375
  triocellatus, 78, 375

rainbow trout, 76
rays, 1
red hake, 198, 201
  king crab, 274
  seabream, 360, 365
  tonguesole, 187, 273, 374
redfishes, 251, 264
Reinhardtius evermanni, 192, 375, 378
  hippoglossoides, 46, 58, 72, 96, 102, 189, 192, 215, 241, 245, 272, 329, 375, 378
  stomias, 102, 145, 173, 187, 192, 272, 328, 336, 375, 378
Remo flounder, 48, 375
rex sole, 71, 97, 110, 186, 188, 272, 277, 284, 374
Rhombosolea leporina, 190, 273, 376
  plebeia, 191, 273, 376
  retiaria, 273, 376
  tapirina, 73, 139, 223, 273, 354, 376
Rhombosoleidae, 12, 13, 16, 17, 19, 22, 23, 26, 27, 47, 58, 190, 191, 373, 375, 376
Rhombosoleinae, 17, 18
ridged-eyed flounder, 273, 349, 375
righteye flounders, 4
Rikuzen flounder, 273, 374
robust tonguefish, 273, 374
rock sole, 68, 86, 99, 168, 188, 190, 912, 220, 225, 228, 272, 277, 281, 282, 284, 328, 336, 374
roughscale sole, 273, 374

Samaridae, 12, 13, 16, 17, 19, 20, 22, 23, 27, 47, 60, 296, 375
Samarinae, 17, 18
Samaris, 47
Samariscus, 47
sand flounder(s), 33, 273, 283, 374
  lances, 193, 202
  shrimp, 197
  sole, 72, 75, 82, 376
sandeel, 247
scaldback, 46, 373

scaldfish, 125, 187, 191, 373
Scomber scombrus, 202
Scophthalmidae, 6, 11, 12, 13, 16, 18, 22, 23, 26, 45, 57, 58, 69, 165, 167, 190, 192, 374, 375, 376
Scophthalmus, 221
   aquosus, 45, 140, 168, 190, 241, 337, 376
   maximus, 74, 139, 169, 190, 192, 215, 245, 347, 376
   rhombus, 169, 192, 215, 245, 376
sculpins, 198, 201, 202
sea raven, 198
Sebastes, 251
   marinus, 202
Senegalese sole, 142, 376
   tonguesole, 142, 145, 296, 297, 299, 374
shoal flounder, 187, 299, 376
shore crab, 197
shortheaded tonguesole, 297, 374
shotted halibut, 187, 273, 349, 374
silver hake, 202
skates, 1, 199, 201, 202
slender sole, 189, 374
slime flounder, 273, 349, 374
small-eyed flounder, 187, 191, 375
smallmouth flounder, 173, 374
sôhachi, 187, 374
sole, 7, 376
Solea, 62, 296, 297
   bleekeri, 169, 191, 297, 376
   elongata, 297, 376
   impar, 72, 376
   lascaris, 72, 376
   ovata, 297, 376
   senegalensis, 142, 376
   solea, 2, 48, 69, 70, 71, 98, 102, 125, 139, 169, 190, 216, 245, 322, 347, 376
   stanalandi, 297, 376
Soleichthys, 29, 48
Soleidae, 4, 6, 12, 13, 14, 15, 16, 17, 19, 20, 21, 22, 23, 27, 28, 35, 48, 55, 58, 60, 62, 69, 70, 102, 152, 167, 165, 190, 191, 214, 221, 240, 245, 250, 296, 301, 306, 308, 373, 374, 375, 376
solenette, 48, 71, 125, 150, 190, 373
southern flounder, 143, 148, 149, 167, 169, 187, 191, 220, 222, 355, 375
   lemon sole, 190, 273, 283, 375
speckled sanddab, 214, 374
   tonguesole, 297, 374
Sphaeramia nematoptera, 97
spiny turbot(s), 296, 300, 305, 308
spottail spiny turbot, 45, 296, 297, 305, 375

spotted codling, 198
   halibut, 189, 273, 348, 349, 362, 363, 376
   turbot, 95, 375
Sprattus sprattus, 247
Stanaland's sole, 297, 376
starry flounder, 1, 71, 97, 140, 145, 189, 190, 192, 194, 195, 196, 220, 225, 272, 375
stingrays, 59, 60
stone flounder, 78, 100, 149, 167, 192, 27, 348, 349, 375
striped searobin, 224
summer flounder, 70, 109, 139, 141, 143, 145, 166, 167, 169 170, 171, 172, 173, 174, 187, 191, 19, 215, 202, 221, 222, 224, 225, 226, 240, 241, 243, 252, 259, 260, 265, 327, 337, 355, 357, 375
swimming crab, 361
Syacium, 46, 60, 296, 297
   gunteri, 187, 299, 376
Symphurus, 20, 30, 32, 49, 58, 62, 296, 306
   civitatium, 166, 376
   plagiusa, 150, 166, 376
Synaptura, 296
   cadenati, 297, 376
   commersonnii, 297, 376
   lusitanica, 297, 376

Taeniopsettinae, 18
tanner crabs, 274
Tarphops, 32, 46
   oligolepis, 150, 376
Tephrinectes, 11, 13, 16, 17, 22, 23, 45, 297
   sinensis, 17, 45, 376
Theragra chalcogramma, 103, 193
thickback sole, 109, 139, 374
Thor's scaldfish, 187, 373
three spotted flounder, 78, 375
tilapias, 353
tongue flatfish, 47, 375
tonguefishes, 6, 20, 30, 49, 293, 296, 297, 299, 300, 306, 307, 308, 309
toothed flounders, 17
topknot, 220, 376
Trinectes, 296
   maculatus, 68, 107, 166, 191, 214, 376
tropical flounder, 355, 375
Tudor's flounder, 273, 373
Turahbuglossus cuvillieri, 14
turbot, 74, 76, 79, 139, 141, 142, 143, 146, 148, 169, 174, 190, 192, 193, 215, 220, 228, 245, 248, 347, 348, 351, 352, 353, 354, 355, 356, 357, 365, 375
tusk, 202

Urophycis chuss, 198
   regia, 198
   tenuis, 198

Vanstralenia, 46
Verasper moseri, 349, 376, 378
   variegatus, 189, 273, 349, 376, 378

wedge sole, 190, 257, 374
west coast sole, 250, 258, 267, 373
white hake, 198
wide-eyed flounder, 141, 149, 187, 218, 220, 250, 297, 373
willowy flounder, 273, 348, 374
windowpane, 45, 140, 145, 168, 174, 190, 200, 201, 203, 241, 243, 260, 337, 376
winter flounder, 46, 68, 78, 84, 86, 99, 103, 139, 140, 141, 143, 144, 145, 148, 167, 173, 174, 186, 189, 191, 196, 198, 200, 201, 202, 203, 214, 215, 217, 221, 222, 224, 225, 226, 229, 240, 241, 243, 251, 252, 259, 260, 328, 329, 330, 338, 340, 357, 375
witch, 58, 80, 110, 112, 143, 145, 153, 167, 186, 187, 193, 200, 203, 240, 241, 245, 247, 252, 254, 260, 261, 263, 264, 265, 328, 329, 331, 338, 340, 374
witch flounder, 374
wolf-fish, 202

Xystreurys rasile, 248, 250, 376

yellowbelly flounder, 190, 273, 283, 376
yellowfin sole, 107, 109, 186, 188, 192, 225, 272, 273, 277, 279, 280, 281, 282, 328, 336, 374
yellowtail flounder, 6, 78, 110, 141, 143, 144, 167, 168, 169, 186, 188, 193, 200, 202, 203, 240, 241, 243, 251, 252, 254, 259, 260, 263, 265, 322, 323, 326, 328, 335, 338, 340, 374

Zanzibar tonguefish, 250, 374
Zebrias, 48, 296, 300
   captivus, 297, 376
   fasciatus, 273, 376
   synapturoides, 297, 376
Zeugopterus punctatus, 220, 376

Subject index

aboriginal fisheries, 273, 283
age and growth, 138–63
    age estimation, 139–41
        adults, 140–41
        larvae and juveniles, 139–40
    growth during metamorphosis, 142–6
    growth of adults, 149–50
        factors affecting growth, 149–50
        trade-off between growth and reproduction, 150
    growth of larvae, 141–2
        factors affecting growth, 141–2
        variation, 141
    growth on nursery grounds, 146–9
        growth compensation/depensation, 148–9
        growth models/experiments, 146–7
        maximum achievable growth, 147–8
    longevity, 150–53
ageing, 3, 5
albinism, 349, 352, 353
ambicolouration, 352
aquaculture, 3, 7, 141, 146, 347–71
    hatchery production, 348–54
        egg production, 348–50
        egg quality, 350
        food and feeding, 350–52
        formulated feeds, 351–2
        genetic improvement, 353–4
        juvenile quality, 352–3
        live food, 350–51
        manipulation of spawning time, 348–50
        microbial environment, 352–4
    intensive farming, 354–6
artisanal fisheries, 6, 293, 297, 298, 299, 300, 310, 312, 393
assessment and management, 319–46
    concepts and terms, 319–21
    population dynamics, 321–34
        recruitment and environment, 332–4
        stock and recruitment, 323–32
        uncertainty, 334
    summary, 335–41
        northeast Atlantic, 340–41
        northeast Pacific, 335
        northwest Atlantic, 335–40
asymmetry, 1, 2, 10, 14, 16, 17, 18, 134, 140
Atlantic fisheries, 240–71
    economic importance, 258–62
        northeast Atlantic, 261–2
        northwest Atlantic, 258–61
        southern Atlantic, 262
    history of exploitation, 251–8
        northeast Atlantic, 255–7
        northwest Atlantic, 251–5
        southern Atlantic, 257–8
    main species and nature, 240–51
        northeast Atlantic, 244–8
        northwest Atlantic, 240–44
        southern Atlantic, 248–51
    management, 263–8
        northeast Atlantic, 265–8
        northwest Atlantic, 263–5
        southern Atlantic, 267
Australian Fisheries Management Authority (AFMA), 286

behaviour, 3, 5, 34, 62, 80, 84, 86, 97, 98, 100, 103, 104, 105, 106, 109, 111, 122, 138, 144, 146, 148, 166, 169, 171, 172, 173, 175, 186, 196, 213–39, 353
    aquaculture, 230
    burying, 215–16
    colour change, 216
    feeding, 219–223
        factors modifying behaviour, 222–3
        feeding behaviour, 219–22
        feeding types, 219

fishing, 228–9
locomotion, 213–5
movements, migration and rhythms, 226–8
predation and reactions to predators, 223–6
    avoidance, 223–5
    burial and substratum selection, 224–5
    effect of size, 225–6
    escape, 225
    reducing encounter/detection, 223–4
reproduction, 216–18
    spawning behaviour, 216–18
rheotaxis and station holding, 215
stock enhancement, 230
biodiversity, 293, 303, 312, 313
biogeography, 4, 42–67
biological clocks, 228
burying, 1, 215–16, 224, 353
by-catch, 6, 247, 248, 251, 264, 266, 277, 282, 283, 298, 301, 303, 308, 309, 310, 312, 313

carrying capacity, 124, 125, 131, 148, 363
cladistics, 15, 17, 18, 62
cladograms, 15, 19, 44,
closed areas, 199, 263, 265, 266, 267, 284, 335, 341
colour change, 1, 2, 3, 5, 213, 216
Common Fisheries Policy (CFP), 266
competition, 120, 165, 175, 186, 194, 198–9, 201–202, 203, 204
competitors, 170, 176, 186, 198–9
concentration hypothesis, 131
conservation, 7, 165, 168, 264, 265, 266, 267, 285, 313, 319, 321, 334, 335
critical habitat, 313

density-dependent growth, 131, 147, 148, 149, 150
mortality, 5, 123, 125, 128, 131
predation, 123
recruitment, 113
Department of Fisheries and Oceans Canada (DFO), 264
development, 3, 96, 98, 99, 109
diel cycles, 171–2
discarding, 185, 247, 248, 264, 266, 267, 277, 285, 296, 298, 301, 303, 308, 312, 320, 332, 334, 340
dispersal, 4, 62, 121, 129, 130, 167, 170, 175
distribution and biogeography, 42–67
    geographic distribution, 45–9
    historical biogeography, 58–63
        Achiridae, 59–60

        Indo-west Pacific region, 61–3
        New World tropical flatfishes, 60–61
        Paralichthyidae, 60
        Pleuronectidae, 59
    patterns of species richness, 49–54
        continental shelves, 52
        continental vs oceanic islands, 54
        insular vs continental regions, 52–4
        latitudinal gradients in species richness, 49–50
        temperate regions, 51
        tropical/subtropical regions, 50
    species richness, 54–8
        Antarctic Ocean, 55–6
        Arctic Ocean, 57
        freshwater, 54–5
        shallow vs deep-sea, 57–8
distribution limits, 121, 122, 123, 127

ecology, planktonic stages, 94–119
    reproduction, 68–87
    trophic, 185–212
economic importance, 258–62, 283–4, 308–309
Economic Return Rate (ERR), 359, 364, 366
ecosystems, 174, 312, 323
eggs, 4, 5, 7, 62, 69–70, 72, 75–9, 81, 82, 86, 87, 94, 96–9, 100, 103–7, 109, 111–13, 120–31, 142, 144, 145, 150
Ekman currents, 99, 100, 101, 102, 110, 111, 169
episodic events, 174–5
essential fish habitat, 165, 176
evolution, 12, 14, 15, 16, 34, 35, 36, 44, 55, 61, 68, 80, 113, 142, 171
Extended Economic Zones (EEZ), 244, 263, 265
eye migration, 10, 131, 139, 140

FAO fishing areas, 240, 241, 250, 294, 295, 302, 303, 304, 305, 306, 307
farming, 7, 348, 354, 356, 358
fecundity, 68, 74, 75–7, 78, 86
feeding, 2. 5, 70, 77, 78, 81, 82, 83, 84, 94, 98, 112, 129, 141, 142, 143, 145, 146, 147, 150, 164, 169, 171, 172, 173, 175, 185–212, 219–223
feeding guilds, 196, 197, 199, 201, 202
FishBase, 8, 22
fisheries, Atlantic, 240–71
    Pacific, 272–91
    tropical, 292–318
Fishery Coordination Committees, 284, 285
Fishery Management Councils, 265, 285
fishing mortality, 320, 322, 332, 335, 340

flatfish assemblages, 21, 35, 42, 44, 49, 50, 51, 52, 54, 57, 58, 167, 199, 200, 292
Flatfish Symposia, 3, 121, 164
fossil flatfishes, 4, 14, 15, 44, 55, 56, 59, 61

gametogenesis, 75
growth, 2, 3, 5, 73, 138–63, 165, 174, 175, 201
habitat associations, 164–84
Heincke's law, 166
history of flatfish research, 2–3
hypermelanosis, 353

ICES divisions, 244, 245, 247, 256, 266, 323, 332, 334, 340, 341
indigenous fishers, 273, 282, 299
Individual Transferable Quotas (ITQ), 267, 282
International Commission for the Northwest Atlantic Fisheries (ICNAF), 263, 264, 265
International Council for the Exploration of the Sea (ICES), 3, 244, 266, 268
International North Pacific Fisheries Commission, 287, 288
International Pacific Halibut Commission (IPHC), 3, 6, 278, 279, 286, 287

juvenile and adult ecology, 164–84
    distribution and ontogeny, 165–74
        dynamics of habitat associations, 171–5
        early juvenile habitat associations, 168–70
        late juvenile and adult habitat associations, 170–71
larvae, 4, 5, 7, 70, 79, 94–119
larval dispersal, 60, 61, 62, 86, 87, 94–119, 129, 130
    drift, 95, 112, 113, 120, 121, 129, 128
    growth, 141–6
    size, 86, 97, 103
    supply, 166
    transport, 94–119, 129
laterality, 2
locomotion, 5, 213–6
longevity, 5, 150–3

management, 263–8, 284–8, 319–46
marine reserves, 204, 313
Market Return Rate (MRR), 363, 364, 365
match-mismatch hypothesis, 95
maternal effects, 142, 144
mating, 1, 218
maturation, 72, 74, 75, 76, 79, 80, 81, 84, 86, 150
maximum growth/optimal feeding conditions hypothesis, 147

Maximum Sustainable Yield (MSY), 319, 320, 335
member-vagrant hypothesis, 131
metamorphosis, 10, 74, 97, 98, 99, 109, 121, 126, 131, 139, 140, 141, 142, 144, 145, 146, 149, 165, 166, 230, 231, 325, 347, 351, 353
metapopulations, 112
migration, 2, 6, 86, 101, 102, 105, 110, 122, 166, 168–74, 176, 213, 226–8, 288, 299
monophyly, 15, 16, 17, 18, 19, 20, 36, 44, 60, 63

National Fisheries Act, Korea, 284
nomenclature, 7–8, 22
North Pacific Fisheries Management Council, 336
Northeast Atlantic Fisheries Commission (NEAFC), 265
Northwest Atlantic Fisheries Organisation (NAFO), 240, 241, 242, 263, 264
nursery grounds, 2, 4, 5, 87, 94, 95, 98, 99, 100–114, 121, 123, 124, 125, 126, 128, 129, 130, 131, 138, 139, 146, 147, 148, 164, 165, 168, 169, 170, 197, 222, 226, 227, 325, 365
nursery size hypothesis, 95, 125

otoliths, 76, 109, 138, 139, 140, 141, 142, 145, 146, 364
overfishing, 266, 279, 282, 300, 302, 308, 309, 310, 311, 312, 334, 335, 357
oxygen levels, 165, 170, 171, 174

Pacific fisheries, 272–91
    economic importance, 283–4
    history of exploitation, 278–83
        Australia, 283
        Canada, 280–81
        general account, 278–9
        Japan, 280
        New Zealand, 282–3
        Republic of Korea, 279
        Russia, 280
        USA, 281–2
    main species and nature, 272–7
    management, 284–8
        Australia and New Zealand, 286–7
        data collection, 287–8
        eastern North Pacific, 285–6
        western North Pacific, 284–5
pigmentation, 10, 216, 352, 353, 361, 364
planktonic stages, interactions in transport processes, 94–119
    adaptations to transport conditions, 105–111

Subject index

congeneric comparisons, 109–110
conspecific comparisons, 110–111
local adaptations, 111
species within a region, 106–109
physical mechanisms of transport/retention, 99–105
    estuarine circulation, 100–103
    fronts and eddies, 103
    models, 103–105
    tidal currents/stream transport, 100
    wind-forcing/Ekman transport, 99–100
transport and population biology, 112–14
    population genetics, 112
    recruitment, 112–114
eggs and larvae, variations, 96–9
    form and function, 96–8
    time and space, 98–9
pollution, 293
polyunsaturated fatty acids (PUFA), 351
population dynamics, 7, 68, 186, 321, 341
precautionary approach, 7, 334, 341
predation, 96, 98, 129, 130, 165, 166, 167, 169, 172, 175, 186, 194, 195, 196, 197–8, 202, 203, 204, 223–6, 347, 352, 353, 357, 364, 365
    cycle, 219
    mortality, 364
predator avoidance, 5, 84, 85, 170, 172, 223–5, 365

rearing, 3, 347, 348, 353, 354, 355, 356, 357, 365
recessus orbitalis, 15, 220
recreational fishing, 259, 273, 283, 365
recruitment, 3, 4, 87, 94, 95, 96, 97, 100, 107, 109, 112, 113, 114, 120–37, 164, 165, 170, 194, 254, 287, 299, 300, 323–34, 360, 361
    variability, 125–32
        processes, 127–30
        relative to other marine fish species, 130–31
    average levels, 123–5
    range of distribution, 121–3
reproduction, 2, 4, 5, 174, 175, 213, 216, 217, 218, 223, 320, 350, 356, 357, 359, 360, 361
reproduction ecology, 68–87
    age and size at first maturation, 79–81
    contaminants, 85–7
    energetics, 81–5
        non-annual spawning, 84
        reproduction and growth, 81–3
        sexual dimorphism, 84–5
        spawning fast, 84
    gonad development, 73–79

batch spawning, 78
egg and sperm quality, 79
fecundity, 75–7
geographic pattern in fecundity, 77–8
ovary, 73–5
testis, 73
spawning, 68–73
    duration, 72–3
    egg size, 69–70
    behaviour, 68
    mode, 68–9
    season, 70–72
seasonal cycles, 172–4
sediment selection, 224, 225
Selective Tidal Stream Transport (STST), 98, 100, 109, 170, 173, 227, 228
settlement, 10, 98, 99, 100, 101, 102, 109, 110, 111, 140, 144, 145, 146, 147, 148, 149, 353
sex determination, 86, 87, 348, 353, 354
    reversal, 354
sexual dimorphism, 84, 138, 149, 218
shrimp fisheries, 297, 298, 300, 301, 308
social behaviour, 68, 175, 218
sound sensitivity, 228, 229,
spawning behaviour, 68, 80, 86, 216–8, 348, 350
    grounds, 72, 87, 98, 99, 100, 103, 111, 112, 216, 227
    mode, 68–9, 142
    season, 70–2, 75, 76, 77, 83, 247, 299
Spawning Stock Biomass (SSB), 266, 267, 268, 332
species accumulation curves, 23, 26, 27, 28
    complex, 29, 30, 44, 45, 165
    diversity, 16, 20, 21, 23, 25, 26, 28, 32, 35, 36, 43, 46, 47, 49, 50, 51, 52, 53, 61,
    range hypothesis, 126, 130
    richness, 42, 45, 49, 50, 52, 54, 55, 57
stage duration hypothesis, 129, 130
stock enhancement, 2, 3, 6, 7, 356–65
    Europe and USA, 357
    Japan, 357–65
        background, 357–8
        objectives and effectiveness, 358–63
        technology, 363–5
Stock-Recruitment (S-R) relationship, 323, 325, 327, 328, 329
Stocking Impact Index (SII), 361
submersibles, 29, 168, 175
subsistence fisheries, 293, 296, 297, 298, 299, 300, 301, 308, 310
sustainable development, 312
Sustainable Fisheries Act, 335
swimming speed, 98, 215

systematic diversity, 10–41
    factors contributing to new species
            discovery, 26–34
        depth, 31–2
        geographic region, 30–1
        size, 32–4
        systematic activities, 28–30
    history of species discovery, 26–8
    intrarelationships, 15–16
    overview, 21
    relative diversity, 25
    species diversity, 21–2
    species within families, 23
    standing diversity estimate for species, 23–5
    synopses of suborders and families, 17–20
    systematic profile, 14–15
systematics, 3, 4

tagging, 3, 138, 146, 149, 361, 364
tidal cycles, 171–2
tidal rhythm, 109, 228
Total Allowable Catch (TAC), 254, 256, 264, 265, 266, 267, 340
transplantation experiments, 3, 138, 139
trophic ecology, 5, 12, 185–212
    feeding groups, 186–97
        anthropogenically produced food, 196
        ontogeny, 196–7
        piscivores, 193–5
        polychaete/crustacean eaters, 186–93
        seasonality, 196
        spatial factors, 197
        specialists, 195–6
    competitors, 198–9

Georges bank case study, 199–203
    influence of changes in flatfish
            populations on ecosystem, 203
    potential competitive interactions, 201–202
    predation by flatfishes, 202
    shifts in abundance/species composition, 199–201
predators, 197–8
tropical fisheries, 292–318
    importance, 308–309
        economic, 308–309
        human, 309
    history of exploitation, 301–308
        commercial landings, 301–303
        geographic occurrence and historical
                landings, 303–308
    main species and nature, 293–301
        commercial species/taxa, 296–7
        gear types, 298–9
        habitats, 293–6
        spawning concentrations, migrating
                stocks, recruitment impacts, 299–300
        industrial vs artisanal fisheries, 300
        nature of fisheries, 297–8
    conservation, 312–13
    conflicts, regulations and management, 310–12

vertical migration, 100, 103, 109, 110, 112, 173

water movements, 4, 100, 103, 113, 120, 124, 127, 131, 170, 194, 197, 198, 204

year-class strength, 95, 99, 100, 104, 113, 114, 120, 124, 127, 131, 170, 194, 197, 198, 204